Lecture Notes in Computer Science 10262

Commenced Publication in 1973
Founding and Former Series Editors:
Gerhard Goos, Juris Hartmanis, and Jan van Leeuwen

More information about this series at http://www.springer.com/series/7407

Fengyu Cong · Andrew Leung
Qinglai Wei (Eds.)

Advances in Neural Networks – ISNN 2017

14th International Symposium, ISNN 2017
Sapporo, Hakodate, and Muroran, Hokkaido, Japan, June 21–26, 2017
Proceedings, Part II

 Springer

Editors
Fengyu Cong
Dalian University of Technology
Dalian
China

Qinglai Wei
Chinese Academy of Sciences
Beijing
China

Andrew Leung
City University of Hong Kong
Kowloon Tong
Hong Kong

ISSN 0302-9743   ISSN 1611-3349   (electronic)
Lecture Notes in Computer Science
ISBN 978-3-319-59080-6   ISBN 978-3-319-59081-3   (eBook)
DOI 10.1007/978-3-319-59081-3

Library of Congress Control Number: 2017941494

LNCS Sublibrary: SL1 – Theoretical Computer Science and General Issues

Printed on acid-free paper

This Springer imprint is published by Springer Nature
The registered company is Springer International Publishing AG
The registered company address is: Gewerbestrasse 11, 6330 Cham, Switzerland

Preface

The twin volumes of *Lecture Notes in Computer Science* constitute the proceedings of the 14th International Symposium on Neural Networks (ISNN 2017) held during June 21–26, 2017, in Sapporo, Hakodate, and Muroran, Hokkaido, Japan. Building on the success of the previous events, ISNN has become a well-established series of popular and high-quality conferences on the theory and methodology of neural networks and their applications. This year's symposium was held for the third time outside China, in Hokkaido, a beautiful island in Japan. As usual, it achieved great success. ISNN aims at providing a high-level international forum for scientists, engineers, educators, and students to gather so as to present and discuss the latest progress in neural network research and applications in diverse areas. It encouraged open discussion, disagreement, criticism, and debate, and we think this is the right way to push the field forward.

Based on the rigorous peer-reviews by the Program Committee members and reviewers, 135 high-quality papers from 25 countries and regions were selected for publication in the LNCS proceedings. These papers cover many topics of neural network-related research including intelligent control, neurodynamic analysis, memristive neurodynamics, computer vision, signal processing, machine learning, optimization etc. Many organizations and volunteers made great contributions toward the success of this symposium. We would like to express our sincere gratitude to City University of Hong Kong and Hokkaido University for their sponsorship, the IEEE Computational Intelligence Society, the International Neural Network Society, and the Japanese Neural Network Society for their technical co-sponsorship. We would also like to sincerely thank all the committee members for all their great efforts in organizing the symposium. Special thanks go to the Program Committee members and reviewers whose insightful reviews and timely feedback ensured the high quality of the accepted papers and the smooth flow of the symposium. We would also like to thank Springer for their cooperation in publishing the proceedings in the prestigious *Lecture Notes in Computer Science* series. Finally, we would like to thank all the speakers, authors, and participants for their support.

April 2017

Fengyu Cong
Andrew C.-S. Leung
Qinglai Wei

Organization

Honorary Chair

Shun'ichi Amari   RIKEN Brain Science Institute, Japan

General Chairs

Hidenori Kawamura   Hokkaido University, Japan
Jun Wang   City University of Hong Kong, SAR China

Advisory Chairs

Kunihiko Fukushima   Fuzzy Logic Systems Institute, Japan
Takeshi Yamakawa   Fuzzy Logic Systems Institute, Japan

Steering Chairs

Haibo He   University of Rhode Island, USA
Derong Liu   University of Illinois, Chicago, USA
Jun Wang   City University of Hong Kong, SAR China

Organizing Committee Chairs

Andrzej Cichocki   RIKEN Brain Science Institute, Japan
Min Han   Dalian University of Technology, China
Bao-Liang Lu   Shanghai Jiao Tong University, China
Masahito Yamamoto   Hokkaido University, Japan

Program Chairs

Fengyu Cong   Dalian University of Technology, China
Andrew C.-S. Leung   City University of Hong Kong, SAR China
Qinglai Wei   CAS Institute of Automation, China

Special Sessions Chairs

Long Cheng   CAS Institute of Automation, China
Satoshi Kurihara   University of Electro-Communications, Japan
Qingshan Liu   Huazhong University of Science and Technology, China
Tomohisa Yamashita   National Institute of Advanced Industrial Science
     and Technology, Japan
Nian Zhang   University of District of Columbia, USA

Tutorial Chairs

Hitoshi Matsubara	Future University Hakodate, Japan
Keiji Suzuki	Future University Hakodate, Japan

Workshop Chairs

Mianxiong Dong	Muroran Institute of Technology, Japan
Jay Kishigami	Muroran Institute of Technology, Japan
Yasuo Kudo	Muroran Institute of Technology, Japan

Publicity Chairs

Jinde Cao	Southeast University, China
Hisao Ishibuchi	Osaka Prefecture University, Japan
Zhigang Zeng	Huazhong University of Science and Technology, China
Huaguang Zhang	Northeastern University, China
Jun Zhang	South China University of Technology, China

Publications Chairs

Jin Hu	Chongqing Jiaotong University, China
He Huang	Soochow University, China
Xinyi Le	Shanghai Jiao Tong University, China
Yongming Li	Liaoning University of Technology, China

Registration Chairs

Shenshen Gu	Shanghai University, China
Hiroyuki Iizuka	Hokkaido University, Japan
Ka Chun Wong	City University of Hong Kong, SAR China

Local Arrangements Chairs

Takashi Kawakami	Hokkaido University of Science, Japan
Koji Nishikawa	Hokkaido University of Science, Japan

Secretaries

Miki Kamata	Hokkaido University, Japan
Ying Qu	Dalian University of Technology, China

Program Committee

Xuhui Bu	Henan Polytechnic University, China
Long Cheng	Chinese Academy of Sciences, China

Fengyu Cong	Dalian University of Technology, China
Ruxandra Liana Costea	Polytechnic University of Bucharest, Romania
Jisheng Dai	Jiangsu University, China
Wai-Keung Fung	Robert Gordon University, UK
Shenshen Gu	Shanghai University, China
Zhishan Guo	Missouri University of Science and Technology, USA
Zhenyuan Guo	Hunan University, China
Chengan Guo	Dalian University of Technology, China
Wei He	Beijing University of Science and Technology, China
Sanqing Hu	Hangzhou Dianzi University, China
Long-Ting Huang	Wuhan University of Technology, China
Min Jiang	Xiamen University, China
Danchi Jiang	University of Tasmania, Australia
Shunshoku Kanae	Fukui University of Technology, Japan
Rhee Man Kil	Korean Advanced Institute of Science and Technology, South Korea
Chiman Kwan	Signal Processing, Inc., Singapore
Chi-Sing Leung	City University of Hong Kong, SAR China
Michael Li	Central Queensland University, Australia
Shoutao Li	Jilin University, China
Cheng Dong Li	Shandong Jianzhu University, China
Jie Lian	Dalian University of Technology, China
Jinling Liang	Southeast University, China
Meiqin Liu	Zhejiang University, China
Ju Liu	Shandong University, China
Wenlian Lu	Fudan University, China
Biao Luo	Chinese Academy of Sciences, China
Dazhong Ma	Northeastern University, China
Tiedong Ma	Chongqing University, China
Jinwen Ma	Peking University, China
Kim Fung Man	City University of Hong Kong, SAR China
Seiichi Ozawa	Kobe University, Japan
Sitian Qin	Harbin Institute of Technology at Weihai, China
Ruizhuo Song	Beijing University of Science and Technology, China
Qiankun Song	Chongqing Jiaotong University, China
John Sum	National Chung Hsing University, China
Weize Sun	Shenzhen University, China
Norikazu Takahashi	Okayama University, Japan
Christos Tjortjis	International Hellenic University, Greece
Kim-Fung Tsang	City University of Hong Kong, SAR China
Jun Wang	City University of Hong Kong, SAR China
Jian Wang	China University of Petroleum, China
Zhanshan Wang	Northeastern University, China
Jing Wang	Beijing University of chemical Technology, China
Shenquan Wang	Changchun University of technology, China
Dianhui Wang	La Trobe University, Australia

Contents – Part II

Signal, Image and Video Processing

Bio-signal and Medical Image Analysis

Contents – Part I

Cognition Computation and Neural Networks

Intelligent Control

Human-Like Robot Arm Robust Nonlinear Control Using a Bio-inspired Controller with Uncertain Properties

Yiping Chang, Aihui Wang$^{(\boxtimes)}$, Shengjun Wen, and Wudai Liao

School of Electronic Information, Zhongyuan University of Technology,
Zhengzhou, China
a.wang@zut.edu.cn

Abstract. This work focuses on a nonlinear robust control of a human arm-like robot arm by using a bio-inspired method based human arm musculoskeletal characteristics, mainly consisting of multi-joint viscosity and multi-joint stiffness. The multi-joint viscosity and multi-joint stiffness are used in designing a bio-inspired operator controller, and the time-varying on estimated human arm multi-joint viscoelasticity (HAMV) data is fed to the designed controller in simulation. Using the designed control architecture, the sufficient robust stable conditions are derived in the presence of uncertainties of modelling and measurement errors, and the control output tracking performance is also realized.

Keywords: HMJA viscoelasticity · Bio-inspired controller · RRCF approach · Robot arm control

1 Introduction

In the last 60 years, the robot arms have always played the some roles of "replacing" and "confronting" the human being, and been mainly focused on industrial fields. However, with the evolution of technology, applications of robot rams research have been broadened taking interest in not only to aiding humans in repetitive tasks, but aiding them in our everyday lives, medical rehabilitation, and social services [1], and these robot arms have always been inspired by the human or animal bodies [2]. Because there are many potential and practical applications, some robots with human arm-inspired motion characteristics which can perform actions smoothly and dexterous as the human arms are still a hot research point in both academic and industrial fields [3,4].

During the last decade, the bio-inspired robot arms have become more and more agile like the human multi-joint arm (HMJA) by measuring or capturing the HMJA motion and converting it to motion of the robot arm [5,6]. Studies on control strategis imitating the movement principle of HMJAs can be considered in developing some bio-inspired robot arms as the humans can control their HMJAs flexibly and robustly. Many concepts that describe the motion principle

© Springer International Publishing AG 2017
F. Cong et al. (Eds.): ISNN 2017, Part II, LNCS 10262, pp. 3–10, 2017.
DOI: 10.1007/978-3-319-59081-3_1

of HMJA have been applied in the robot arms control, for example equilibrium-point control hypothesis, cost functions, and electromyography (EMG) signal-based methods. The various given trajectories can be obtained by using the optimal principles of the above methods based the learned and obtained information pairs in advance. However, in order to generate multifarious movement mode to the same movement assignment, the information pairs are needed to achieve online control. This is difficult in practice.

Assuming that the HMJA has a model similar to a regular connected robot arm, the mechanical properties of the HMJA musculoskeletal system can be mainly modelled by using a called HMJA viscoelasticity [7,8]. The HMJA viscoelasticity includes the multi-joint stiffness and the multi-joint viscosity, which are regulated by the central nervous system (CNS) to make the HMJA can move Arbitrarily to the external different environments or various movements (see Fig. 1). The HMJA viscoelasticity has been widely used in diseases diagnosis, vehicle driving system, and rehabilitation training fields by measuring viscoelastic properties of HMJA [9]. Similarly, if the viscoelastic properties during HMJA can be adapted effectively in the robot arm control, many skillful strategies of the HMJA may be embed into the robot arm control.

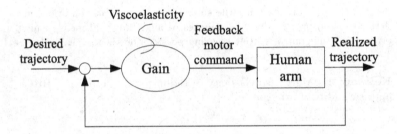

Fig. 1. A human motion control system

Moreover, there not only exist measurement errors from HMJA viscoelasticity estimating data, and but also the robot arms have highly nonlinear, disturbances and model uncertainties. Therefore, it is very difficult to achieve the robustness and output tracking performances. Address these issues, many approaches, such as, state or disturbance observer methods, Lyapunov-based methods, and cost function are used. However, the most existing approaches require that the controlled objects have the precise state space equations in designing a controller, whilst in many cases, the existing methods are used to obtain the approximation model based on a real system. To improve this question, and also for the practical application consideration, the operator-based robust right coprime factorisation (RRCF) approach [10] is becoming an effective and practicable method in linear or nonlinear control system analysis and design.

Addressing the existing challenges, the paper focus on a human-like robot arm robust nonlinear control using a bio-inspired controller with uncertain properties. We propose a new control approach that is inspired by the biological model of HMJA viscoelasticity. The bio-inspired controller is design by using HMJA viscoelastic properties, and there not only exist measurement uncertainties in HMJA viscoelasticity estimating. The objective is that the robot arm can perform a random wide variety of dexterous operations based on the remote motions by the human arm in unstructured environments. Addressing the designed control system, we will discuss the controller design, investigate the robustness and tracking performance.

2   Preliminaries

2.1   Robot Arm

The two-link robot arm dynamics can be modeled as [9],

$$\mathbf{M}(\theta)\ddot{\theta} + \mathbf{H}(\dot{\theta}, \theta) = \tau \tag{1}$$

where, θ is angular, and $\theta = (\theta_1, \theta_2)^T$, $\theta_i(t)$ (i = 1, 2) is the ith link joint angle. $\tau = (\tau_1, \tau_2)^T$, $\tau_i(t)$ is the ith link control input torque. $\mathbf{H}$ and $\mathbf{M}$ are the Coriolis-Centrifugal force vector and inertial matrix, and

$$\mathbf{M} = \begin{bmatrix} Z_1 + 2Z_2 \cos\theta_2 & Z_3 + Z_2 \cos\theta_2 \\ Z_3 + Z_2 \cos\theta_2 & Z_3 \end{bmatrix}, \quad \mathbf{H} = \begin{bmatrix} -Z_2 \sin\theta_2(\dot{\theta}_2^{\,2} + 2\dot{\theta}_1\dot{\theta}_2) \\ Z_2\dot{\theta}_1^{\,2} \sin\theta_2 \end{bmatrix} \tag{2}$$

where $Z_1 = m_1 l_{g1}^2 + m_2(l_1^2 + l_{g2}^2) + I_1 + I_2$, $Z_2 = m_2 l_1 l_{g2}$, and $Z_3 = m_2 l_{g2}^2 + I_2$ are the structural parameters.

2.2   Human Multi-joint Arm

Imitating the robot arm, the two-link HMJA dynamics can also be modeled,

$$\mathbf{M}_A(\mathbf{q})\ddot{\mathbf{q}} + \mathbf{H}_A(\dot{\mathbf{q}}, \mathbf{q}) = \tau_A(\dot{\mathbf{q}}, \mathbf{q}, \mathbf{u}) \tag{3}$$

where, $\mathbf{q}$ is angular, and $\mathbf{q} = [\theta_s(t), \theta_e(t)]^T$, $\theta_s(t)$ is the shoulder joint angle and $\theta_e(t)$ is the elbow joint angle, the subscripts s and e denote the shoulder joint and the elbow joint, respectively. $\tau_A = [\tau_s, \tau_e]^T$ denote the multi-joint torque. $\mathbf{M}_A$ and $\mathbf{H}_A$ have the same definition and structure as $\mathbf{M}$ and $\mathbf{H}$ in (1).

From (3), we have,

$$\delta\tau_A = -\mathbf{R}_A(t)\delta\dot{\mathbf{q}} - \mathbf{K}_A(t)\delta\mathbf{q} + \frac{\partial\tau_A}{\partial\mathbf{u}}\delta\mathbf{u} \tag{4}$$

where, $\mathbf{R}_A(t)$ and $\mathbf{K}_A(t)$ represent human arm multi-joint viscosity and multi-joint stiffness, and

$$-\frac{\partial\tau_A}{\partial\dot{\mathbf{q}}} \equiv \mathbf{R}_A(t) = \begin{bmatrix} R_{A-ss} & R_{A-se} \\ R_{A-es} & R_{A-ee} \end{bmatrix}, \quad -\frac{\partial\tau_A}{\partial\mathbf{q}} \equiv \mathbf{K}_A(t) = \begin{bmatrix} K_{A-ss} & K_{A-se} \\ K_{A-es} & K_{A-ee} \end{bmatrix} \tag{5}$$

3   Control System Design and Analysis

The proposed control system based on HMJA viscoelastic properties is given in Fig. 2, where the component units, consisting of the robot arm dynamics $P+\Delta P$, the controller operator A, the bio-inspired controller operator B with measurement uncertainties ΔB, and the tracking controllers operator C are connected. $r = (\theta_{1d}, \theta_{2d})$ and $y = (\theta_1, \theta_2)$ are the control reference angular inputs and the plant control angular outputs.

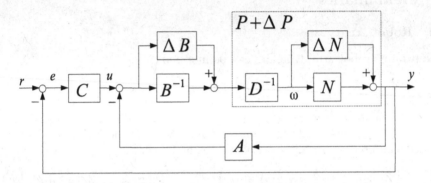

Fig. 2. The proposed robust nonlinear tracking control system

To control the robot arm joint angular, an robust nonlinear control architecture shown in Fig. 3 is designed firstly based operator-RRCF approach. For the robot arm dynamics with uncertainties, the operator model $\tilde{P} = (\tilde{P}_1, \tilde{P}_2)$, includes two parts, the nominal plant $P = (P_1, P_1)$ and the uncertain plant $\Delta P = (\Delta P_1, \Delta P_2)$, namely, $\tilde{P} = P + \Delta P$. The nominal plant P and the real plant $\tilde{P}$ are assumed to have right factorization as $P_i = N_i D_i^{-1}$ (i = 1, 2) and $\tilde{P}_i = P_i + \Delta P_i = (N_i + \Delta N_i)D_i^{-1}$ (i = 1, 2), respectively, N_i, ΔN_i, and D_i (i = 1, 2) are the stable operators, D_i is invertible, ΔN_i is unknown. Addressing the robot arm dynamic in (1), the right factorizations N_1 and D_1 can be modeled as

$$D_i(\omega)(t) = \mathbf{M}_i(\omega(t))\ddot{\omega}(t) + \mathbf{H}_i(\dot{\omega}(t), \omega(t)) \tag{6}$$
$$N_i(\omega)(t) = \omega(t) \tag{7}$$

For the proposed control architecture shown in Fig. 2, the controllers operator A, B, ΔB are designed controller operators to ensure the robustness and stability. B is a bio-inspired operator controller which is designed to obtain the expected motion mechanism of HMJA by using time-varying on estimating HMJA viscoelasticity. In order to satisfy that $T_i = A_i N_i + B_i D_i$ (i = 1, 2), the controller operator A is designed,

$$A(y)(t) = I(y)(t) \tag{8}$$

And the controller operator B is,

$$B^{-1}(s)(t) = -\mathbf{R}(t)\dot{e}_1(t) - \mathbf{K}(t)e_1(t) \tag{9}$$

here $\mathbf{R}(t)$, $\mathbf{K}(t)$ are the expected two-joint viscosity and two-joint stiffness, respectively, and which can be modelled and obtained by using the robot arm dynamic model, like Eqs. (4) and (5). Here, the two-joint viscosity $\mathbf{R}(t)$ and two-joint stiffness $\mathbf{K}(t)$ are replaced by the estimating viscoelasticity data $\mathbf{R}_A(t)$ and $\mathbf{K}_A(t)$ of a HMJA.

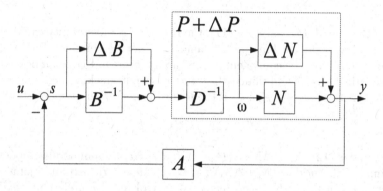

Fig. 3. The proposed robust control architecture

Based on the designed $N_i(\omega)(t)$, $D_i(\omega)(t)$, $A(y)(t)$ and $B^{-1}(s)(t)$, we can find

$$T_i = A_i N_i + B_i D_i = 2\omega(t) \tag{10}$$

is an unimodular operator. Addressing the designed control architecture in Fig. 2, if there does not exist measurement errors of HMJA viscoelasticity and conditon (10) is satisfied, and the robust stability is guaranteed [10]. However, there are the measurement errors or uncertainties of HMJA viscoelasticity, which usually can be described as ΔB. In order to ensure the system robustness and stability, a new condition related to ΔB is discussed.

Theorem 1. For the Fig. 3, the Bezout identity of the nominal model and the real model are $A_i N_i + B_i D_i = T_i \in u(W, U)$, $A_i(N_i + \Delta N_i) + (B_i + \Delta B_i)D_i = \hat{T}_i \in u(W, U)$, respectively. If

$$\left\| (\Delta N_i + \Delta B_i D_i) T_i^{-1} \right\| < 1, \quad i = 1, 2 \tag{11}$$

is satisfied, and the robustness and stability can be guaranteed.

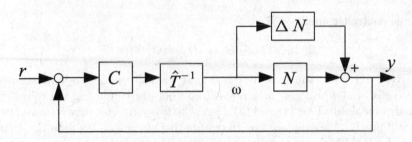

Fig. 4. The equivalent control architecture of Fig. 2

Based on the above condition (11), the robustness and stability can be guaranteed. However, the tracking performance does not be obtained. Based on the proposed conditions, the equivalent control architecture of Fig. 3 can be obtained and shown in Fig. 4. To obtained the control output tracking performance, an controller operator $\mathbf{C}$ is designed in the following condition, namely,

$$(N_i + \Delta N_i)\hat{T}_i^{-1}\mathbf{C} = I \tag{12}$$

Namely, under (12), $N_i + \Delta N_i + (B_i + \Delta B_i)D_i = \hat{T}_i$ are unimodular operators, which implies that $y(t) = (N_i + \Delta N_i)\,\hat{T}_i^{-1}\mathbf{C}r(t)$. Hence, the expected joint angular output y can track the given reference control input r under the condition $(N_i + \Delta N_i)\hat{T}_i^{-1}\mathbf{C} = I$. However, because the ΔN_i is unknown, and ΔB_i has also uncertainties. Therefore, based on the condition of (12), we can not design directly the perfect tracking controller $\mathbf{C}$ in expected tracking control performance. SO, in the paper, we design a tracking controller C is to improve control tracking performance, it is,

$$C = \Gamma_{\alpha i}e_i(t) + \Gamma_{\beta i}\int_0^t e_i(\tau)d\tau \tag{13}$$

where $\Gamma_{\alpha i}$, $\Gamma_{\beta i}$) are the designed controller parameters.

4   Simulation Results

According to the presented HMJA viscoelasticity online estimating method in [11], the stiffness data and viscosity data of a multi-joint HMJA are measured in Fig. 5(a) and (b), where, the HMJA moves from the starting point $(x, y) = [-41.7596, 33.8013]$ (cm) to the end point $(x, y) = [20.4489, 42.4762$ (cm).

In control simulation, the controlled objet is assumed as a HMJA model, the expected movement trajectory is the above experimental path in estimating HMJA viscoelasticity. The measured HMJA multi-joint stiffness data and multi-joint viscosity data are fed to the controller operator B. The robot arm structural parameters uncertainties is to be $Z_i = Z_i^* + \Delta Z_i$, $\Delta = 0.05$, where Z_i^*

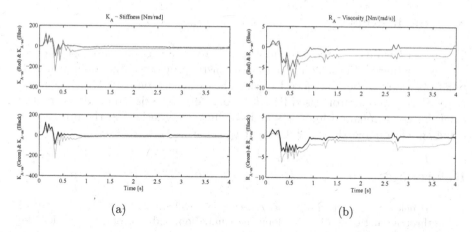

Fig. 5. Measured experimental data: (a) Stiffness; (b) Viscoesity (Color figure online)

can be assumed to be real value. The unknown external disturbances are to be $\tau_d = 0.5 + 0.05 * sin(2\pi t)$. The effect of structural uncertainties and disturbances can be as ΔN. Moreover, the uncertainties of controller operator ΔB is be $\Delta B = \Delta B^* + \sigma \Delta B$, $\sigma = 0.05$, where ΔB^* can be assumed to be a real value. The control parameters are $K_{\alpha 1} = K_{\alpha 2} = 50$, and $K_{\beta 1} = K_{\beta 2} = 0.02$. Using the proposed architecture, the tracking simulation results consisting of joint angles movement and endpoint position motion of robot arm are shown in Fig. 6(a) and (b), respectively. From Fig. 6(a) and (b), the expected results can be achieved, namely robustness, stability, and tracking can be obtained.

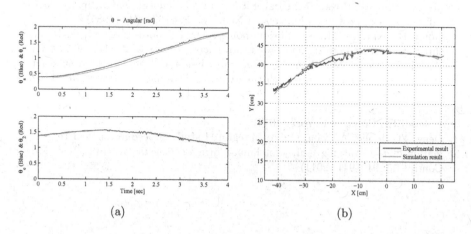

Fig. 6. Simulation results: (a) Angles; (b) Position (Color figure online)

5   Conclusions

This paper has investigated a robot arm robust nonlinear control by using a new bio-inspired method based on HMJA viscoelastic properties, the robot arm endpoint position can be controlled by using the estimated online HMJA viscoelasticity. Based on operator-based RRCF theory, for the designed architecture, the sufficient conditions of robustness and stability were derived in the presence of coupling effects, and the control output tracking was also realized.

References

1. Wittmeier, S., Alessandro, C., Bascarevic, N., Dalamagkidis, K., et al.: Toward anthropomimetic robotics: development, simulation, and control of a musculoskeletal torso. Artif. Life **19**, 171–193 (2014)
2. Zlotowski, J., Proudfoot, D., Yogeeswaran, K., Bartneck, C.: Anthropomorphism: opportunities and challenges in human robot interaction. Int. J. Soc. Robot. **7**, 347–360 (2015)
3. Campos, F., Calado, J.: Approaches to human arm movement control—a review. Annu. Rev. Control **33**, 69–77 (2009)
4. Berret, B., Gauthier, J.P., Papaxanthis, C.: How humans control arm movements. Proc. Steklov Inst. Math. **261**, 44–58 (2008)
5. Jagodnik, K.M., Thomas, P.S., Bogert, A.J.V.D., Branicky, M.S., Kirsch, R.F.: Human-like rewards to train a reinforcement learning controller for planar arm movement. IEEE Trans. Hum.-Mach. Syst. **46**, 723–733 (2016)
6. Zhang, Z., Beck, A., Magnenat-Thalmann, N.: Human-like behavior generation based on head-arms model for robot tracking external targets and body parts. IEEE Trans. Cybern. **45**, 1390–1400 (2015)
7. Mussa-Ivaldi, F., Hogan, N., Bizzi, E.: Neural, mechanical, and geometric factors subserving arm posture in humans. J. Neurosci. **5**, 2732–2743 (1985)
8. Gomi, H., Kawato, M.: Equilibrium-point control hypothesis examined by measured arm-stiffness during multi-joint movement. Science **272**, 117–120 (1996)
9. Wang, A., Deng, M., Wang, D.: Operator-based robust control design for a human arm-like manipulator with time-varying delay measurements. Int. J. Control Autom. Syst. **11**, 1112–1121 (2013)
10. Deng, M.: Operator-Based Nonlinear Control System Design and Applications. Wiley-IEEE Press, New York (2014)
11. Deng, M., Inoue, A., Zhu, Q.: An integrated study procedure on real time estimation of time varying multijoint human arm viscoelasticity. Trans. Inst. Meas. Control **33**, 919–941 (2011)

Adaptive NNs Fault-Tolerant Control for Nonstrict-Feedback Nonlinear Systems

Guowei Dong[1($\boxtimes$)], Yongming Li[1], Duo Meng[2], Fuming Sun[1], and Rui Bai[1]

[1] College of Science, Liaoning University of Technology, Jinzhou, China
dongguowei1992@163.com
[2] School of Civil and Architectural Engineering, Liaoning University of Technology,
121001 Jinzhou, People's Republic of China

Abstract. In this paper, the problem of fault-tolerant control (FTC) is investigated for a class of nonlinear single input and single output (SISO) systems in the non-strict feedback form. The considered system possess unknown nonlinear functions, unmeasured states, unknown time-varying delays, unknown control direction and actuator faults (bias and gain faults). Neural networks (NNs) are adopted to approximate the unknown nonlinear functions. Then, a state observer is constructed to solve the problem of unmeasured states. In the frame of adaptive backstepping design technique, by combining with Nussbaum gain function and Lyapunov-Krasobskii functional theory, an adaptive NNs output feedback FTC method is developed. It is shown that all signals in the closed-loop system are proved to be bounded, and the system output can follow the given reference signal well.

Keywords: Nonstrict-feedback nonlinear systems · Fault-tolerant control · Adaptive NNs control

1 Introduction

In the past decades, fuzzy systems and NNs have been popularly used in fuzzy modeling and controller design for uncertain nonlinear systems [1,2]. However, the results obtained in [1,2] are only suitable for those systems that all the components of the considered systems are in good operating conditions, i.e., the faults did not occur in the considered systems. In practical control systems, there are usually some faults [1]. These faults will make the stability of the system decreased, and even affect the safety and reliability of the control system. Thus, some researches have been done on the problem of FTC for the controlled system, and a deal of effective adaptive neural networks (NNs) or fuzzy FTC design methods have been developed [4–6]. Among, [4] investigated the adaptive NNs FTC problem under the assumption that the states of the systems can be measured directly. Adaptive fuzzy backstepping output-feedback-based fault-tolerant method is developed in [5,6] with unmeasured states. It is worth to be

F. Cong et al. (Eds.): ISNN 2017, Part II, LNCS 10262, pp. 11–19, 2017.
DOI: 10.1007/978-3-319-59081-3_2

noticed that the above-mentioned FTC problems are aiming at the systems in the pure-feedback or strict-feedback forms.

Therefore, the above method cannot be used for non-strict feedback systems [7]. In general, compared with nonlinear strict-feedback systems (or pure-feedback systems), non-strict feedback systems have the unknown nonlinear functions, which contain the whole state vector of each subsystems. And also, the intermediate control functions are the function including whole state vector. If the control method for strict-feedback systems (or pure-feedback systems) were adopted with the aim to solve the control design problem for non-strict feedback systems, the algebraic loop problem may occur. In order to avoid this problem, the study for non-strict feedback systems has gained considerable interest in the past years and some considerable efforts have been developed, for example [8–10]. In addition, the work in [8–10] did not consider the problem of time-varying delay and unknown control direction. Therefore, they cannot be utilized to deal with the control design problem considered in this paper.

In this paper, by using NNs and fuzzy state observer to approximate the unknown nonlinear functions and estimate the unmeasured states, respectively. Combining with Nussbaum gain function methods, and in the frame of adaptive backstepping design technique, an adaptive NNs output feedback FTC method is developed. The proposed method can not only guarantee that all the signals in the closed-loop system are bounded, but also the system output can follow the given reference signal well.

2   Problem Formulations and Preliminaries

2.1   Nonlinear System and Actuator Fault Model

Consider an uncertain SISO nonlinear system with actuator faults.

$$\begin{cases} \dot{\tau}_i = f_i(\bar{\tau}) + \tau_{i+1} + h_i(y(t - \sigma_i(t))), \quad i = 1, \ldots, n-1, \\ \quad \vdots \\ \dot{\tau}_n = f_n(\bar{\tau}) + gu^q + h_n(y(t - \sigma_n(t))), \\ y = \tau_1 \end{cases} \tag{1}$$

where $\bar{\tau} = [\tau_1, \cdots, \tau_n]^T$ is a state vector, g denotes an unknown constant, while $h_i(y(t - \sigma(t)))$ and $f_i(y)$ are unknown nonlinear functions, u^q denotes the control input of the system.

Therefore, according to [2,8], The bias and gain faults are as the following form:

$$u^q(t) = (1 - m)u(t) + \omega(t) \tag{2}$$

where $\omega(t)$ denotes a bounded function, which can be given in the next section. $0 \leq m \leq 1$ denotes the lost control rate, which is an unknown constant.

In this paper, the control objective is to develop an observer-based adaptive NNs backstepping FTC strategy for the system (1) with bias and gain faults (2), which can not only validate the boundeness of the whole signals y_r in the closed-loop system, but also ensure that the system output can follow the given reference signal y well.

To achieve the above objective, several assumptions are given.

Assumption 1: There exist known constants d_i, $(1 \leq i \leq n)$, such that the time delays $|\sigma_i(t)| \leq d_i$

Assumption 2: $h_i(\cdot)$ is a nonlinear function, and it satisfies the following inequality:

$$|h_i(y(t))|^2 \leq z_1(t)H_i(z_1(t)) + \bar{h}_i(y_r(t)) + \varpi_i \quad (1 \leq i \leq n) \qquad (3)$$

where $h_i(\cdot)$ is a bounded function and $h_i(\cdot) = 0$, $H_i(\cdot)$ is a known function, ϖ_i are unknown constants.

2.2   Neural Network System

In this paper, the unknown nonlinear functions existed in controlled system are approximated by employing NNs. The general form of neural network system is $f(\tau) = \xi^T \phi(\tau)$, where $\xi \in R^\upsilon$, the NN node number $\upsilon > 1$ and ξ is the parameter estimation vector. $\phi(\tau)$ are chosen as the form of Gaussian functions, i.e. Then $\xi^T \phi(\tau)$ can approximate any given function $f(\tau)$ in a compact set, i.e.

$$f(\tau) = \xi^T \phi(\tau) + \delta \qquad (4)$$

where δ is the approximation error with $|\delta| \leq \delta^*$ and δ^* is an unknown positive parameter.

2.3   Nussbaum-Type Function

A Nussbaum gain technique-based design method is adopted in this paper, and Nussbaum-type function $N(\varsigma)$ owns the following characteristics:

$$\lim_{m \to \infty} \sup \frac{1}{m} \int_0^m N(\varsigma)d\varsigma = \infty$$
$$\lim_{m \to \infty} \sup \frac{1}{m} \int_0^m N(\varsigma)d\varsigma = -\infty \qquad (5)$$

Nussbaum common features are $\varsigma^2 \cos(\varsigma)$, $\varsigma^2 \sin(\varsigma)$ and $\exp(\varsigma^2) \cos(\varsigma^2)$. In this paper, the form of $\exp(\varsigma^2) \cos(\varsigma^2)$ is adopted.

Lemma 1: For system (1), define $N(\varsigma) = \exp(\varsigma^2) \cos(\varsigma^2), 0 \leq \varsigma < t$, there exists a function $V(t) \geq 0$, positive constants C and D, such that the following inequality holds:

$$\dot{V}(t) \leq -CV(t) + \sum_{i=1}^{n} \ell_j[\beta_i N'(\varsigma_i) + 1]\dot{\varsigma}_i + D \qquad (6)$$

3   Design of Fuzzy State Observer

Since the states of the considered systems are partial measurable, a state observer is needed with the aim to estimate the unmeasured states.

Let $\eta = g(1 - m)$, $x_i = \bar{\tau}/\eta = [\tau_1/\eta, \tau_2/\eta, \cdots \tau_n/\eta]^T$, thus the system (1) becomes

$$\begin{cases} \dot{x}_i = x_{i+1} + \frac{f_i(\bar{\tau})}{\eta} + \frac{1}{\eta}h_i(y(t - \sigma_i(t))), i = 1, 2, \cdots n - 1 \\ \vdots \\ \dot{x}_n = u(t) + \frac{f_i(\bar{\tau})}{\eta} + \frac{1}{\eta}h_n(y(t - \sigma_n(t))) \\ \dot{y} = f_1(\bar{\tau}) + \eta x_2 + h_1(y(t - \sigma_1(t))) \end{cases} \tag{7}$$

According to the transformation from (1) to (7), the coefficient of $u(t)$ becomes 1, while the coefficient of x_2, which is in the last equation of (7), is η, rather than 1. Hence, the Nussbaum technique should be adopted in this paper with the aim to erase the effect of η.

Constructing a state observer for system (7) as

$$\begin{cases} \dot{\hat{x}}_i = -k_i\hat{x}_1 + \hat{x}_{i+1} + k_i y, \ i = 1, 2, \cdots, n - 1 \\ \dot{\hat{x}}_n = -k_n\hat{x}_1 + u(t) + k_i y \end{cases} \tag{8}$$

Let $e_i = x_i - \hat{x}_i$ be the observer errors, where $\hat{x}_i = [\hat{x}_1, \cdots \hat{x}_n]^T$. Based on (7) and (8), we have

$$\dot{e}_i = \dot{x}_i - \dot{\hat{x}}_i = Ae_i + B\frac{\omega(t)}{\eta} + \frac{F}{\eta} + \frac{h}{\eta} \tag{9}$$

where $F = [f_1(\bar{\tau}), \cdots, f_n(\bar{\tau})]^T$, $h = [h_1(y(t - \sigma_1(t))), \cdots h_n(y(t - \sigma_n(t)))]$, $B = [0, \cdots, 0 \ 1]^T$ and $A = \begin{bmatrix} -k_1 & & \\ \vdots & I_{(n-1)\times(n-1)} & \\ -k_n \ 0 & \cdots & 0 \end{bmatrix}$

$\underbrace{}_{n-1}$

According to selecting the appropriate vector $[k_1, \cdots, k_n]^T$, thus the matrix A can be guaranteed a Hurwitz form. And also, for any given $Q = Q^T > 0$, there exists $P = P^T > 0$ such that

$$A^T P + PA = -Q \tag{10}$$

Consider a Lyapunov function candidate:

$$V_0 = e^T Pe/2 + W_0 \tag{11}$$

where

$$W_0 = \frac{1}{2b(1 - \sigma^*)}\|P\|^2 e^{-rt} \sum_{i=1}^{n} \int_{t-\sigma(t)}^{t} e^{rm} z_1(m)(H_i(z_1(m)))dm \tag{12}$$

where b is a known constant. The time derivative of V_0 is

$$\dot{V}_0 = -\frac{1}{2}e^T Qe + e^T P(B\frac{\omega(t)}{\eta} + \frac{F}{\eta} + \frac{h}{\eta}) + \dot{W}_0 \tag{13}$$

Thus, from mean value theorem, the function $f_i(\bar{\tau})$ can be represented as the following formula

$$f_i(\bar{\tau}) = y\bar{f}_i(\bar{\tau}) \tag{14}$$

According to (4), the nonlinear function $\|P\|^2 \sum_{i=1}^{n} y\bar{f}_i^2(\bar{\tau})/\eta^2$ can be approximated by NNs, one has

$$\|P\|^2 \sum_{i=1}^{n} \frac{y\bar{f}_i^2(\bar{\tau})}{\eta^2} = \Psi^{*T}\phi(\bar{\tau}) + \varepsilon \tag{15}$$

where $b' = be^{r\sigma}$ and $|\omega(t)/\eta| \leq \kappa$ where κ is an unknown constant and $|\varepsilon| \leq \varepsilon^*$, then we can obtain

$$\dot{V}_0 \leq -(\lambda_{\min}(Q) - \frac{1}{2} - \frac{1}{2b'} - \frac{b'}{2\eta})\|e\|^2 + y(\Psi^{*T}\phi(\tau) + \varepsilon)$$
$$+ \frac{b'}{2}\|P\|^2 \sum_{i=1}^{n} \kappa^2 + \frac{1}{2b(1-\sigma^*)}\|P\|^2 \sum_{i=1}^{n} z_1 H_1(z_1) - rW_0 + d_0^* \tag{16}$$

where d_0^* is a constant and $d_0^* \geq \left\|\|\|^2(\bar{h}_i(y_r(t)) + \varpi_i)\right./ 2b'$

4   Neural Networks Control Design

In this section, according to the backstepping technique, an adaptive fuzzy output feedback fault tolerate controller design method will be presented, and the Lyapunov function stability theory is adopted to verify the stability of the considered system. The coordinate transformation of n-step backstepping control design is chosen as

$$z_1 = y - y_r, \ z_i = \hat{x}_i - \alpha_{i-1}, (i = 2, \cdots n) \tag{17}$$

where z_1 is the system's tracking error. α_{i-1} denotes the virtual control input.

Step 1: From (7) and (17), we have

$$\dot{z}_1 = f_1(\bar{\tau}) + \eta x_2 + h_1(y(t - \sigma_i(t))) - \dot{y}_r \tag{18}$$

Consider a Lyapunov function candidate:

$$V_1 = \frac{1}{2}z_1^2 + \frac{1}{2\gamma_1}\tilde{\theta}_1^2 + \frac{1}{2\gamma_2}\tilde{\theta}_2^2 + W_1 + V_0 \tag{19}$$

where $\gamma_1 > 0$ and $\gamma_2 > 0$ are design constants, and

$$W_1 = \frac{1}{2b(1-\sigma^*)}e^{-rt}\int_{t-\sigma_1(t)}^{t} e^{rm}z_1(m)(H_1(z_1(m)))dm \tag{20}$$

According to (4), we use NN to approximate the unknown nonlinear function $f_1(\bar{\tau})$ as:

$$f_1(\bar{\tau}) = \Phi^{*T}\xi(\bar{\tau}) + \mu_1(\bar{\tau}) \tag{21}$$

Where $|\mu_1(x)| \leq \mu^*$. Define $\theta_1^* = \Psi_1^{*T}\Psi_1^*$, $\theta_2^* = \Phi_1^{*T}\Phi_1^*$, $\hat{\theta}_1$ and $\hat{\theta}_2$ are used to estimate θ_1^* and θ_2^*, respectively. The estimation error is $\tilde{\theta}_i = \theta_i^* - \hat{\theta}_i$ $(i = 1, 2)$. The time derivative of V_1 is

$$
\begin{aligned}
\dot{V}_1 \leq &-(\lambda_{\min}(Q) - \tfrac{1}{2b'} - \tfrac{b'}{2\eta} - 1 - \tfrac{\eta^2}{b'})\|e\|^2 + \tfrac{1}{\gamma_2}\tilde{\theta}_2^T(\tfrac{\gamma_2^2 z_1^2}{4\lambda} - \dot{\hat{\theta}}_2) \\
&+ z_1(\tfrac{z_1}{2} + \tfrac{b'\bar{\eta}z_1}{4} + \tfrac{b'z_1}{2} + \tfrac{\hat{\theta}_1 z_1}{4\lambda} + \tfrac{\hat{\theta}_2 z_1}{4\lambda} + \tfrac{1}{2b(1-\sigma^*)}z_1 H_1(z_1)) \\
&+ z_1\tfrac{1}{2b(1-\sigma^*)}\|P\|^2\sum_{i=1}^{n}z_1 H_i(z_1) + \tfrac{1}{\gamma_1}\tilde{\theta}_1^T(\tfrac{\gamma_1^2 z_1^2}{4\lambda} - \dot{\hat{\theta}}_1) \\
&+ \eta z_1 z_2 + \eta z_1\alpha_1 - z_1\dot{y}_r + d_0^* + \bar{d}_1 + D_1 - rW_0 - rW_1
\end{aligned}
\tag{22}
$$

where $D_1 = b'\|P\|^2\sum_{i=1}^{n}\kappa^2\left/2 + y^2/2 + \theta_1^* + 2\varepsilon^2 + \mu_1^2(\bar{\tau}) + 2\lambda\right.$ and $\bar{d}_1 = d_0^*\left/\|P\|^2\right.$.

The virtual control α_1 and the parameters adaptive functions θ_i $(i = 1, 2)$ as:

$$
\begin{aligned}
\alpha_1 = &\dot{N}(\varsigma)[c_1 z_1 - \dot{y}_r + \tfrac{z_1}{2} + \tfrac{b'\bar{\eta}z_1}{4} + \tfrac{b'z_1}{2} + \tfrac{\hat{\theta}_1 z_1}{4\lambda} + \tfrac{\hat{\theta}_2 z_1}{4\lambda} \\
&+ \tfrac{n}{2b(1-\sigma^*)}H_1(z_1) + \tfrac{1}{2b(1-\sigma^*)}\|P\|^2\sum_{i=1}^{n}z_1 H_i(z_1)]
\end{aligned}
\tag{23}
$$

$$
\begin{aligned}
\dot{\vartheta} = &\tfrac{z_1}{\ell}[c_1 z_1 - \dot{y}_r + \tfrac{z_1}{2} + \tfrac{b'\bar{\eta}z_1}{4} + \tfrac{b'z_1}{2} + \tfrac{\hat{\theta}_1 z_1}{4\lambda} + \tfrac{\hat{\theta}_2 z_1}{4\lambda} \\
&+ \tfrac{n}{2b(1-\sigma^*)}H_1(z_1) + \tfrac{1}{2b(1-\sigma^*)}\|P\|^2\sum_{i=1}^{n}z_1 H_i(z_1)]
\end{aligned}
\tag{24}
$$

$$
\dot{\hat{\theta}}_1 = \frac{\gamma_1^2 z_1^2}{4\lambda} - \rho_1\hat{\theta}_1, \dot{\hat{\theta}}_2 = \frac{\gamma_2^2 z_1^2}{4\lambda} - \rho_2\hat{\theta}_2
\tag{25}
$$

Substituting (23)–(25) into (22) results in

$$
\begin{aligned}
\dot{V}_1 \leq &-(\lambda_{\min}(Q) - \tfrac{1}{2b'} - \tfrac{b'}{2\eta} - 1 - \tfrac{\eta^2}{b'})\|e\|^2 + \eta z_1 z_2 - c_1 z_1^2 \\
&+ \ell_1(\eta N'(\varsigma) + 1)\dot{\varsigma} - \tfrac{n-1}{2b(1-\sigma^*)}z_1 H_1(z_1) + \sum_{i=1}^{2}\tfrac{\rho_i}{\gamma_i}\tilde{\theta}_i^T\hat{\theta}_i \\
&+ d_0^* + \bar{d}_1 + D_1 - rW_0 - rW_1
\end{aligned}
\tag{26}
$$

Step i: From (8), (9) and (18), we have

$$
\begin{aligned}
\dot{z}_i = &z_{i+1} + \alpha_i - k_i\hat{x}_1 - \tfrac{\partial\alpha_1}{\partial y}(\Phi_1^T\xi(\tau) + \mu_1(\tau) + \eta\hat{x}_i + \eta e_i \\
&+ h_1(y(t - \sigma_1(t)))) - \sum_{j=1}^{i}\tfrac{\partial\alpha_{i-1}}{\partial y_r^{(j-1)}}y_r^{(j)} - \sum_{i=1}^{2}\tfrac{\partial\alpha_{i-1}}{\partial\theta_i}\dot{\theta}_i
\end{aligned}
\tag{27}
$$

where $i = 2, 3, \cdots n - 1$, then construct a Lyapunov function V_i as

$$
V_i = V_{i-1} + \frac{1}{2}z_i^2 + W_1
\tag{28}
$$

Similar to α_1, the virtual control input α_i as:

$$
\begin{aligned}
\alpha_i = &-c_i z_i + k_i\hat{x}_1 - z_{i-1} - \tfrac{z_i}{2}(\tfrac{\partial\alpha_{i-1}}{\partial y})^2 - \tfrac{z_i}{4\lambda}(\tfrac{\partial\alpha_{i-1}}{\partial y})^2 + \sum_{j=1}^{2}\tfrac{\partial\alpha_{i-1}}{\partial\theta_j}\dot{\theta}_j \\
&- \tfrac{z_i}{4\lambda}(\tfrac{\partial\alpha_{i-1}}{\partial y})^2\hat{x}_2^2 - \tfrac{b'}{2}(\tfrac{\partial\alpha_{i-1}}{\partial y})^2 z_i + \sum_{j=1}^{i}\tfrac{\partial\alpha_{i-1}}{\partial y_r^{(j-1)}}y_r^{(j)}
\end{aligned}
\tag{29}
$$

The time derivative of V_i is

$$\dot{V}_i \leq -(\lambda_{\min}(Q) - \tfrac{1}{2b'} - \tfrac{b'}{2\beta} - 1 - \tfrac{\bar{\eta}^2}{b'} - (i-1)\lambda\bar{\eta}^2)\|e\|^2 + z_i z_{i-1}$$
$$-\tfrac{n-i}{2b(1-\sigma^*)}z_1 H_1(z_1) - \sum_{j=1}^{i} c_j z_j^2 + \sum_{i=1}^{2} \tfrac{\rho_i}{\gamma_i}\tilde{\theta}_i^T\hat{\theta}_i - rW_0 \qquad (30)$$
$$-irW_1 + \ell_1(\eta N'(\varsigma) + 1)\dot{\vartheta} + \eta z_1 z_2 + d_0^* + i\bar{d}_1 + D_i$$

where $D_i = D_{i-1} + \lambda\bar{\eta}^2 + \theta_2^* + \mu_i^{*2}$.

Step n: In this step, the actual control input $u(t)$ appears. From (7), (8) and (17), we have

$$\dot{z}_n = u(t) - k_n\hat{x}_1 - \sum_{i=1}^{n} \tfrac{\partial\alpha_{n-1}}{\partial y_r^{(i-1)}}y_r^{(i)} - \sum_{i=1}^{2}\tfrac{\partial\alpha_{n-1}}{\partial\theta_1}\dot{\theta}_i - \tfrac{\partial\alpha_1}{\partial y}(\Phi_1^T\xi(\tau)$$
$$+\mu_1(\tau) + \eta\hat{x}_2 + \eta e_2 + h_1(y(t-\sigma_1(t)))) \qquad (31)$$

Construct a Lyapunov function V_n as:

$$V_n = V_{n-1} + \frac{1}{2}z_n^2 + W_1 \qquad (32)$$

Design the actual controller $u(t)$ as:

$$u(t) = k_n\hat{x}_1 + \sum_{i=1}^{n}\tfrac{\partial\alpha_{n-1}}{\partial y_r^{(i-1)}}y_r^{(i)} + \sum_{i=1}^{2}\tfrac{\partial\alpha_{n-1}}{\partial\theta_1}\dot{\theta}_i - \tfrac{b'}{2}(\tfrac{\partial\alpha_{n-1}}{\partial y})^2$$
$$-\tfrac{z_n^2}{4\lambda}(\tfrac{\partial\alpha_{n-1}}{\partial y})^2\hat{x}_2^2 - \tfrac{z_n^2}{2}(\tfrac{\partial\alpha_{n-1}}{\partial y})^2 - \tfrac{z_n^2}{4\lambda}(\tfrac{\partial\alpha_{n-1}}{\partial y})^2 - c_n z_n - z_{n-1} \qquad (33)$$

From (33), one has

$$\dot{V}_n \leq -(\lambda_{\min}(Q) - \tfrac{1}{2b'} - \tfrac{b'}{2\eta} - 1 - \tfrac{\bar{\eta}^2}{b'} - (n-1)\lambda\bar{\eta}^2)\|e\|^2$$
$$-(c_n - \tfrac{1}{2}\bar{\eta}^2)z_1^2 - (c_n - \tfrac{1}{2}\bar{\eta}^2)z_2^2 - \sum_{i=1}^{2}\tfrac{\rho_i}{\gamma_i}\tilde{\theta}_i^2 - \sum_{j=3}^{n-1}c_j z_j^2 \qquad (34)$$
$$+ \sum_{i=1}^{2}\tfrac{\rho_i}{2\gamma_i}\theta_i^{*2} + d_0^* + n\bar{d}_1 + D_n - rW_0 - nrW_1 + \ell_1(\eta N'(\varsigma) + 1)\dot{\varsigma}$$

The inequality (34) can be rewritten as

$$\dot{V}_n \leq -CV_n + D \qquad (35)$$

where

$$C = \min\{-(\lambda_{\min}(Q) - 1/2b' - b'/2\eta - 1 - \bar{\eta}^2/b' - (n-1)\lambda\bar{\eta}^2),$$
$$2(c_1 - \bar{\eta}^2/2), 2(c_2 - \bar{\eta}^2/2), 2c_3, 2c_4 \cdots 2c_{n-1}, \rho_1/2\gamma_1, \rho_2/2\gamma_2\} \qquad (36)$$

There exists a constant $\tilde{D}$ such that $\tilde{D} \geq \varsigma_1(\eta N'(\varsigma) + 1)\dot{\vartheta}$, and

$$D = d_0^* + n\bar{d}_1 + D_n - rW_0 - nrW_1 + \tilde{D} + \sum_{i=1}^{2}\tfrac{\rho_i}{2\gamma_i}\theta_i^{*2} \qquad (37)$$

Integrate the differential inequality (35), we have

$$V = V_n \le e^{-ct}(V(0) - D/C) + D/C \tag{38}$$

From (38) and Lemma 1, the boundeness of the whole signals in the closed-loop system can be obtained.

The above design and analysis are summarized in the Theorem 1.

Theorem 1: For system (1) with fault, under Assumptions 1, 2 and Lemma 1, the controller functions (33), state observer (8), the intermediate control functions (23) and (29), and the parameter adaptation functions (25) obtained based on the above derivations, the following properties can hold: (1) The boundeness of the whole signals in the closed-loop system can be validated; (2) The system output can follow the given reference signal well.

5   Conclusions

This paper has presented an observer-based adaptive NNs FTC method. Firstly, NNs have been utilized for approximating the unknown nonlinear functions, and the states observers have been constructed for estimating the unmeasured states. Then, by using the properties of Nussbaum gain function and Lyapunov-Krasobskii functional theory, and combining with adaptive backstepping design technique, the problem of FTC with unknown time-varying delays, unmeasured states, and unknown control direction has been solved. It is shown that not only all signals in the closed-loop system are proved to be bounded, but the system output can follow the given reference signal well.

Acknowledgments. This work was supported by the National Natural Science Foundation of China (Nos. 61573175, 61572244) and Liaoning BaiQianWan Talents Program.

References

1. Kwan, C., Lewis, F.L.: Robust backstepping control of nonlinear systems using neural networks. IEEE Syst. Man Cybern. Part A **30**, 753–766 (2000)
2. Kuljaca, O., Swamy, N., Lewis, F.L., Kwan, C.: Design and implementation of industrial neural network controller using backstepping. IEEE Trans. Ind. Electron. **50**, 193–201 (2003)
3. Polycarpou, M., Zhang, X.D., Xu, R., Yang, Y.L., Kwan, C.: A neural network based approach to adaptive fault tolerant flight control. In: Intelligent Control, Proceedings of the 2004 IEEE International Symposium, pp. 61–66 (2004)
4. Chen, M., Tao, G.: Adaptive fault-tolerant control of uncertain nonlinear large-scale systems with unknown dead zone. IEEE Trans. Cybern. **46**, 1851–1862 (2016)
5. Shen, Q.K., Jiang, B., Cocquempot, V.: Adaptive fuzzy observer-based active fault-tolerant dynamic surface control for a class of nonlinear systems with actuator faults. IEEE Trans. Fuzzy Syst. **22**, 338–349 (2014)

6. Tong, S.C., Huo, B.Y., Li, Y.M.: Observer-based adaptive decentralized fuzzy fault-tolerant control of nonlinear large-scale systems with actuator failures. IEEE Trans. Fuzzy Syst. **22**, 1–15 (2014)
7. Chen, B., Liu, X.P., Ge, S.S., Lin, C.: Adaptive fuzzy control of a class of nonlinear systems by fuzzy approximation approach. IEEE Trans. Fuzzy Syst. **20**, 1012–1021 (2012)
8. Wang, H.Q., Chen, B., Liu, K.F., Liu, X.P., Lin, C.: Adaptive neural tracking control for a class of nonstrict-feedback stochastic nonlinear systems with unknown backlash-like hysteresis. IEEE Trans. Neural Netw. Learn. Syst. **25**, 947–958 (2014)
9. Li, Y.M., Tong, S.C.: Adaptive fuzzy output-feedback stabilization control for a class of switched nonstrict-feedback nonlinear systems. IEEE Trans. Cybern. doi:10.1109/TCYB.2016.2536628
10. Chen, B., Zhang, H.G., Lin, C.: Observer-based adaptive neural network control for nonlinear systems in nonstrict-feedback form. IEEE Trans. Neural Netw. Learn. Syst. **27**, 89–98 (2016)
11. Li, Y.M., Tong, S.C.: Adaptive neural networks decentralized FTC design for nonstrict-feedback nonlinear interconnected large-scale systems against actuator faults. IEEE Trans. Neural Netw. Learn. Syst. doi:10.1109/TNNLS.2016.2598580

Neural Adaptive Dynamic Surface Control of Nonlinear Systems with Partially Constrained Tracking Errors and Input Saturation

Hairong Dong, Xiaoyu Wang, Shigen Gao$^{(\boxtimes)}$, and Yubing Wang

State Key Laboratory of Rail Traffic Control and Safety,
Beijing Jiaotong University, Beijing 100044, China
gaoshigen@bjtu.edu.cn

Abstract. This paper considers the neural adaptive dynamic surface control with partially constrained tracking errors and input saturation for a class of strict-feedback nonlinear systems with uncertain parameters. An error transformation method is utilized to guarantee the prescribed performance control of the partially constrained states, which restricts the partial states located in the prescribed bounds all through. Reduced-order interceptive signals are used to solve the problem of input saturation. Neural networks are utilized to online estimate the uncertainties of the system, and dynamic surface control technique is incorporated to circumvent the complexity explosion problem. The stability of the resulted system and all the signals in the system are proved by the Lyapunov stability theorem. At last, a simulation is presented to demonstrate the effectiveness of this control scheme.

Keywords: Neural adaptive control · Dynamic surface control · Partial tracking error constrained · Prescribed performance control · Input saturation · Uncertain nonlinear system

1 Introduction

In recent years, considerable developments on adaptive control for uncertain nonlinear systems to handle the uncertainties from the practical requirements and theoretical challenges. Subsequently, neural networks (NNs) have been incorporated into the adaptive control relying on the parallel processing and function approximation capacities [2–4], then, NNs-based adaptive backstepping control becomes an research hotpot for various high-order nonlinear system for unknown parameters and modeling uncertainties [7,12,15].

However, undesirable transient behaviors appear widely in the NNs-based approximations methods, caused by the convergence rate of precise approximation of NNs. Moreover, although the NNs-based control approaches satisfy the

This work is supported jointly by the Fundamental Research Funds for Central Universities (No. 2016RC054) and the State Key Laboratory of Rail Traffic Control and Safety (No. RCS2017ZQ001).

© Springer International Publishing AG 2017
F. Cong et al. (Eds.): ISNN 2017, Part II, LNCS 10262, pp. 20–27, 2017.
DOI: 10.1007/978-3-319-59081-3_3

Lyapunov stability in the infinite time, the partial tracking errors in finite time are difficult to be kept in some prescribed bounds. Then, to solve the problem mentioned above, a smooth adaptive neural controller was proposed in [14] where an integral-type Lyapunov function was introduced to solve the problem of guaranteeing transient performance. Furthermore, the barrier Lyapunov function (BLF)-based control method with full state constraints has been active research area for the nonlinear system, see [11]. An error-transformation (ET) method, which aims at transforming the "constrained" error into an equivalent "unconstrained" one for the strict feedback nonlinear uncertain systems was proposed in [1]. At the same time, many control systems encounter constraints on the control inputs, that is, the control signals implemented are usually limited in magnitude caused by the physical constrains, control design with input saturation is another necessary point, since the input saturation may cause performance degradation even instability of closed-loop systems [8,9,13].

In this paper, we design a neural adaptive dynamic surface control (DSC) for some strict feedback nonlinear systems with unmeasurable states, partially constrained states, and input saturation. NNs-based observer is designed to online estimation the unmeasurable states, the ET-based and dynamic surface control methods proposed [1,10] are utilized to guarantee the partially constrained tracking errors and to circumvent the complexity explosion problem. And, partially inspired by the method in [9], where a full-order auxiliary system is designed to compensate the effect caused by input saturation, this paper presents a reduced-order auxiliary-based design method to compensate the complicated nonlinearities caused by the input saturation, it is also noticeable that the method in [9] requires all the states in the controller while in this paper, only output information is required.

2   Preliminaries and Problem Formulation

The considered nonlinear SISO system with uncertainties is given as follows:

$$\begin{cases} \dot{x}_1 = x_2 + f_1(x_1) + d_1(t), \\ \dot{x}_i = x_{i+1} + f_i(\underline{x}_i) + d_i(t), \ i = 2, 3, \cdots, n-1 \\ \dot{x}_n = \text{sat}(u(t)) + f_n(X) + d_n(t) \end{cases} \quad (1)$$

where $\underline{x}_i = [x_1, \cdots, x_i]^T \in R^i, (i = 2, 3, \cdots, n)$ is state vector, and $\underline{x}_n = X$ is full state vector; $f_i(\cdot)$ is the smooth nonlinear function with unknown parameters; $d_i(t)$ is the bounded external disturbances where $|d_i(t)| \leqslant d_i^*$ and d_i^* is an unknown constant; u is the control input, $\text{sat}(u)$ is the input with saturation. The states are partitioned into two parts $\underline{x}_j = [x_1, \cdots, x_j] \in R^l, (j = 1, 2, \cdots, l)$ is the constrained part for the states error are constrained while $\underline{x}_k = [x_{l+1}, \cdots, x_k] \in R^r (k = l+1, \cdots, n)$ is the free part, where $l + r = n$. The constrained errors for the state $\underline{x}_j$ is constrained in the set $|z_j(t)| < \rho_j(t), j = 1, 2, \cdots, l$. $\text{sat}(u(t)) = u$ when $|u(t)| < u_M$, and $\text{sat}(u(t)) = \text{sign}(u(t))u_M$ when $|u(t)| \geq u_M$, where u_M is the bound of $u(t)$. And $y = x_1$ is the output of the system. It is noticed that the partially constrained errors problem becomes the full-state constraints problem studied in [6] if $\underline{x}_j = [x_1, \cdots, x_n]$ and the output constraint problem studied in [16] if $\underline{x}_j = x_1$.

For the given system (1), our goal is to design a neural adaptive dynamic surface control $u(t)$ such that $i)$ the states in the closed-loop system are uniformly bounded, $ii)$ the tracking error of the output and the reference signal remain in the certain of the prescribed bounds.

Assumption 1. *The desired trajectory $y_r(t)$ and its derivatives $y_r^{(1)}(t)$, $y_r^{(2)}(t)$ are given and both are bounded functions.*

Assumption 2. *There exists a set of constant b_i $(i = 1, 2, \cdots, n)$, $\forall X_1, X_2 \in R^i$, the inequality is satisfied as follows: $|F_i(X_1) - F_i(X_2)| \leqslant b_i \| X_1 - X_2 \|$.*

A performance function $\rho_j(t)$ is defined to guarantee the tracking error remaining in the range of prescribed constraints: $\rho_j(t) = (\rho_{0j} - \rho_{\infty j}) e^{-a_j t} + \rho_{\infty j}$ where ρ_{0j}, $\rho_{\infty j}$ and a_j are positive constants. Then, the transient tracking errors can be guaranteed with the prescribed range as follows:

$$-\delta_j \rho_j(t) < z_j(t) < \rho_j(t), \qquad \text{if } z_j(0) \geq 0 \tag{2}$$

$$\rho_j(t) < z_j(t) < \delta_j \rho_j(t), \qquad \text{if } z_j(0) < 0 \tag{3}$$

where $0 < \delta_j \leq 1$ is parameter to adjust the prescribed constraint scale. $\rho_{\infty j}$ restrain the tracking error in a pretty small size when the system is nearly steady. a_j adjusts the decreasing rate of the performance function, which requires the performance of convergence. Then, the transformed tracking error ξ_j can be defined as: $\xi_j = \frac{z_j}{\eta_j(t)}, \eta_j = q\bar{\eta}_j + (1 - q)\underline{\eta}_j$ where $q = 1$ if $z_j(t) \geqslant 0$, and $q = 0$ if $z_j(t) < 0$. The parameters $\bar{\eta}_j$, $\underline{\eta}_j$ are defined as follows:

$$\begin{cases} \bar{\eta}_j = \rho_j \\ \underline{\eta}_j = -\delta_j \rho_j(t) \end{cases} \text{if } z_j(0) \geqslant 0, \qquad \begin{cases} \bar{\eta}_j = \delta_j \rho_j(t) \\ \underline{\eta}_j = -\rho_j(t) \end{cases} \text{if } z_j(0) < 0 \tag{4}$$

Lemma 1. [5]: *Eq. (4) holds if and only if ρ_{0j}, $\rho_{\infty j}$, a_j and δ_j are selected satisfying (2) and (3): $0 < \xi < 1, \forall t > 0$.*

3   Observer Design Using Neural Networks

In this part, a state observer is established to estimate the states in system (1) which are not available to measure. The system (1) can be written in another way:

$$\dot{X} = AX + Ky + \sum_{i=1}^{n-1} B_i f_i(\hat{\underline{x}}_i) + B_n[f_n(X) + \text{sat}(u)] + \Delta F + d \tag{5}$$

And $y = CX$. $\hat{\underline{x}}_i$, $\hat{X}$ stand for the estimates of $\underline{x}_i$ and X respectively. $\Delta F_i = f_i(\underline{x}_i) - f_i(\hat{\underline{x}}_i)$, $\Delta F_n = f_n(X) - f_n(\hat{X})$, $(i = 1, 2, \cdots, n-1)$, $X = [x_1, \cdots, x_n]^\mathrm{T}$,

$$A = \begin{bmatrix} -k_1 & \\ \vdots & I \\ -k_n & 0 \cdots 0 \end{bmatrix}, \; K = \begin{bmatrix} k_1 \\ \vdots \\ k_n \end{bmatrix}, \; B_i = [\underbrace{0 \cdots 0}_{i} 1 \cdots 0]^{\mathrm{T}}, \; B_n = [0, \cdots, 1]^{\mathrm{T}}, \; \Delta F =$$

$[\Delta F_1, \cdots, \Delta F_n]^{\mathrm{T}}$, $C = [1 \cdots 0 \cdots 0]$, and $d = [d_1, \cdots, d_n]^{\mathrm{T}}$. There exists a vector K ensuring the matrix A a strict Hurwitz one. Then, there exists a symmetric matrix P that satisfies $A^{\mathrm{T}}P + PA = -Q$ when given a matrix $Q = Q^{\mathrm{T}}$. Then, the function can be written as follows: $f_i(\hat{\underline{x}}_i) = \hat{f}_i(\hat{\underline{x}}_i \mid \theta_i^*) + \varepsilon_i^* = \theta_i^{\mathrm{T}}\varphi_i(\hat{\underline{x}}) + \varepsilon_i^*$, $(i = 1, 2, \cdots, n)$, where ε_i^* is the neural approximation error and $|\varepsilon_i^*| < \varepsilon_m$, θ_i^* is the value of one θ that makes neural approximation error minimum. θ_i^* is defined as follows: $\theta_i^* = \arg \min\limits_{\theta_i \in U_i} \left[\sup\limits_{\hat{\underline{x}}_i \in \Omega_i} |\hat{f}_i(\hat{\underline{x}}_i \mid \theta_i^*) - f_i(\hat{\underline{x}}_i)| \right]$, where U_i is compact region for θ_i.

Design a state observer as:

$$\begin{cases} \dot{\hat{x}}_i = \hat{x}_{i+1} + \hat{f}_i(\hat{\underline{x}}_i) + k_i(y - \hat{x}_1), & i = 1, 2, \cdots, n-1 \\ \dot{\hat{x}}_n = \mathrm{sat}(u(t)) + f_n(\hat{X}) + k_n(y - \hat{x}_1) \end{cases} \tag{6}$$

The observer error e is defined: $e = [e_1, e_2, \cdots, e_n]^{\mathrm{T}} = X - \hat{X}$. Then, $\dot{e}$ can be written as $\dot{e} = Ae + \varepsilon + \sum_{i=1}^{n} B_i \tilde{\theta}_i^{\mathrm{T}} \varphi_i + \Delta F + d(t)$, where $\varepsilon = [\varepsilon_1, \cdots, \varepsilon_n]^{\mathrm{T}}$, and $\tilde{\theta}_i = \theta^* - \theta_i$, $i = 1, \cdots, n$.

Selecting the following Lyapunov function: $V_0 = e^{\mathrm{T}}Pe$, then the time derivative form of V_0 can be get: $\dot{V}_0 \leqslant -\lambda_{min}(Q)\|e\|^2 + 2e^{\mathrm{T}}P(\varepsilon + \sum_{i=1}^{n} B_i \tilde{\theta}_i^{\mathrm{T}} \varphi_i + \Delta F + d(t))$. where $\lambda_{min}(Q)$ is the smallest eigenvalue of matrix Q. Using the fact that $\varphi_i \varphi_i^{\mathrm{T}} \leqslant I$ and some inequalities, we can turn the above equation into the following form:

$$\dot{V}_0 \leqslant -P_0\|e\|^2 + \sum_{i=1}^{n} \tilde{\theta}_i^{\mathrm{T}} \theta_i + L_0 \tag{7}$$

where $r = 2\sum_{i=1}^{n} b_i\|P\|$, $P_0 = \lambda_{min}(Q) - r - 1 - n\sigma\|P\|^2 - n\|P\|$, $L_0 = \|P\|^2\|\varepsilon^*\|^2 + \sum_{i=1}^{n} d_i^{*2}$.

Remark 1. From (7), it can be concluded that the observation errors cannot be guaranteed to convergence and is not sufficient for a stable system, thus needing to be considered next.

4 Partial Tracking Error Constrained Adaptive Controller Design

The controller design is divided into three parts including n steps. The first part includes the first nth states with error constrained and input saturation, the other part deals the input saturation only. For the first part, we change the coordinates as follows: $z_1 = y - y_r$, $z_j = \hat{x}_j - \vartheta_j$, $\chi_j = \vartheta_j - \alpha_{j-1}$, $\xi_j = \frac{z_j}{\eta_j}$, $S_j = \frac{\xi_j}{1 - \xi_j}$, $j = 1, \cdots, l$. For the second part, we change the coordinates as follows:

$z_k = \hat{x}_k - \vartheta_k, \chi_k = \vartheta_k - \alpha_{k-1}, S_k = \vartheta_k - \alpha_{k-1}, \dot{h}_k = -c_k h_k + h_{k+1}, \dot{h}_n = -c_n h_n + (\text{sat}(u) - u), k = l+1, \cdots, n-1$ where z_i is error surface; ϑ_i is obtained through a first-order filter on intermediate control function α_{i-1}, and χ_i is the output error of the first-order filter; S_j is the transformed error surfaces.

Step 1: First we can get the time derivative form of the error surface S_1:

$$\dot{S}_1 = q_1(e_2 + z_2 + \alpha_1 + \chi_2 + \theta_1^T \varphi_1 + \tilde{\theta}_1^T \varphi_1 + \varepsilon_1 + \Delta F_1 + d_1 - \dot{y}_r - \dot{\eta}_1 \xi_1) \quad (8)$$

where $q_1 = \frac{1}{(1-\xi_1)^2 \eta_1}$. Choose the Lyapunov function candidates as $V_1 = V_0 + \frac{1}{2}S_1^2 + \frac{1}{2\gamma_1}\tilde{\theta}_1^T \tilde{\theta}_1$ where γ_1 is the design constant matrice.

Considered the Lyapunove theory, we choose the intermediate control α_1 and the adaptive law θ_1 as $\alpha_1 = -\beta_1 S_1 \eta_1 - \beta_1 \frac{S_1}{q_1} - z_2 - \theta_1^T \varphi_1 + \dot{y}_r + \dot{\eta}_1 \xi_1, \dot{\theta} = \gamma_1 S_1 q_1 \varphi_1 - \tau_1 \theta_1$, where β_1 and τ_1 are design parameters. To avoid the explosion of complexity, variable ϑ_2 is introduced and is obtained as: $\varsigma_2 \dot{\vartheta}_2 + \vartheta_2 = \alpha_1, \vartheta_2(0) = \alpha_1(0)$. By the definition of $\chi_2 = \vartheta_2 - \alpha_1$, it can be obtained that $\dot{\vartheta}_2 = -\frac{\chi_2}{\varsigma_2}$, and $\dot{\chi}_2 = \dot{\vartheta}_2 - \dot{\alpha}_1 = \frac{\chi_2}{\varsigma_2} + H_2$, where $H_2(\cdot)$ is a continuous function of variables $S_1, S_2, \chi_2, y_r, \dot{y}_r, \ddot{y}_r$ and θ_1, its expression can be described as: $H_2(\cdot) = \beta_1 \dot{S}_1 \eta_1 + \beta_1 \frac{\dot{S}_1}{q_1} + \xi_2 \dot{\eta}_2 + \frac{\theta_1^T \partial \varphi_1(\hat{x}_1)}{\partial \hat{x}_1} \dot{\hat{x}}_1 - \ddot{y}_r - \ddot{\eta}_1 \xi_1$. The first-order filter before each step is similar, and we omit it in the next following steps.

Step $j(2 \leqslant j \leqslant l-1)$: Similar to step 1, we have

$$\dot{S}_j = q_j(\dot{z}_j - \dot{\eta}_j \xi_j) = q_j(\theta_j^T \varphi_j + z_{j+1} + \alpha_j + \chi_{j+1} - \dot{\vartheta}_j + k_j e_1 - \dot{\eta}_j \xi_j) \quad (9)$$

where $q_j = \frac{1}{(1-\xi_j)^2 \eta_j}$. The Lyapunov function is choosen as $V_j = V_{j-1} + \frac{1}{2}S_j^2 + \frac{1}{2\gamma_j}\tilde{\theta}_j^T \tilde{\theta}_j + \frac{1}{2}\chi_j^2$ where $\gamma_j > 0$ is the constant matrix. Noticing that the Lyapunov function in the following steps are similar, we omit them in the next writting.

To maintain the stability of the system, the intermediate control α_j and the adaptive function θ_j are chosen as: $\alpha_j = -\beta_j S_j \eta_j - \beta_j \frac{S_j}{q_j} - z_{j+1} + \dot{\vartheta}_j - k_j e_1 + \dot{\eta}_j \xi_j, \dot{\theta}_j = \gamma_j S_j q_j \varphi_1 - \tau_j \theta_j$ where β_j and τ_j are design parameters.

Step l: The lth state is the last one with error constrained, and we have

$$\dot{S}_l = q_l(\dot{z}_l - \dot{\eta}_l \xi_l) = q_l(\theta_l^T \varphi_l + z_{l+1} + \alpha_l + \chi_{l+1} + h_{l+1} - \dot{\vartheta}_l + k_l e_1 - \dot{\eta}_l \xi_l) \quad (10)$$

where $q_l = 1/(1-\xi_l)^2 \eta_l$.

Similar to step j, the intermediate control α_l and the adaptive law θ_l are as follows: $\alpha_l = -\beta_l S_l \eta_l - \beta_l \frac{S_l}{q_l} - z_{l+1} + \dot{\vartheta}_l - k_l e_1 + \dot{\eta}_l \xi_l - h_{l+1}, \dot{\theta}_l = \gamma_l S_l q_l \varphi_l - \tau_l \theta_l$. where β_l and τ_l are design parameters.

Step $k(l+1 \leqslant k \leqslant n-1)$: The time derivative form of S_k can be expressed as

$$\dot{S}_k = \dot{\hat{x}}_k - \dot{\vartheta}_k - \dot{h}_k = S_{k+1} + \chi_{k+1} + \alpha_k + \theta_k^T \varphi_k + k_k e_1 - \dot{\vartheta}_k + c_k h_k \quad (11)$$

Similar to the former steps, α_k and θ_k can be chosen as follows: $\alpha_k = -c_k(\hat{x}_k - \vartheta_k) - \sigma S_k - \theta_k^T \varphi_k + \dot{\vartheta}_k - k_k e_1, \dot{\theta}_k = \gamma_k S_k \varphi_k - \tau_k \theta_k$ where c_k and τ_k are design parameters.

Step n: In this step, the control input u will be designed. First give the time derivative of S_n as

$$\dot{S}_n = \dot{\hat{x}}_n - \dot{\vartheta}_n - \dot{h}_n = \theta_k^T \varphi_k + k_n e_1 - \dot{\vartheta}_n + c_n h_n + u \tag{12}$$

To make the Lyapunov function negative definite, the input u and the adaptive law θ_l are chosen as: $u = -c_n(\hat{x}_n - \vartheta_n) - \sigma S_n - \theta_n^T \varphi_n + \dot{\vartheta}_n - k_n e_1, \dot{\theta}_n = \gamma_n S_n \varphi_n - \tau_n \theta_n$. where c_n and τ_n are design parameters. Then we can get

$$\dot{V}_n \leqslant -P_1 \|e\|^2 + \frac{1}{\sigma} \sum_{i=1}^{n} \tilde{\theta}_i^T \hat{\theta}_i + L_1 - (2\beta_1 - 2q_1^*)S_1^2 + \sum_{i=1}^{l} S_i q_i \chi_{i+1} - \sum_{i=2}^{l} (2\beta_i - \sigma$$

$$q_i^*)S_i^2 + \sum_{i=1}^{n} \frac{\tau_i}{r_i} \tilde{\theta}_i^T \theta_i + \sum_{i=2}^{n} \chi_k(-\frac{\chi_k}{\varsigma} + H_k) + \frac{1}{\sigma} \sum_{i=2}^{n} \tilde{\theta}_i^T \hat{\theta}_i - \sum_{i=l+1}^{n} c_i S_i^2 + \sum_{i=l+1}^{n-1} S_i \chi_{i+1}$$

$$\tag{13}$$

where V_n is the final Lyapunov function concluding all the signals in each step.

Let $\Omega_i = \{(e, S_i, \theta_i, \chi_i) : [e^T e + \frac{1}{2} \sum_{i=1}^{n} S_i^2 + \sum_{i=1}^{n} \frac{1}{2r_i} \tilde{\theta}_i^T \tilde{\theta}_i + \frac{1}{2} \sum_{i=2}^{n} \chi_i^2] \leqslant D_i\}$ where D_i is a known positive constant. Since Ω_i is a compact set and H_{i+1} is a continuous function, there exists a positive constant M_{i+1} such that $|H_{i+1}| \leqslant M_{i+1}$ on Ω_i. Consequently, we have $|\chi_{i+1} H_{i+1}| \leqslant \frac{1}{2}\chi_{i+1}^2 + \frac{1}{2}M_{i+1}^2$. Then, (13) can be written as

$$\dot{V}_n \leqslant -P_1 \|e\|^2 + L_2 - \sum_{i=1}^{n} (\frac{\tau_i}{2r_i} - \frac{2}{\sigma})\tilde{\theta}_i^T \theta_i + \frac{1}{\sigma}\tilde{\theta}_1^T \theta_1 - \sum_{i=2}^{l} (2\beta_i - \sigma q_i^* - \frac{1}{2}q_i^*)S_i^2$$

$$+ \frac{1}{2}c_n^2 - (2\beta_1 - \frac{5}{2}q_1^*)S_1^2 - \sum_{i=l+1}^{n} (c_i - \frac{1}{2})S_i^2 - \sum_{i=1}^{l} (\frac{1}{\varsigma} - \frac{1}{2} - \frac{1}{2}q_i^*)\chi_{i+1}^2$$

$$- \sum_{i=l+1}^{n-1} (\frac{1}{\varsigma} - 1)\chi_{i+1}^2$$

where $L_2 = L_1 + \sum_{i=1}^{n} \frac{\tau_i}{2r_i}\|\theta_i^*\|^2 + \sum_{i=2}^{n} \frac{1}{2}M_i^2$.

Choose $P_1 > 0$, $\frac{\tau_i}{2r_i} - \frac{2}{\sigma} > 0 (i = 1, 2, 3, \cdots, n)$, $\frac{\tau_i}{2r_i} - \frac{1}{\sigma} > 0 (i = 1, 2, 3, \cdots, n)$, $2\beta_i - \sigma q_i^* - \frac{1}{2}q_i^* > 0 (i = 2, 3, \cdots, l)$, $2\beta_1 - \frac{5}{2}q_1^* > 0$, $c_i - \frac{1}{2} > 0 (i = l+1, \cdots, n)$, $\frac{1}{\varsigma} - \frac{1}{2} - \frac{1}{2}q_i^* > 0 (i = 1, \cdots, l)$, $\frac{1}{\varsigma} - 1 > 0 (i = l+1, \cdots, n-1)$, and define $C = min\{P_1 \lambda_{min}(P), \frac{\tau_i}{2r_i} - \frac{2}{\sigma}(i = 1, \cdots, n), \frac{\tau_i}{2r_i} - \frac{1}{\sigma} > 0(i = 1, \cdots, n), 2\beta_i - \sigma q_i^* - \frac{1}{2}q_i^* > 0(i = 2, \cdots, l), 2\beta_1 - \frac{5}{2}q_1^*, c_i - \frac{1}{2} > 0(i = l+1, \cdots, n), \frac{1}{\varsigma} - \frac{1}{2} - \frac{1}{2}q_i^* > 0(i = 1, \cdots, l), \frac{1}{\varsigma} - 1 > 0(i = l+1, \cdots, n-1)\}$ and $D = L_2$. As a result, the inequality $\dot{V}_n \leqslant -CV_n + D$ can be obtained, the remainder of the detailed proof is similar to [1], which is not studied in details here.

5   Simulation Example and Results Analysis

Consider the second-order system in (1), where $f_i(\underline{x}_i) = x_1^2 \sin x_1$, $f_n(X) = -a - b - cx_2^2$, and a, b are known constants. The nominal values of the states are $x_1(0) = 0.2$, $x_2(0) = 0$. The reference signal is $y_r(t) = \sin(t)$. Our goal is to make the tracking error convergent subject to the performance function as follows: $\rho_1(t) = (\rho_{1_0} - \rho_{1_\infty})e^{-a_1 t} + \rho_{1_\infty}$, where $\rho_{1_0} = 0.7$, $\rho_{1_\infty} = \pi/20$. The controller u, intermediate control α_1 and adaptive laws are chosen as follows: $\alpha_1 = -c_1 S_1 - z_2 + \dot{y}_r - h(t) - \hat{f}_1$, $u = -(c2 + 0.5)S_2/q_2 - \hat{f}_2 - k_2 e_1 + \dot{\theta}_2 - h(t) + \dot{\eta}_2 \xi_2 - c_2 S_2 q_2$, $\dot{\theta}_1 = -\tau_1 \theta_1 + \gamma_1 S_1 q_1 \varphi_1$, $\dot{\theta}_2 = -\tau_2 \theta_2 + \gamma_2 S_2 \varphi_2$; The design parameters are chosen as $\tau_1 = 0.1$, $\gamma_1 = 10$, $\tau_2 = 0.1$, $\gamma_2 = 10$, $k_1 = 2$, $k_2 = 2$, $\varsigma_1 = 0.01$, $\beta_1 = \beta_2 = 12$, and the input saturation is set to be 10.

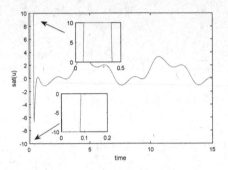
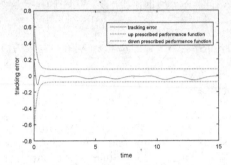

Fig. 1. sat(u)   **Fig. 2.** Constrained tracking error

The simulation results with the use of the mentioned adaptive control method are shown in Figs. 1 and 2, where Fig. 1 shows the trajectory of control input with saturation, and Fig. 2 shows the track error $z_1(t)$ with error constrained. The proposed control method is proved to guarantee the convergence of all variables. What's more, the transient performance of tracking error with constrained is guaranteed all the time.

6   Concluding Remarks

In this paper, a neural adaptive output-feedback DSC method with partial prescribed tracking errors and input saturation has been proposed for a class of strict feedback nonlinear systems. By applying the constraint bounds to the transformation of the tracking errors into new error variables, the transient performance is guaranteed for all the time and the tracking errors within the prescribed bounds. Using the Lyapunov stability theorem, the states in the whole system are proved to be stable.

References

1. Bechlioulis, C.P., Rovithakis, G.A.: Adaptive control with guaranteed transient and steady state tracking error bounds for strict feedback systems. Automatica **45**(2), 532–538 (2009)

2. Gao, S., Dong, H., Chen, Y., Ning, B., Chen, G., Yang, X.: Approximation-based robust adaptive automatic train control: An approach for actuator saturation. IEEE Trans. Intell. Transp. Syst. **14**(4), 1733–1742 (2013)
3. Gao, S., Dong, H., Lyu, S., Ning, B.: Truncated adaptation design for decentralized neural dynamic surface control of interconnected nonlinear systems under input saturation. Int. J. Control **89**(7), 1447–1466 (2016)
4. Gao, S., Dong, H., Ning, B., Sun, X.: Neural adaptive control for uncertain mimo systems with constrained input via intercepted adaptation and single learning parameter approach. Nonlinear Dyn. **82**(3), 1109–1126 (2015)
5. Han, S.I., Lee, J.M.: Partial tracking error constrained fuzzy dynamic surface control for a strict feedback nonlinear dynamic system. IEEE Trans. Fuzzy Syst. **22**(5), 1049–1061 (2014)
6. He, W., Chen, Y., Yin, Z.: Adaptive neural network control of an uncertain robot with full-state constraints. IEEE Trans. Cybern. **46**(3), 620–629 (2016)
7. Kwan, C., Lewis, F.L.: Robust backstepping control of nonlinear systems using neural networks. IEEE Trans. Syst. Man Cybern.-Part A: Syst. Hum. **30**(6), 753–766 (2000)
8. Li, T., Li, R., Li, J.: Decentralized adaptive neural control of nonlinear interconnected large-scale systems with unknown time delays and input saturation. Neurocomputing **74**(14), 2277–2283 (2011)
9. Polycarpou, M., Farrell, J., Sharma, M.: On-line approximation control of uncertain nonlinear systems: issues with control input saturation. In: Proceedings of the 2003 American Control Conference, 2003, vol. 1, pp. 543–548. IEEE (2003)
10. Swaroop, D., Hedrick, J.K., Yip, P.P., Gerdes, J.C.: Dynamic surface control for a class of nonlinear systems. IEEE Trans. Autom. Control **45**(10), 1893–1899 (2000)
11. Tee, K.P., Ge, S.S., Tay, E.H.: Barrier lyapunov functions for the control of output-constrained nonlinear systems. Automatica **45**(4), 918–927 (2009)
12. Tong, S., Li, Y.: Observer-based fuzzy adaptive control for strict-feedback nonlinear systems. Fuzzy Sets Syst. **160**(12), 1749–1764 (2009)
13. Wen, C., Zhou, J., Liu, Z., Su, H.: Robust adaptive control of uncertain nonlinear systems in the presence of input saturation and external disturbance. IEEE Trans. Autom. Control **56**(7), 1672–1678 (2011)
14. Zhang, T., Ge, S.S., Hang, C.C.: Adaptive neural network control for strict-feedback nonlinear systems using backstepping design. Automatica **36**(12), 1835–1846 (2000)
15. Zhou, Q., Shi, P., Lu, J., Xu, S.: Adaptive output-feedback fuzzy tracking control for a class of nonlinear systems. IEEE Trans. Fuzzy Syst. **19**(5), 972–982 (2011)
16. Zhou, Q., Wang, L., Wu, C., Li, H., Du, H.: Adaptive fuzzy control for nonstrict-feedback systems with input saturation and output constraint. IEEE Trans. Syst. Man Cybern.: Syst. **47**(1), 1–12 (2017)

An Application of Master-Slave ADALINE for State Estimation of Power System

Zhanshan Wang$^{(\boxtimes)}$, Haoyuan Gao, and Huaguang Zhang

School of Information Science and Engineering,
Northeastern University, Shenyang 110819, China
zhanshan_wang@163.com

Abstract. This paper presents two-fold adaptive linear neural networks (ADALINE) to gain the current operating state of power system for a fast and accurate estimation. On the one hand, the Slave-ADALINE applies the fixed and larger step-size least mean square algorithm to accelerate the convergence speed of weights. On the other hand, the Master-ADALINE follows least mean square with a variable step-size factor to achieve the minimum of steady-state error. In this paper the IEEE-30 network of power system is used to verify the effectiveness of the proposed method, and comparisons of simulation results with Particle Swarm Optimization algorithm and single ADALINE are also provided.

Keywords: State estimation · Master-Slave ADALINE · Least mean square (LMS) · Power system

1 Introduction

In the last years, a rapid progress from the conventional electrical grids toward the new smart grids has happened to deal with the increasing requirements of customers [1]. In fact, various power system applications such as optimal power flow, economic dispatch, and security assessment rely on the state variables of power systems under management that are filtered initially by state estimation [2]. Real time monitoring of power systems has therefore become very important, and the timely detection of contingencies has also become important in order to allow the undertaking ofremedial actions to avoidany potentially dangerous situation [3].

F.C. Schweppe, in the 1970s, firstly presented the concept of the power system state estimation and applied weighted least squares (WLS) method to solve this problem [4]. But, with the high development of the Distributed Generations, the complexity of power system, operation and communication will also affect the optimal state estimation. In response to these challenges, various methods especially based on the evolutionary algorithms have been proposed in many

---

Z. Wang—This work was supported by the National Natural Science Foundation of China (Grant Nos. 61473070, 61433004, 61627809), and SAPI Fundamental Research Funds (Grant No. 2013ZCX01).

F. Cong et al. (Eds.): ISNN 2017, Part II, LNCS 10262, pp. 28–35, 2017.
DOI: 10.1007/978-3-319-59081-3_4

literatures. In the 2015, Reza used the firefly algorithm to solve state estimation problems [1]. A hybrid method based on Particle Swarm Optimization (PSO) was proposed [5,6] for distribution state estimation with the Distributed Generations. Specially for the PSO, many researchers have tried to improve the performance of PSO, focusing on the individual best position (Pbest) and global best position (Gbest) [7,8]. In order to improve the performance of PSO, a new PSO is proposed in [9]. The algorithm can adaptively change the initial trajectory of a particle to make the particle explore a new region. Nevertheless, the above methods still need to use more memory resources.

In recent years, adaptive linear neural network (ADALINE) has been widely used in harmonic analysis [10–14]. In 2014, a new algorithm minimizes an objective function based on weighted square of the error and using a modified recursive Gauss Newton (MRGN) method was introduced by Nanda [15]. The method in [10–15] can minimize the tracking error, and has a faster convergence rate. Meanwhile, its multi-input and single-output structure can reduce the complexity of the system design. However, the ADALINE technique prematurely converges during the estimation of the signal with time-varying parameters, affecting the accuracy of estimation. Therefore, in 2009, G.W. Chang presented a two-stage ADALINE for harmonics and interharmonics measurement [16], but the computing time is double. In this paper, the authors will use a two-fold ADALINE structure, i.e. applying the Master-Slave ADALINE to solve the state estimation of power system. Compared with the reference [16], the proposed method has parallel processing characteristics, which can improve the speed of computation. The IEEE-30 network of power system is used to verify the achievability of the way, and comparisons of simulation results with PSO algorithm [9] and single ADALINE [15] is tested.

The rest of this paper is organized as follows. Section 2 shows the power system state estimation of specific implementation. Section 3 presents the MS ADALINE structure and algorithm. Section 4 presents the simulation results of IEEE-30 network of power system and the simulation results are compared with PSO algorithm and single ADAINE. Section 5 draws some conclusions of the present paper.

2 Specific Accomplishment of State Estimation

Before presenting the master-slave adaptive linear neural network structure and algorithm in detail, we need to get the mathematical model of the state estimation of voltage. So, in this section will introduce the common mathematical model of power lines and branch power flow calculation formula, and the specific processes of state estimation of voltage.

In the steady-state analysis of power system, mathematical model of power lines is based on the resistance, reactance, and admittance, serial or parallel conductance through the equivalent circuits. Figure 1 shows the π-type equivalent circuit of transmission line. Among them, $Z = R + jX$, where R is the resistor of power line, X is the inductance of power lines. $Y = jB$ is the admittance

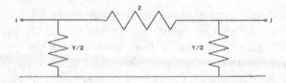

Fig. 1. The π-equivalent circuit of transmission line

of power lines. The active and reactive power calculation formula of the branch from node i to node j is defined as follows,

$$P_{ij} = |V_i|^2 |y_{ij}| \cos(-\alpha_{ij}) - |V_i||V_j||y_{ij}| \cos(\delta_i - \delta_j - \alpha_{ij}) \tag{1}$$

$$Q_{ij} = |V_i|^2 |y_{ij}| \sin(-\alpha_{ij}) - |V_i||V_j||y_{ij}| \sin(\delta_i - \delta_j - \alpha_{ij}) \tag{2}$$

where, $|V_i|$ and δ_i are the amplitude and phase of the voltage node i, respectively. $|V_j|$, δ_j are the amplitude and phase of the voltage node j. $|y_{ij}|$, α_{ij} are the admittance modulus and phase of the branch from node i to node j ($y_{ij} = |y_{ij}| \angle \alpha_{ij}$), respectively.

By comparing Eq. (1) with (2), let

$$W_1 = 1, W_2 = |V_j| \cos(\delta_j), W_3 = |V_j| \sin(\delta_j) \tag{3}$$

Therefore, from the (1)–(2),we can derive the following formulas,

$$|V_j| = \sqrt{W_2^2 + W_3^2} \tag{4}$$

$$\delta_j = \arctan(W_3/W_2) \tag{5}$$

3 Structure of MS ADALINE

This section will introduce ADALINE method to solve the power system state estimation problem. Figure 1 is the structure diagram of MS ADA-LINE. The structure is formed by two conventional master ADALINE and slave ADALINE, whose weights are denoted as $\{\hat{w}_{1M}(n), \hat{w}_{2M}(n), \hat{w}_{3M}(n)\}$ and $\{\hat{w}_{1S}(n), \hat{w}_{2S}(n), \hat{w}_{3S}(n)\}$. At the same time, the master and slaver ADA-LINE have the same reference signal of input and desired output, which is $\{I_1(n), I_2(n), I_3(n)\}$ and $D(n)$, and the corresponding feedback signal of error is $\{E_M(n), E_S(n)\}$. The error feedback signal is transferred to the decision controller to adjust the real-time weights. The Slave-ADALINE applies fixed, larger step-size least mean square (LMS) algorithm to weights for accelerating the speed of convergence. At the moment, the Master-ADALINE follows least mean square with a variable step-size factor, in order to accomplish the minimum of steady-state error. Finally, after some iterations MS ADALINE weights can be obtained to calculate amplitude and phase of the node j, the formulas are as follows,

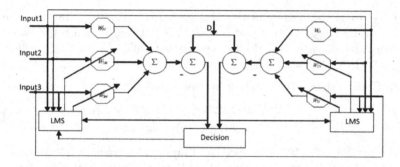

Fig. 2. The framework of MS ADALINE for power state estimation

$$|V_j| = \sqrt{\hat{w}_{2M}(n)^2 + \hat{w}_{3M}(n)^2} \tag{6}$$
$$\delta_j = \arctan(\hat{w}_{3M}(n)/\hat{w}_{2M}(n)) \tag{7}$$

symbols S_i and δ_i are the amplitude and phase of the harmonic i, respectively. Weights of MS ADALINE are adjusted as follows.

Step-1: The adjustment of weights $\{\hat{w}_{1S}(n), \hat{w}_{2S}(n), \hat{w}_{3S}(n)\}$ of the Slave-ADALINE.

$$\hat{w}_{1S}(n) = 1 \tag{8}$$
$$\hat{w}_{2S}(n+1) = \hat{w}_{2S}(n) + \mu_S E_S(n) I_2(n) \tag{9}$$
$$\hat{w}_{3S}(n+1) = \hat{w}_{3S}(n) + \mu_S E_S(n) I_3(n) \tag{10}$$
$$E_S(n) = D(n) - Y_S(n) \tag{11}$$
$$Y_S(n) = [\hat{w}_{1S}, \hat{w}_{2S}, \hat{w}_{3S}][I_1, I_2, I_3]^T \tag{12}$$

symbol $D(n)$ is the desired output, $Y_S(n)$ is the output of the Slave-ADALINE respectively.

Step-2: The adjustment of weights $\{\hat{w}_{1M}(n), \hat{w}_{2M}(n), \hat{w}_{3M}(n)\}_{i=1}^{L}$ of the Master-ADALINE.

$$\hat{w}_{1M}(n+1) = 1 \tag{13}$$
$$hatw_{2M}(n+1) = \begin{cases} \hat{w}_{2S}(n+1), if(A_S(m) < A_M(m)) \\ \hat{w}_{2M}(n) + \mu_M E_M(n) I_2(n), else \end{cases} \tag{14}$$
$$\hat{w}_{3M}(n+1) = \begin{cases} \hat{w}_{3S}(n+1), if(A_S(m) < A_M(m)) \\ \hat{w}_{3M}(n) + \mu_M E_M(n) I_3(n), else \end{cases} \tag{15}$$
$$A_S(m) = \sum_{m=0}^{Q} E_S^2(m) \qquad A_M(m) = \sum_{m=0}^{Q} E_M^2(m) \tag{16}$$
$$E_M(n) = D(n) - Y_M(n) \tag{17}$$
$$Y_M(n) = [\hat{w}_{1M}(n), \hat{w}_{2M}(n), \hat{w}_{3M}(n)][I_1(n), I_2(n), I_3(n)]^T \tag{18}$$

symbol $Y_M(n)$ is the output of the Master-ADALINE. After the end of each iteration, $A_S(m)$ and $A_M(m)$ will be calculated. Decision controller based on the results of comparison of the two calculated values is used to predict the Master-ADALINE updated weights.

Step-3: Update the variable step of Master-ADALINE.

$$\mu_M(n+1) = \begin{cases} \frac{\mu_M(n)+\mu_S}{2}, if(A_S(m) < A_M(m)) \\ \max[C_1\mu_M(n), \mu_{\min}], else \end{cases} \tag{19}$$

From the formula (19), one can see taht, if the tracking performance of Master-ADALINE is better, Master-ADALINE step value is the average value of Master-ADALINE step and Slaver-ADALINE step, which makes the Master-ADALINE, converges faster. In order to obtain small steady-state error, the step value of Master-ADALINE should be further reduced.

Step-4: According to the formula (1) and (2), the amplitude and phase of the voltage of the node j can be calculated, respectively.

In order to obtain a good convergence efficiency, the values of C_1, μ_{min}, μ_S and μ_M need to be chosen. The above discussions show that μ_S determines the global convergence of MS ADALINE, μ_M determines the accuracy of convergence, therefore, the selection of these two values plays a key in the performance of the network. These two main values can be determined based on previous experience.

The weights of the Master-ADALINE are updated by the expected outputs until them no long changed obviously, or the maximum number of iterations is reached. The active power and reactive power are alternating as the expected input of the MS ADALINE. The reference input signals as shown in Table 1.

Table 1. The reference input signals

Input signals	Pij	Qij
Input1	$|V_i|^2|y_{ij}|\cos(-\alpha_{ij})$	$|V_i|^2|y_{ij}|\cos(-\alpha_{ij})$
Input2	$-|V_i||y_{ij}|\cos(\delta_i - \alpha_{ij})$	$-|V_i||y_{ij}|\sin(\delta_i - \alpha_{ij})$
Input3	$-|V_i||y_{ij}|\sin(\delta_i - \alpha_{ij})$	$|V_i||y_{ij}|\cos(\delta_i - \alpha_{ij})$

Remark 1. The proposed method used to deal with the state of power system needs less memory space compared with previous method, like PSO [9]. So, this needs less time to compute the results.

Remark 2. Compared with [16], the proposed method has parallel processing characteristics, which can improve the speed of computation.

Remark 3. Compared with the single ADALINE [15], the proposed method has a two-fold structure, i.e. master ADALINE and slave ADALINE. The slave ADA-LINE mainly is used to improve the speed of convergence, at the same time, the master ADALINE could accomplish the minimum of steady-state error.

4   Simulation Results

A IEEE-30 network of power system is used to verify the achievability of the proposed method. Meanwhile, some comparisons of simulation results with PSO [9] and single ADALINE [15] are presented. Then the simulation results indicate that the proposed method has better accuracy than the PSO algorithm and the single ADALINE, and convergence rate of which is faster than the single ADALINE.

Figure 3 shows the comparisons between the results of the MS ADALINE and PSO algorithms and the single ADALINE. MS ADALINE and single ADAINE have better performance than PSO for the ability of voltage amplitude and phase estimation. What's more, MS ADALINE voltage amplitude estimated average error is 0.0015769, phase estimated average error is 0.0047077. PSO voltage amplitude estimated average error is 0.035515, phase estimated average error is 0.022969. The single ADALINE voltage amplitude estimated average error is 0.0022536, phase estimated average error is 0.021897. So MS ADALINE results are better than the PSO algorithm and single ADALINE. MS ADALINE results are more accurate, and MS ADALINE model has obvious advantages on simulation time, whose value is 0.015 s, and PSO is 0.103 s (CPU 887 1.5 GHz).

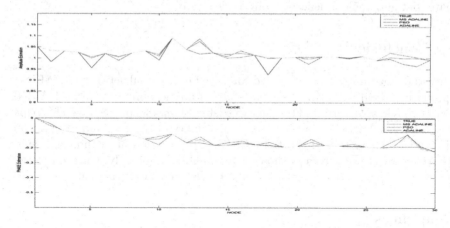

Fig. 3. The comparisons of estimated voltage amplitude and phase

Figure 4 shows the comparisons of tracking performance of MS ADALINE, Single ADALINE and PSO. The PQ_{12} is the actual measured value. It can be seen from the right simulation diagram that MS ADALINE coincides with the expected waveform after 16 iterations, PSO converges after the 37th iteration, and single ADALINE converges to the expected value after the 25th iteration. The left diagram is a comparison of the MS ADALINE and PSO algorithms of node 2, the horizontal axis is the number of iterations, and the vertical axis is the error degree. The error degree is defined as $\Delta = (\hat{V}_i(k) - V_{imeas})^2 + (\hat{\delta}_i(k) - \delta_{imeas})^2$, where, $\hat{V}_i(k)$, $\hat{\delta}_i(k)$ are the estimated values of each iteration

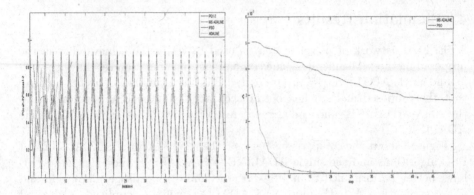

Fig. 4. The comparisons of tracking performance of MS ADALINE, Single ADALINE and PSO

and V_{imeas}, δ_{imeas} are the actual measured values of the node. Above, the PSO algorithm has a lower convergence rate, and the estimation accuracy is worse. The estimation precision of MS ADALINE is better than PSO and single ADA-LINE. Therefore, MS ADALINE not only can improve the accuracy of the estimate, but also could ameliorate convergence rate.

5　Conclusion

This study introduces a master-slave adaptive linear neural network (ADALINE) approach to deal with power system state estimation problem. MS ADALINE has a two-fold structure, and the characteristics of parallel processing. This paper uses a IEEE-30 network to verify the achievability of the way, and comparisons of simulation results with Particle Swarm Optimization algorithm and single ADAINE. Simulation results shows MS ADALINE not only can improve the accuracy of the estimate, but also could ameliorate convergence rate.

References

1. Khorshidi, R., Shabaninia, F., Niknam, T.: A new smart approach for state estima-tion of distribution grids considering renewable energy sources. Energy **94**, 29–37 (2016)
2. Choi, S., Sakis Meliopoulos, A., Ratnesh, K.: Autonomous state estimation based diagnostic system in smart grid. In: IEEE PES Innovative Smart Grid Technologies Conference (ISGT), Washington, DC, pp. 1–6 (2013)
3. Asprou, M., Kyriakides, E., Chakrabarti, S.: The use of a PMU-based state esti-mator for tracking power system dynamics. In: IEEE PES General Meeting - Con-ference & Exposition, National Harbor, MD, pp. 1–5 (2014)
4. Schweppe, F.C., Wildes, J.: Power system static-state estimation. Part I-III: Exact Model IEEE Trans. Power Appar. Syst. **89**(1), 120–135 (1970)

5. Naka, S., Genji, T., Yura, T., Fukuyama, Y.: A hybrid particle swarm optimization for distribution state estimation. In: IEEE Power Engineering Society General Meeting (IEEE Cat. No. 03CH37491) 2 (2003)
6. Liu, Z., Ji, T., Tang, W.H., Wu, Q.H.: Optimal harmonic estimation using a particle swarm optimizer. IEEE Trans. Power Deliv. 23(2), 1166–1174 (2008)
7. Liu, Z., Zhang, J., Zhou, S., Li, X., Liu, K.: Coevolutionary particle swarm optimization using AIS and its application in multiparameter estimation of PMSM. IIEEE Trans. Cybern. 43(6), 1921–1935 (2013)
8. Liu, Z., Li, X., Wu, L., Zhou, S., Liu, K.: GPU-accelerated parallel coevolutionary algorithm for parameters identification and temperature monitoring in permanent magnet synchronous machines. IEEE Trans. Ind. Inform. 11(5), 1220–1230 (2015)
9. Liu, Z., Wei, H., Zhong, Q.C., Liu, K., Xiao, X., Wu, L.: Parameter estimation for VSI-Fed PMSM based on a dynamic PSO with learning strategies. IEEE Trans. Power Electron., to be published. doi:10.1109/TPEL.2016
10. Sarkar, A., Choudhury, S., Sengupta, S.: A self-synchronized ADALINE network for on-line tracking of power system harmonics. Meas 44(4), 784–790 (2011)
11. Marei, M., El-Saadany, E., Salama, M.: A processing unit for symmetrical components and harmonics estimationbased on a new adaptive linear combiner structure. IEEE Trans. Power Deliv. 19(3), 1245–1252 (2004)
12. Joorabian, M., Mortazavi, S., Khayyami, A.: Harmonic estimation in a power system using a novel hybrid least square-Adaline algorithm. Electr. Power Syst. Res. 79(1), 107–116 (2009)
13. Garanayak, P., Panda, G.: Fast and accurate measurement of harmonic parameters employing hybrid adaptive linear neural network and filtered-x least mean square algorithm. IET Gener. Transm. Distrib. 10(2), 421–436 (2016)
14. Dash, P., Swain, D., Routray, A., Liew, A.: Harmonic estimation in a power system using adaptive perceptrons. IEEE Proc. Gener. Transm. Distrib. 143(6), 565–574 (1996)
15. Nanda, S., Biswal, M., Dash, P.K.: Estimation of time varying signal parameters using an improved Adaline learning algorithm. Int. J. Electron. Commun. 68(2), 115–129 (2014)
16. Chang, G.W., Chen, C.I., Liang, Q.W.: A two-stage ADALINE for harmonics and interharmonics measurement. IEEE Trans. Ind. Electron. 56(6), 2220–2228 (2009)

Motion and Visual Control for an Upper-Limb Exoskeleton Robot via Learning

Jian-Bin Huang[1], I-Yu Lin[1], Kuu-Young Young[1(✉)],
and Chun-Hsu Ko[2]

[1] Department of Electrical Engineering,
National Chiao Tung University, Hsinchu, Taiwan
kyoung@mail.nctu.edu.tw
[2] Department of Electrical Engineering, I-Shou University, Kaohsiung, Taiwan

Abstract. The arrival of an aging society brings up many challenges, including the demanding need in medical resources. In responding, the exoskeleton robot becomes one of the focuses, which provides assistance for people with loco-motive problems. Motivated by it, our laboratory has developed a wearable upper-limb exoskeleton robot, named as HAMEXO. It is of 2 DOF and intended to provide motion assistance for users in their daily activities. To serve the purpose, HAMEXO is equipped with a visual system to detect objects in the environment, and also a motion controller for its governing. To deal with the coupling involved during the movements of the two joints and the need to adapt to various users, we adopted the learning approach for controller design. Experiments are performed to demonstrate its effectiveness.

Keywords: Upper-limb exoskeleton robot · Motion and visual control · Learning

1 Introduction

Along with the coming of an aging society, the number of people with limb mobility is increasing, Consequently, medical staffs, caregivers, and medical resources are highly demanded for providing assistance in walking, nursing care, and daily lives. It solicits the introduction of robots to relieve the workloads from their human counterparts. Among them, the exoskeleton robot, which can be worn on the human body directly and operated in concert with the wearer, has received much attention [1–3]. The exoskeleton robots can basically be classified into three types: upper-body, lower-body, and full-body [2–7]. Among previous research, they have been applied for rehabilitation, daily activities, and others. NTUH-ARM [6] and ETS-MARSE [7] were developed for full-arm rehabilitation, which were heavy and fixed to a base. TTL-Exo [5], a light and portable 6-DOF dual arm, was also developed for rehabilitation. Being mounted on a base or wheelchair, they can be applied for eating, drinking, brushing, etc. [1, 4]. Meanwhile, EMAS II [2] and HAL-UL [3] were designed to be light for higher portability.

When used for assistance, the exoskeleton robot can operate in either passive, active-assisted, or active-resistive mode [6, 7]. In the passive mode, the robot dictates

© Springer International Publishing AG 2017
F. Cong et al. (Eds.): ISNN 2017, Part II, LNCS 10262, pp. 36–43, 2017.
DOI: 10.1007/978-3-319-59081-3_5

the entire motion without any force from the user. It is generally adopted for the cases that the user was almost unable to move his/her arm. In the active-assisted or active-resisted mode, the user joins force with the robot to move. The robot usually takes a supporting role when the user executes the task. For these active types of assistance, it is crucial for the robot to come up with proper assistive force. For that, biological signals from the user, such as electromyography (EMG) and electroencephalography (EEG) [1], are frequently used to detect user's motion intention. Another approach is to determine the assistive force by sensing the applied force from the user [8]. Meanwhile, these two approaches can also be combined together by using both biological and force information [1].

Motivated by the demand of motion assistance for people with weak mobility, our laboratory has developed a wearable 2-DOF upper-limb exoskeleton robot, named as HAMEXO. For the use in daily life, such as object picking or drinking, we equip the HAMEXO with a visual system for detecting the objects in the working environment. To execute the motion solicited via the visual system, we develop a motion controller for its governing. As the coupling is present during the movements of the two joints and the adaptability is demanded in applying it for different users, we propose using the learning approach for controller design. The adaptive network-based fuzzy inference system (ANFIS) is adopted for its execellence at adaptation [9]. In this stage of research, we focus on the passive mode of assistance. Meanwhile, the effectiveness of the proposed motion and visual control system is demonstrated via the experiments for object fetching.

2 Design and Development for HAMEXO

HAMEXO (Human and Machine Exoskeleton) is developed to be a 2-DOF upper-body wearable exoskeleton robot. It is designed based on the human upper-body anatomy and dynamics for better fitting in wearing [10, 11]. The two DOFs are intended for the flexion and extension of the shoulder (θ_1) and elbow (θ_2), which should provide the freedoms for simple picking and reaching tasks in daily activities. Referring to the actual range of motion of human body, the ranges of θ_1 and θ_2 are designed to be

$$0^\circ \leq \theta_1 \leq 90^\circ, 0^\circ \leq \theta_2 \leq 135^\circ \tag{1}$$

The 3D CAD modeling of HAMEXO is as shown in Fig. 1. Its frame is made of aluminum for providing the demanded strength and lightness. For each of the two links, there is a PLA (polylactide) 3D printed platform together with a strap belt for securing user's arm to the exoskeleton. The upper-arm, forearm, shoulder, and backpack are all equipped with sliding parts to accommodate to variations in human bodies. The brushless DC motors (BLDCMs) were adopted as the actuators, coupled with reduction gears and also incremental encoders for position feedback. Other designs include: hard foam as padding between the user and exoskeleton for comfort and power-kill switch for safety concern. Note that, as HAMEXO is designed to be wearable, it can also be fixed to a work station to relieve the user from its load. As shown in Fig. 2, HAMEXO can be hung on the rack of the work station and the casters allow it to move. Figure 3(a) shows the photo of the developed HAMEXO and Fig. 3(b) a user wearing it.

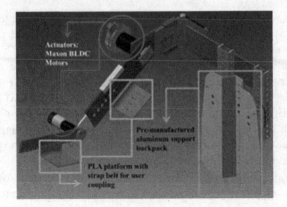

Fig. 1. 3D CAD modeling of HAMEXO.

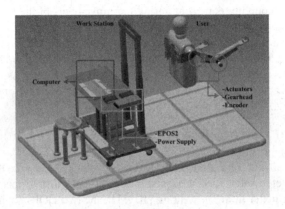

Fig. 2. HAMEXO with the work station.

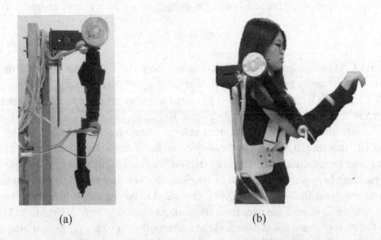

(a)   (b)

Fig. 3. (a) The HAMEXO and (b) a user wearing HAMEXO.

3   Proposed Motion and Visual Control System

The proposed system consists of mainly a visual system for object detection and an ANFIS PID position controller for motion governing. Figure 4(a) shows the setup of the visual system, which includes two cameras for locating the objects in the 3D workspace. Their locations are arranged according to the task, so that they well observe the objects involved. To be portable to go along with HAMEXO, we adopted the CMUcam5 pixy (shown in Fig. 4(b)) as the camera [12], which is light and also with the ability of color recognition. The calibration procedure has been performed to derive accurate parameters for the two cameras. The 2D imagines obtained by them can then be used to determine the 3D object location.

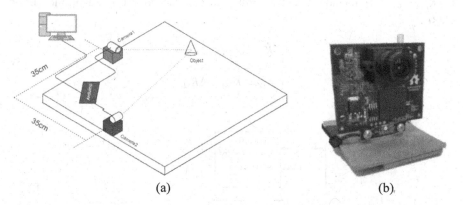

(a)   (b)

Fig. 4. The visual system for object detection: (a) the arrangement of the two cameras and (b) the CMUcam5 pixy.

After both the locations of the object and HAMEXO are identified, the ANFIS PID controller, shown in Fig. 5, is applied to move HAMEXO to reach the object. In Fig. 5, according to the relative locations between the object and HAMEXO, the motion planner first generates a path (θ_d) for execution. For smoothness consideration, we utilize the B-spline method to generate the path. The planned path (θ_d) is forwarded to

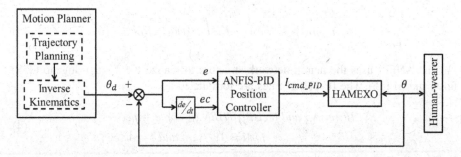

Fig. 5. The proposed motion controller based on ANFIS.

the ANFIS-PID position controller for execution. The controller then derives proper control commands in current (I_{cmd_PID}) according to the feedbacks of position error (e) and position error rate (ec), which shall drive the motors to move HAMEXO to follow the path (θ_d).

The ANFIS, famous for its excellence on adaptation, has been applied for speed control of the BLDCM [13]. In our previous work, it has been applied to determine system parameters for a multi-DOF robot control system based on EMG signals, and achieved desirable performance [14]. Figure 6 shows the system block diagram of the proposed ANFIS-PID position controller, equipped on each of the two links of HAMEXO. It is basically a PID controller with adjustable K_p, K_I, K_D gains tuned by the ANFIS. The controller starts with a set of initial gains (K_{P0}, K_{I0}, K_{D0}). Through a learning process, the ANFIS shall determine proper amount of ($\Delta K_P, \Delta K_I, \Delta K_D$) added to ($K_{p0}, K_{I0}, K_{D0}$) for adjustment according to position error (e) and position error rate (ec):

$$\begin{cases} K_P = K_{P0} + \Delta K_P \\ K_I = K_{I0} + \Delta K_I \\ K_D = K_{D0} + \Delta K_D \end{cases} \tag{2}$$

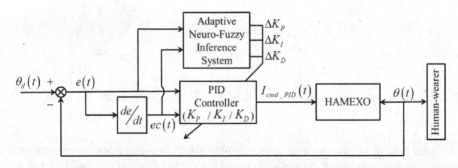

Fig. 6. Block diagram of the proposed ANFIS-PID position controller.

Current control signal $I_{cmd\_PID}(t)$ generated by the ANFIS-PID position controller will drive the motors to move HAMEXO, formulated as

$$I_{cmd\_PID}(t) = K_P e(t) + K_I \int_0^t e(t)dt + K_D \dot{e}(t) \tag{3}$$

The ANFIS uses the neural network structure to realize the Takagi-Sugeno (T-S) fuzzy model [15]. The IF-THEN rules are formulated as

$$R^i : IF\ (e\,is\,A_j)\ and\ (ec\,is\,B_j)\ THEN\ (f_i = p_i e + q_i ec + r_i)$$
$$for\ i = 1, \cdots, m\ and\ j = 1, \cdots, n \tag{4}$$

where R^i is the i's rule of the ANFIS, f_i the output variable (ΔK_p), (A_j, B_j) fuzzy sets characterized by the membership function in the antecedent, and (p_i, q_i, r_i) inference parameter sets in the consequent, respectively. The architecture of ANFIS for deriving ΔK_p, ΔK_I and ΔK_D can be constructed by referring to [16].

4 Experiment

To evaluate the performance of the proposed motion and visual system, we invited three young subjects, two males and one female (shown in Fig. 7), to conduct the experiments. They were all right-handed with the height of 160 (female), 165, and 171 cm and weight of 50, 70, and 62 kg, respectively. For safety concern, the maximum motor speeds for the shoulder and elbow were set to be 300 and 250 rpm, respectively. Evaluation on the proposed ANFIS PID position controller, including its ability in tackling the coupling effect between joints and in adapting to various users, has been reported in our previous work [16]. Here, we concentrate on how it can be linked with the visual system for object fetching. During the experiments, we applied the visual system to locate the object first and the motion controller to move HAMEXO in carrying the arm to fetch the object. We arranged the object to appear in an arbitrary manner, so that the subject did not know where it would be in advance. Figure 8 show the experimental setup for subject A, in which the cup was put on the desk first (Fig. 8(a)), lifted up to the air (Fig. 8(b)), and then put back to the desk (Fig. 8(c)). Figure 9 shows the trajectories of both the shoulder and elbow joints during the motion, in which the blue dots 1, 2, and 3 represent the three object locations, the red line the trajectory designed by the motion planner based on these locations, and the blue line the actual trajectory executed by HAMEXO. In Fig. 9, the actual joint trajectories followed the planned ones quite well, and all three target locations were reached. Similar results were also observed for the experiments conducted by subjects B and C, indicating the effectiveness of the proposed system.

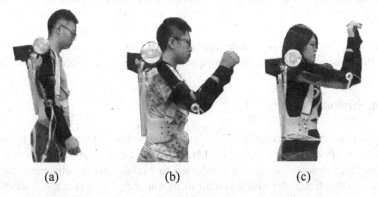

| (a) | (b) | (c) |

Fig. 7. Photos of subjects A, B, and C invited for the experiments.

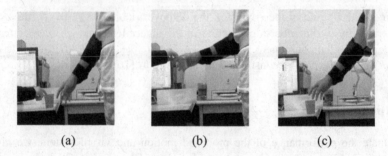

(a)   (b)   (c)

Fig. 8. Setup for the experiment of object feching: (a) reach point 1, (b) reach point 2, and (c) back to point 1.

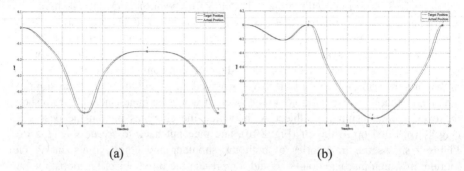

(a)   (b)

Fig. 9. Experimental results (subject A): trajectories for (a) shoulder and (b) elbow. (Color figure online)

To further investigate the effect of learning for the proposed motion controller, we also used a pure fuzzy system, i.e., not a neural-fuzzy type of system, to tune K_p, K_I, K_D gains for the PID controller shown in Fig. 6. For this object-fetching task involving only two joints, the fuzzy system was able to derive suitable gains that led to satisfactory performance at the expense of time. In fact, the derived gains were quite close to those tuned by the proposed ANFIS. Meanwhile, to be more effective on gain tuning and also able to deal with more complicated tasks, we consider the proposed ANFIS PID position controller is more appropriate for future system development.

5   Conclusion

In this paper, we have proposed a motion and visual control system for the upper-limb exoskeleton robots, and applied it to HAMEXO, a such kind of robot developed in our laboratory. Experiments have been conducted to evaluate its effectiveness. In future works, we will enhance the visual system in its portability and also the ANFIS-based motion controller in its learning, including further study on the transferability for different wearers, so that HAMEXO can be applied for more complicated tasks and more adaptive to various users.

Acknowledgment. This work was supported in part by the Ministry of Science and Technology, Taiwan, under Grant NSC 102-2221-E-009-138-MY3.

References

1. Kiguchi, K., Hayashi, Y.: An EMG-based control for an upper-limb power-assist exoskeleton robot. IEEE Trans. Syst. Man Cybern. B Cybern. **42**(4), 1064–1071 (2012)
2. Hasegawa, Y., Oura, S., Takahashi, J.: Exoskeletal meal assistance system (EMAS II) for patients with progressive muscular disease. Adv. Robot. **27**(18), 1385–1398 (2013)
3. Otsuka, T., Kawaguchi, K., Kawamoto, H., Sankai, Y.: Development of upper-limb type HAL and reaching movement for meal-assistance. In: Proceedings of the 2011 IEEE International Conference on Robotics and Biomimetics, Phuket, Thailand, pp. 883–888 (2011)
4. Huete, A.J., Victores, J.G., Martinez, S., Gimenez, A., Balaguer, C.: Personal autonomy rehabilitation in home environments by a portable assistive robot. IEEE Trans. Syst. Man Cybern. C Appl. Rev. **42**(4), 561–570 (2012)
5. Ugurlu, B., Nishimura, M., Hyodo, K., Kawanishi, M., Narikiyo, T.: Proof of concept for robot-aided upper limb rehabilitation using disturbance observers. IEEE Trans. Hum.-Mach. Syst. **45**(1), 110–118 (2015)
6. Wang, W.W., Tsai, B.C., Hsu, L.C., Fu, L.C., Lai, J.S.: Guidance-control-based exoskeleton rehabilitation robot for upper limbs: application to circle drawing for physiotherapy and training. J. Med. Biol. Eng. **34**(3), 284–292 (2014)
7. Rahman, M.H., Saad, M., Ochoa-Luna, C., Kenné, J.P., Archambault, P.S.: Cartesian trajectory tracking of an upper limb exoskeleton robot. In: Proceedings of the 38th Annual Conference on IEEE Industrial Electronics Society, Montreal, Canada, pp. 2668–2673 (2012)
8. Lee, H.D., Lee, B.K., Kim, W.S., Han, J.S., Shin, K.S., Han, C.S.: Human-robot cooperation control based on a dynamic model of an upper limb exoskeleton for human power amplification. Mechatronics **24**(2), 168–176 (2014)
9. Jang, J.-S.R.: ANFIS: adaptive-network-based fuzzy inference system. IEEE Trans. Syst. Man Cybern. **23**(3), 665–685 (1993)
10. Karner, J., Reichenfelser, W., Gfoehler, M.: Kinematic and kinetic analysis of human motion as design input for an upper extremity bracing system. In: Proceedings of the 9th IASTED International Conference on Biomedical Engineering, Innsbruck, Austria, pp. 376–383 (2012)
11. Masjedi, M., Duffell, L.D.: Dynamic analysis of the upper limb during activities of daily living: comparison of methodologies. Inst. Mech. Eng. H, J. Eng. Med. **227**(12), 1275–1283 (2013)
12. CMUcam5 Pixy Camera. http://www.cmucam.org/projects/cmucam5. Accessed 10 Dec 2016
13. Premkumara, K., Manikandanb, B.V.: Adaptive neuro-fuzzy inference system based speed controller for brushless DC motor. Neurocomputing **138**(22), 260–270 (2014)
14. Liu, H.J., Young, K.Y.: An adaptive upper-arm EMG-based robot control system. Int. J. Fuzzy Syst. **12**(3), 181–189 (2010)
15. Takagi, T., Sugeno, M.: Fuzzy identification of systems and its applications to modeling and control. IEEE Trans. Syst. Man Cybern. **1**(1), 116–132 (1985)
16. Huang, Y.B., Young, K.Y., Ko, C.H.: Effective control for an upper-extremity exoskeleton robot using ANFIS. In: Proceeding of 2016 IEEE International Conference on System Science and Engineering, Nantou, Taiwan (2016)

Approximation-Based Adaptive Neural Tracking Control of an Uncertain Robot with Output Constraint and Unknown Time-Varying Delays

Da-Peng Li[1(✉)], Yan-Jun Liu[2], Dong-Juan Li[3], Shaocheng Tong[2], Duo Meng[3], and Guo-Xing Wen[4]

[1] School of Electrical Engineering, Liaoning University of Technology, Jinzhou 121001, China
li_dapengsir@163.com
[2] College of Science, Liaoning University of Technology, Jinzhou 121001, China
liuyanjun@live.com, jztongsc@163.com
[3] School of Chemical and Environmental Engineering, Liaoning University of Technology, Jinzhou 121001, China
lidongjuan@live.com, mengduo2004@163.com
[4] Department of Mathematics, Binzhou University, Binzhou 256600, China
gxwen@live.cn

Abstract. This paper presents an adaptive neural control design for an n-link rigid robot with both output constraint and unknown time-varying delays. The main design difficulties caused by both the output constraint and unknown time-varying delayed states. In order to overcome these difficulties, the novel Barrier Lyapunov Functions (BLF) and iterative backstepping procedures are employing to guarantee constraints satisfaction of the position of the robot. The Lyapunov-krasovskii functionals (LKFs) are utilized to eliminate and compensate the effect of unknown functions with time-varying delayed states in communication channels. By using the Lyapunov analysis, the stability of closed-loop systems is proven.

Keywords: Neural networks · Adaptive control · Backstepping · Barrier Lyapunov functions · Time-varying delay systems

1 Introduction

In recent year, adaptive control designs have been got much attention on the nonlinear systems with unknown function. Based on the approximation characteristic of fuzzy logic systems and neural networks, the adaptive tracking control schemes were proposed for nonlinear SISO systems [1–4] and MIMO system [5, 6] with unknown function, the adaptive controllers have been proposed. In [7–9], the early researches

F. Cong et al. (Eds.): ISNN 2017, Part II, LNCS 10262, pp. 44–51, 2017.
DOI: 10.1007/978-3-319-59081-3_6

were studied for robot by employing the adaptive robust neural networks controllers. However, the above research results omitted the effect of constrains.

It is indispensable that constraints often appear in the real systems, such as the flexible crane systems [10], the stirred tank reactor systems [11] and the robotic manipulator systems [12, 13]. Adaptive controller designs for several classes of SISO nonlinear systems with output constraint in [14] and state constraints in [15, 16] have been studied using the BLFs. Adaptive NN control design was presented for nonlinear MIMO systems with state constraints in [17]. Some subsequent studies have extended constant constraints to time-varying constraints. As the main factor effecting system performance, time delays are not considered in the above-mentioned works.

To meet the needs of practical systems, the handling of time delays in the real systems have become an active research domain, for example magnetic levitation systems in [18], chemical systems in [19] and crane systems in [20]. The stabilization analysis and adaptive controllers were studied for nonlinear systems in [21–23] to compensate for the unknown time-delay based on backstepping technique and LKFs. Based on the LKFs and robust control, [24] proposed an tracking control for the n-link flexible-joint manipulator with unknown time-delay states. It is a field worthy of further study that how to control both the time-varying states and constraints in an uncertain robot.

In this paper, we try to deal with the problem of adaptive neural tracking control for the nonlinear uncertain robot with both time-varying states and output constraint. The main contributions of the present method are summarized that: Based on the BLFs, the transgression of constraints is overcome in the uncertain robot; the unknown tine-varying functions are eliminated by the LKFs. Finally, the proposed control method can guarantee that the semiglobal uniform ultimate boundedness (SGUUB) of the closed-loop signals and output constraint is not violated.

2 System Descriptions

The motion equation of an n-link rigid robotic system with time-varying delayed states can be described by

$$M(q)\ddot{q} + C(q,\dot{q})\dot{q} + H(q(t-\tau_1(t)), \dot{q}(t-\tau_2(t))) + G(q) = u - J^T(q)f(t) \quad (1)$$

where $q, \dot{q}, \ddot{q} \in R^n$ are the position, velocity and acceleration vectors, respectively. $M(q) \in R^{n \times n}$ stands for the symmetric positive definite inertia matrix, $C(q,\dot{q})\dot{q} \in R^n$ is the unknown Centripetal and Coriolis torques, $J^T(q)$ is the unknown reversible Jacobian matrix, $f(t)$ represents the constrained force with being bounded uniformly, $G(q)$ denotes the unknown gravitational force, $u \in R^n$ is the applied torques, $H(\cdot)$ represents the unknown time-delayed function, τ_1 and τ_2 are the unknown time-varying states in communication channels which satisfy $\tau_i(t) \leq \tau_{\max}$ and $\dot{\tau}_i(t) \leq \tau \leq 1$ with $\tau_{\max}$ and τ being known constants.

For the definitions of $x_1 = q$ and $x_2 = \dot{q}$, the dynamic of n-link rigid robotic systems are transformed into the state-space expressions as

$$\begin{cases} \dot{x}_1 = x_2 \\ \dot{x}_2 = M^{-1}(x_1)(-H(x_1(t-\tau_1(t)), x_2(t-\tau_2(t))) - C(x_1, x_2)x_2 \\ \qquad -G(x_1) - J^{*T}(x_1)f(t) + u) \\ y_1 = x_1 \end{cases} \qquad (2)$$

The control objective of this paper is to design an adaptive NN controller u for system (1) to ensure that link position $y_1 = x_1 = [q_1, \ldots, q_n]^T$ tracks the design reference trajectory $y_d = [y_{d_1}, \ldots, y_{d_n}]^T$, while the states and all signals in the close-loop system is SGUUB and all the output constrain are not violated.

Assumption 1 [17]: For all $t > 0$, there are positive constants k_{c_1} and $A_0, A_1, \cdots A_n$, such that the desired trajectory $y_d(t)$ satisfies $|y_d(t)| \leq A_0 \leq k_{c_1}$, and its time derivative $y_d^{(j)}(t)$ satisfies $\left|y_d^{(j)}(t)\right| \leq A_j, j = 1, 2, \cdots, n$.

Assumption 2 [24]: For the unknown nonlinear continuous function $H(\cdot)$ is bounded by the positive continuous function $\bar{H}(\cdot)$, the inequality hold $\|H(x_1, x_2)\| \leq \bar{H}(x_1, x_2)$.

Assumption 3 [24]: There are some positive continuous functions $q_1(\cdot)$ and $q_2(\cdot)$, the inequality holds $\bar{H}(x_1, x_2) \leq q_1(x_1)\|x_1\| + q_2(x_1, x_2)\|x_2\|$ with functions $q_1(x_1)$ and $q_2(x_1, x_2)$ are abbreviated to q_1 and q_2.

3 The Controller Design and Stability Analysis

Based on the first equation of system (1), define the tracking error as $z_1 = x_1 - y_d = [x_{11} - y_{d_1}, \ldots, x_{1n} - y_{d_n}]^T$ and $z_2 = x_2 - \alpha_1 = [x_{21} - \alpha_{11}, \ldots, x_{2n} - \alpha_{1n}]^T$, it is easy to get $\dot{z}_1 = \dot{x}_1 - \dot{y}_d = x_2 - \dot{y}_d = z_2 + \alpha_1 - \dot{y}_d$ and $\dot{z}_{1i} = \dot{x}_{1i} - \dot{y}_{d_i} = z_{2i} + \alpha_{1i} - \dot{y}_{d_i}$ with the adaptive neural tracking controller α_1 will be defined liter on.

Choose the BLF as $V_{B1} = 1/2 \sum_{i=1}^{n} \log k_{b_i}^2 / k_{b_i}^2 - z_{1i}^2$, where k_{b_i}, $i = 1, \ldots, n$ is a design constant. Based on the tracking error z_1, the differentiating of V_{B1} yields

$$\dot{V}_{B1} = \sum_{i=1}^{n} \frac{z_{1i}}{k_{b_i}^2 - z_{1i}^2} (z_2 + \alpha_1 - \dot{y}_d) \qquad (3)$$

Choose the virtual controller α_1 as

$$\alpha_1 = -k_1 z_1 + \dot{y}_d \qquad (4)$$

where $k_1 = \text{diag}(k_{1i})$ with k_{1i}, $i = 1, \ldots, n$ is the design positive constant.
Based on (4), the equality (3) becomes

$$\dot{V}_{B1} = -\sum_{i=1}^{n} \frac{k_i z_{1i}^2}{k_{b_i}^2 - z_{1i}^2} + \sum_{i=1}^{n} \frac{z_{1i} z_{2i}}{k_{b_i}^2 - z_{1i}^2} \qquad (5)$$

Define the tracking error $z_2 = x_2 - \alpha_1$, and its derivative is given as $\dot{z}_2 = \dot{x}_2 - \dot{\alpha}_1$ and consider the second equation of system (1), we have

$$
\begin{aligned}
\dot{z}_2 = {}& M^{-1}(x_1)(-H(x_1(t - \tau_1(t)), x_2(t - \tau_2(t))) \\
& -G(x_1) - C(x_1, x_2)x_2 - J^{*T}(x_1)f(t) + u) - \dot{\alpha}_1
\end{aligned}
\tag{6}
$$

where $\dot{\alpha}_1$ is defined as $\dot{\alpha}_{1i} = \partial\alpha_{1i}/\partial x_{1i}x_{2i} + \sum_{k=0}^{1}\partial\alpha_{1i}\big/\partial y_{d_i}^k y_{d_i}^{(k+1)}$.

Choose the BLF as $V_{B2} = V_{B1} + 1/2z_2^T M(x_1)z_2 + 1/2\delta \sum_{i=1}^{n}\tilde{\theta}_i\Gamma_i^{-1}\tilde{\theta}_i$, where $\tilde{\theta}_i$ is weight estimation error. Based on (6), the differentiating of V_{B2} yields

$$
\begin{aligned}
\dot{V}_{B2} = {}& z_2^T(-C(x_1,x_2)x_2 - G(x_1) - M(x_1)\dot{\alpha}_1 - J^{*T}(x_1)f(t)) \\
& + \frac{1}{\delta}\sum_{i=1}^{n}\tilde{\theta}_i\Gamma_i^{-1}\dot{\tilde{\theta}}_i + z_2^T u + \dot{V}_{B1} - z_2^T H(x_1(t-\tau_1(t)), x_2(t-\tau_2(t)))
\end{aligned}
\tag{7}
$$

Form Assumption 2, using the Young's inequality, the following inequality holds

$$
-z_2^T H(x_1(t-\tau_1(t)), x_2(t-\tau_2(t))) \le \frac{1}{4\bar{v}}z_2^T z_2 + \bar{v}\bar{H}^2(x_1(t-\tau_1(t)), x_2(t-\tau_2(t)))
\tag{8}
$$

where the variable $\bar{v} = 1 - \tau$.

The unknown function $U(Z)$ is defined by

$$
\begin{aligned}
U(Z) = {}& -C(x_1,x_2)x_2 - G(x_1) - M(x_1)\dot{\alpha}_1 \\
& - J^{*T}(x_1)f(t) + \frac{1}{z_2^T}\sum_{k=1}^{2}\exp(\tau_k(t))\left(q_k^2\|x_k(t)\|^2\right)
\end{aligned}
\tag{9}
$$

The NN approximation $U(Z)$ are defined as $U(Z) = \theta^{*T}S(Z) + \varepsilon(Z)$, where $U(Z)$ is a function of $x_1, x_2, y_d, \cdots, \dot{y}_d$ and $Z = \left[x_1^T, x_2^T, y_d, \cdots, \dot{y}_d\right]^T \in \Omega_1$, there is a design positive constant $\bar{\delta}$ which is upper bounded of the approximation error $\varepsilon(Z)$, i.e., $|\varepsilon(Z)| \le \bar{\varepsilon}$.

The signal function is defined as following:

$$
d(z_2) = \begin{cases} 0, & z_2 = [0,0,\ldots,0]^T \\ 1, & \textit{Otherwies} \end{cases}
\tag{10}
$$

Consider $z_2 = [0,0,\ldots,0]^T$, From a practical point of view, once the system reaches its origin, control performance is best, i.e., no control action should be taken for less power consumption. According to the definitions of V_{B1} and V_{B2}, we can get $\dot{V}_{B2} = -\sum_{i=1}^{n}k_i z_{1i}^2\big/k_{bi}^2 - z_{1i}^2 \le 0$ with neural networks is not need to be added in V_{B2}.

For $z_2 \neq [0, 0, \ldots, 0]^T$, the unknown function $U(Z)$ can be approximated by RBF neural works. By substituting (8), (9) and (10) into (7), we obtain

$$
\begin{aligned}
\dot{V}_{B2} \leq &- \sum_{k=1}^{2} \exp(\tau_k(t)) \left(q_k^2 \|x_k(t)\|^2 \right) + \frac{1}{\delta} \sum_{i=1}^{n} \tilde{\theta}_i \Gamma_i^{-1} \dot{\hat{\theta}}_i + z_2^T u + \frac{1}{4\bar{v}} z_2^T z_2 \\
&+ \dot{V}_{B1} + z_2^T \theta^{*T} S(Z) + z_2^T \varepsilon(Z) + \bar{v} \bar{H}^2 (x_1(t - \tau_1(t)), x_2(t - \tau_2(t)))
\end{aligned}
\tag{11}
$$

Using the Young's inequality, we obtain

$$
z_2^T \sum_{i=1}^{n} \varepsilon(Z) \leq \frac{1}{2\eta_i} z_2^T z_2 + \frac{1}{2} \sum_{i=1}^{n} \eta_i \bar{\varepsilon}_i^2
\tag{12}
$$

where η_i is a design constant.

Introduce the actual controller u and the NN adaptation law as

$$
u = d(z_2) \left(-k_2 z_2 - \Xi - \hat{\theta}^T S(Z) - \frac{1}{2\eta} z_2 - \frac{1}{4\bar{v}} z_2 \right)
\tag{13}
$$

$$
\dot{\hat{\theta}}_i = d(z_2) \Gamma_i \delta_i \left[z_2^T S(Z_i) - \sigma_i \hat{\theta}_i \right]
\tag{14}
$$

where σ_i is a design constant. The weight estimation error by $\tilde{\theta}_i = \hat{\theta}_i - \theta_i^*$. The unknown optimal weight vector θ_i^* is estimated by the estimation weight vector $\hat{\theta}_i$ and $\Xi = \left[z_{11} \big/ k_{b_1}^2 - z_{21}^2, \ldots, z_{1n} \big/ k_{b_n}^2 - z_{2n}^2 \right]^T$, $k_2 = \mathrm{diag}(k_{2i})$ with k_{2i}, $i = 1, \ldots, n$ is design positive constant.

For $z_2 \neq [0, 0, \ldots, 0]^T$, substituting (5), (12), (13) and (14) into (11) leads to

$$
\begin{aligned}
\dot{V}_{B2} \leq &- \sum_{k=1}^{2} \exp(\tau_k(t)) \left(q_k^2 \|x_k(t)\|^2 \right) - \sum_{i=1}^{n} \sigma_i \tilde{\theta}_i^T \hat{\theta}_i - \sum_{i=1}^{n} \frac{k_i z_{1i}^2}{k_{b_i}^2 - z_{1i}^2} \\
&+ \frac{1}{2} \sum_{i=1}^{n} \eta_i \bar{\varepsilon}_i^2 - z_2^T k_2 z_2 + \bar{v} \bar{H}^2 (x_1(t - \tau_1(t)), x_2(t - \tau_2(t)))
\end{aligned}
\tag{15}
$$

Notion the term $\sum_{i=1}^{n} \sigma_i \tilde{\theta}_i^T \hat{\theta}_i$ in (15) and the equality $\hat{\theta}_i = \tilde{\theta}_i + \theta_i^*$, we have

$$
-\sum_{i=1}^{n} \sigma_i \tilde{\theta}_i^T \hat{\theta}_i \leq -\frac{1}{2} \sum_{i=1}^{n} \sigma_i \|\tilde{\theta}_i\|^2 + \frac{1}{2} \sum_{i=1}^{n} \sigma_i \|\theta_i^*\|^2
\tag{16}
$$

Noting Assumption 3, the time-delay function $\bar{H}(\cdot)$ in (15) can be rewritten as

$$
\bar{v} H^2 (x_1(t - \tau_1(t)), x_2(t - \tau_2(t))) \leq \bar{v} (q_1^2 x_1^2 (t - \tau_1(t)) + q_2^2 x_2^2 (t - \tau_2(t)))
\tag{17}
$$

Substituting (16) and (17) into (15) leads to

$$\dot{V}_{B2} \le -\sum_{k=1}^{2} \exp(\tau_k(t))\left(q_k^2\|x_k(t)\|^2\right) - \sum_{i=1}^{n}\frac{k_i z_{1i}^2}{k_{b_i}^2 - z_{1i}^2} + \frac{1}{2}\sum_{i=1}^{n}\eta_i\bar{\varepsilon}_i^2 - z_2^T k_2 z_2$$
$$-\frac{1}{2}\sum_{i=1}^{n}\sigma_i\|\tilde{\theta}_i\|^2 + \frac{1}{2}\sum_{i=1}^{n}\sigma_i\|\theta_i^*\|^2 + q_1^2 x_1^2(t - \tau_1(t)) + q_2^2 x_2^2(t - \tau_2(t)) \tag{18}$$

Choose the Lyapunov-Krasovskii function candidates

$$V_K = d(z_2)\sum_{k=1}^{2}\left(\exp(-(t - \tau_k(t)))\int_{t-\tau_k(t)}^{t}\exp(s)(Q(x_k(s)))ds\right) \tag{19}$$

where the function $Q(\cdot) = q_1^2\|x_1\|^2 + q_2^2\|x_2\|^2$.

For $z_2 \ne [0, 0, \ldots, 0]^T$, the time derivative of V_K is given by

$$\dot{V}_K = \sum_{k=1}^{2}\exp(\tau_k(t))\left(q_k^2\|x_k(t)\|^2\right) - \bar{v}(q_1^2 x_1^2(t - \tau_1(t)) + q_2^2 x_2^2(t - \tau_2(t))) - \bar{v}V_K \tag{20}$$

Define the Lyapunov function as $V = V_{B2} + V_K$. For $z_2 \ne [0, 0, \ldots, 0]^T$, based on (18) and (20), the time derivative of the Lyapunov function V can be rewritten as

$$\dot{V} \le -V_K - \frac{1}{2}\sum_{i=1}^{n}\sigma_i\|\tilde{\theta}_i\|^2 + \frac{1}{2}\sum_{i=1}^{n}\sigma_i\|\theta_i^*\|^2 - \sum_{i=1}^{n}\frac{k_i z_{1i}^2}{k_{b_i}^2 - z_{1i}^2} + \frac{1}{2}\sum_{i=1}^{n}\eta_i\bar{\varepsilon}_i^2 - z_2^T k_2 z_2 \tag{21}$$

The equality (21) can be rewritten as

$$\dot{V} = -\rho V + C \tag{22}$$

where $\rho = \min\{2min(k_2 - I)\lambda_{\min}(M(x_1)), \min(2k_{1i}), \sigma_i\lambda_{\min}(\Gamma_i), i = 1, \ldots, n\}$

$$C = \frac{1}{2}\sum_{i=1}^{n}\eta_i\bar{\varepsilon}_i^2 + \frac{1}{2}\sum_{i=1}^{n}\sigma_i\|\theta_i^*\|^2$$

Theorem 1: Consider the n-link rigid robotic systems (1) under Assumptions 1 and 2. Under the set Ω_z, the virtual controller α_1 in (4), the actual controller u in (13) and the adaptive law in (14), if the design parameters are chosen appropriately, the designed adaptive control strategy can ensure that: (1) the output constraint is not never violated; (2) the tracking errors will converge in a compact set about zero; (3) the closed–loop systems is SGUUB.

Proof: The proof process is similar to previous Lyapunov analysis method. Thus, the proof process is omitted here.

4 Conclusion

In order to stabilize an n-link rigid robotic system with unknown time-varying delayed states and output constraint, an adaptive neural tracking control scheme has been developed. The appropriate BLF and iterative backstepping design are employed to prevent violation of the output constraint. The unknown time-varying delayed states have been compensated by LKF. By using the Lyapunov analysis, it can be proved that the boundedness of all signals in the closed-loop, tracking errors converge to a bounded compact set and the output constraint is not violated.

Acknowledgements. The work is supported by the National Natural Science Foundation of China (61473139, 61622303 and 61603164), the Doctoral Scientific Research Staring Fund of Binzhou University under Grant 2016Y14 and the project for Distinguished Professor of Liaoning Province.

References

1. Chen, C.L.P., Wen, G.X., Liu, Y.J., Liu, Z.: Observer-based adaptive backstepping consensus tracking control for high-order nonlinear semi-strict-feedback multiagent systems. IEEE Trans. Cybern. **46**, 1591–1601 (2016)
2. Lai, G.Y., Liu, Z., Zhang, Y., Chen, C.L.P.: Adaptive fuzzy tracking control of nonlinear systems with asymmetric actuator backlash based on a new smooth inverse. IEEE Trans. Cybern. **46**, 1250–1262 (2016)
3. Wen, G.X., Chen, C.L.P., Liu, Y.J., Liu, Z.: Neural-network-based adaptive leader-following consensus control for second-order nonlinear multi-agent systems. IET Control Theory Appl. **9**, 1927–1934 (2015)
4. Liu, Z., Chen, C., Zhang, Y., Chen, C.L.P.: Adaptive neural control for dual-arm coordination of humanoid robot with unknown nonlinearities in output mechanism. IEEE Trans. Cybern. **45**, 507–518 (2015)
5. Chen, M., Chen, W.H., Wu, Q.X.: Adaptive fuzzy tracking control for a class of uncertain MIMO nonlinear systems using disturbance observer. Sci. China Inf. Sci. **57**, 1–13 (2014)
6. Yang, Q.M., Jagannathan, S., Sun, Y.X.: Robust integral of neural network and error sign control of MIMO nonlinear systems. IEEE Trans. Neural Netw. Learn. Syst. **26**, 3278–3286 (2015)
7. Kwan, C., Lewis, F.L., Dawson, D.M.: Robust neural-network control of rigid-link electrically driven robots. IEEE Trans. Neural Netw. **9**, 581–588 (1998)
8. Kwan, C., Lewis, F.L., Kim, Y.H.: Robust neural network control of flexible-joint robots. Asian J. Control **1**, 188–197 (1999)
9. Kwan, C., Lewis, F.L.: Robust backstepping control of nonlinear systems using neural networks. IEEE Trans. Syst. Man Cybern. Part A: Syst. **30**, 753–766 (2000)
10. He, W., Zhang, S., Ge, S.S.: Adaptive control of a flexible crane system with the boundary output constraint. IEEE Trans. Ind. Electron. **64**, 4126–4133 (2014)
11. Li, D.J., Li, D.P.: Adaptive controller design-based neural networks for output constraint continuous stirred tank reactor. Neurocomputing **153**, 159–163 (2015)
12. He, W., Nie, S.X., Meng, T.T., Liu, Y.J.: Modeling and vibration control for a moving beam with application in a drilling riser. IEEE Trans. Control Syst. Technol., to be published. doi:10.1109/TCST.2016.2577001

13. He, W., Ge, W.L., Li, Y.C., Liu, Y.J., Yang, C.G., Sun, C.Y.: Model identification and control design for a humanoid robot. IEEE Trans. Syst. Man Cybern.: Syst. **47**, 45–57 (2017)
14. Liu, Z., Lai, G.Y., Zhang, Y., Chen, C.L.P.: Adaptive neural output feedback control of output-constrained nonlinear systems with unknown output nonlinearity. IEEE Trans. Neural Netw. Learn. Syst. **26**, 1789–1802 (2015)
15. Tee, K.P., Ge, S.S.: Control of nonlinear systems with partial state constraints using a barrier Lyapunov function. Int. J. Control **84**, 2008–2023 (2011)
16. Liu, Y.J., Tong, S.C.: Barrier Lyapunov functions for Nussbaum gain adaptive control of full state constrained nonlinear systems. Automatica **76**, 143–152 (2017)
17. Liu, Y.J., Tong, S.C., Chen, C.L.P., Li, D.J.: Adaptive NN control using Integral Barrier Lyapunov Functionals for uncertain nonlinear block-triangular constraint systems. IEEE Trans. Cybern., to be published. doi:10.1109/TCYB.2016.2581173
18. Choi, J.S., Baek, Y.S.: A single dof magnetic levitation system using time delay control and reduced-order observer. J. Mech. Sci. Technol. **16**, 1643–1651 (2002)
19. Mehrkanoon, S., Shardt, Y.A.W., Suykens, J.A.K.: Estimating the unknown time delay in chemical processes. Eng. Appl. Artif. Intell. **55**, 219–230 (2016)
20. Delgado, E., Diaz-Cacho, M., Bustelo, D., Barreiro, A.: Generic approach to stability under time-varying delay in teleoperation: application to the position-error control of a gantry crane. IEEE/ASME Trans. Mechatron. **18**, 1581–1591 (2013)
21. Wang, M., Chen, B., Liu, X.Y., Shi, P.: Adaptive fuzzy tracking control for a class of perturbed strict-feedback nonlinear time-delay systems. Fuzzy Sets Syst. **159**, 949–967 (2008)
22. Chen, B., Liu, X.P., Liu, K.F., Lin, C.: Novel adaptive neural control design for nonlinear MIMO time-delay systems. Automatica **45**, 1554–1560 (2009)
23. Li, D.P., Li, D.J.: Adaptive neural tracking control for nonlinear time-delay systems with full state constraints. IEEE Trans. Syst. Man Cybern.: Syst., to be published. doi:10.1109/TSMC.2016.2637063
24. Chang, Y.C., Wu, M.F.: Robust tracking control for a class of flexible-joint time-delay robots using only position measurements. Int. J. Syst. Sci. **47**, 3336–3349 (2016)

Neural Network Based Power Tracking Control of Wind Farm

Liyuan Liang[1], Yongduan Song[1,2(✉)], and Mi Tan[2]

[1] School of Electronic and Information Engineering, Beijing Jiaotong University,
Beijing 100044, China
15120218@bjtu.edu.cn
[2] The Key Laboratory of Dependable Service Computing in Cyber Physical Society,
Ministry of Education and School of Automation, Chongqing University,
Chongqing 400044, China
{ydsong,tanmi1989}@cqu.edu.cn

Abstract. This paper investigates the power tracking control problem of wind farm consisting of a large number of wind turbines, each of which is to deliver certain amount of power so that the combined power from the wind farm is able to meet the total power demand. For such power tracking control problem, the precise total demanded power is unavailable and there involve modeling uncertainties as well as external disturbances. To address the issue of unknown power trajectory, an analytical model is proposed to reconstruct the unknown desired power profile. Neural network based control scheme is developed to ensure stable power tracking.

Keywords: Wind farm · Unknown desired power profile · Neural network · Power tracking

1   Introduction

As one of the most attractive renewable energy resources due to its environmental friendly and economically competitive nature, wind energy has become more and more prolific rendering control technology an enabling one for utilizing this kind of resource to its full efficiency [1], while various speed wind turbine control is challenging. These challenges can be attributed to the energy resource itself (the wind), and its stochastic nature. How to design an intelligent and robust control scheme to keep power trajectory close along with uncertain power demand remains an interesting yet challenging problem. A common practice in addressing the nonlinear natures of WT is based on linear system theory [2], and this inevitably causes approximation error. Also various nonlinear controllers have been proposed. For example, Song *et al.* [3] have presented a nonlinear and adptive controller to track asymptotically a desire rotor speed. A sling mode observer is adopted to estimate the aerodynamic torque in spite of system uncertainties [4]. She *et al.* [5] have proposed a improved strategy

---

L. Liang—This work was supported in part by technology transformation program of Chongqing higher education university (KJZH17102).

F. Cong et al. (Eds.): ISNN 2017, Part II, LNCS 10262, pp. 52–59, 2017.
DOI: 10.1007/978-3-319-59081-3_7

by using adaptive backstepping controller based on neural networks. However, the resultant solutions in the aforementioned literature are not only structurally complicated and computationally expensive, but also is based on the assumption that the required tracking curve is known a priori precisely. In this paper we focus on a low-cost neuro-adaptive proportional (P) control for wind farm to maintain satisfactory power delivery to the load side and the literature as most closely related to our work, and alongside such control has salient feature of simplicity in structure and low complexity in computation.

In power system planning and dispatching, its often required to control power output of the power plant to meet (balance) the short-term or long-term power demand from customer side. Most previous power control schemes are based on the assumption that such demand (required power trajectory) is known a prior precisely. However, the power demand from power user side is always uncertain and hard to be precisely predicted. So it is more crucial to investigate the power tracking problem in wind power system where the desired power trajectory is unavailable. To our best knowledge, it has not been well addressed by using existing wind power controllers. This paper aims to synthesize and implement a robust neural-adaptive feedback control based torque control architecture in the presence of unknown power demand for variable speed wind turbines (VSWTs) during below-rated operation in wind field.

The remainder of this paper is organized as follows. Section 2 gives the problem formulates the problem and reviews some preliminaries for the control of wind turbine systems. Section 3 elucidates neuroadaptive feedback control design for wind energy conversion systems with detail stability analysis, where power tracking is achieved, followed by some concluding remarks in Sect. 4.

2   Problem Formulation and Preliminaries

2.1   Overall Control Scheme for Wind Farm

The main objective of the distributed control system is to adjust the out power of each WT so that the output power of entire wind farm can track desired power asymptotically. In this work, we use proportional distributed algorithm to generate the reference power of WTs. Further a neuro-adative feedback controller is proposed without using the uncertain information of WT parameters and known power demands, and the control scheme is easy for implementation.

The proportional power distribution scheme therefore given as follows,

$$P_{sum}^* = P_1^* + P_2^* + \cdots + P_n^*$$
$$= \alpha_1 P_{sum}^* + \alpha_2 P_{sum}^* + \cdots + \alpha_n P_{sum}^* \tag{1}$$

where $\alpha_1 + \alpha_2 + \cdots + \alpha_n = 1$, P_{sum}^* is the total output electric power that the entire wind farm should generate, and P_i^* ($i = 1, 2, \cdots, n$) is the reference distributed power for the ith WT, and α_i denotes the distribution ratio.

2.2   Neural Networks

NNs have been widely applied to many engineering fields as approximation model for the any continuous function over a compact set with sufficient accuracy [6–8]. For any continuous function, it holds that

$$f(z) = W^{*\mathrm{T}}S(z) + \eta(z) \tag{2}$$

where $z = [z_1, z_2, \cdots, z_q]^{\mathrm{T}} \in R^q$ is the input vector of the approximator, $W^* \in R^P$ is the ideal weight matrix, $S(z) = [s_1(z), s_2(z), \cdots, s_p(z)]^{\mathrm{T}} \in R^P$ is the basic function vector, $\eta(z)$ is the approximation error which satisfies $|\eta(z)| < \bar{\eta}$, and $\bar{\eta}$ is an unknown bound parameter. If the number of nodes in the NN p is chosen large enough, then the approximation error can be made arbitrarily small. Also it is worth noting that $\|S(z)\| \le S_m < \infty$ and $\|W^*\| \le W_m < \infty$, where and are some positive constants. A NN unit will be embedded to the controller to compensate the uncertainties in the system as detailed in follows.

2.3   System Description

Here, we consider the ith $(i = 1, 2, \cdots, n)$ VSWT with the help of [1] in the wind farm whose dynamics shown in Fig. 1 is described by

$$J_{ti}\dot{\omega}_{ri} = T_{ai} - K_{ti}\omega_{ri} - n_{gi}T_{emi} - \xi_i(\cdot) \tag{3}$$

where $T_{ai} = K_{ai}(\lambda)\omega_r^2$ with $K_{ai}(\lambda_i) = \frac{1}{2}\rho\pi R^5 \frac{C_{Pi}}{\lambda_i^3}$ being a time-varying and uncertain coefficient. For a WT, $K_{ai}(\lambda_i)$ is a single peak and bounded continuous function of λ_i, $\xi_i(\cdot) = c_i(\omega_{r1}, \omega_{r2}, \cdots \omega_{r(i-1)}, \omega_{r(i+1)}, \cdots \omega_{rn})$ denotes all the possible external bounded disturbances associated with rotate speeds from other wind turbines.

Note that the output power is described by

$$P_{gi} = T_{emi}\omega_{gi} = n_{gi}T_{emi}\omega_{ri} \tag{4}$$

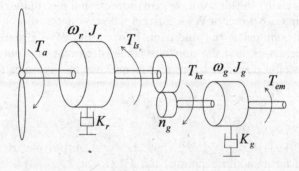

Fig. 1. Drive train dynamics.

Then by (3) we get

$$
\begin{aligned}
\dot{P}_{gi} &= n_{gi}\dot{T}_{emi}\omega_{ri} + n_{gi}T_{emi}\dot{\omega}_{ri} \\
&\triangleq n_{gi}\dot{T}_{emi}\omega_{ri} + \frac{1}{J_{ti}}K_{ai}\left(\lambda_i\right)\omega_{ri}P_{gi} - \frac{K_{ti}}{J_{ti}}P_{gi} \\
&\quad - \frac{1}{J_{ti}\omega_{ri}^2}P_{gi}^2 - \frac{d_i(\cdot)}{J_{ti}\omega_{ri}}P_{gi}
\end{aligned}
\tag{5}
$$

which can be expressed in a matrix form as

$$
\dot{P}_{gi} = b_i(t)u_i + f_i(\cdot)
\tag{6}
$$

where

$$
\begin{aligned}
b_i\left(t\right) &= n_{gi}\omega_{ri} \\
u_i &= \dot{T}_{emi} \\
f_i\left(\cdot\right) &= \frac{1}{J_{ti}}K_{ai}\left(\lambda_i\right)\omega_{ri}P_{gi} - \frac{K_{ti}}{J_{ti}}P_{gi} \\
&\quad - \frac{1}{J_{ti}\omega_{ri}^2}P_{gi}^2 - \frac{\xi_i(\cdot)}{J_{ti}\omega_{ri}}P_{gi}
\end{aligned}
$$

The output power form a wind farm can then be expressed as

$$
\dot{P} = B(t)U + F(\cdot)
\tag{7}
$$

where $P = (P_{g1}, P_{g2}, \cdots P_{gn})^{\mathrm{T}} \in R^n$ is output electric power that every wind turbine generates. $B(t) = diag\left(b_1(t), b_2(t), \cdots b_n(t)\right)$ is the unknown and time-varying control gain matrix of the system, and $F(\cdot) = (f_1(\cdot), f_2(\cdot), \cdots f_n(\cdot))^{\mathrm{T}} \in R^n$ can be treated as the lumped bounded disturbances. $U = (u_1, u_2, \cdots u_n)^{\mathrm{T}} \in R^n$ presents the derivation of control vector of the wind energy conversion systems.

Remark 1: For a WT, $K_{ai}(\lambda_i)$ is a single peak and bounded continuous function of λ_i. Also note that since $\omega_{ri} > 0$, it holds that $\min\{eig(B(t))\} > 0$ and $\|B(t)\|$ is bounded as long as ω_{ri} is bounded as seen clearly from the definition of $b_i(t)$.

2.4   Reconstructing the Unknown Power Trajectory

The desired power set points from the load side is very difficult, if not impossible, to be obtained in advance, which has been widely overlooked in the literature. Now we address this issue. As a first step, we reconstruct the unknown desired power trajectory P^* using P_d estimated power curve [9] via

$$
\begin{cases}
P_d(t) = d_0(\cdot)P^*(t) + \varepsilon_{d0}(\cdot) \\
0 < \underline{d_0} \leq \|d_0(\cdot)\| \leq \overline{d_0} < \infty \\
\|\varepsilon_{d0}(\cdot)\| \leq \varepsilon_0^d < \infty
\end{cases}
\tag{8}
$$

where $P_d(t) = (P_{d1}(t), P_{d2}(t), \cdots P_{dn}(t))^{\mathrm{T}} \in R^n$ is the estimation of the power target trajectory $P^*(t) = (P_1^*(t), P_2^*(t), \cdots P_n^*(t))^{\mathrm{T}} \in R^n$, $d_0(t) \in R^{n \times n}$ is an unknown and time-varying diagonal matrix and $\varepsilon_{d0} \in R^n$ is the estimation error. Here, $\underline{d_0}$, $\overline{d_0}$, ε_0^d are some unknown positive constants. Similarly,

$$\begin{cases} \dot{P}_d(t) = d_1(\cdot)\dot{P}^*(t) + \varepsilon_{d1}(\cdot) \\ 0 < \underline{d_1} \leq \|d_1(\cdot)\| \leq \overline{d_1} < \infty \\ \|\varepsilon_{d1}(\cdot)\| \leq \varepsilon_1^d < \infty \end{cases} \tag{9}$$

where $d_1(t) \in R^{n \times n}$ is an unknown and time-varying diagonal matrix and $\varepsilon_{d1} \in R^n$ is the estimation error. Here, $\underline{d_1}$, $\overline{d_1}$, ε_1^d are some unknown positive constants.

In this paper, we define $e = P - P^* \in R^n$ as the output tracking error, the control objective is to desire adaptive tracking controller for wind system, such that the actual P closely follows the desired trajectory P*. However the precise P* is not available for control design, most existing ways are invalid, calling for more dedicated approach for control design.

For later technical development, the following assumption is required:

Assumption: The desired target trajectory P^* and $\dot{P}^*$ are uncertain yet bounded, and the estimation of the target state P_d and $\dot{P}_d$ are available for control design.

To facilitate the design, $e_m = P - P_d \in R^n$ is defined, in which e_m is computable error. Now, we establish the relation between e and e_m, $\dot{e}$ and $\dot{e}_m$. First, it is straightforward to derive

$$\begin{aligned} e_m &= P - P_d = P - P^* + P^* - P_d \\ &= e + P^* - d_0(t)P^* - \varepsilon_{d0}(t) \\ &= e + (I - d_0(t))P^* - \varepsilon_{d0} = e + \delta_1 \end{aligned} \tag{10}$$

with

$$\delta_1 = (I - d_0(t))P^* - \varepsilon_{d0}. \tag{11}$$

Similarly, it holds that

$$\begin{aligned} \dot{e}_m &= \dot{P} - \dot{P}_d = \dot{P} - \dot{P}^* + \dot{P}^* - \dot{P}_d \\ &= \dot{e} + \dot{P}^* - d_1(t)\dot{P}^* - \varepsilon_{d1}(t) \\ &= \dot{e} + (I - d_1(t))\dot{P}^* - \varepsilon_{d1} = \dot{e} + \delta_2 \end{aligned} \tag{12}$$

with

$$\delta_2 = (I - d_1(t))\dot{P}^* - \varepsilon_{d1}. \tag{13}$$

Remark 2: In view of Assumption and according to (8), (9), (11) and (13), it holds that δ_1 and δ_2 are bounded. Hence, from (10), we can see that if $e_m \in \ell_\infty$ then $e \in \ell_\infty$.

Now we reexpress (12) as

$$\begin{aligned} \dot{e}_m &= \dot{e} + \delta_2 = \dot{P} - \dot{P}^* + \delta_2 \\ &= B(\cdot)U + F(\cdot) - \dot{P}^* + \delta_2 \end{aligned} \tag{14}$$

Note that in (14), $B(\cdot)$ and $F(\cdot)$ are all quite complex and time varying, meanwhile, uncertain desired power demand is involved. Here in this work, we exploit a neural-adaptive proportional control scheme to deal with this problem with no need the specific information on system model and parameters, as detailed in the next section.

3   Control Design and Stability Analysis

Note that matrix $B(\cdot)$ is time-dependent but positive definite, and then there exists an unknown positive constant b_0 satisfying

$$0 < b_0 \leq \min\left\{eig\left(B(\cdot)\right)\right\} \tag{15}$$

From (14), we have

$$\dot{e}_m = B(\cdot)U + F(\cdot) - \dot{P}^* + \delta_2 = B(\cdot)U + Q(\cdot) \tag{16}$$

in which

$$Q(\cdot) = F(\cdot) - \dot{P}^* + \delta_2 \leq \|F(\cdot)\| + \left\|\dot{P}^*\right\| + \|\delta_2\| = \phi(P, \dot{P}^*) \tag{17}$$

As we can see that $\phi(\cdot)$ contains significant nonlinearities and uncertainties. Since NN has been proven effective in approximating unknown nonlinear function in compact set with given precision [1,3]. Hence, we reconstruct $\phi(\cdot)$ via a NN estimator as

$$\phi(\cdot) = W^{*\mathrm{T}}S(z) + \eta(z) \tag{18}$$

where $W^* \in R^L$ is the weight vector and $\eta(z)$ is the NN approximation error, where $|\eta(z)| < \eta_N < \infty$. Here, η_N is some unknown constant. $S(z) \in R^L$ is the basic function with $z = \sum_{i=1}^{n}\left(|P_i| + \left|\dot{P}_{di}\right|\right)$ being the actual training signal to the NN. Then

$$\phi(\cdot) = W^{*\mathrm{T}}S(z) + \eta(z) \leq \|W^*\| \cdot \|S(z)\| + \eta_N \leq a\Phi(z) \tag{19}$$

where $\Phi(z) = \|S(z)\| + 1$, $a = \max\{\|W^*\|, \eta_N\}$.

The corresponding control strategy turns out to be simpler since the required computation cost is relatively inexpensive and it is of the following P structure,

$$U = -\left(k_{P1} + \Delta k_{P1}(\cdot)\right)e_m \tag{20}$$

where $(\Delta k_{P1}(\cdot))$ is time varying, and is updated adaptively by

$$\Delta k_{P1}(\cdot) = c_1\hat{a}\Phi^2(z) \tag{21}$$

Further, the adaptive law for $\hat{a}$ to approximate its actual value a is given as

$$\dot{\hat{a}} = -\gamma\hat{a} + c_1\Phi^2(z)\|e_m\|^2 \tag{22}$$

where k_{P1}, c_1 and γ are positive parameters chosen by the designer. Now we are ready to present the following result on the stability of the proposed P control for the power system:

Theorem: Consider the nonlinear system (6) with the error dynamics defined by (14). If the control algorithm specified in (20) is implemented, the tracking error e is ensured to be ultimately uniformly bounded (UUB).

Proof: To show the stability, we define $\tilde{a} = a - b_0\hat{a}$, and integrate it into the following Lyapunov function candidate

$$V = \frac{1}{2}e_m{}^T e_m + \frac{1}{2b_0}\tilde{a}^2 \tag{23}$$

Considering (16) and (19), and differentiating V yields:

$$
\begin{aligned}
\dot{V} &= e_m{}^T \dot{e}_m - \dot{\hat{a}}\tilde{a} \\
&= e_m{}^T \left(B(\cdot)U + Q(\cdot)\right) - \dot{\hat{a}}\tilde{a} \\
&= e_m{}^T B(\cdot)U + e_m{}^T Q(\cdot) - \dot{\hat{a}}\tilde{a} \\
&\leq e_m{}^T B(\cdot)U + \|e_m\|\,\|Q(\cdot)\| - \dot{\hat{a}}\tilde{a} \\
&\leq e_m{}^T B(\cdot)U + \|e_m\|\,a\Phi(\cdot) - \dot{\hat{a}}\tilde{a}
\end{aligned}
\tag{24}
$$

Taking into account U derived in (20) and $B(\cdot)$ as fore-mentioned, (24) becomes

$$\dot{V} \leq a\Phi(z)\|e_m\| - \left(k_{P1} + \Delta k_{P1}(\cdot)\right)b_0\|e_m\|^2 - \tilde{a}\dot{\hat{a}} \tag{25}$$

Using Young's inequality and Inserting $\Delta k_{P1}(\cdot)$ and $\dot{\hat{a}}$ given in (21) and (22) respectively, one immediately gets that for any $c_1 > 0$,

$$
\dot{V} \leq a\left(c_1\Phi^2(z)\|e_m\|^2 + \tfrac{1}{4c_1}\right) - \left(k_{P1} + c_1\hat{a}\Phi^2(z)\right)b_0\|e_m\|^2 - \tilde{a}\left(-\gamma\hat{a} + c_1\Phi^2(z)\|e_m\|^2\right) \tag{26}
$$

Note that

$$\tilde{a}\hat{a} = \frac{1}{b_0}(a\tilde{a} - \tilde{a}^2) \leq \frac{1}{2b_0}(a^2 - \tilde{a}^2) \tag{27}$$

Then, (26) can be reexpressed as

$$
\begin{aligned}
\dot{V} &\leq ac_1\Phi^2(z)\|e_m\|^2 + \tfrac{a}{4c_1} - k_{P1}b_0\|e_m\|^2 \\
&\quad - c_1\hat{a}\Phi^2(z)b_0\|e_m\|^2 + \gamma\tilde{a}\hat{a} - c_1\tilde{a}\Phi^2(z)\|e_m\|^2 \\
&\leq -k_{P1}b_0\|e_m\|^2 - \tfrac{\gamma}{2b_0}\tilde{a}^2 + \tfrac{\gamma}{2b_0}a^2 + \tfrac{a}{4c_1} \\
&\leq -\Lambda_1 V + \Theta_1
\end{aligned}
\tag{28}
$$

with $\Lambda_1 = \min\left(2k_{P1}b_0, \gamma\right)$, $\Theta_1 = \frac{\gamma}{2b_0}a^2 + \frac{a}{4c_1}$ being some constants. Clearly (33) implies that $V \in \ell_\infty$, thus $\tilde{a} \in \ell_\infty$ (also $\hat{a}$) and $e_m \in \ell_\infty$, then $e \in \ell_\infty$, $\dot{\hat{a}} \in \ell_\infty$, and $U \in \ell_\infty$.

Remark 3: As reflected in the error residual set Θ_1, the developed neoraadaptive algorithm offers the obvious recipe for improving the control precision by enlarging k_{P1} and c_1 and reducing γ properly.

Remark 4: An analytical model linking the desired power curve with the estimated power trajectory is established, and a neuroadaptive proportional control scheme is developed to ensure stable power tracking, which is simple in structure and inexpensive in computation.

Remark 5: It is worth noting that the proportional gain in the strategy is tuned automatically using the proposed algorithm without the need for human interference.

4 Conclusion

This work explicitly addressed the power tracking problem in wind farms in the presence of uncertain dynamics and unknown power target trajectory. A low-cost neuroadative tracking control scheme with no need for precise target trajectory is developed. The proposed method allows P gains to tune adaptively, avoiding the trail and error process.

References

1. Meng, W.C., Yang, Q.M., Ying, Y., Sun, Y., Yang, Z.Y., Sun, Y.X.: Adaptive power capture control of variable-speed wind energy conversion systems with guaranteed transient and steady-state performance. IEEE Trans. Energy Convers. **28**(3), 716–725 (2013)
2. Liu, F.: Stabilization of switched linear system with bounded disturbances and unobservable switchings. Sci. China Ser. F Inf. Sci. **50**(5), 711–718 (2007)
3. Song, Y.D., Dhinakaran, B., Bao, X.Y.: Variable speed control of wind turbines using nonlinear and adaptive algorithms. J. Wind Eng. Ind. Aerodyn. **85**(3), 293–308 (2000)
4. Beltran, B., Ahmed-Ali, T., Benbouzid, M.: High-order sliding mode control of variable-speed wind turbines. IEEE Trans. Ind. Electron. **56**(9), 3314–3321 (2009)
5. She, Y., She, X., Baran, M.: Universal tracking control of wind conversion system for purpose of maximum power acquisition under hierarchical control structure. IEEE Trans. Energy Convers. **26**(3), 766–775 (2011)
6. Ge, S.S., Lee, T.H., Wang, J.: Adaptive control of non-affine nonlinear systems using neural networks. Proc. IEEE Int. Symp. Intell. Control **180**(17), 13–18 (2000)
7. Battilotti, S., De Santis, A.: Robust output feedback control of nonlinear stochastic systems using neural networks. IEEE Trans. Neural Netw. Learn. Syst. **14**(1), 103–116 (2003)
8. MohammadZadeh, S., Masoumi, A.A.: Modeling residential electricity demand using neural network and econometrics approaches. In: International Conference on CIE, pp. 1–6 (2010)
9. Song, Y.D., Zhang, B.B., Zhao, K.: Neuroadaptive control of unknown MIMO systems tracking uncertain target under sensor failures. Automation (2016, under review)

A Generalized Policy Iteration Adaptive Dynamic Programming Algorithm for Optimal Control of Discrete-Time Nonlinear Systems with Actuator Saturation

Qiao Lin, Qinglai Wei[✉], and Bo Zhao

University of Chinese Academy of Sciences, Beijing 100190, China
{linqiao2014,qinglai.wei,zhaobo}@ia.ac.cn

Abstract. In this study, a nonquadratic performance function is introduced to overcome the saturation nonlinearity in actuators. Then a novel solution, generalized policy iteration adaptive dynamic programming algorithm, is applied to deal with the problem of optimal control. To achieve this goal, we use two neural networks to approximate control vectors and performance index function. Finally, this paper focuses on an example simulated on Matlab, which verifies the excellent convergence of the mentioned algorithm and feasibility of this scheme.

Keywords: Adaptive dynamic programming · Neural network · Optimal control · Saturating actuators

1 Introduction

In the control field, saturation nonlinearity of the actuators is universal phenomenon. So optimizing control of systems in which actuators have problem of saturating nonlinearity, is a major and increasing concern [1,2]. However, these traditional methods were proposed without considering the optimal control problem. In order to overcome this shortcoming, Lewis et al. [3] used adaptive dynamic programming (ADP) algorithm. The ADP algorithm [4–6], an effective brain-like method, which can give the solution to Hamilton-Jacobi-Bellman (HJB) equation forward-in-time, provides an important way of obtaining policy of optimizing control. The value and policy iteration algorithms [7,8] are key of the ADP algorithms. Considering the superiority of ADP algorithm, growing researchers chose ADP algorithm in terms of optimal control. Zhang et al. [9] used greedy ADP algorithm to design the infinite-time optimal tracking controller. Qiao et al. [10] applied ADP algorithm to a large wind farm and a STATCOM, with focusing on Coordinated reactive power control. Liu et al. [11] developed an optimizing controller for some systems which were discrete-time nonlinear and had control constraints by DHP. As mentioned in [12], ADP algorithm is also suitable for time-delay systems with the same saturation challenge

© Springer International Publishing AG 2017
F. Cong et al. (Eds.): ISNN 2017, Part II, LNCS 10262, pp. 60–65, 2017.
DOI: 10.1007/978-3-319-59081-3_8

as above. However, in order to realize constrained optimal control, there is still no research using the generalized policy iteration ADP algorithm.

This paper focuses on the generalized policy iteration ADP algorithm. The present algorithm has i-iteration and j-iteration. When j is equal to zero, the proposed algorithm will be a value iteration algorithm, while becoming a policy iteration algorithm when j approaches the infinity. Firstly, the nonquadratic performance function is introduced to overcome the saturation nonlinearity. Then, the process of the generalized policy iteration algorithm is given. Lastly, the simulation results verify the efficiency of the developed method.

2   Problem Statement

We will study the following discrete-time nonlinear systems:

$$x_{k+1} = F(x_k, u_k)$$
$$= f(x_k) + g(x_k)u_k \tag{1}$$

where $u_k \in \mathbb{R}^m$ is control vector, $x_k \in \mathbb{R}^n$ is the state vector, $f(x_k) \in \mathbb{R}^n$ and $g(x_k) \in \mathbb{R}^{n \times m}$ are system functions. We denote $\Omega_u = \{u_k | u_k = [u_{1k}, u_{2k}, \ldots, u_{mk}]^\mathsf{T} \in \mathbb{R}^m, |u_{ik}| \leq \overline{u}_i, i = 1, 2, \ldots, m\}$, where $\overline{u}_i$ can be regarded as the saturating bound. Let $\overline{U} = diag[\overline{u}_1, \overline{u}_2, \ldots, \overline{u}_m]$.

The generalized nonquadratic performance index function is $J(x_k, \underline{u}_k) = \sum_{i=k}^{\infty} \{x_i^\mathsf{T} Q x_i + W(u_i)\}$, where $\underline{u}_k = \{u_k, u_{k+1}, u_{k+2}, \ldots\}$, the weight matrix Q and $W(u_i) \in \mathbb{R}$ are positive definite.

Inspired by the paper [3], we can introduced $W(u_i) = 2 \int_0^{u_i} \Lambda^{-\mathsf{T}}(\overline{U}^{-1}s)\overline{U}Rds$, where R is positive definite, $s \in \mathbb{R}^m$, $\Lambda \in \mathbb{R}^m$, $\Lambda^{-\mathsf{T}}$ denotes $(\Lambda^{-1})^\mathsf{T}$, and $\Lambda(\cdot)$ can choose $\tanh(\cdot)$.

Then we can use $J^*(x_k) = \min_{\underline{u}_k} J(x_k, \underline{u}_k)$ to stand for the optimal performance index function and use u_k^* to be the optimal control vector. So from the principle of discrete-time Bellman's optimality, we can obtain the optimal performance index function as

$$J^*(x_k) = \min_{u_k} \left\{ x_k^\mathsf{T} Q x_k + 2 \int_0^{u_k} \Lambda^{-\mathsf{T}}(\overline{U}^{-1}s)\overline{U}Rds + J^*(x_{k+1})) \right\}. \tag{2}$$

And we can use the following equation to stand for the optimal control vector:

$$u_k^* = \arg\min_{u_k} \left\{ x_k^\mathsf{T} Q x_k + 2 \int_0^{u_k} \Lambda^{-\mathsf{T}}(\overline{U}^{-1}s)\overline{U}Rds + J^*(x_{k+1}) \right\}. \tag{3}$$

The goal of this paper is to get the optimal control vector u_k^* and the optimal performance index function $J^*(x_k)$.

3   Derivation of the Generalized Policy Iteration ADP Algorithm

From [16], it's known that the traditional ADP algorithm just have one iteration procedure. However, the generalized policy iteration ADP algorithm has i-iteration and j-iteration. Specially, for i-iteration, the generalized policy iteration ADP algorithm doesn't need to solve the HJB equation, which speed the convergence rate of the developed ADP algorithm.

According to [17], if a control vector can stabilize the system (1) and make the performance index function finite at the same time, it can be concluded that the control vector is admissible.

Next, we will get that the control vector and cost function of the developed generalized policy iteration ADP algorithm are updated in each iteration. First, the cost function $V_0(x_k)$ can be initialed as follows:

$$V_0(x_k) = x_k^\mathsf{T} Q x_k + 2 \int_0^{v_0(x_k)} \Lambda^{-\mathsf{T}} (\overline{U}^{-1} s) \overline{U} R ds + V_0(F(x_k, v_0(x_k))), \quad (4)$$

where the $v_0(x_k)$ is an initial admissible control vector. Then, for $i = 1$, the control vector $v_1(x_k)$ can be gained by:

$$v_1(x_k) = \arg\min_{u_k} \left\{ x_k^\mathsf{T} Q x_k + 2 \int_0^{u_k} \Lambda^{-\mathsf{T}} (\overline{U}^{-1} s) \overline{U} R ds + V_0(F(x_k, u_k)) \right\}. \quad (5)$$

Then, we will introduced the second iteration procedure. Define an arbitrary non-negative integer sequence, that is $\{L_1, L_2, L_3, \ldots\}$. L_1 is the upper boundary of j_1. When j_1 increases from 0 to L_1, we can have the iterative cost function by

$$V_{1,j_1+1}(x_k) = x_k^\mathsf{T} Q x_k + 2 \int_0^{v_1(x_k)} \Lambda^{-\mathsf{T}} (\overline{U}^{-1} s) \overline{U} R ds + V_{1,j_1}(F(x_k, v_1(x_k))), (6)$$

where

$$V_{1,0}(x_k) = x_k^\mathsf{T} Q x_k + 2 \int_0^{v_1(x_k)} \Lambda^{-\mathsf{T}} (\overline{U}^{-1} s) \overline{U} R ds + V_0(F(x_k, v_1(x_k))). \quad (7)$$

In the second iteration, the cost function changes to be $V_1(x_k) = V_{1,L_1}(x_k)$. For $i = 2, 3, 4, \ldots$, the control vector and cost function of the developed ADP algorithm are updated by:

(1) i-iteration

$$v_i(x_k) = \arg\min_{u_k} \left\{ x_k^\mathsf{T} Q x_k + 2 \int_0^{u_k} \Lambda^{-\mathsf{T}} (\overline{U}^{-1} s) \overline{U} R ds + V_{i-1}(F(x_k, u_k)) \right\}, \quad (8)$$

(2) j-iteration

$$V_{i,j_i+1}(x_k) = x_k^\mathsf{T} Q x_k + 2 \int_0^{v_i(x_k)} \Lambda^{-\mathsf{T}} (\overline{U}^{-1} s) \overline{U} R ds + V_{i,j_i}(F(x_k, v_i(x_k))), \quad (9)$$

where $j_i = 0, 1, 2, \ldots, L_i,$

$$V_{i,0}(x_k) = x_k^{\mathsf{T}} Q x_k + 2 \int_0^{v_i(k)} \Lambda^{-\mathsf{T}} (\overline{U}^{-1} s) \overline{U} R ds + V_{i-1}(F(x_k, v_i(x_k))) \quad (10)$$

and we can get the iterative cost function by

$$V_i(x_k) = V_{i,L_i}(x_k). \tag{11}$$

From (4)–(11), we make use of $V_{i,j_i}(x_k)$ to approximate $J^*(x_k)$ and $v_i(x_k)$ to approximate u_k^*. In the following, an example is applied to illustrate the convergence and feasibility of the presented ADP algorithm.

4 Simulation Example

The following nonlinear system is mass-spring system:

$$x(k+1) = f(x_k) + g(x_k)u(k), \tag{12}$$

where

$$x_k = \begin{bmatrix} x_{1k} \\ x_{2k} \end{bmatrix},$$

$$f(x_k) = \begin{bmatrix} x_{1k} + 0.05x_{2k} \\ -0.0005x_{1k} - 0.0335x_{1k}^3 + x_{2k} \end{bmatrix},$$

$$g(x_k) = \begin{bmatrix} 0 \\ 0.05 \end{bmatrix},$$

and the system is controlled with control constraint of $|u| \le 0.6$. The cost function is defined by

$$J(x_k) = \sum_{i=k}^{\infty} \left\{ x_i^{\mathsf{T}} Q x_i + 2 \int_0^{u_i} \tanh^{-\mathsf{T}} (\overline{U}^{-1} s) \overline{U} R ds \right\},$$

where $Q = \begin{bmatrix} 1 & 0 \\ 0 & 1 \end{bmatrix}$, $R = 0.5$, $\overline{U} = 0.6$.

The developed iteration ADP algorithm is implemented by NNs. The hidden layers of the critic network and action network both are 10 neurons. For each iteration step, we train the networks for 4000 training steps so as to make the training error become minimum. The learning rate of the above two networks both are 0.01.

From Fig. 1(a) and (b), we can get the convergent process of the cost function $V_{i,j_i}(x_k)$ and the subsequence $V_i(x_k)$. Next, we use the optimal control vectors to control the system (12) with the initial state $x(0) = [1, -1]^{\mathsf{T}}$ for 200 time steps. Figure 1(c) and (d) display the changing curves of the state x and the control u. The effective of the presented ADP algorithm in handling optimal control problem for discrete-time nonlinear systems with actuator saturation is verified through the simulation results.

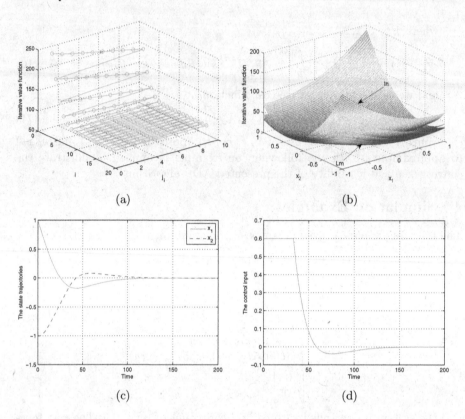

Fig. 1. Simulation results (a) Convergence of V_{i,j_i} (b) Convergence of V_i (c) State trajectories (d) Control vectors

5 Conclusion

In this paper, a novel ADP algorithm is chosen to treat the optimal control problem for discrete-time nonlinear systems with control constraint. One example demonstrates the convergence and feasibility of the presented iteration ADP algorithm. Since the time-delay problem is another hot topic in the control field, it's significant to use the developed ADP algorithm to handle the time-delay systems in the future.

Acknowledgments. This work was supported partly by the National Natural Science Foundation of China (Nos. 61374105, 61374051, 61533017, 61233001, 61273140, 61304086 and U1501251).

References

1. Saberi, A., Lin, Z., Teel, A.: Control of linear systems with saturating actuators. IEEE Trans. Autom. Control **41**(3), 368–378 (1996)
2. Sussmann, H., Sontag, E., Yang, Y.: A general result on the stabilization of linear systems using bounded controls. IEEE Trans. Autom. Control **39**(12), 2411–2425 (1994)
3. Abu-Khalaf, M., Lewis, F.: Nearly optimal control laws for nonlinear systems with saturating actuators using a neural network HJB approach. Automatica **41**(5), 779–791 (2005)
4. Werbos, P.: Approximate dynamic programming for real-time control and neural modeling. In: White, D.A., Sofge, D.A. (eds.) Handbook of Intelligent Control: Neural, Fuzzy, and Adaptive Approaches (1992)
5. Liu, D., Wang, D., Zhao, D., et al.: Neural-network-based optimal control for a class of unknown discrete-time nonlinear systems using globalized dual heuristic programming. IEEE Trans. Autom. Sci. Eng. **9**(3), 628–634 (2012)
6. Wei, Q., Song, R., Yan, P.: Data-driven zero-sum neuro-optimal control for a class of continuous-time unknown nonlinear systems with disturbance using ADP. IEEE Trans. Neural Netw. Learn. Syst. **27**(2), 444–458 (2016)
7. Wei, Q., Liu, D., Shi, G., et al.: Optimal multi-battery coordination control for home energy management systems via distributed iterative adaptive dynamic programming. IEEE Trans. Industr. Electron. **42**(7), 4203–4214 (2015)
8. Bhasin, S., Kamalapurkar, R., Johnson, M., et al.: A novel actorcritic- identifier architecture for approximate optimal control of uncertain nonlinear systems. Automatica **49**(1), 82–92 (2013)
9. Zhang, H., Wei, Q., Luo, Y.: A novel infinite-time optimal tracking control scheme for a class of discrete-time nonlinear systems via the greedy HDP iteration algorithm. IEEE Trans. Syst. Man Cybern.-Part B: Cybern. **38**(4), 937–942 (2008)
10. Qiao, W., Harley, R.G., Venayagamoorthy, G.K.: Coordinated reactive power control of a large wind farm and a STATCOM using heuristic dynamic programming. IEEE Trans. Energy Convers. **24**(2), 493–503 (2009)
11. Liu, D., Wang, D., Yang, X.: An iterative adaptive dynamic programming algorithm for optimal control of unknown discrete-time nonlinear systems with constrained inputs. Inf. Sci. **220**(1), 331–342 (2013)
12. Song, R., Zhang, H., Luo, Y., et al.: Optimal control laws for time-delay systems with saturating actuators based on heuristic dynamic programming. Neurocomputing **73**, 3020–3027 (2010)
13. Vrabie, D., Vamvoudakis, K., Lewis, F.: Adaptive optimal controllers based on generalized policy iteration in a continuous-time framework. In: 17th Mediterranean Conference on Control & Automation, Thessaloniki, Greece, pp. 1402–1409 (2009)
14. Lin, Q., Wei, Q., Liu, D.: A novel optimal tracking control scheme for a class of discrete-time nonlinear systems using generalized policy iteration adaptive dynamic programming algorithm. Int. J. Syst. Sci. **48**(3), 525–534 (2017)
15. Apostol, T.: Mathematical Analysis, 2nd edn. Addison-Wesley Press
16. Wang, F., Zhang, H., Liu, D.: Adaptive dynamic programming: an introduction. IEEE Comput. Intell. Mag. **4**(2), 39–47 (2009)
17. Liu, D., Wei, Q.: Policy iteration adaptive dynamic programming algorithm for discrete-time nonlinear systems. IEEE Trans. Neural Netw. Learn. Syst. **25**(3), 621–634 (2014)

Exponential Stability of Neutral T-S Fuzzy Neural Networks with Impulses

Shujun Long[1] and Bing Li[2(✉)]

[1] College of Mathematics and Information Science, Leshan Normal University,
Leshan 614004, China
[2] College of Mathematics and Statistics, Chongqing Jiaotong University,
Chongqing 400074, China
libingcnjy@163.com

Abstract. In this paper, the stability of neutral T-S fuzzy neural networks with impulses is considered. By extending a singular impulsive differential inequality to the fuzzy version, some new criteria are established for the exponential stability of network under consideration. The results obtained improve some related works in previous literature. A numerical example is given to illustrate the effectiveness of the theoretical methods.

Keywords: Fuzzy neural network · Exponential stability · Impulse · Neutral delays

1 Introduction

The neural network has become the focus of research over the last decades because of its wide applications in many areas such as associative memory, pattern classification, and optimization. Due to the finite speed of switching of neuron amplifiers and signal propagation, time delays are unavoidable and may lead to some more complicated dynamic behaviors including oscillation, divergence, chaos, instability and so on. In particular, many practical systems are modeled as the delayed differential systems of neutral type, which depend on not only the derivative term of the current state but also the derivative term of the past state. In recent years, a great deal of research interest has been focused on the stability of neural networks with neutral delays [1–4].

In real world, impulsive phenomena are encountered in many fields such as biological neural networks, bursting rhythm models and optimal control models, etc. [5]. Therefore, it is of prime importance to consider both impulses and delays on the dynamical behaviors of the system. There have been many papers to consider dynamic behaviors of the differential systems with both impulses and delays [6,7].

T-S fuzzy systems are nonlinear systems described by a set of IF-THEN rules. Some nonlinear dynamical systems can be approximated by the overall fuzzy linear T-S model for the purpose of stability analysis [8]. The stability

© Springer International Publishing AG 2017
F. Cong et al. (Eds.): ISNN 2017, Part II, LNCS 10262, pp. 66–74, 2017.
DOI: 10.1007/978-3-319-59081-3_9

analysis of T-S fuzzy neural networks with delays has been widely considered [9–15]. To the best of authors' knowledge, the exponential stability of neutral T-S fuzzy neural networks with impulses has not been fully investigated, which is very challenging and remains as an open issue.

Motivated by the above discussion, we will investigate the exponential stability of neutral T-S fuzzy neural networks with impulses. By using the property of $\mathcal{M}$-matrix and fuzzy logic method, we extend the inequality in [3] to a new fuzzy version which enables us to study the exponential stability of the network under consideration. The results obtained in this paper improve some related results in literature. An example is given to illustrate the effectiveness of the theoretical results.

2   Model Description and Preliminaries

Let R^n be the space of n-dimensional real column vectors, $\mathcal{N} \triangleq \{1, 2, \ldots, n\}$, $\mathbb{N} \triangleq \{1, 2, \ldots\}$, and $R^{m \times n}$ denote the set of $m \times n$ real matrices. For $\mathbb{A}, \mathbb{B} \in R^{m \times n}$ or $\mathbb{A}, \mathbb{B} \in R^n$, the notation $\mathbb{A} \leq \mathbb{B}$ means that each pair of corresponding elements of $\mathbb{A}$ and $\mathbb{B}$ satisfies the inequality " $\leq$ ". E denotes the identity matrix with compatible dimension. Let $C[\mathbb{X}, \mathbb{Y}]$ be the space of continuous mappings from the space $\mathbb{X}$ to the space $\mathbb{Y}$. $PC[I, R^n] = \{\psi : I \to R^n | \psi(s)$ is continuous for all but at most countable points $s \in I$ and at these points $s \in I, \psi(s^+)$ and $\psi(s^-)$ exist and $\psi(s) = \psi(s^+)\}$. $PC^1[I, R^n] = \{\psi : I \to R^n | \psi(s)$ is continuous differentiable for all but at most countable points $s \in I$ and at these points $s \in I, \psi(s^+), \psi(s^-), \psi'(s^+)$ and $\psi'(s^-)$ exist and $\psi(s) = \psi(s^+), \psi'(s) = \psi'(s^+)\}$. Here, $I \subset R$ is an interval, $\psi(s^+)$ and $\psi(s^-)$ denote the right-hand and left-hand limits of the function $\psi(s)$ at time s, respectively, $\psi'(s)$ denotes the derivative of the function $\psi(s)$. In particular, we denote $\mathcal{C} \triangleq C[[-\tau, 0], R^n]$, $\mathcal{PC} \triangleq PC[[-\tau, 0], R^n]$ and $\mathcal{PC}^1 \triangleq PC^1[[-\tau, 0], R^n]$.

For $x \in R^n, \mathbb{A}, \mathbb{B} \in R^{n \times n}, \varphi \in PC[I, R^n]$, we define

$$[x]^+ = (|x_1|, \ldots, |x_n|)^T, \quad [\mathbb{A}]^+ = (|a_{ij}|)_{n \times n}, \quad \mathbb{A} \circ \mathbb{B} = (a_{ij}b_{ij})_{n \times n}$$

$$[\varphi(t)]_\tau^+ = (|\varphi_1(t)|_\tau, \ldots, |\varphi_n(t)|_\tau)^T, \quad |\varphi_i(t)|_\tau = \sup_{-\tau \leq s \leq 0} |\varphi_i(t+s)|, \ i \in \mathcal{N}.$$

Let $D^+\varphi(t)$ be the upper-right-hand derivative of $\varphi(t)$ at time t. For $\varphi \in \mathcal{PC}^1$ and $\psi \in \mathcal{PC}^1$, we introduce the following norms $\|\varphi\|_\tau = \max_{1 \leq i \leq n} \{ \max_{-\tau \leq s \leq 0} |\varphi_i(s)| \}$, $\|\psi\|_{1\tau} = \max_{1 \leq i \leq n} \{ \max_{-\tau \leq s \leq 0} |\psi_i(s)|, \max_{-\tau \leq s \leq 0} |\psi'_i(s)| \}$. For an $\mathcal{M}$-matrix D, we denote $\Omega_{\mathcal{M}}(D) \triangleq \{z \in R^n | Dz > 0, z > 0\}$.

In this paper, we consider the following neutral T-S fuzzy neural networks with impulses

Plant Rule l.

IF $\theta_1(t)$ is η_1^l and $\cdots$ and $\theta_p(t)$ is η_p^l, **THEN**

$$\begin{cases} \frac{dx(t)}{dt} = -D_l x(t) + (A_l \circ F(x(t)))e_n + (B_l \circ G(x(t-\tau(t))))e_n \\ \qquad + (C_l \circ H(x'(t-r(t))))e_n, \ t \ge t_0, \ t \ne t_k, \\ x(t_k) = I_k(x(t_k^-)), \ t = t_k, \\ x(t_0 + s) = \phi(s), \quad -\tau \le s \le 0, \end{cases} \tag{1}$$

where $x(t) = (x_1(t), \cdots, x_n(t))^T$ is the state vector associated with the neurons, $e_n = (1, 1, \ldots, 1)^T$. For any $l = 1, \ldots, r$, $\eta_s^l (s = 1, \ldots, p)$ is the fuzzy set, $\theta(t) = (\theta_1(t), \ldots, \theta_p(t))^T$ is the premise variable vector, r is the number of fuzzy IF-THEN rules. $D_l = \text{diag}\{d_1^l, \ldots, d_n^l\}$ with $d_i^l > 0 (i = 1, \ldots, n, l = 1, \ldots, r)$, $A_l = (a_{ij}^l)_{n\times n}, B_l = (b_{ij}^l)_{n\times n}, C_l = (c_{ij}^l)_{n\times n} (l = 1, \ldots, r)$ are the connection weight matrices. $F(x(t)) = (f_{ij}(x_j(t)))_{n\times n}, G(x(t-\tau(t))) = (g_{ij}(x_j(t - \tau_{ij}(t))))_{n\times n}, H(x'(t-r(t))) = (h_{ij}(x_j'(t-r_{ij}(t))))_{n\times n}$ are activation functions. For some constant $\tau > 0$, the time-varying delays $\tau_{ij}(t), r_{ij}(t)$ satisfy that $0 \le \tau_{ij}(t) \le \tau, 0 < r_{ij}(t) \le \tau$ for $i, j \in \mathcal{N}$. $\phi(s) = (\phi_1(s), \phi_2(s), \ldots, \phi_n(s))^T \in \mathcal{PC}^1$ is the initial function vector. $I_k = (I_{1k}, \ldots, I_{nk})^T \in C[R^n, R^n]$ represents the the impulsive function. The fixed impulsive moments $t_k (k \in \mathbb{N})$ satisfy $t_0 < t_1 < t_2 < \cdots$ and $\lim_{k\to\infty} t_k = \infty$.

The defuzzified output of neural network (1) is represented as follows:

$$\begin{cases} \frac{dx(t)}{dt} = \sum_{l=1}^r h_l(\theta(t)) \times [-D_l x(t) + (A_l \circ F(x(t)))e_n \\ \qquad + (B_l \circ G(x(t-\tau(t))))e_n + (C_l \circ H(x'(t-r(t))))e_n], t \ne t_k, \\ x(t_k) = I_k(x(t_k^-)), \ t = t_k, \\ x(t_0 + s) = \phi(s), \quad -\tau \le s \le 0, \end{cases} \tag{2}$$

where $h_l(\theta(t)) = \frac{\nu_l(\theta(t))}{\sum_{l=1}^r \nu_l(\theta(t))}$, $\nu_l(\theta(t)) = \prod_{s=1}^p \eta_s^l(\theta_s(t))$.

The following definition will be used in later discussion.

Definition 1. [3] The zero solution of (2) is said to be globally exponentially stable if there are constants $\lambda > 0$ and $M \ge 1$ such that for any solution $x(t, t_0, \phi)$ with the initial functions $\phi \in \mathcal{PC}^1$

$$\|x(t, t_0, \phi)\|_{1\tau} \le M\|\phi\|_{1\tau} e^{-\lambda(t-t_0)}, \ t \ge t_0.$$

Let $y(t) = x'(t)$, then neural network (2) can be transformed to the following $2n$-dimensional singular T-S fuzzy delayed neural network with impulses

$$\begin{cases} \frac{dx(t)}{dt} = \sum_{l=1}^r h_l(\theta(t)) \times [-D_l x(t) + (A_l \circ F(x(t)))e_n \\ \qquad + (B_l \circ G(x(t-\tau(t))))e_n + (C_l \circ H(y(t-r(t))))e_n], \\ 0 = \sum_{l=1}^r h_l(\theta(t)) \times [-y(t) - D_l x(t) + (A_l \circ F(x(t)))e_n \\ \qquad + (B_l \circ G(x(t-\tau(t))))e_n + (C_l \circ H(y(t-r(t))))e_n], t \ne t_k, \\ x(t_k) = I_k(x(t_k^-)), \ y(t_k) = y(t_k^+), \ t = t_k, \\ x(t_0 + s) = \phi(s), \ y(t_0 + s) = \phi'(s), \quad -\tau \le s \le 0. \end{cases} \tag{3}$$

It is clear to conclude that the global exponential stability of the zero solution of the network (2) is evidently equivalent to the global exponential stability of the zero solution of the network (3).

3   Main Results

Theorem 1. Let $u(t) = (u_1(t), \ldots, u_r(t))^T \geq 0$ be a solution of the following fuzzy singular delay differential inequality

$$\begin{cases} KD^+u(t) \leq \sum_{l=1}^{m} e_l(t)[P_l u(t) + Q_l[u(t)]_\tau], \ t \in [\sigma, b), \\ u(s) \in PC[[\sigma - \tau, \sigma], R_+^r], \end{cases} \tag{4}$$

where $\sigma < b \leq +\infty$, $u_i(t) \in C[[\sigma, b), R_+]$ for $i \in S \subset \mathcal{N}^* \overset{\Delta}{=} \{1, \ldots, r\}$, $u_i(t) \in PC[[\sigma, b), R_+]$ for $i \in S^* \overset{\Delta}{=} \mathcal{N}^* - S$ and

(P_1) $K = \text{diag}\{k_1, \ldots, k_r\}$ satisfies $k_i > 0$, $i \in S$ and $k_i = 0$, $i \in S^*$.

(P_2) $\Pi_l = -(P_l + Q_l)$ are $\mathcal{M}$−matrices for $l = 1, \ldots, m$ and $\bigcap_{l=1}^{m} \Omega_{\mathcal{M}}(\Pi_l)$ is nonempty, where $Q_l = (q_{ij}^l)_{r \times r} \geq 0$, $P_l = (p_{ij}^l)_{r \times r}$ satisfy $p_{ij}^l \geq 0$, $i \neq j$, $i, j \in \mathcal{N}^*$, $e_l(t)$ satisfy $e_l(t) \geq 0$ and $\sum_{l=1}^{m} e_l(t) = 1$ for all $t \in [\sigma, b)$.

Then we have

$$u(t) \leq z e^{-\lambda(t-\sigma)}, \ t \in [\sigma, b), \tag{5}$$

provided that the initial function satisfies

$$u(s) \leq z e^{-\lambda(s-\sigma)}, \ \sigma - \tau \leq s \leq \sigma, \tag{6}$$

in which $\lambda > 0$ and $z = (z_1, \ldots, z_r)^T \in \bigcap_{l=1}^{m} \Omega_{\mathcal{M}}(\Pi_l)$ satisfy that for $l = 1, \ldots, m$

$$\left(\lambda K + P_l + Q_l e^{\lambda \tau}\right) z \leq 0 \tag{7}$$

Proof. The proof is similar to that of [3]. So we omit it here.

In order to study the stability issue, we further introduce the following assumptions.

(H_1) There exist nonnegative constants u_{ij}, v_{ij} and w_{ij} such that

$$|f_{ij}(z)| \leq u_{ij}|z|, \ |g_{ij}(z)| \leq v_{ij}|z|, \ |h_{ij}(z)| \leq w_{ij}|z|, \ i, j \in \mathcal{N}, \ z \in R.$$

(H_2) Let $\widehat{\Pi}_l = -(\widehat{P}_l + \widehat{Q}_l)$ $(l = 1, \ldots, r)$ be $\mathcal{M}$−matrices, where

$$\widehat{P}_l = \begin{pmatrix} -D_l + [A_l]^+ \circ U & 0 \\ D_l + [A_l]^+ \circ U & -E \end{pmatrix} \overset{\Delta}{=} (\widehat{p}_{ij}^l)_{2n \times 2n} \in R^{2n \times 2n},$$

$$\widehat{Q}_l = \begin{pmatrix} [B_l]^+ \circ V & [C_l]^+ \circ W \\ [B_l]^+ \circ V & [C_l]^+ \circ W \end{pmatrix} \triangleq (\hat{q}_{ij}^l)_{2n \times 2n} \in R^{2n \times 2n},$$

with $U = (u_{ij})_{n \times n}$, $V = (v_{ij})_{n \times n}$, $W = (w_{ij})_{n \times n}$.

(H_3) There exist nonnegative matrices $R_k = (r_{ij}^{(k)})_{n \times n}$ such that

$$[I_k(x)]^+ \leq R_k[x]^+, \ x \in R^n, \ k \in \mathbb{N}.$$

(H_4) The set $\Omega = \bigcap_{l=1}^{r} \Omega_{\mathcal{M}}(\widehat{\Pi}_l)$ is nonempty (i.e., $\Omega \neq \emptyset$), for a given $z = (z_1, \ldots, z_{2n})^T \in \Omega$, the scalar $\lambda > 0$ satisfies

$$(\lambda \widehat{K} + \widehat{P}_l + \widehat{Q}_l e^{\lambda \tau})z < 0 \, (l = 1, \ldots, r), \tag{8}$$

where

$$\widehat{K} = \begin{pmatrix} E & 0 \\ 0 & 0 \end{pmatrix} \triangleq \mathrm{diag}\{\hat{k}_1, \ldots, \hat{k}_{2n}\}. \tag{9}$$

(H_5) Let $z = (z_x^T, z_y^T)^T$, $z_x = (z_1, \ldots, z_n)^T$, $z_y = (z_{n+1}, \ldots, z_{2n})^T$. There exists a constant γ such that

$$\frac{\ln \gamma_k}{t_k - t_{k-1}} \leq \gamma < \lambda, \ k \in \mathbb{N}, \tag{10}$$

where $\gamma_k \geq 1$ and $R_k z_x \leq \gamma_k z_x$, $k \in \mathbb{N}$.

Theorem 2. Assume that $(H_1) - (H_5)$ hold. Then the zero solution of the neural network (3) is globally exponentially stable in $\mathcal{PC}$ with the exponentially convergent rate $\lambda - \gamma$.

Proof. By calculating the upper-right-hand derivative $D^+[x(t)]^+$ along the solution of (3), we get for $t \in (t_{k-1}, t_k)$, $k \in \mathbb{N}$

$$D^+[x(t)]^+ \leq \sum_{l=1}^{r} h_l(\theta(t))\{(-D_l + [A_l]^+ \circ U)[x(t)]^+$$

$$+([B_l]^+ \circ V)[x(t)]_\tau^+ + ([C_l]^+ \circ W)[y(t)]_\tau^+\}. \tag{11}$$

From the second formula in (3) and (H_1), we have

$$|y_i(t)| \leq \sum_{l=1}^{r} h_l(\theta(t)) \left[d_i^l |x_i(t)| + \sum_{j=1}^{n} |a_{ij}^l| |u_{ij}| |x_j(t)| \right.$$

$$\left. + \sum_{j=1}^{n} |b_{ij}^l| |v_{ij}| |x_j(t - \tau_{ij}(t))| + \sum_{j=1}^{n} |c_{ij}^l| |w_{ij}| |y_j(t - r_{ij}(t))| \right], \tag{12}$$

which implies that

$$0 \le \sum_{l=1}^{r} h_l(\theta(t))\{-[y(t)]^+ + (D_l + [A_l]^+ \circ U)[x(t)]^+$$
$$+([B_l]^+ \circ V)[x(t)]_\tau^+ + ([C_l]^+ \circ W)[y(t)]_\tau^+\}, \ t \in (t_{k-1}, t_k), \ k \in \mathbb{N}. \quad (13)$$

Let

$$u(t) = (x^T(t), y^T(t))^T \in R^{2n}. \quad (14)$$

Then from (9), (11), (13), (14) and (H_2), we get

$$\widehat{K}D^+[u(t)]^+ \le \sum_{l=1}^{r} h_l(\theta(t))\left\{\widehat{P}_l[u(t)]^+ + \widehat{Q}_l[u(t)]_\tau^+\right\}, t \in (t_{k-1}, t_k). \quad (15)$$

Let $\mathcal{N}^* = \{1, \dots, 2n\}$, $S = \{1, \dots, n\} = \mathcal{N}$ and $S^* = \mathcal{N}^* - S = \{n+1, \dots, 2n\}$. Then we get

$$\hat{k}_i > 0, \ i \in S \text{ and } \hat{k}_i = 0, \ i \in S^*, \ \widehat{Q} = (\hat{q}_{ij})_{2n \times 2n} \ge 0,$$

$$\hat{p}_{ij} \ge 0, \ i \ne j \text{ and } \hat{p}_{ij} = 0, \ i \ne j, \ i \in \mathcal{N}^*, \ j \in S^*,$$

$$u_i(t) = x_i(t) \in C[[t_{k-1}, t_k), R], \ i \in S, \ k \in \mathbb{N},$$

$$u_i(t) = y_{i-n}(t) \in PC[[t_{k-1}, t_k), R], \ i \in S^*, \ k \in \mathbb{N}.$$

Combining (15) with the initial condition, it follows from Theorem 1 that

$$[u(t)]^+ \le z\|\phi\|_{1\tau}e^{-\lambda(t-t_0)}, \ t_0 \le t < t_1. \quad (16)$$

Suppose that for all $m = 1, 2, \dots, k$ the inequalities

$$[u(t)]^+ \le \gamma_0\gamma_1 \cdots \gamma_{m-1}z\|\phi\|_{1\tau}e^{-\lambda(t-t_0)}, \ t_{m-1} \le t < t_m. \quad (17)$$

hold, where $\gamma_0 = 1$.

It is easy to conclude that

$$[u(t_k)]^+ \le \gamma_0 \cdots \gamma_{k-1}\gamma_k z\|\phi\|_{1\tau}e^{-\lambda(t_k-t_0)}. \quad (18)$$

Let

$$\tilde{z} = (\tilde{z}_1, \dots, \tilde{z}_{2n})^T = \gamma_0 \cdots \gamma_{k-1}\gamma_k\|\phi\|_{1\tau}e^{-\lambda(t_k-t_0)}z.$$

Then, by the property of $\Omega_\mathcal{M}(\widehat{\Pi}_l)$, we get the vector $\tilde{z} \in \Omega$. Thus, one has

$$[u(t)]^+ \le \tilde{z}e^{-\lambda(t-t_k)}, \ t \in [t_k - \tau, t_k]. \quad (19)$$

Therefore, from (H_2), (8), (19) and Theorem 1, we get

$$[u(t)]^+ \le \tilde{z}e^{-\lambda(t-t_k)} = \gamma_0 \cdots \gamma_{k-1}\gamma_k\|\phi\|_{1\tau}ze^{-\lambda(t-t_0)}, \ t \in [t_k, t_{k+1}), \quad (20)$$

Applying induction gives that

$$[u(t)]^+ \leq e^{\gamma(t_1-t_0)} \cdots e^{\gamma(t_{k-1}-t_{k-2})} \|\phi\|_{1\tau} z e^{-\lambda(t-t_0)}$$
$$= \|\phi\|_{1\tau} z e^{-(\lambda-\gamma)(t-t_0)}, \ \forall t \in [t_0, t_k), \ k \in \mathbb{N}, \tag{21}$$

which means

$$[u(t)]^+ \leq \|\phi\|_{1\tau} z e^{-(\lambda-\gamma)(t-t_0)}, \ t \geq t_0.$$

The proof is completed.

Remark. In particular, when $r = 1$ in Theorem 2, we get the main results in [3]. Furthermore, Theorem 2 has a wider range of applications than those results in [1,4,9,11,15] due to the simultaneous presence of the fuzzy factor, the impulses and the neutral delay. In addition, Theorem 2 removes the monotonicity of active functions and the boundedness of the derivative of the delay which usually required in previous literature [1,9,11].

4   Numerical Example

Example. Consider the neutral T-S fuzzy neural network with impulses as follows:

Plant Rules
Rule 1: IF $\{\theta_1(t)$ is $\eta^1\}$, THEN

$$\begin{cases} x_1'(t) = -5x_1(t) + \frac{1}{2}\sin x_2(t) + \sin x_2(t - \tau_{12}(t) + \frac{1}{4}x_1'(t - r_{11}(t)) \\ x_2'(t) = -6x_2(t) - \frac{1}{3}\sin x_1(t) + x_1(t - \tau_{21}(t)) + \frac{1}{4}x_2'(t - r_{22}(t)) \end{cases} \tag{22}$$

Rule 2: IF $\{\theta_2(t)$ is $\eta^2\}$, THEN

$$\begin{cases} x_1'(t) = -6x_1(t) - \sin x_2(t) + x_1(t - \tau_{11}(t) + \frac{1}{5}x_2'(t - r_{12}(t)) \\ x_2'(t) = -5x_2(t) + \sin x_1(t) + x_2(t - \tau_{22}(t)) - \frac{1}{5}x_1'(t - r_{21}(t)) \end{cases} \tag{23}$$

for $t \neq t_k$, with the membership functions for Rule 1 and Rule 2 are $\eta^1 = \frac{1}{e^{-2\theta_1(t)}}$, $\eta^2 = 1 - \eta^1$. When $t = t_k$, the impulses

$$\begin{cases} x_1(t_k) = \alpha_{1k}x_1(t_k^-) + \beta_{1k}x_2(t_k^-) \\ x_2(t_k) = -\beta_{2k}x_1(t_k^-) + \alpha_{2k}x_2(t_k^-) \end{cases}, \ t = t_k, \tag{24}$$

where $\alpha_{1k} = \frac{2}{5}e^{0.1k}$, $\beta_{1k} = \frac{3}{5}e^{0.1k}$, $\alpha_{2k} = \frac{3}{4}e^{0.1k}$, $\beta_{2k} = \frac{1}{4}e^{0.1k}$ and $t_k - t_{k-1} = 2k$, $\tau_{ij}(t) = \frac{1}{2}|\sin(i+j)t| \leq \frac{1}{2} \overset{\triangle}{=} \tau$, $r_{ij}(t) = \frac{1}{2} - \frac{1}{3}|\cos(i+j)t|$ for $i, j = 1, 2, .$

We calculate that $\widehat{\Pi}_1$ and $\widehat{\Pi}_2$ are nonsingular $\mathcal{M}$-matrix. By choosing $z = (1, 1, 12, 10)^T \in \Omega_{\mathcal{M}}(\widehat{\Pi}_1) \bigcap \Omega_{\mathcal{M}}(\widehat{\Pi}_2)$, $z_x = (1, 1)^T$ and $\lambda = 0.095$, we get the inequalities

$$(\lambda \widehat{K} + \widehat{P}_1 + \widehat{Q}_1 e^{\lambda\tau})z = (-0.2259, -1.9156, -2.3159, -0.0056)^T < (0, 0, 0, 0)^T,$$

$$(\lambda \widehat{K} + \widehat{P}_2 + \widehat{Q}_2 e^{\lambda\tau})z = (-1.7719, -0.3535, -1.8619, -0.4435)^T < (0, 0, 0, 0)^T.$$

Let $\gamma_k = \max\{\alpha_{1k} + \beta_{1k}, \alpha_{2k} + \beta_{2k}\}$, then we derive

$$\gamma_k = e^{0.1k} \geq 1 \text{ and } \frac{\ln\gamma_k}{t_k - t_{k-1}} = \frac{\ln e^{0.1k}}{2k} = 0.05 = \gamma < \lambda = 0.095.$$

By following from Theorem 2, we deduce that the zero solution is globally exponentially stable with the exponential convergence rate 0.045. The simulation result is presented in Fig. 1.

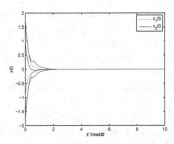

Fig. 1. The simulation result for example

Acknowledgments. This work was supported in part by the National Natural Science Foundation of China under Grant 11501065, the Natural Science Foundation of Chongqing under Grant cstc2015jcyjA00033, the Scientific Research Fund of Sichuan Provincial Education Department under Grant 16TD0029, the Natural Science Foundation of Chongqing Municipal Education Commission under Grants KJ1600504 and KJ1705138, the Research Foundation of Chongqing Jiaotong University under Grant 2014kjc-II-019, and the Project of Leshan Normal University under Grant Z1324.

References

1. Rakkiyappan, R., Balasubramaniam, P., Cao, J.D.: Global exponential stability results for neutral-type impulsive neural networks. Nonlinear Anal.: RWA **11**, 122–130 (2010)
2. Orman, Z.: New sufficient conditions for global stability of neutral-type neural networks with time delays. Neurocomputing **97**, 141–148 (2012)
3. Xu, D.Y., Yang, Z.G., Yang, Z.C.: Exponential stability of nonlinear impulsive neutral differential equations with delays. Nonlinear Anal. **67**, 1426–1439 (2007)
4. Tu, Z.W., Wang, L.W.: Global Lagrange stability for neutral type neural networks with mixed time-varying delays. Int. J. Mach. Learn. Cybern. 1–11 (2016)
5. Lakshmikantham, V., Bainov, D., Simeonov, P.: Theory of Impulsive Differential Equations. World Scientific, Singapore (1989)
6. Yang, Z.C., Xu, D.Y.: Stability analysis of delay neural networks with impulsive effects briefs. IEEE Trans. Circuits Syst.-II Express **52**, 517–521 (2005)
7. Gopalsamy, K.: Stability of artificial neural networks with impulses. Appl. Math. Comput. **154**, 783–813 (2004)

8. Takagi, T., Sugeno, M.: Stability analysis and design of fuzzy control systems. Fuzzy Sets Syst. **45**, 135–156 (1993)

9. Balasubramaniam, P., Ali, M.S.: Stability analysis of Takagi-Sugeno fuzzy Cohen-Grossberg BAM neural networks with discrete and distributed time-varying delays. Math. Comput. Model. **53**, 151–160 (2011)

10. Ahn, C.: Some new results on stability of Takagi-Sugeno fuzzy Hopfield neural networks. Fuzzy Sets Syst. **179**, 100–111 (2011)

11. Muralisankar, S., Gopalakrishnan, N.: Robust stability criteria for Takagi-Sugeno fuzzy Cohen-Grossberg neural networks of neutral type. Neurocomputing **144**, 516–525 (2014)

12. Chen, B., Liu, X.P., Tong, S.C.: New delay-dependent stabilization conditions of TCS fuzzy systems with constant delay. Fuzzy Sets Syst. **158**, 2209–2224 (2007)

13. Lin, C., Wang, Q.G., Lee, T.H.: Delay dependent LMI conditions for stability and stabilization of Takagi-Sugeno fuzzy systems with bounded time-delays. Fuzzy Sets Syst. **157**, 1229–1247 (2006)

14. Ahn, C.: Passive and exponential filter design for fuzzy neural networks. Inf. Sci. **238**, 126–137 (2013)

15. Long, S.J., Xu, D.Y.: Global exponential p-stability of stochastic nonautonomous Takagi-Sugeno fuzzy cellular neural networks with time-varying delays and impulses. Fuzzy Sets Syst. **253**, 82–100 (2014)

Robust NN Control of the Manipulator in the Underwater Vehicle-Manipulator System

Weilin Luo[✉] and Hongchao Cong

School of Mechanical Engineering and Automation,
Fuzhou University, Fuzhou 350116, China
wlluo@fzu.edu.cn, 15695905563@163.com

Abstract. Neural networks (NN) are applied to the tracking control of a three-link manipulator attached to an autonomous underwater vehicle (AUV). Lyapunov design is employed to obtain the NN based robust controller. The interaction between the AUV and the manipulator is considered. Nonlinearity in the plant is compensated by NN based identification. To illustrate the validity of the proposed controller, numerical simulation is performed and the comparison between the NN based controller and a conventional proportional-derivative (PD) controller is conducted.

Keywords: Underwater vehicle-manipulator system · Neural networks · Robust control · Uncertainties

1 Introduction

Underwater vehicles play an supporting role in the undersea exploration by human beings. Usually three kinds of underwater vehicle are available in performing underwater missions, which refer to as HOVs (human occupied vehicles), ROVs (remotely operated vehicles) and AUVs (autonomous underwater vehicles). A underwater vehicle-manipulator system (UVMS) is formed when a n-link manipulator is attached to an underwater vehicle such as ROV and AUV. By means of UVMS, complicated underwater missions can be fulfilled for instance undersea sampling, fixing or repairing. In the meantime, the safety and operability of an UVMS become more important, comparing an UVMS with an AUV or ROV. A control system with high performance can guarantee the safety and operability requirements.

One of the difficulties in obtaining a high-performance controller for an UVMS is how to deal with the uncertainties in the plant. Such uncertainties result from the disturbance by environment (e.g., current, oceanic internal wave), variation of payload and tether force, noises in the mechanical equipments, parametric perturbation, etc. Consequently, the robustness of the control system of an UVMS is of great importance. During the past years, developing an adaptive robust controller for an UVMS is of interest for many researchers. For example,

© Springer International Publishing AG 2017
F. Cong et al. (Eds.): ISNN 2017, Part II, LNCS 10262, pp. 75–82, 2017.
DOI: 10.1007/978-3-319-59081-3_10

Mohan and Kim presented a coordinated motion control scheme using a disturbance observer for an autonomous UVMS [1]. Xu et al. proposed a neuro-fuzzy approach to the control of UVMS in the presence of payload variation and hydrodynamic disturbance [2]. Han and Chung presented the use of restoring moments to the motion control of an UVMS with uncertainties caused by external disturbances [3]. Korkmaz et al. proposed the inverse dynamics control algorithm for the trajectory tracking of an underactuated UVMS in the presence of parameter uncertainty and disturbance [4]. Antonelli and Cataldi addressed the low level control of an UVMS affected by external disturbance using a recursive adaptive control scheme [5] and a virtual decomposition approach [6]. Barbalata et al. proposed a proportional-integral limited controller for a lightweight UVMS affected by disturbance [7]. Ji et al. presented the motion control of an UVMS affected by uncertainties and disturbances using zero moment point equation [8]. Woolfrey et al. studied the kinematic control of an UVMS affected by wave disturbance using model predictive control [9].

This paper presents a neural network approach to the tracking control of the manipulator of a UVMS. A three-link manipulator is considered to be attached to an AUV. In the study, it is assumed that the AUV is hovering when the manipulation is being performed. The interaction between the AUV and the manipulator is viewed as an external disturbance to the control of manipulator. An on-line forward neural network is applied to identify the nonlinearity that involves the disturbance. The overall control system is obtained using Lyapunov design to guarantee the stability of controller.

2　Problem Formulations

For an underwater vehicle with a n-link manipulator, a general model can be described as

$$M(q)\dot{\zeta} + C(q,\zeta)\zeta + D(q,\zeta)\zeta + g(q, R_B^I) = \tau. \tag{1}$$

where $q \in \mathbb{R}^n$ is the vector of joint position, $\zeta \in \mathbb{R}^{(6+n)}$ is the vector of generalized velocity, R_B^I is the transformation from the body-fixed frame to the inertia frame, $M(q) \in \mathbb{R}^{(6+n)\times(6+n)}$ is the inertia matrix, $C(q,\zeta)\zeta \in \mathbb{R}^{(6+n)}$ is the vector of Coriolis and centripetal terms, $D(q,\zeta)\zeta \in \mathbb{R}^{(6+n)}$ is the vector of dissipative effects, $g(q, R_B^I) \in \mathbb{R}^{(6+n)}$ is the vector of gravity and buoyancy effects, $\tau \in \mathbb{R}^{(6+n)}$ is the vector of generalized forces [10]. An UVMS with a 3-link manipulator and the coordinate system can be depicted as Fig. 1, where $o - xyz$ denotes the inertia (or earth-fixed) reference frame while $o_i - x_i y_i z_i$ denote the body-fixed reference frames.

When the vehicle is kept hovering while the manipulator is working, the body-fixed frame $o_0 - x_0 y_0 z_0$ can be viewed as an inertia frame. The dynamics of the manipulator can be described as

$$M(q)\ddot{q} + C(q,\dot{q})\dot{q} + D(q,\dot{q})\dot{q} + g(q) + w = \tau_m. \tag{2}$$

where w represents the interaction between the vehicle and the manipulator. Since it is difficult to express such an interaction in precise mathematical equation, w is taken as an external disturbance to the plant. In the study, it is assumed

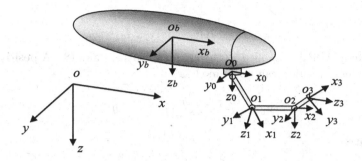

Fig. 1. An UVMS with 3-link manipulator

that the first link of the manipulator rotates around axis $o_0 z_0$ (as shown in Fig. 1) with the angle q_1; the second and the third links rotate in the plane $o_0 z_0 x_1$ with q_2 and q_3 respectively. Such a motion allocation guarantees the end effector can reach the object through the rotation of the three links.

Suppose the three links are driven by DC motors, the force vector τ_m in (2) can be calculated by

$$\tau_m = K_m I. \tag{3}$$

where K_m is the conversion coefficient matrix from electrical current to torque, I is the vector of electrical current.

The dynamics of the electrical circuit can be described as

$$L\dot{I} + RI + K\dot{q} = u_e. \tag{4}$$

where L is the matrix of armature inductance, R is the matrix of armature resistance, K is the voltage constant matrix, u_e is the vector of armature voltage.

Equations (2) and (4) constitute a cascaded system involving a mechanical subsystem and an electrical one. As can be recognized, the output of the cascaded system is the vector $q = [q_1, q_2, q_3]^T$ while the controller input is u_e.

3   Control Design

3.1   NN Based Controller

To obtain the torque in the mechanical subsystem, a desired signal is designed as

$$I_d = K_m^{-1}(M(q)\ddot{q}_d + C(q,\dot{q})\dot{q}_d + D(q,\dot{q})\dot{q} + g(q) + u_1). \tag{5}$$

where q_d is the desired trajectory, u_1 is introduced as an auxiliary controller. Similarly, another auxiliary controller can be introduced in the electrical subsystem:

$$u_e = RI_d + K\dot{q}_d + u_2. \tag{6}$$

Define $e = q_d - q$, $\eta = I_d - I$, $\xi = \dot{e} + \alpha e (\alpha > 0)$, one has

$$M(q)\dot{\xi} = (\alpha M(q) - C(q,\dot{q}))\dot{e} + w + K_m \eta - u_1, \tag{7}$$

and

$$L\dot{\eta} = -R\eta - K(\xi - \alpha e) + L\dot{I}_d - u_2. \tag{8}$$

Lyapunov design is applied to the error systems (7) and (8). A positive definite Lyapunov function can be defined as

$$V_1 = \frac{1}{2}(e^T e + \xi^T M \xi + \eta^T L \eta). \tag{9}$$

Its derivative is

$$\dot{V}_1 = -\alpha e^T e + \xi^T (K_m \eta + \alpha M(q)\dot{e} + \alpha C(q, \dot{q})e + e + w - u_1) \\ + \eta^T (-R\eta - K(\xi - \alpha e) + L\dot{I}_d - u_2). \tag{10}$$

On-line forward neural networks are applied to identify the nonlinearities in the above equation. In detail, let:

$$f_1 = \alpha M(q)\dot{e} + \alpha C(q, \dot{q})e + e + w = W_1^T \phi(X_1) + \varepsilon_1, \tag{11}$$

$$f_2 = K_m \xi - R\eta - K(\xi - \alpha e) + L\dot{I}_d = W_2^T \phi(X_2) + \varepsilon_2. \tag{12}$$

where W_i are weight matrices, X_i are net inputs, ε_i are approximation errors [11].

The two auxiliary controllers u_1 and u_2 can be designed as:

$$u_1 = W_{1e}^T \phi(X_1) + \alpha_1 M(q)\xi, \tag{13}$$

and

$$u_2 = W_{1e}^T \phi(X_2) + \alpha_2 L\eta. \tag{14}$$

where W_{ie} are updated weight matrices, α_i are control gains.

3.2 Stability Analysis

Substituting Eqs. (13) and (14) into (10) yields:

$$\dot{V}_1 = -\alpha e^T e + (\xi^T \varepsilon_1 + \eta^T \varepsilon_2) \\ + \xi^T \tilde{W}_1^T \phi(X_1) + \eta^T \tilde{W}_2^T \phi(X_2) - \alpha_1 \xi^T M(q)\xi - \alpha_2 \eta^T L\eta. \tag{15}$$

where $\tilde{W}_i = W_i - W_{ie}$ are weight error matrices. To guarantee the robustness of the net weight, the algorithms of updated weight matrices are designed as:

$$\dot{W}_{1e} = k_1 \phi(X_1)\xi^T - k_2 \|\xi\| W_{1e}, \tag{16}$$

$$\dot{W}_{2e} = k_1 \phi(X_2)\eta^T - k_2 \|\zeta\| W_{2e}. \tag{17}$$

where $\zeta = [\xi, \ \eta]^T$ is an augmented error vector. A stepping Lyapunov function is defined as

$$V_2 = V_1 + \frac{1}{2k_1}(\left\|\tilde{W}_1\right\|_F^2 + \left\|\tilde{W}_2\right\|_F^2). \tag{18}$$

Its derivative is

$$\dot{V}_2 \leq -2\alpha_0 V_2 + (\xi^T \varepsilon_1 + \eta^T \varepsilon_2) - a(\alpha_1 \xi^T M(q)\xi + \alpha_2 \eta^T L\eta)$$
$$+ \frac{k_2}{k_1} \|\zeta\| \left((\tilde{W}_1, W_1)_F + (\tilde{W}_2, W_2)_F - a \left\|\tilde{W}_1\right\|_F^2 - a \left\|\tilde{W}_2\right\|_F^2\right), \quad (19)$$

where $a \in [0, 1]$, $\alpha_0 = \min\{(1-a)\alpha_1, (1-a)\alpha_2, (1-a)k_2 \|\xi\|, (1-a)k_2 \|\zeta\|\}$. Define a compact set $U(\zeta) = \{\zeta| \|\zeta\| \leq b, b > 0\}$, the following inequality can be obtained:

$$\dot{V}_2 \leq -2\alpha_0 V_2 + \lambda, (\lambda > 0) \tag{20}$$

when ζ is within the set $U(\zeta)$. Otherwise, if $\|\zeta\| > b$, the following inequality can be derived

$$\dot{V}_2 \leq -2\alpha_0 V_2, \tag{21}$$

by selecting appropriate controller parameters.

It can be concluded from inequalities (20) and (21) that the tracking system is stable. Moreover, it should be noted that UUB (uniformly ultimately bounded) stability rather than asymptotic stability can be guaranteed.

4   Numerical Simulation

To demonstrate the validity of the proposed controller, an example is studied with respect to an AUV-manipulator system. The particulars of the manipulator are: $l_1 = l_2 = l_3 = 1$m, $m_1 = m_2 = 1$kg, $m_3 = 2m_1$, $L = diag(0.1, 0.1, 0.1)$H, $R = diag(1, 1, 1)\Omega$, $K = diag(0.5, 0.5, 0.5)$V·s/rad, $K_m = diag(1, 1, 1)$N·m/A. The interaction between the vehicle and the manipulator is assumed as a disturbance acted on the first link of the manipulator. In the study, it is selected as an uniformly distributed pseudorandom signals in the interval [0, 200N]. Moreover, a disturbance is assumed to act on the end effector at the time $t = 1.5$–1.7s with the amplitude 200N.

The controller parameters are selected as: $\alpha_1 = \alpha_2 = 200$, $k_1 = 50, k_2 = 0.8$. Sigmoid function is employed as the activation function of hidden layers. The initial weights are set as zeros. To confirm the advantage of the proposed controller over conventional controllers, a PD controller is taken as:

$$u_e = \lambda_1 \dot{e} + \lambda_2 e, \tag{22}$$

by selecting $\lambda_1 = \lambda_2 = 300$.

Figure 2 presents the tracking results of the joint positions. Figures 3 and 4 show the results of the trajectory tracking of the end effector. As can be recognized from the comparison results, better performance of the NN controller proposed in the study is achieved.

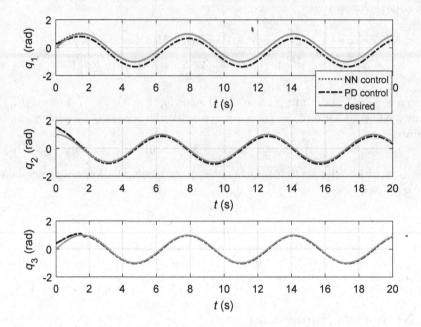

Fig. 2. Tracking results of the joint positions

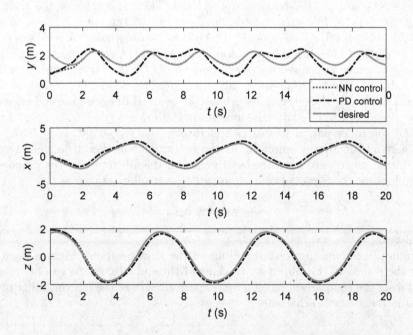

Fig. 3. Tracking results of the trajectory of the end effector

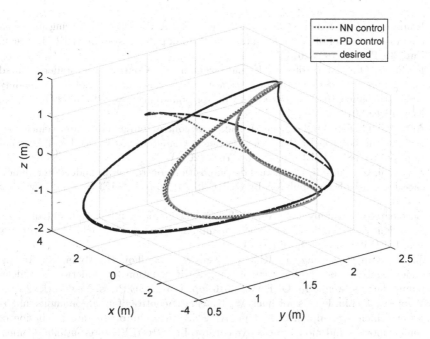

Fig. 4. Space tracking results of the trajectory of the end effector

5   Conclusions

Neural networks are applied to the control of the manipulator of an UVMS. To guarantee the robustness of the controller, Lyapunov design is incorporated. Simulation results show the validity of the proposed control scheme. It should be noted however that a specific case is studied in the paper. The vehicle is assumed to be hovering while the manipulation is being performed, which means that the manipulator can be dealt with as a ground robot. The validity of the proposed control scheme should be verified in a more complicated case, e.g., both the vehicle and the manipulator are under controlled.

Acknowledgments. This work was partially supported by the Special Item supported by the Fujian Provincial Department of Ocean and Fisheries (No. MHGX-16), the Special Item for University in Fujian Province supported by the Education Department (No. JK15003), and the Special Item supported by Fuzhou University (No. 2014-XQ-16).

References

1. Mohan, S., Kim, J.: Coordinated motion control in task space of anautonomous underwater vehicle-manipulator system. Ocean Eng. **104**, 155–167 (2015)
2. Xu, B.R., Pandian, S., Sakagami, N., Petry, F.: Neuro-fuzzy control of underwater vehicle-manipulator systems. J. Franklin. Inst. **349**, 1125–1138 (2012)

3. Han, J., Chung, W.K., Sakagami, N., Petry, F.: Active use of restoring moments for motion control of an underwater vehicle-manipulator system. IEEE J. Ocean. Eng. **39**(1), 100–109 (2014)

4. Korkmaz, O., Kemal Ider, S., Kemal Ozgoren, M.: Control of an underactuated underwater vehicle manipulator system in the presence of parametric uncertainty and disturbance. In: 2013 American Control Conference, pp. 578–584. IEEE Press, New York (2013)

5. Antonelli, G., Cataldi, E.: Recursive adaptive control for an underwater vehicle carrying a manipulator. In: 22nd Mediterranean Conference on Control and Automation, pp. 847–852. IEEE Press, New York (2014)

6. Antonelli, G., Cataldi, E.: Virtual decomposition control for an underwater vehicle carrying a n-DoF manipulator. In: OCEANS 2015, pp. 1–9. IEEE Press, New York (2015)

7. Barbalata, C., Dunnigan, M.W., Ptillot, Y.: Dynamic coupling and control issues for a lightweight underwater vehicle manipulator system. In: OCEANS 2014, pp. 1–6. IEEE Press, New York (2014)

8. Ji, D., Kim, D., Kang, J., Kim, J., Nguyen, N., Choi, H., Byun, S.: Redundancy analysis and motion control using ZMP equation for underwater vehicle-manipulator systems. In: OCEANS 2016, pp. 1–6. IEEE Press, New York (2016)

9. Woolfrey, J., Liu, D., Carmichael, M.: Kinematic control of an autonomous underwater vehicle - manipulator system (AUVMS) using autoregressive prediction of vehicle motion and model predictive control. In: 2016 IEEE International Conference on Robotics and Automation, pp. 4591–4596. IEEE Press, New York (2016)

10. Antonelli, G.: Underwater Robots. Springer, Heidelberg (2014)

11. Kwan, C., Lewis, F.L., Dawson, D.M.: Robust neural network control of rigid-link electrically-driven robots. IEEE Trans. Neural Networks. **9**(4), 581–588 (1998)

Nonsingular Terminal Sliding Mode Based Trajectory Tracking Control of an Autonomous Surface Vehicle with Finite-Time Convergence

Shuailin Lv[1], Ning Wang[1(✉)], Yong Wang[1], Jianchuan Yin[2],
and Meng Joo Er[1,3]

[1] Marine Engineering College, Dalian Maritime University, Dalian, China
lvshuailin@gmail.com, n.wang.dmu.cn@gmail.com, 15241192269@163.com
[2] Navigation College, Dalian Maritime University, Dalian, China
yinjianchuan@gmail.com
[3] School of EEE, Nanyang Technological University, Singapore, Singapore
emjer@ntu.edu.sg

Abstract. In this paper, a nonsingular terminal sliding mode (NTSM) based tracking control (NTSMTC) scheme for an autonomous surface vehicle (ASV) subject to unmodelled dynamics and unknown disturbances is proposed. The salient features of the NTSMTC scheme are as follows: (1) The NTSMTC scheme is designed by combining the NTSM technique with an established finite-time unknown observer (FUO) which enhances the system robustness significantly and achieves accurate tracking performance; (2) By virtue of the NTSMTC scheme, not only that unknown estimation errors are controlled to zero but also tracking errors can be stabilized to zero in a finite time; (3) The finite-time convergence of the entire closed-loop control system can be ensured by the Lyapunov approach. Simulation studies are further provided to demonstrate the effectiveness and remarkable performance of the proposed NTSMTC scheme for trajectory tracking control of an ASV.

Keywords: Nonsingular terminal sliding mode (NTSM) · Finite-time stability · Finite-time unknown observer (FUO) · Trajectory tracking control · Autonomous surface vehicle (ASV)

1 Introduction

In last decades, autonomous surface vehicles (ASVs) have drawn more and more attention mainly due to important roles in military and civilian applications.

N. Wang—This work is supported by the National Natural Science Foundation of P.R. China (under Grants 51009017 and 51379002), Applied Basic Research Funds from Ministry of Transport of P.R. China (under Grant 2012-329-225-060), China Postdoctoral Science Foundation (under Grant 2012M520629), the Fund for Dalian Distinguished Young Scholars (under Grant 2016RJ10), the Innovation Support Plan for Dalian High-level Talents (under Grant 2015R065), and the Fundamental Research Funds for the Central Universities (under Grant 3132016314).

© Springer International Publishing AG 2017
F. Cong et al. (Eds.): ISNN 2017, Part II, LNCS 10262, pp. 83–92, 2017.
DOI: 10.1007/978-3-319-59081-3_11

However, suffering from a variety of external disturbance variations including winds, waves and currents, ASVs are highly nonlinear and the exact ASV model can hardly be known, which makes it much challenging and difficult when designing a controller for ASVs.

Traditionally, fuzzy logic systems (FLS) [1] and fuzzy neural networks (FNN) [2] are usually employed for tracking control of an ASV, which can explicitly take into account complicated unknowns including external disturbances and even unmodelled dynamics. As a result, the previous approximation-based methods can achieve many good properties including disturbance rejection capacity and high steady-state accuracy. However, it should be pointed out that only asymptotic or exponential convergence can be obtained in the previous works rather than finite-time convergence.

Recently, finite-time control theorems have been increasingly studied; for example, nonsingular terminal sliding mode (NTSM) technique [3], homogeneity [4] and adding a power integrator (API) [5] approaches. Note that fast convergence rate and high robustness can be achieved pertaining to the foregoing finite-time based methods. Motivated by the above observations, finite-time trajectory tracking and heading controller have been established by Wang in [6] and [7], respectively. However, finite-time control problems of an ASV in the presence of complicated unknowns is still largely open. It is mainly for this reason that finite-time convergence is pursued in this paper in order to achieve fast and precise tracking performance.

In this paper, a nonsingular terminal sliding mode (NTSM) based tracking control (NTSMTC) scheme is proposed. To be more specific, the NTSM technique and the designed finite-time unknown observer (FUO) are integrated to preserve the advantages of each method, i.e., fast convergence and high robustness. Moreover, rigorously proof has been given to ensure the overall closed-loop system to be finite-time stable and it has been proven that tracking errors can be stabilized to zero in a finite time, which as a result leads to accurate tracking performance.

2   Problem Formulation

The kinematics and dynamics of an ASV moving in a planar space can be expressed as follows:

$$\begin{cases} \dot{\boldsymbol{\eta}} = \mathbf{J}(\psi)\boldsymbol{v} \\ \mathbf{M}\dot{\boldsymbol{v}} = \mathcal{N}(\boldsymbol{\eta}, \boldsymbol{v}) + \boldsymbol{\tau} + \boldsymbol{\tau}_\delta \end{cases} \tag{1}$$

where

$$\mathcal{N}(\boldsymbol{\eta}, \boldsymbol{v}) = -\mathbf{C}(\boldsymbol{v})\boldsymbol{v} - \mathbf{D}(\boldsymbol{v})\boldsymbol{v} - \mathbf{g}(\boldsymbol{\eta}, \boldsymbol{v}) \tag{2}$$

Here, $\boldsymbol{\eta} = [x, y, \psi]^{\mathrm{T}}$ is the 3-DOF position (x, y) and heading angle (ψ) of the ASV, $\boldsymbol{v} = [u, v, r]^{\mathrm{T}}$ is the corresponding linear velocities (u, v), i.e., surge and sway velocities, and angular rate (r), i.e., yaw, in the body-fixed frame,

$\boldsymbol{\tau} = [\tau_1, \tau_2, \tau_3]^{\mathrm{T}}$ and $\boldsymbol{\tau}_\delta := \mathbf{M}\mathbf{R}^{\mathrm{T}}(\psi)\boldsymbol{\delta}(t)$ with $\boldsymbol{\delta}(t) = [\delta_1(t), \delta_2(t), \delta_3(t)]^{\mathrm{T}}$ denote control input and mixed external disturbance, and $\mathbf{J}(\psi)$ is a rotation matrix governed by

$$\mathbf{J}(\psi) = \begin{bmatrix} \cos\psi & -\sin\psi & 0 \\ \sin\psi & \cos\psi & 0 \\ 0 & 0 & 1 \end{bmatrix} \tag{3}$$

with the following properties:

$$\mathbf{J}^{\mathrm{T}}(\psi)\mathbf{J}(\psi) = \mathbf{I}, \text{ and } \|\mathbf{J}(\psi)\| = 1, \ \forall\,\psi \in [0, 2\pi] \tag{4a}$$

$$\dot{\mathbf{J}}(\psi) = \mathbf{J}(\psi)\mathbf{S}(r) \tag{4b}$$

$$\mathbf{J}^{\mathrm{T}}(\psi)\mathbf{S}(r)\mathbf{J}(\psi) = \mathbf{J}(\psi)\mathbf{S}(r)\mathbf{J}^{\mathrm{T}}(\psi) = \mathbf{S}(r) \tag{4c}$$

where $\mathbf{S}(r) = \begin{bmatrix} 0 & -r & 0 \\ r & 0 & 0 \\ 0 & 0 & 0 \end{bmatrix}$, the inertia matrix $\mathbf{M} = \mathbf{M}^{\mathrm{T}} > 0$, the skew-symmetric matrix $\mathbf{C}(\mathbf{v}) = -\mathbf{C}(\mathbf{v})^{\mathrm{T}}$, and the damping matrix $\mathbf{D}(\mathbf{v})$ can be written as follows:

$$\mathbf{M} = \begin{bmatrix} m_{11} & 0 & 0 \\ 0 & m_{22} & m_{23} \\ 0 & m_{32} & m_{33} \end{bmatrix} \tag{5a}$$

$$\mathbf{C}(\mathbf{v}) = \begin{bmatrix} 0 & 0 & c_{13}(\mathbf{v}) \\ 0 & 0 & c_{23}(\mathbf{v}) \\ -c_{13}(\mathbf{v}) & -c_{23}(\mathbf{v}) & 0 \end{bmatrix} \tag{5b}$$

$$\mathbf{D}(\mathbf{v}) = \begin{bmatrix} d_{11}(\mathbf{v}) & 0 & 0 \\ 0 & d_{22}(\mathbf{v}) & d_{23}(\mathbf{v}) \\ 0 & d_{32}(\mathbf{v}) & d_{33}(\mathbf{v}) \end{bmatrix} \tag{5c}$$

where $m_{11} = m - X_{\dot{u}}$, $m_{22} = m - Y_{\dot{v}}$, $m_{23} = mx_g - Y_{\dot{r}}$, $m_{32} = mx_g - N_{\dot{v}}$, $m_{33} = I_z - N_{\dot{r}}$; $c_{13}(\mathbf{v}) = -m_{11}v - m_{23}r$, $c_{23}(\mathbf{v}) = m_{11}u$; $d_{11}(\mathbf{v}) = -X_u - X_{|u|u}|u| - X_{uuu}u^2$, $d_{22}(\mathbf{v}) = -Y_v - Y_{|v|v}|v|$, $d_{23}(\mathbf{v}) = -Y_r - Y_{|v|r}|v| - Y_{|r|r}|r|$, $d_{32}(\mathbf{v}) = -N_v - N_{|v|v}|v| - N_{|r|v}|r|$ and $d_{33}(\mathbf{v}) = -N_r - N_{|v|r}|v| - N_{|r|r}|r|$. Here, m is the mass of the ASV, I_z is the moment of inertia about the yaw rotation, $Y_{\dot{r}} = N_{\dot{v}}$, and symbols X_*, Y_*, N_* represent corresponding hydrodynamic derivatives.

Consider the desired trajectory generated by

$$\begin{cases} \dot{\boldsymbol{\eta}}_d = \mathbf{J}(\psi_d)\mathbf{v}_d \\ \mathbf{M}\dot{\mathbf{v}}_d = \boldsymbol{\mathcal{N}}_d(\boldsymbol{\eta}_d, \mathbf{v}_d) + \boldsymbol{\tau}_d \end{cases} \tag{6}$$

where

$$\boldsymbol{\mathcal{N}}_d(\boldsymbol{\eta}_d, \mathbf{v}_d) = -\mathbf{C}(\mathbf{v}_d)\mathbf{v}_d - \mathbf{D}(\mathbf{v}_d)\mathbf{v}_d \tag{7}$$

Here, $\boldsymbol{\eta}_d = [x_d, y_d, \psi_d]^{\mathrm{T}}$ and $\mathbf{v}_d = [u_d, v_d, r_d]^{\mathrm{T}}$ represent the desired position and velocity vectors.

The objective in this context is to design a control law such that the actual trajectory in (1)–(2) can track exactly the desired targets generated by (6)–(7) in a finite time $0 < T < \infty$, i.e., $\boldsymbol{\eta}(t) \equiv \boldsymbol{\eta}_d(t)$ and $\mathbf{v}(t) \equiv \mathbf{v}_d(t), \forall\, t > T$.

3   Controller Design and Stability Analysis

3.1   Controller Design

Consider the following transformations on $\boldsymbol{v}$ and $\boldsymbol{v}_d$:

$$\boldsymbol{\omega} = \mathbf{J}\boldsymbol{v} \tag{8a}$$

$$\boldsymbol{\omega}_d = \mathbf{J}_d\boldsymbol{v}_d \tag{8b}$$

where $\boldsymbol{\omega} = [\omega_1, \omega_2, \omega_3]^T$, $\boldsymbol{\omega}_d = [\omega_{d1}, \omega_{d2}, \omega_{d3}]^T$, $\mathbf{J} = \mathbf{J}(\psi)$ and $\mathbf{J}_d = \mathbf{J}(\psi_d)$.
   Combining (1)–(2) with (8a) yields

$$\begin{cases} \dot{\boldsymbol{\eta}} = \boldsymbol{\omega} \\ \dot{\boldsymbol{\omega}} = \mathbf{J}\mathbf{M}^{-1}\boldsymbol{\tau} + \mathcal{H}(\boldsymbol{\eta}, \boldsymbol{\omega}) + \boldsymbol{\delta}(t) \end{cases} \tag{9}$$

where

$$\mathcal{H}(\boldsymbol{\eta}, \boldsymbol{\omega}) = \mathbf{S}(\underset{3}{\omega})\boldsymbol{\omega} + \mathbf{J}\mathbf{M}^{-1}\mathcal{N}(\boldsymbol{\eta}, \mathbf{J}^T\boldsymbol{\omega}). \tag{10}$$

From (6)–(7) and (8b) yields

$$\begin{cases} \dot{\boldsymbol{\eta}}_d = \boldsymbol{\omega}_d \\ \dot{\boldsymbol{\omega}}_d = \mathbf{J}_d\mathbf{M}^{-1}\boldsymbol{\tau}_d + \mathcal{H}_d(\boldsymbol{\eta}_d, \boldsymbol{\omega}_d) \end{cases} \tag{11}$$

where

$$\begin{aligned} \mathcal{H}_d(\boldsymbol{\eta}_d, \boldsymbol{\omega}_d) = &-\mathbf{J}_d\mathbf{M}^{-1}\left(\mathbf{C}\left(\mathbf{J}_d^T\boldsymbol{\omega}_d\right) + \mathbf{D}\left(\mathbf{J}_d^T\boldsymbol{\omega}_d\right)\right)\mathbf{J}_d^T\boldsymbol{\omega}_d \\ &+\mathbf{S}(\underset{d3}{\omega})\boldsymbol{\omega}_d. \end{aligned} \tag{12}$$

Using (9)–(10) and (11)–(12), we have

$$\begin{cases} \dot{\boldsymbol{\eta}}_e = \boldsymbol{\omega}_e \\ \dot{\boldsymbol{\omega}}_e = \mathbf{J}\mathbf{M}^{-1}\boldsymbol{\tau} + \mathcal{H}_e(\boldsymbol{\eta}, \boldsymbol{\omega}, \boldsymbol{\eta}_d, \boldsymbol{\omega}_d) + \boldsymbol{f}_u(\boldsymbol{\eta}, \boldsymbol{\omega}, \boldsymbol{\delta}, t) \end{cases} \tag{13}$$

where

$$\mathcal{H}_e(\cdot) = (\mathbf{J}_d\mathbf{M}^{-1}(\mathbf{C}(\mathbf{J}_d^T\boldsymbol{\omega}_d) + \mathbf{D}(\mathbf{J}_d^T\boldsymbol{\omega}_d))\mathbf{J}_d^T)\boldsymbol{\omega}_d$$
$$+ \mathbf{S}\boldsymbol{\omega} - \mathbf{S}_d\boldsymbol{\omega}_d - \mathbf{J}_d\mathbf{M}^{-1}\boldsymbol{\tau}_d \tag{14a}$$

$$\boldsymbol{f}_u(\cdot) = \boldsymbol{\delta} + \mathbf{J}\mathbf{M}^{-1}\mathcal{N}(\boldsymbol{\eta}, \mathbf{J}^T\boldsymbol{\omega}) \tag{14b}$$

Here, $\mathbf{S} = \mathbf{S}(\omega_3)$, $\mathbf{S}_d = \mathbf{S}(\omega_{d3})$, $\boldsymbol{\eta}_e = \boldsymbol{\eta} - \boldsymbol{\eta}_d := [\eta_{e1}, \eta_{e2}, \eta_{e3}]^T$ and $\boldsymbol{\omega}_e = \boldsymbol{\omega} - \boldsymbol{\omega}_d := [\omega_{e1}, \omega_{e2}, \omega_{e3}]^T$.

Assumption 1. *The unknown term $\boldsymbol{f}_u$ in (13)–(14) satisfies*

$$\left\| \ddot{\boldsymbol{f}}_u \right\| \leq L_{fu} \tag{15}$$

for a bounded constant $L_{fu} < \infty$.

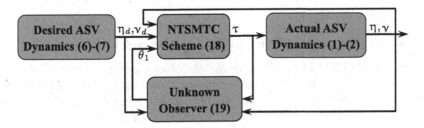

Fig. 1. Control system diagram.

In the light of (13)–(14), we define the nonsingular terminal sliding mode (NTSM) manifold as follows:

$$\boldsymbol{\sigma}(t) = \boldsymbol{\eta}_e(t) + \frac{1}{\beta}\boldsymbol{\omega}_e^{p/q}(t) \tag{16}$$

where $\boldsymbol{\sigma}(t) = [\sigma_1(t), \sigma_2(t), \sigma_3(t)]^{\mathrm{T}}$.

Differentiating $\boldsymbol{\sigma}(t)$ with respect to time, we obtain

$$\dot{\boldsymbol{\sigma}} = \boldsymbol{\omega}_e + \frac{p}{q\beta}\mathrm{diag}(\boldsymbol{\omega}_e^{(p/q)-1})\dot{\boldsymbol{\omega}}_e \tag{17}$$

where $\mathrm{diag}(\boldsymbol{\omega}_e^{(p/q)-1}) := \mathrm{diag}(\omega_{e1}^{(p/q)-1}, \omega_{e2}^{(p/q)-1}, \omega_{e3}^{(p/q)-1})$ and $\dot{\boldsymbol{\omega}}_e := [\dot{\omega}_{e1}, \dot{\omega}_{e2}, \dot{\omega}_{e3}]^{\mathrm{T}}$.

Concerning the ASV tracking error dynamics (13)–(14) and sliding functions (16)–(17), the NTSM based tracking control (NTSMTC) scheme can be designed accordingly

$$\boldsymbol{\tau} = -\mathbf{MJ}^{-1}\left(\beta\frac{q}{p}(\mathbf{J}\boldsymbol{v} - \mathbf{J}_d\boldsymbol{v}_d)^{[2-(p/q)]} + \boldsymbol{\kappa}\mathrm{sgn}(\boldsymbol{\sigma})\right)$$
$$-\mathbf{MS}\boldsymbol{v} + \mathbf{MJ}^{-1}\left(\mathbf{S}_d\mathbf{J}_d\boldsymbol{v}_d + \mathbf{J}_d\mathbf{M}^{-1}\boldsymbol{\tau}_d\right)$$
$$+\mathbf{MJ}^{-1}\left(\mathbf{J}_d\mathbf{M}^{-1}\mathcal{N}_d(\boldsymbol{\eta}_d, \boldsymbol{v}_d) - \boldsymbol{\theta}_1\right) \tag{18}$$

Here, $\beta > 0, p > 0$ and $q > 0$ are positive old integers satisfying $1 < p/q < 2$, $\boldsymbol{\kappa} = \mathrm{diag}(\kappa_1, \kappa_2, \kappa_3)$ with positive constants $\kappa_j (j = 1, 2, 3)$, and $\mathrm{sgn}(\boldsymbol{\sigma}) = [\mathrm{sgn}(\sigma_1), \mathrm{sgn}(\sigma_2), \mathrm{sgn}(\sigma_3)]^{\mathrm{T}}$, with $\boldsymbol{\theta}_1$ derived by the following finite-time unknown observer (FUO):

$$\dot{\boldsymbol{\theta}}_0 = \boldsymbol{\zeta}_0 + \mathbf{JM}^{-1}\boldsymbol{\tau} + \mathcal{H}_e(\cdot)$$
$$\boldsymbol{\zeta}_0 = -\lambda_1\mathcal{L}^{1/3}\mathrm{sig}^{2/3}(\boldsymbol{\theta}_0 - \boldsymbol{\omega}_e) + \boldsymbol{\theta}_1$$
$$\dot{\boldsymbol{\theta}}_1 = \boldsymbol{\zeta}_1$$
$$\boldsymbol{\zeta}_1 = -\lambda_2\mathcal{L}^{1/2}\mathrm{sig}^{1/2}(\boldsymbol{\theta}_1 - \boldsymbol{\zeta}_0) + \boldsymbol{\theta}_2$$
$$\dot{\boldsymbol{\theta}}_2 = -\lambda_3\mathcal{L}\mathrm{sgn}(\boldsymbol{\theta}_2 - \boldsymbol{\zeta}_1) \tag{19}$$

where $\boldsymbol{\theta}_i := [\theta_{i1}, \theta_{i2}, \theta_{i3}]^{\mathrm{T}}, i = 0, 1, 2, \boldsymbol{\zeta}_k := [\zeta_{k1}, \zeta_{k2}, \zeta_{k3}]^{\mathrm{T}}, k = 0, 1, \lambda_j > 0, j = 1, 2, 3$ and $\mathcal{L} = \mathrm{diag}(\ell_1, \ell_2, \ell_3)$. The corresponding control system diagram are also illustrated in Fig. 1.

3.2   Stability Analysis

The key result ensuring finite-time stability of the closed-loop is now stated.

Theorem 1 (NTSMTC). *Consider the closed-loop system composed of (13)–(14), (16)–(17) and (18)–(19), the actual trajectory and velocity of the ASV system (1)–(2) will converge to the desired signals generated by (6)–(7) in a finite time $0 < T < \infty$, i.e., $\boldsymbol{\eta}(t) \equiv \boldsymbol{\eta}_d(t)$ and $\boldsymbol{v}(t) \equiv \boldsymbol{v}_d(t), \forall\, t > T$.*

Proof. Consider the Lyapunov function as follows:

$$V = \frac{1}{2}\boldsymbol{\sigma}^{\mathrm{T}}\boldsymbol{\sigma}. \tag{20}$$

Differentiating V along (13)–(14) yields

$$
\begin{aligned}
\dot{V} &= \boldsymbol{\sigma}^{\mathrm{T}}\dot{\boldsymbol{\sigma}} \\
&= \boldsymbol{\sigma}^{\mathrm{T}}\left(\boldsymbol{\omega}_e + \frac{p}{q\beta}\mathrm{diag}(\boldsymbol{\omega}_e^{(p/q)-1})\dot{\boldsymbol{\omega}}_e \right) \\
&= \boldsymbol{\sigma}^{\mathrm{T}}\left(\boldsymbol{\omega}_e + \frac{p}{q\beta}\mathrm{diag}(\boldsymbol{\omega}_e^{(p/q)-1})\left(\mathbf{RM}^{-1}\boldsymbol{\tau} + \mathcal{H}_e(\cdot) + \boldsymbol{f}_u\right) \right).
\end{aligned} \tag{21}
$$

Substituting (18) into (21) yields

$$
\begin{aligned}
\dot{V} = \boldsymbol{\sigma}^{\mathrm{T}}\Bigg[\boldsymbol{\omega}_e &+ \frac{p}{q\beta}\mathrm{diag}(\boldsymbol{\omega}_e^{(p/q)-1})\Bigg(-\beta\frac{q}{p}\boldsymbol{\omega}_e^{[2-(p/q)]} \\
&- \boldsymbol{\kappa}\,\mathrm{sgn}\,(\boldsymbol{\sigma}) + \boldsymbol{f}_u - \boldsymbol{\theta}_1 \Bigg)\Bigg].
\end{aligned} \tag{22}
$$

Define unknown observation errors as follows:

$$\mathbf{z}_1 = \boldsymbol{\theta}_0 - \boldsymbol{\omega}_e, \;\; \mathbf{z}_2 = \boldsymbol{\theta}_1 - \boldsymbol{f}_u, \;\; \mathbf{z}_3 = \boldsymbol{\theta}_2 - \dot{\boldsymbol{f}}_u \tag{23}$$

Then the error dynamics can be derived as

$$
\begin{aligned}
\dot{\mathbf{z}}_1 &= -\lambda_1 \mathcal{L}^{1/3}\mathrm{sig}^{2/3}(\mathbf{z}_1) + \mathbf{z}_2 \\
\dot{\mathbf{z}}_2 &= -\lambda_2 \mathcal{L}^{1/2}\mathrm{sig}^{1/2}(\mathbf{z}_2 - \dot{\mathbf{z}}_1) + \mathbf{z}_3 \\
\dot{\mathbf{z}}_3 &= -\lambda_3 \mathcal{L}\,\mathrm{sgn}(\mathbf{z}_3 - \dot{\mathbf{z}}_2) - \ddot{\boldsymbol{f}}_u
\end{aligned} \tag{24}
$$

i.e.,

$$
\begin{aligned}
\dot{z}_{1j} &= -\lambda_1 \ell_j^{1/3}\mathrm{sig}^{2/3}(z_{1j}) + z_{2j} \\
\dot{z}_{2j} &= -\lambda_2 \ell_j^{1/2}\mathrm{sig}^{1/2}(z_{2j} - \dot{z}_{1j}) + z_{3j} \\
\dot{z}_{3j} &\in -\lambda_3 \ell_j\,\mathrm{sgn}(z_{3j} - \dot{z}_{2j}) + [-L_{fu}, L_{fu}].
\end{aligned} \tag{25}
$$

According to [8], $\mathbf{z}_1, \mathbf{z}_2$ and $\mathbf{z}_3$ can be stabilized to zero in a finite time, and this yields

$$\boldsymbol{\theta}_0 \equiv \boldsymbol{\omega}_e, \ \boldsymbol{\theta}_1 \equiv \boldsymbol{f}_u, \ \boldsymbol{\theta}_2 \equiv \dot{\boldsymbol{f}}_u. \tag{26}$$

Combining (22) and (26) we have

$$\dot{V} = \boldsymbol{\sigma}^{\mathrm{T}} \left(\boldsymbol{\omega}_e + \frac{p}{q\beta} \mathrm{diag}(\boldsymbol{\omega}_e^{(p/q)-1}) \left(-\beta \frac{q}{p} \boldsymbol{\omega}_e^{[2-(p/q)]} - \kappa \mathrm{sgn}(\boldsymbol{\sigma}) \right) \right)$$

$$\leq - \min_{j=1,2,3} \left\{ \frac{p}{q\beta} \omega_{ej}^{(p/q)-1} \kappa_j \right\} \sum_{j=1}^{3} |\sigma_j| \tag{27}$$

Define

$$\rho = \sqrt{2} \cdot \min_{j=1,2,3} \left\{ \frac{p}{q\beta} \omega_{ej}^{(p/q)-1} \kappa_j \right\}. \tag{28}$$

Clearly, when $\omega_{ej} \neq 0$, since p and q are positive old integers and $1 < p/q < 2$, we have $\rho > 0$. Thus,

$$\dot{V} \leq -\rho V^{1/2}. \tag{29}$$

Then finite-time stability can be ensured according to [9, Theorem 1]. When $\omega_{ej} = 0$, substituting control law (18) into (13)–(14), we have

$$\dot{\boldsymbol{\omega}}_e = -\beta \frac{q}{p} \boldsymbol{\omega}_e^{[2-(p/q)]} - \kappa \mathrm{sgn}(\boldsymbol{\sigma}) + \boldsymbol{f}_u - \boldsymbol{\theta}_1 \tag{30}$$

Hence,

$$\dot{\omega}_{ej} = -\beta \frac{q}{p} \omega_{ej}^{[2-(p/q)]} - \kappa_j \mathrm{sgn}(\sigma_j) \tag{31}$$

And this yields

$$\dot{\omega}_{ej} = -\kappa_j \mathrm{sgn}(\sigma_j) \tag{32}$$

with $j = 1, 2, 3$.

Therefore, $\dot{\omega}_{ej} < 0$ when $\sigma_j > 0$, and $\dot{\omega}_{ej} > 0$ when $\sigma_j < 0$. Clearly, $\dot{\omega}_{ej} = 0$ is not an attractor. It can be concluded that manifold (16) can be reached in a finite time $t_1^* > 0$.

Next, we will prove that once the manifold is reached, tracking errors $\boldsymbol{\eta}_e$ and $\boldsymbol{\omega}_e$ will converge to zero along the manifold in a finite time.

When $\boldsymbol{\sigma} = 0$, from (16), we have

$$\boldsymbol{\eta}_e + \frac{1}{\beta} \boldsymbol{\omega}_e^{p/q} = 0 \tag{33}$$

i.e.,

$$\eta_{ej} + \frac{1}{\beta} \dot{\eta}_{ej}^{p/q} = 0, \ j = 1, 2, 3. \tag{34}$$

It follows that tracking errors η_{ej} and ω_{ej} can be stabilized to zero along $\sigma_j = 0$ at time $t_2^* = p\beta^{(-q/p)} \cdot \eta_{ej}^{[1-(q/p)]}(t_1^*)/(p-q) + t_1^*$.

Now we can get the conclusion that the closed-loop system (13)–(14), (16)–(17) and (18)–(19) is finite-time stable. This completes the proof.

Remark 1. If $p = q = 1$, the NTSMTC scheme (18) will degrade to a sliding mode control SMC scheme (τ_{SMC}) accordingly

$$
\begin{aligned}
\tau_{\mathrm{SMC}} = &- \mathbf{M}\mathbf{J}^{-1}\left(\beta\left(\mathbf{J}\boldsymbol{\nu} - \mathbf{J}_d\boldsymbol{\nu}_d\right) + \boldsymbol{\kappa}\mathrm{sgn}\left(\boldsymbol{\sigma}\right)\right) \\
&- \mathbf{M}\mathbf{S}\boldsymbol{\nu} + \mathbf{M}\mathbf{J}^{-1}\left(\mathbf{S}_d\mathbf{J}_d\boldsymbol{\nu}_d + \mathbf{J}_d\mathbf{M}^{-1}\boldsymbol{\tau}_d\right) \\
&+ \mathbf{M}\mathbf{J}^{-1}\left(\mathbf{J}_d\mathbf{M}^{-1}\boldsymbol{\mathcal{N}}_d(\boldsymbol{\eta}_d, \boldsymbol{\nu}_d) - \boldsymbol{\theta}_1\right)
\end{aligned}
\tag{35}
$$

with $\boldsymbol{\theta}_1$ derived by (19).

Remark 2. The chattering can be reduced by replacing the $\mathrm{sgn}(\sigma_j)$ function with a saturation function described by

$$
fsat(\underset{j}{\sigma}; \varepsilon, \vartheta) = \begin{cases} \mathrm{sgn}(\sigma_j), |\sigma_j| > \varepsilon \\ \frac{\mathrm{sig}^\vartheta(\sigma_j)}{\varepsilon^\vartheta}, |\sigma_j| \le \varepsilon \end{cases}
\tag{36}
$$

with $\varepsilon > 0$ and $0 < \vartheta < 1$.

4 Simulation Studies

This section assesses the control performance of the proposed NTSMTC law in terms of trajectory tracking of an ASV. Simulations studies are conducted on a well-known ASV named CyberShip II [10].

Assume external disturbances in (1) are governed by

$$
\boldsymbol{\delta}(t) = \begin{bmatrix} 3\cos(0.1\pi t - \pi/3) \\ 4\cos(0.2\pi t + \pi/4) \\ 6\cos(0.3\pi t + \pi/6) \end{bmatrix}.
\tag{37}
$$

Consider the desired trajectory generated by (6)–(7), assume $\boldsymbol{\tau}_d = [4, 3\cos^2(0.1\pi t), \sin^2(0.1\pi t)]^{\mathrm{T}}$, the initial conditions are $\boldsymbol{\eta}(0) = [15.5, 8, \pi/4]^{\mathrm{T}}$, $\boldsymbol{\nu}(0) = [0, 0, 0]^{\mathrm{T}}$, $\boldsymbol{\eta}_d(0) = [16, 7.8, \pi/3]^{\mathrm{T}}$ and $\boldsymbol{\nu}_d(0) = [1, 0, 0]^{\mathrm{T}}$.

Correspondingly, parameters of the FUO are: $\lambda_1 = 2.2$, $\lambda_2 = 1.1$, $\lambda_3 = 0.8$, $\mathcal{L} = \mathrm{diag}(30, 30, 30)$; and parameters of the NTSMTC scheme are: $\beta = 1$, $p = 5$, $q = 3$, $\boldsymbol{\kappa} = \mathrm{diag}(3.6, 3.6, 3.6)$, $\varepsilon = 6.8$, $\vartheta = 0.58$.

In comparison with the traditional SMC approach $\boldsymbol{\tau}_{\mathrm{SMC}}$ in (35), it can be clearly seen from Fig. 2 that the actual trajectory (solid line) can track the desired (dashed line) one with faster convergence rate. Correspondingly, the actual position $\boldsymbol{\eta} = [x, y, \psi]^{\mathrm{T}}$ and the desired signal $\boldsymbol{\eta}_d = [x_d, y_d, \psi_d]^{\mathrm{T}}$ are shown in Fig. 3, which exhibits the higher tracking accuracy. In addition to precise position tracking, actual velocity vector $\boldsymbol{\nu} = [u, v, r]^{\mathrm{T}}$ can track the desired target $\boldsymbol{\nu}_d = [u_d, v_d, r_d]^{\mathrm{T}}$ very quickly, as shown in Fig. 4. The time-varying unknowns

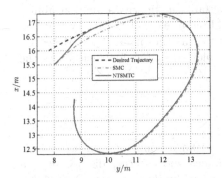

Fig. 2. Desired and actual trajectories.

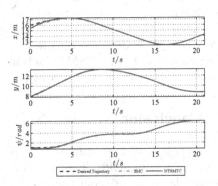

Fig. 3. Curves of position tracking.

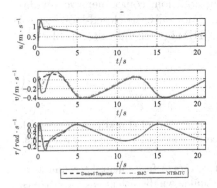

Fig. 4. Curves of velocity tracking.

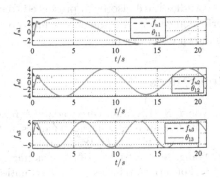

Fig. 5. Curves of unknown estimation.

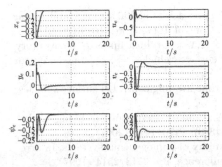

Fig. 6. Curves of zero tracking errors.

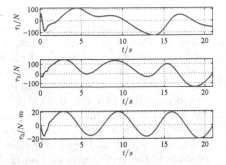

Fig. 7. Curves of smooth control inputs.

$f_u = [f_{u1}, f_{u2}, f_{u3}]^T$ and the finite-time identification results $\theta_1 = [\theta_{11}, \theta_{12}, \theta_{13}]^T$ are shown in Fig. 5, which shows the remarkable unknown estimation ability associated with the proposed FUO. It should be noted that trajectory tracking errors can be rendered to zero in a finite time, as shown in Fig. 6. Moreover, control inputs of the NTSMTC scheme are plotted in Fig. 7, which shows the smooth control actions dynamically.

5   Conclusion

In this paper, to pursue finite-time control of an autonomous surface vehicle (ASV) in the presence of unmodelled dynamics and external disturbances, a nonsingular terminal sliding mode (NTSM) based tracking control (NTSMTC) scheme has been proposed. Under the NTSMTC scheme, not only that unmodelled dynamics and unknown disturbances can be completely identified but also finite-time convergence property can be achieved, and thereby contributing to fast convergence rate and high robustness. In addition, comprehensive simulation studies have also been presented to confirm not only the closed-loop control performance but also the effectiveness of the NTSMTC scheme in terms of exact unknown observation.

References

1. Wang, N., Er, M.J.: Direct adaptive fuzzy tracking control of marine vehicles with fully unknown parametric dynamics and uncertainties. IEEE Trans. Control Syst. Technol. **24**(5), 1845–1852 (2016)
2. Wang, N., Er, M.J.: Self-constructing adaptive robust fuzzy neural tracking control of surface vehicles with uncertainties and unknown disturbances. IEEE Trans. Control Syst. Technol. **23**(3), 991–1002 (2015)
3. Feng, Y., Yu, X.H., Man, Z.H.: Non-singular terminal sliding mode control of rigid manipulators. Automatica **38**(12), 2159–2167 (2002)
4. Hong, Y.G., Xu, Y.S., Huang, J.: Finite-time control for robot manipulators. Syst. Control Lett. **46**(4), 243–253 (2002)
5. Qian, C.J., Lin, W.: A continuous feedback approach to global strong stabilization of nonlinear systems. IEEE Trans. Autom. Control **46**(7), 1061–1079 (2002)
6. Wang, N., Qian, C.J., Sun, J.C., Liu, Y.C.: Adaptive robust finite-time trajectory tracking control of fully actuated marine surface vehicles. IEEE Trans. Control Syst. Technol. **24**(4), 1454–1462 (2016)
7. Wang, N., Lv, S.L., Liu, Z.Z.: Global finite-time heading control of surface vehicles. Neurocomputing **175**, 662–666 (2016)
8. Shtessel, Y.B., Shkolnikov, I.A., Levant, A.: Smooth second-order sliding modes: missile guidance application. Automatica **43**(8), 1470–1476 (2007)
9. Bhat, S.P., Bernstein, D.S.: Finite-time stability of homogeneous systems. In: Proceedings of the 1997 American Control Conference, pp. 2513–2514 (1997)
10. Skjetne, R., Fossen, T.I., Kokotovi'c, P.V.: Adaptive maneuvering with experiments for a model ship in a marine control laboratory. Automatica **41**(2), 289–298 (2005)

Saturated Kinetic Control of Autonomous Surface Vehicles Based on Neural Networks

Zhouhua Peng[1,2]([✉]), Jun Wang[2], and Dan Wang[1]

[1] School of Marine Engineering, Dalian Maritime University,
Dalian 116026, People's Republic of China
dwangdl@gmail.com, zhpeng@dlmu.edu.cn
[2] Department of Computer Science, City University of Hong Kong,
Kowloon Tong, Hong Kong
jwang.cs@cityu.edu.hk

Abstract. This paper investigates the saturated kinetic control of autonomous surface vehicles subject to unknown kinetics and limited control torques. The unknown kinetics stems from parametric model uncertainty, unmodelled hydrodynamics, and environmental forces due to wind, waves and ocean currents. By approximating the unknown kinetics using neural networks, a bounded kinetic control law is proposed based on a saturated function, with the main advantage being that the control input is known as a *priori*. The resulting closed-loop kinetic control system is proved to be input-to-state stable.

Keywords: Neural networks · Autonomous surface vehicles · Unknown kinetics · Saturated control

1 Introduction

One of the key challenges in motion control of marine vehicles comes from the model uncertainty, unmodelled dynamics, and environmental forces due to wind, waves, and ocean currents [1]. Numerous control methods have been proposed to address the uncertainty existing in vehicle kinetics, ranging from sliding model

The work of Z. Peng was supported in part by the National Natural Science Foundation of China under Grant 51579023, and in part by High Level Talent Innovation and Entrepreneurship Program of Dalian under Grant 2016RQ036, and in part by the Hong Kong Scholars Program under Grant XJ2015009, and in part by the China Post-Doctoral Science Foundation under Grant 2015M570247.
The work of J. Wang was supported in part by the National Natural Science Foundation of China under Grant 61673330, and in part by the Research Grants Council of the Hong Kong Special Administrative Region, China, under Grant 14207614.
The work of D. Wang was supported in part by the National Natural Science Foundation of China under Grants 61673081, and in part by the Fundamental Research Funds for the Central Universities under Grant 3132016313, and in part by the National Key Research and Development Program of China under Grant 2016YFC0301500.

© Springer International Publishing AG 2017
F. Cong et al. (Eds.): ISNN 2017, Part II, LNCS 10262, pp. 93–100, 2017.
DOI: 10.1007/978-3-319-59081-3_12

control [2,3], adaptive control [4,5], neural network control [8–13,15–17,19], fuzzy control [6,7], L_1 adaptive control [20], disturbance-observer-based control [18], model reference adaptive control [6], to extended-state-observer-based control [14,21]. In particular, neural networks and fuzzy systems are widely explored for approximating vehicle kinetics [8,9,11–13,15–17,19]; however, the constraint problem in vehicle kinetics is not touched.

Constraints are ubiquitous in a practical system. Violating constraints may degrade performance, and even lead to instability in some circumstances. In the existing works, three types of constraint problem are mainly targeted, namely, input constraint [22–24], state constraint [24], and output constraint [25,26].

In this paper, a practical design method is presented for saturated kinetic control of autonomous surface vehicles with unknown kinetics and limited control torques. The unknown kinetics may be caused by parametric model uncertainty, unmodelled hydrodynamics, and environmental forces due to wind, waves, and ocean currents. By approximating the unknown kinetics using neural networks, a bounded kinetic control law is proposed based on a saturated function. A key advantage of the developed control law is that the control input is known as a priori. The resulting closed-loop kinetic control system is input-to-state stable and the error signals are proved to be uniformly ultimately bounded via Lyapunov analysis. The developed saturated kinetic control law can be used in a variety of motion control scenario such as target tracking, trajectory tracking, path following, and formation control of marine vehicles subject to unknown kinetics and limited control torques.

In the following, the problem formulation is stated in Sect. 2. Then, the saturated kinetic controller design and analysis is presented in Sect. 3. Finally, conclusions are drawn in Sect. 4.

2   Preliminaries and Problem Formulation

2.1   Neural Networks

Given a continuous function $f(x) : \mathbb{R}^n \to \mathbb{R}$, there exists an ideal wight W such that the function can be approximated by a neural network as

$$f(x) = W^T \beta(x) + \varepsilon(x), \ x \in \Omega, \tag{1}$$

where $\beta(x)$ is a known activation function; $\varepsilon(x)$ is the approximation error; Ω is a compact set; Besides, there exist positive constants ε^*, β^*, and W^* such that $\|\varepsilon(x)\| \leq \varepsilon^*, \|\beta(x)\| \leq \beta^*$, and $\|W\|_F \leq W^*$, where $\|\cdot\|$ denotes the 2-norm and $\|\cdot\|_F$ denotes the Frobenius norm.

2.2   Problem Formulation

According to [1], the kinetics of surface vehicles can be expressed by (Fig. 1)

$$M\dot{\nu} = \tau - C(\nu)\nu - D(\nu)\nu + g(\nu, \eta) + \tau_w(t), \tag{2}$$

where $M = M^T \in \mathbb{R}^{3 \times 3}$ is a known inertial matrix; $\nu = [u, v, r]^T \in \mathbb{R}^3$ is a vector denoting surge velocity, sway velocity, and angular velocity expressed in the body-fixed reference frame; $C(\nu) = -C(\nu) \in \mathbb{R}^{3 \times 3}$ is a centrifugal and coriolis matrix; $D(\nu) \in \mathbb{R}^{3 \times 3}$ is a damping matrix; $\tau_w = [\tau_{wu}, \tau_{wv}, \tau_{wr}]^T \in \mathbb{R}^3$ is a vector of environmental forces that are assumed to be bounded; $\tau = [\tau_u, \tau_v, \tau_r]^T \in \mathbb{R}^3$ denotes the control input to be designed;

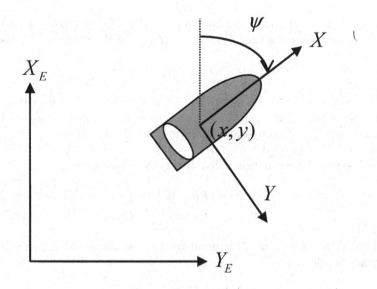

Fig. 1. Reference frames.

The objective is to develop a saturated kinetic control law for vehicle kinetics (2) with unknown kinetics and limited control torques to track a velocity set-point $\nu_r \in \Re^3$.

3   Saturated Kinetic Control Law Design and Analysis

3.1   Estimation of Unknown Kinetics

Rewrite the vehicle kinetics (2) as

$$M\dot{\nu} = \tau + f(\cdot), \tag{3}$$

where $f(\cdot) = -C(\nu)\nu - D(\nu)\nu + g(\nu, \eta) + \tau_w(t)$.

Similar to [27,28], the following lemma is introduced.

Lemma 1. Given $\varepsilon_k^* > 0$ for $k = 1, 2, 3$, there exists a set of bounded weights $W = [W_1, W_2, W_3] \in \Re^{n \times 3}$ with $W_k \in \Re^n$ such that the continuous function $f(\vartheta)$ is approximated by using a neural network as

$$f(\vartheta) = W^T \beta(\vartheta) + \varepsilon, \vartheta \in \Omega, \tag{4}$$

where $\vartheta = [\bar{\nu}^T(t), \tau^T(t)]^T \in \Re^6$ with $\bar{\nu}(t) = \nu(t) - \nu(t - t_d^*)$, t_d^* is the sample period, and $\varepsilon = [\varepsilon_1, \varepsilon_2, \varepsilon_3]^T \in \Re^3$ satisfies $|\varepsilon_k| \leq \varepsilon_k^*$.

Let $\hat{\nu} = [\hat{u}, \hat{v}, \hat{r}]^T \in \Re^3$ be an estimate of ν, and then an estimator for the vehicle kinetics (3) is devised as

$$M\dot{\hat{\nu}} = -F(\hat{\nu} - \nu) + \tau_a + \tau, \tag{5}$$

where $\tau_a = \hat{W}^T \beta(\vartheta)$ and $F = \text{diag}\{k_1, k_2, k_3\} \in \Re^{3 \times 3}$ is a control gain. $k_1, k_2,$ and k_3 are positive constants. $\hat{W}_k$ is an estimate of W_k that updated as

$$\dot{\hat{W}}_k(t) = -\Gamma \, \text{Proj}[\hat{W}_k(t), \beta(\vartheta)\tilde{\nu}_k], \tag{6}$$

where $\tilde{\nu} = \hat{\nu} - \nu = [\tilde{\nu}_1, \tilde{\nu}_2, \tilde{\nu}_3]^3$ denotes the velocity estimation error; $\Gamma \in \Re$ denotes an adaptation gain; $Proj[\cdot, \cdot]$ denotes the projection operation [29]. According to [29], the projection operation assures that there exists a constant ϵ satisfying $\|\hat{W}(t)\|_F \leq W^* + \epsilon$.

As a result, the estimation subsystem is expressed by

$$\begin{cases} M\dot{\tilde{\nu}} = -F\tilde{\nu} + \tilde{W}^T \beta(\vartheta) - \varepsilon, \\ \dot{\hat{W}}_k = -\Gamma \, Proj[\hat{W}_k, \beta(\vartheta)\tilde{\nu}_k]. \end{cases} \tag{7}$$

Lemma 2. The subsystem (7) is input-to-state stable provided that the control parameter is selected as

$$\lambda_{\min}(K) - \tfrac{1}{2} > 0. \tag{8}$$

Proof. Construct the Lyapunov function as $V = \frac{1}{2}\{\tilde{\nu}^T M\tilde{\nu} + \Gamma^{-1} \sum_{k=1}^{3}(\tilde{W}_k^T \tilde{W}_k)\}$, which is bounded by $\lambda_{\min}(P)\|E\|^2/2 \leq V \leq \lambda_{\max}(P)\|E\|^2/2$ with $E = [\tilde{\nu}^T, \tilde{W}_1^T, \tilde{W}_2^T, \tilde{W}_3^T]^T$ and $P = \text{diag}\{M, \Gamma^{-1}\}$.

Tacking the time derivative of V and using (7), it leads to

$$\dot{V} = -\tilde{\nu}^T F\tilde{\nu} - \tilde{\nu}^T \varepsilon. \tag{9}$$

Substituting the inequality $-\tilde{\nu}^T \varepsilon \leq 1/2\|\tilde{\nu}\|^2 + 1/2\|\varepsilon\|^2$ into (9), one has

$$\dot{V} \leq -\left(\lambda_{\min}(F) - \frac{1}{2}\right)\|\tilde{\nu}\|^2 + \frac{1}{2}\|\varepsilon\|^2. \tag{10}$$

Letting $\alpha = \lambda_{\min}(F) - \frac{1}{2}$, it follows that $\dot{V} \leq -\alpha\|E\|^2 + \frac{1}{2}\|\varepsilon\|^2 + \alpha\|\tilde{W}\|_F^2$ under the condition (8). Note that $\|E\| \geq \|\varepsilon\|/\sqrt{\alpha} + \sqrt{2}\|\tilde{W}\|_F$ renders

$$\dot{V} \leq -\frac{\alpha}{2}\|E\|^2. \tag{11}$$

As a consequence, the subsystem (7) is input-to-state stable, and

$$\|E(t)\| \leq \kappa_1(E(0), t) + \kappa_2(\|\varepsilon\|) + \kappa_3(\|\tilde{W}\|_F), \tag{12}$$

where κ_1 is a class $\mathcal{KL}$ function and

$$\kappa_2(s) = \sqrt{\frac{\lambda_{\max}(P)}{\lambda_{\min}(P)}} \frac{s}{\sqrt{\alpha}}, \kappa_3(s) = \sqrt{\frac{\lambda_{\max}(P)}{\lambda_{\min}(P)}} \sqrt{2}s. \tag{13}$$

The boundenss of $\tilde{W}$ is guaranteed by projection operation [29]. Besides, the upper bound for $\tilde{W}$ is given by $\|\tilde{W}\| \leq 2W^* + \epsilon$.

3.2   Kinetic Control Law Design

Let the velocity tracking error be denoted by $e = \nu - \nu_r$ and an estimated velocity tracking error be denoted by $\hat{e} = \hat{\nu} - \nu_r$.

Taking the time derivative of $\hat{e}$ along (16) yields

$$M\dot{\hat{e}} = \tau + \tau_a - F(\hat{\nu} - \nu). \tag{14}$$

To stabilize $\hat{e}$, a *saturated kinetic control law* is taken as

$$\tau = -\frac{K\hat{e}}{\sqrt{\|\hat{e}\|^2 + \Delta^2}} - \tau_a, \tag{15}$$

where $K = \text{diag}\{k_4, k_5, k_6\}$ with k_4, k_5, k_6 being constants; $\Delta \in \Re$ is a positive constant.

Substituting (15) into (16), it follows that

$$M\dot{\hat{e}} = -\frac{K\hat{e}}{\sqrt{\|\hat{e}\|^2 + \Delta^2}} - F\tilde{\nu}, \tag{16}$$

A key feature of the kinetic control law (15) is that the control torques are bounded, and the bounds are known as a priori to a designer. Since $\|\hat{e}\|/\sqrt{\|\hat{e}\|^2 + \Delta^2} < 1$, the explicit bound for τ is given by

$$\|\tau\| \leq K^* + (W^* + \epsilon)\beta^*, \tag{17}$$

where $\|K\|_F \leq K^*$.

Lemma 3. The subsystem (16) is input-to-state stable provided that

$$\frac{\lambda_{\min}(K)}{c_0} - \Delta \geq \xi, \tag{18}$$

where ξ is defined in (24).

Proof. Letting $\Xi = \sqrt{\|\hat{e}\|^2 + \Delta^2}$, construct the Lyapunov function

$$V = \Xi - \Delta, \tag{19}$$

which is bounded as $\frac{\|\hat{e}\|^2}{2\Xi} \leq \Xi - \Delta \leq \frac{\|\hat{e}\|^2}{\Xi}$.

The time derivative of V along (16) is given by

$$\dot{V} = -\frac{\hat{e}^T K \hat{e}}{\Xi^2} - \frac{\hat{e}^T F \tilde{\nu}}{\Xi},$$

$$\leq -c_0 \frac{\|\hat{e}\|^2}{\Xi} + (c_0 - c) \frac{\|\hat{e}\|^2}{\Xi} + \|F\tilde{\nu}\|, \tag{20}$$

where $c = \lambda_{\min}(K)/\Xi$. Provided that

$$\frac{\lambda_{\min}(K)}{c_0} \geq \Delta + \|\hat{e}\|, \tag{21}$$

it follows that $\dot{V} \leq -c_0 \frac{\|\hat{e}\|^2}{\Xi} + \|F\tilde{\nu}\|$ since $\Delta + \|\hat{e}\| \geq \sqrt{\Delta^2 + \|\hat{e}\|^2}$.

Noting that $\frac{\|\hat{e}\|}{\sqrt{\Xi}} \geq \sqrt{\frac{2\|F\tilde{\nu}\|}{c_0}}$ renders $\dot{V} \leq -\frac{c_0}{2} \frac{\|\hat{e}\|^2}{\Xi}$, it is concluded that subsystem (16) is input-to-state stable and

$$\|\hat{e}(t)\| \leq \max\{\kappa_4(\hat{e}(0), t), \kappa_5(\|\tilde{\nu}\|)\} \tag{22}$$

where κ_4 is a $\mathcal{KL}$ function and

$$\kappa_5(s) = \mu^{-1}\left(2\sqrt{\frac{\|F\|_F s}{c_0}}\right). \tag{23}$$

where $\mu(\|\hat{e}\|) = \|\hat{e}\|/\sqrt{\Xi}$. Letting

$$\xi = \max\{\kappa_4(\hat{e}(0), 0), \kappa_5(\|\tilde{\nu}^*\|)\}, \tag{24}$$

where $\tilde{\nu}^*$ is a constant satisfying $\|\tilde{\nu}\| \leq \tilde{\nu}^*$, it follows that a sufficient condition for (21) is given in (18).

The stability of cascade system formed by (16) and (7) is presented in Theorem 1.

Theorem 1. Under the conditions (8) and (18), the cascade system formed by (16) and (7) is input-to-state stable. Besides, the error signals of e and $\hat{e}$ are uniformly ultimately bounded.

Proof. By Lemma C.4 in [30], it follows that the cascade system formed by (16) and (7) is input-to-state stable, implying the boundedness of $\hat{e}$. Besides, noting that $\|e\| = \| -\tilde{\nu} + \hat{e}\| \leq \|\hat{e}\| + \|\tilde{\nu}\|$, it follows that the kinetic tracking error e is uniformly ultimately bounded. This completes the proof.

4   Conclusions

In this paper, a design method is presented for bounded kinetic control of autonomous surface vehicles subject to unknown kinetics and limited control torques. The unknown kinetics comes from parametric model uncertainty, unmodelled hydrodynamics, and environmental forces. A bounded kinetic control law is proposed based on a saturated function and neural networks. A key advantage of the proposed saturated kinetic control law is that the control input is known as a *priori*. It is proved that the resulting closed-loop kinetic control system is input-to-state stable and the error signals are uniformly ultimately bounded. The proposed design method can be used in various motion control scenario such as target tracking, trajectory tracking, path following, and formation control of marine vehicles subject to unknown kinetics and limited control torques.

References

1. Fossen, T.: Handbook of Marine Craft Hydrodynamics and Motion Control. Wiley, Hoboken (2011)
2. Ashrafiuon, H., Muske, K.R., McNinch, L.C., Soltan, R.A.: Sliding-mode tracking control of surface vessels. IEEE Trans. Ind. Electron. **55**(11), 4004–4012 (2008)
3. Cui, R., Zhang, X., Cui, D.: Adaptive sliding-mode attitude control for autonomous underwater vehicles with input nonlinearities. Ocean Eng. **123**, 45–54 (2016)
4. Skjetne, R., Fossen, T.I., Kokotovic, P.V.: Adaptive maneuvering, with experiments, for a model ship in a marine control laboratory. Automatica **41**(2), 289–298 (2005)
5. Yin, S., Xiao, B.: Tracking control of surface ships with disturbance and uncertainties rejection capability. IEEE/ASME Trans. Mechatron. (2016). doi:10.1109/TMECH.2016.2618901
6. Yang, Y., Zhou, C., Ren, J.: Model reference adaptive robust fuzzy control for ship steering autopilot with uncertain nonlinear systems. Appl. Soft Comput. **3**(4), 305–316 (2003)
7. Xiang, X., Yu, C., Zhang, Q.: Robust fuzzy 3D path following for autonomous underwater vehicle subject to uncertainties. Comput. Oper. Res. (2016). doi:10.1016/j.cor.2016.09.017
8. Tee, K., Ge, S.: Control of fully actuated ocean surface vessels using a class of feedforward approximators. IEEE Trans. Control Syst. Technol. **14**(4), 750–756 (2006)
9. Chen, M., Ge, S.S., How, B.V.E., Choo, Y.S.: Robust adaptive position mooring control for marine vessels. IEEE Trans. Control Syst. Technol. **21**(2), 395–409 (2013)
10. Dai, S.L., Wang, M., Wang, C., Li, L.: Learning from adaptive neural network output feedback control of uncertain ocean surface ship dynamics. Int. J. Adapt. Control Signal Process. **28**(3–5), 341–365 (2012)
11. Peng, Z., Wang, D.: Robust adaptive formation control of underactuated autonomous surface vehicles with uncertain dynamics. IET Control Theory A. **5**(12), 1378–1387 (2011)
12. Peng, Z., Wang, D., Chen, Z., Hu, X., Lan, W.: Adaptive dynamic surface control for formations of autonomous surface vehicles with uncertain dynamics. IEEE Trans. Control Syst. Technol. **21**(2), 513–520 (2013)
13. Zheng, Z., Sun, L.: Path following control for marine surface vessel with uncertainties and input saturation. Neurocomputing **177**, 158–167 (2016)

14. Liu, L., Wang, D., Peng, Z.: ESO-based line-of-sight guidance law for path following of underactuated marine surface vehicles with exact sideslip compensation. IEEE J. Oceanic Eng. (2016, in press). doi:10.1109/JOE.2016.2569218

15. Peng, Z., Wang, D., Shi, Y., Wang, H., Wang, W.: Containment control of networked autonomous underwater vehicles with model uncertainty and ocean disturbances guided by multiple leaders. Inf. Sci. **316**(20), 163–179 (2015)

16. Peng, Z., Wang, J., Wang, D.: Containment maneuvering of marine surface vehicles with multiple parameterized paths via spatial-temporal decoupling. IEEE/ASME Trans. Mechatron. (2016). doi:10.1109/TMECH.2016.2632304

17. Peng, Z., Wang, J., Wang, D.: Distributed containment maneuvering of multiple marine vessels via neurodynamics-based output feedback. IEEE Trans. Ind. Electron. (2016) doi:10.1109/TIE.2017.2652346

18. Peng, Z., Wang, D., Wang, J.: Cooperative dynamic positioning of multiple marine offshore vessels: a modular design. IEEE/ASME Trans. Mechatron. **31**(3), 1210–1221 (2016)

19. Peng, Z., Wang, D., Zhang, H., Sun, G.: Distributed neural network control for adaptive synchronization of uncertain dynamical multiagent systems. IEEE Trans. Neural Netw. Learn. Syst. **25**(8), 1508–1519 (2014)

20. Svendsen, C.H., Hoick, N.O., Galeazzi, R., Blanke, M.: L_1 adaptive manoeuvring control of unmanned high-speed water craft. In: IFAC Conference on Manoeuvring and Control of Marine Craft, vol. 45, no. 27, pp. 144–151 (2012)

21. Lei, Z.L., Guo, C.: Disturbance rejection control solution for ship steering system with uncertain time delay. Ocean Eng. **95**(1), 78–83 (2015)

22. Laghrouche, S., Harmouche, M., Chitour, Y.: Global tracking for underactuated ships with bounded feedback controllers. Int. J. Control **87**(10), 2035–2043 (2014)

23. Wang, H., Wang, D., Peng, Z.: Adaptive dynamic surface control for cooperative path following of marine surface vehicles with input saturation. Nonlinear Dyn. **77**(1), 107–117 (2014)

24. Chwa, D.: Global tracking control of underactuated ships with input and velocity constraints using dynamic surface control method. IEEE Trans. Control Syst. Technol. **19**(6), 1357–1370 (2011)

25. Zhao, Z., He, W., Ge, S.S.: Adaptive neural network control of a fully actuated marine surface vessel with multiple output constraints. IEEE Trans. Control Syst. Technol. **22**(4), 1536–1543 (2014)

26. He, W., Yin, Z., Sun, C.: Adaptive neural network control of a marine vessel with constraints using the asymmetric barrier Lyapunov function. IEEE Trans. Cybern. (2016). doi:10.1109/TCYB.2016.2554621

27. Calise, A.J., Hovakimyan, N., Idan, M.: Adaptive output feedback control of nonlinear systems using neural networks. Automatica **37**(12), 1201–1211 (2001)

28. Peng, Z., Wang, D., Wang, W., Liu, L.: Containment control of networked autonomous underwater vehicles: a predictor-based neural DSC design. ISA Trans. **59**, 160–171 (2015)

29. Lavretsky, E., Gibson, T.E., Projection operator in adaptive systems. arXiv:1112.4232 (2011)

30. Krstic, M., Kokotovic, P.V., Kanellakopoulos, I.: Nonlinear and Adaptive Control Design, 1st edn. Wiley, New York (1995)

Virtual Structure Formation Control via Sliding Mode Control and Neural Networks

Qi Qin[1], Tie-Shan Li[1(✉)], Cheng Liu[1], C.L. Philip Chen[2],
and Min Han[3]

[1] Navigation College, Dalian Maritime University, Dalian, China
1139438104@qq.com, tieshanli@126.com,
lassieliucheng@163.com
[2] Department of Computer and Information Science,
University of Macau, Macau, China
philip.chen@ieee.org
[3] Department of Control Science and Engineering,
Dalian University of Technology, Dalian, China
minhan@dlut.edu.cn

Abstract. In this paper, a sliding mode controller is presented for the trajectory tracking by a group of ships with an established formation along a given parametrized path via neural network and sliding mode control technique. The control objective for each ship is to keep its relative positon in the formation while a virtual Formation Reference Point (FRP) tracks a predefined path. We first solve the virtual structure formation problems via sliding mode control method due to its excellent adaptability to external disturbance and system perturbation. Moreover, a radial basis function NN is considered in the design of the controller to approximate the unknown uncertainties efficiently. Some simulations are given to verify the theoretical results in this paper.

Keywords: Formation control · Virtual structure · Sliding mode control · Neural networks

1 Introduction

Over the past few years, the formation control of multiagent systems attracted great attention. This is mainly due to the increasing demand for utilizing multiple agents to perform difficult tasks to improve efficiency, reduce the energy loss and greater tolerance and adaptability. The ships formation system by applying multiagent theory shows a great prospect of application in military and industrial production areas, including fleet combat, underway ship replenishment, environmental monitoring, oil and natural gas prospecting and so on.

Several methods have been proposed to achieve the desired formation including leader-follower method [1–3], virtual structure strategy [4–6], behavior-based method [7], graph theory-based method, artificial potential field [8] and so on. Using the virtual structure strategy, we regard the formation as a rigid object, and define a virtual point in the rigid object called Formation Reference Point (FRP). So each individual ship will

© Springer International Publishing AG 2017
F. Cong et al. (Eds.): ISNN 2017, Part II, LNCS 10262, pp. 101–108, 2017.
DOI: 10.1007/978-3-319-59081-3_13

have a relative position to the FRP. The control objective can be divided into the geometric task and the dynamic task. Actually, in some real-world practical applications, agents are governed by both position and velocity states, which brings us the problem about second-order dynamics. Therefore, the sliding mode control method [8–10] has been widely investigated due to its easy implementation and high robustness against uncertain disturbances. Moreover, the unknown nonlinear functions which is difficult to obtain are ubiquitous in physical systems. The neural network (NN) technique is a powerful tool to approximate arbitrary functions [11–14].

Thus, in this paper, we choose virtual structure strategy to solve the formation problems, although lack of the overall flexibility and robustness. We focus more on the abilities of formation keeping and transformation.

2   Problem Description

2.1   Vehicle Dynamics

We consider a ship formation system including i ships whose structures are well-regulated and symmetric, and the dynamic of each ship is represented in [15]

$$\begin{cases} \dot{x}_i = u_i \cos \psi_i - v_i \sin \psi_i \\ \dot{y}_i = u_i \sin \psi_i + v_i \cos \psi_i \\ \dot{\psi}_i = r_i \\ \dot{u}_i = f_{ui} + \tau_{1i}/m_{11i} \\ \dot{v}_i = f_{vi} + \tau_{2i}/m_{22i} \\ \dot{r}_i = f_{ri} + \tau_{3i}/m_{33i} \end{cases} \tag{1}$$

In the formula, $[x_i, y_i]$ stands for the actual coordinate of the ship. $[u_i, v_i, r_i]$ stands for the speed along each axis. $m_{jji}, j = 1, 2, 3$ stands for the proper and added mass along each axis of the ship, which is a known item. f_{ui}, f_{vi}, f_{ri} are unknown functions.

2.2   Virtual Structure Formation

Figure 1 shows the basic geometric structure about five ships in a virtual structure formation which is created by a set of formation designation vectors. The idea is for the FRP to follow a given parametrized path η_d with a desired formation speed along it.

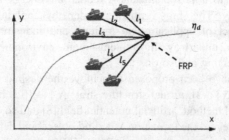

Fig. 1. Illustration of a formation system.

The individual path for ship i is given by

$$\eta_{di} = [x_{di}, y_{di}, \psi_{di}]^{\mathrm{T}} = \eta_d + R(\psi_i)l_i \tag{2}$$

where $l_i = [l_{xi}, l_{yi}, \phi_i]$, l_{xi} and l_{yi} stand for the expected transverse and longitudinal distance between ship i and the FRP in its horizontal body axes system. ϕ_i stands for the difference of the heading angle between ship i and the FRP.

For ships sailing on the sea, the output is the 3 DOF vector $\eta = [x, y, \psi]^{\mathrm{T}}$, where (x, y) is the coordinate and ψ is the heading angle. The desired path is given by $\eta_d = [x_d, y_d, \psi_d]^{\mathrm{T}}$. Then the desired heading angle can be defined as follow

$$\psi_d = \arctan(y_d'/x_d') \tag{3}$$

The orthogonal rotation matrix $R(\psi_i)$ for the ships is given by

$$R(\psi_i) = \begin{bmatrix} \cos\psi_i & -\sin\psi_i & 0 \\ \sin\psi_i & \cos\psi_i & 0 \\ 0 & 0 & 1 \end{bmatrix} \tag{4}$$

Control Objective: The control objective is to design control inputs τ_i for ship i to track the FRP with desired signal η_d and to guarantee the formation tracking error arbitrarily small.

3 Controller Design

To facilitate the controller design, define the transformation of coordinates as

$$\begin{aligned} z_{1i} &= x_i \cos\psi_i + y_i \sin\psi_i \\ z_{2i} &= -x_i \sin\psi_i + y_i \cos\psi_i \\ z_{3i} &= r_i \end{aligned} \tag{5}$$

whose time derivative is given by

$$\begin{aligned} \dot{z}_{1i} &= u_i + z_{2i}r_i \\ \dot{z}_{2i} &= v_i - z_{1i}r_i \\ \dot{z}_{3i} &= r_i \end{aligned} \tag{6}$$

Define the coordinate transformation as

$$\begin{aligned} z_{1di} &= x_{di} \cos\psi_{di} + y_{di} \sin\psi_{di} \\ z_{2di} &= -x_{di} \sin\psi_{di} + y_{di} \cos\psi_{di} \\ z_{3di} &= \psi_{di} \end{aligned} \tag{7}$$

whose time derivative is given by

$$
\begin{aligned}
\dot{z}_{1di} &= \dot{x}_{di}\cos\psi_{di} + \dot{y}_{di}\sin\psi_{di} + \dot{\psi}_{di}z_{2di} \\
\dot{z}_{2di} &= -\dot{x}_{di}\sin\psi_{di} + \dot{y}_{di}\cos\psi_{di} - \dot{\psi}_{di}z_{1di} \\
\dot{z}_{3di} &= \dot{\psi}_{di}
\end{aligned}
\tag{8}
$$

Define the error variables as

$$
\begin{aligned}
z_{1ei} &= z_{1i} - z_{1di} \\
z_{2ei} &= z_{2i} - z_{2di} \\
z_{3ei} &= z_{3i} - z_{3di}
\end{aligned}
\tag{9}
$$

whose time derivative is given by

$$
\begin{aligned}
\dot{z}_{1ei} &= \dot{z}_{1i} - \dot{z}_{1di} \\
\dot{z}_{2ei} &= \dot{z}_{2i} - \dot{z}_{2di} \\
\dot{z}_{3ei} &= \dot{z}_{3i} - \dot{z}_{3di}
\end{aligned}
\tag{10}
$$

The first sliding-mode manifold can be designed in the form as follows:

$$
S_{1i} = c_{1i}z_{3ei} + \dot{z}_{3ei}, c_{1i} > 0
\tag{11}
$$

Consider a scalar function $V_1 = S_{1i}^2/2$, whose time derivative satisfies

$$
\dot{V}_1 = S_{1i}\dot{S}_{1i} = S_{1i}(c_{1i}\dot{z}_{3ei} - \ddot{z}_{3di} + f_{ri} + \tau_{3i}/m_{33i})
\tag{12}
$$

where f_{ri} is an unknown function, the radial basis function neural network (RBFNN) is used to approximate the unknown function f_{ri}

$$
f_{ri} = W_{ri}^T H_{ri}(\eta_i, \dot{\eta}_i) + \varepsilon_{ri}
\tag{13}
$$

Define the estimation of the unknown function f_{ri} as $\hat{f}_{ri}$. The estimation of f_{ri} can be written as

$$
\hat{f}_{ri} = \hat{W}_{ri}^T H_{ri}(\eta_i, \dot{\eta}_i)
\tag{14}
$$

Choose the control law as

$$
\tau_{3i} = m_{33i}[-c_{1i}\dot{z}_{3ei} + \ddot{z}_{3di} + \hat{f}_{ri} - k_1 S_{1i} - \eta_{1i}\mathrm{sgn}(S_{1i})]
\tag{15}
$$

then, (12) becomes

$$
\dot{V}_1 \le -k_1 S_{1i}^2 - \eta_{1i}|S_{1i}| + S_{1i}(\tilde{f}_{ri})
$$

The second sliding-mode manifold can be designed in the form as follows:

$$S_{2i} = c_{2i}z_{1ei} + \dot{z}_{1ei}, c_{2i} > 0 \tag{16}$$

Consider a scalar function $V_2 = S_{2i}^2/2$, whose time derivative satisfies

$$\dot{V}_2 = S_{2i}\dot{S}_{2i} = S_{2i}[c_{2i}\dot{z}_{1ei} - \ddot{z}_{1di} + f_{ui} + \tau_{1i}/m_{11i} + \dot{z}_{2i}r_i + z_{2i}(f_{ri} + \tau_{3i}/m_{33i})] \tag{17}$$

Define the estimation of the unknown function f_{ui} as $\hat{f}_{ui}$. The estimation of f_{ui} can be written as

$$\hat{f}_{ui} = \hat{W}_{ui}^T H_{ui}(\eta_i, \dot{\eta}_i) \tag{18}$$

Choose the control law as

$$\tau_{1i} = m_{11i}(-c_{2i}\dot{z}_{1ei} + \ddot{z}_{1di} - \hat{f}_{ui} - \dot{z}_{2i}r_i - z_{2i}\hat{f}_{ri} - z_{2i}\tau_{3i}/m_{33i} - k_{2i}S_{2i} - \eta_i\mathrm{sgn}(S_{2i})) \tag{19}$$

then, (17) becomes

$$\dot{V}_2 \le - k_{2i}S_{2i}^2 - \eta_{2i}|S_{2i}| + S_{2i}[\tilde{f}_{ui} + z_{2i}\tilde{f}_{ri}]$$

The third sliding-mode manifold can be designed in the form as follows:

$$S_{3i} = c_{3i}z_{2ei} + \dot{z}_{2ei}, c_{3i} > 0 \tag{20}$$

Consider a scalar function $V_3 = S_{3i}^2/2$, whose time derivative satisfies

$$\dot{V}_3 = S_{3i}\dot{S}_{3i} = S_{3i}(c_{3i}\dot{z}_{2ei} - \ddot{z}_{2di} + f_{vi} + \tau_{2i}/m_{22i} + \dot{z}_{1i}r_i + z_{1i}(f_{ri} + \tau_{3i}/m_{33i})) \tag{21}$$

Define the estimation of the unknown function f_{vi} as $\hat{f}_{vi}$. The estimation of f_{vi} can be written as

$$\hat{f}_{vi} = \hat{W}_{vi}^T H_{vi}(\eta_i, \dot{\eta}_i) \tag{22}$$

Choose the control law as

$$\tau_{2i} = m_{22i}(-c_{3i}\dot{z}_{2ei} + \ddot{z}_{2di} - \hat{f}_{vi} - \dot{z}_{1i}r_i - z_{1i}\hat{f}_{ri} - z_{1i}\tau_{3i}/m_{33i} - k_{3i}S_{3i} - \eta_{3i}\mathrm{sgn}(S_{3i})) \tag{23}$$

then, (21) becomes

$$\dot{V}_3 \le - k_{3i}S_{3i}^2 - \eta_{3i}|S_{3i}| + S_{3i}[\tilde{f}_{vi} + z_{1i}\tilde{f}_{ri}]$$

Construct the following Lyapunov function candidate

$$V = V_1 + V_2 + V_3 + \tilde{W}_{ui}^T \tilde{W}_{ui}/2\gamma_{1i} + \tilde{W}_{vi}^T \tilde{W}_{vi}/2\gamma_{2i} + \tilde{W}_{ri}^T \tilde{W}_{ri}/2\gamma_{3i} \tag{24}$$

whose time derivative is given by

$$
\begin{aligned}
\dot{V} &= \dot{V}_1 + \dot{V}_2 + \dot{V}_3 - \tilde{W}_{ui}^T \dot{\hat{W}}_{ui}/\gamma_{1i} - \tilde{W}_{vi}^T \dot{\hat{W}}_{vi}/\gamma_{2i} - \tilde{W}_{ri}^T \dot{\hat{W}}_{ri}/\gamma_{3i} \\
&\leq \sum_{n=1}^{3} -k_{ni}S_{ni}^2 - \eta_{ni}|S_{ni}| + \tilde{f}_{ri}(S_{1i} + z_{2i}S_{2i} + z_{1i}S_{3i}) + S_{2i}\tilde{f}_{ui} + S_{3i}\tilde{f}_{vi} \\
&\quad - \tilde{W}_{ui}^T \dot{\hat{W}}_{ui}/\gamma_{1i} - \tilde{W}_{vi}^T \dot{\hat{W}}_{vi}/\gamma_{2i} - \tilde{W}_{ri}^T \dot{\hat{W}}_{ri}/\gamma_{3i} \\
&\leq \sum_{n=1}^{3} -k_{ni}S_{ni}^2 - \eta_{ni}|S_{ni}| - \tilde{W}_{ui}^T(\dot{\hat{W}}_{ui}/\gamma_{1i} - S_{2i}H_{ui}) - \tilde{W}_{vi}^T(\dot{\hat{W}}_{vi}/\gamma_{2i} - S_{3i}H_{vi}) \\
&\quad - \tilde{W}_{ri}^T[\dot{\hat{W}}_{ri}/\gamma_{3i} - H_{ri}(S_{1i} + z_{2i}S_{2i} + z_{1i}S_{3i})]
\end{aligned}
\tag{25}
$$

Choose the adaptive laws as

$$
\begin{aligned}
\dot{\hat{W}}_{ui} &= \gamma_{1i} S_{2i} H_{ui} \\
\dot{\hat{W}}_{vi} &= \gamma_{2i} S_{3i} H_{vi} \\
\dot{\hat{W}}_{ri} &= \gamma_{3i}(S_{1i} + z_{2i}S_{2i} + z_{1i}S_{3i}) H_{ri}
\end{aligned}
\tag{26}
$$

then, (26) with (25) is

$$\dot{V} \leq \sum_{n=1}^{3} -k_{ni}S_{ni}^2 - \eta_{ni}|S_{ni}|$$

Which means the proposed control law can guarantee all the signals in the closed-loop system uniformly ultimately bounded. And consider the control laws in (15), (19), (23), adaptive laws in (26), the formation tracking error can be arbitrarily small. Therefore, we can conclude the main result of this paper as follows.

Theorem 1: Consider the closed loop system (1) under the control laws in (15), (19), (23), adaptive laws in (26). Given any positive number p, for bounded initial conditions satisfying $V(0) < p$, there exist a set of designed parameters, which guarantees the formation tracking error can be made arbitrarily small.

4   Simulation

To verify the theoretical analysis in this paper, consider a multiagent system with 3 same ships whose parameters are given by $m_{11} = 200$, $m_{22} = 250$, $m_{33} = 80$, $d_{11} = 70$, $d_{22} = 100$, $d_{33} = 50$, the reference signal of the Formation Reference

Point(FRP) is given by $x_d = 0.01t$, $y_d = r_0 \sin(w_0 t)$, $\psi_d = a \tan[r_0 w_0 \cos(w_0 t)]$, the expected formation is given by $l_1 = [0.15, 0.2, 0]$, $l_2 = [-0.2, 0.2, 0]$, $l_3 = [0, -0.5, 0]$.

Figure 2 shows the formation trajectories in the process of sailing, which is well-regulated and stable.

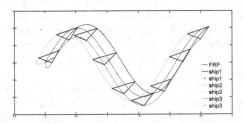

Fig. 2. The formation of three ships

5 Conclusion

In this paper, an NN-based sliding mode control approach for ships by a specified formation without any leaders has been proposed, which solves the strict formation keeping, the bounded disturbances and parameter variations and the unknown terms problems. The stability of the system is proved based on Lyapunov stability theory, which guarantees all signals in the closed-loop system are bounded. Simulation results are given to demonstrate the effectiveness of the controller.

Acknowledgements. This work is supported in part by the National Natural Science Foundation of China (Grant Nos: 61572540, 51179019, 51279106, 61374114), the Macau Science and Technology Development under Grant 008/2010/A1 and UM Multiyear Research Grants, the Fundamental Research Program for Key Laboratory of the Education Department of Liaoning Province (LZ2015006), the Fundamental Research Funds for the Central Universities under Grants 3132016313 and 3132016311.

References

1. Yiguang, H., Guanrong, C., Linda, B.: Distributed observers design for leader-following control of multi-agent networks. Automatica **44**, 846–850 (2008)
2. Zhouhua, P., Dan, W., Zhiyong, C., Xiaojing, H., Weiyao, L.: Adaptive dynamic surface control for formations of autonomous surface vehicles with uncertain dynamics. IEEE Trans. Control Syst. Technol. **21**(2), 513–519 (2013)
3. Ren, C.-E., Chen, C.L.P.: Sliding mode leader-following consensus controllers for second-order non-linear multi-agent systems. IET Control Theory Appl. **9**(10), 1544–1552 (2015)
4. Randal, W.B., Jonathan, L., Fred, Y.H.: A coordination architecture for spacecraft formation control. IEEE Trans. Control Syst. Technol. **9**(6), 777–790 (2001)

5. Roger, S., Sonja, M., Thor, I.F.: Nonlinear formation control of marine craft. In: Proceedings of the 41st IEEE Conference on Decision and Control, pp. 1699–1703. IEEE, Las Vegas (2002)
6. Jawhar, G., Hasan, M., Maarouf, S., Faical, M.: Formation path following control of unicycle-type mobile robots. Robot. Auton. Syst. 58(5), 727–736 (2010)
7. Filippo, A., Stefano, C., Thor, I.F.: Formation control for underactuated surface vessels using the null-space-based behavioral control. In: Proceedings of IEEE/RSJ International Conference on Intelligent Robots and Systems, pp. 5942–5947. Beijing (2006)
8. Mabrouk, M.H., McInnes, C.R.: Solving the potential field local minimum problem using internal agent states. Robot. Auton. Syst. 56(12), 1050–1060 (2008)
9. Sharma, A., Panwar, V.: Control of mobile robot for trajectory tracking by sliding mode control technique. In: ICEEOT, pp. 3988–3994 (2016)
10. Wenwu, Y., He, W., Fei, C., Xinghuo, Y., Guanghui, W.: Second-order consensus in multiagent systems via distributed sliding mode control. IEEE Trans. Cybern. 1–10 (2016)
11. Zengguang, H., Long, C., Min, T.: Decentralized robust adaptive control for the multiagent system consensus problem using neural networks. IEEE Trans. Syst. Man Cybern. Part B (Cybern.) 39(3), 636–647 (2009)
12. Chen, C.L.P., Wen, G.X., Liu, Y.J., Wang, F.Y.: Adaptive consensus control for a class of nonlinear multiagent time-delay systems using neural networks. IEEE Trans. Neural Netw. Learn. Syst. 25(6), 1217–1226 (2014)
13. Anmin, Z., Krishna, D.K., Zengguang, H.: Distributed consensus control for multi-agent systems using terminal sliding mode and Chebyshev neural networks. Int. J. Robust Nonlinear Control 23(3), 334–357 (2013)
14. Hongwen, M., Zhuo, W., Ding, W., Derong, L., Pengfei, Y., Qinglai, W.: Neural-network-based distributed adaptive robust control for a class of nonlinear multiagent systems with time delays and external noises. IEEE Trans. Syst. Man Cybern.: Syst. 46(6), 750–758 (2016)
15. Fossen, T.I.: Handbook of Marine Craft Hydrodynamics and Motion Control. Wiley, Norway (2011)

Neuro Adaptive Control of Asymmetrically Driven Mobile Robots with Uncertainties

Zhixi Shen, Yaping Ma, and Yongduan Song$^{(\boxtimes)}$

Key Laboratory of Dependable Service Computing in Cyber Physical Society
of Ministry of Education, Chongqing University, Chongqing 400044, China
{shenzhixi,mayaping,ydsong}@cqu.edu.cn

Abstract. This paper presents an adaptive tracking control scheme
for asymmetrically actuated wheeled mobile robot (WMR) with uncer-
tain/unknown mass center. First, we establish the WMR dynamic model
with consideration of the fact that its center of mass is normally unknown
or even shifting due to dynamic loading and/or load shifting. Second,
a structurally simple controller is developed to deal with time-varying
unknown control gain and parametric/non-parametric uncertainties of
WMR, where the asymmetric and non-smooth input saturation with no
a prior knowledge of bounds of input saturation is addressed.

Keywords: Adaptive control · Inputs saturation · WMR · Neural
network · Uncertain load

1 Introduction

Significant progress has been made in control developments of WMR during the
past decades [1–9]. However, there exists some drawbacks, such as sliding mode
control involves the notorious chattering problem and feedback lineralization
schemes require exact kinematic and dynamic model.

Furthermore, most existing works commonly assume that the center of mass
(CM) of WMR is certain and known. However, in many practical applications,
CM might not be precisely available or even be shifting with time, due to, for
instance, payload loading/unloading and shifting during WMR operation. As a
result, most well-known dynamic models established in literature that are based
on known and fixed CM are no longer valid, making the underlying control
problem more challenging as the skew-symmetric property of the matrix $\dot{\bar{M}} - 2\bar{V}$
(definitions of $\bar{M}$ and $\bar{V}$ can be found in [10]) has changed.

In addition, it is important to design controller for systems with input saturation, that is because any actuator always has a limitation of the physical
inputs. Unfortunately, only a few researchers have focused on the design of controllers with saturated inputs for WMR [11,12], however, the bounds of input
constraints is known for control design and non-linear functions in the saturation
internal must be known and smooth, which is rather restrictive in practice.

© Springer International Publishing AG 2017
F. Cong et al. (Eds.): ISNN 2017, Part II, LNCS 10262, pp. 109–117, 2017.
DOI: 10.1007/978-3-319-59081-3_14

The main contributions of the work can be summarized as follows:

(1) A new WMR model that accounts for uncertain/time-varying shifting CM due to payload loading/unloading is developed in which slipping/skidding uncertainties are reflected.
(2) By using a well-defined smooth function, the actuator inputs saturation are handled without requiring the prior knowledge of bounds of input saturation.
(3) Introducing the virtual parameter based neural network adaptive control approach, it leads to a low-cost and user-friendly control scheme for WMR.

2   Modeling and Problem Formulation

2.1   Kinematic Model of WMR

To proceed, we give the variable definition: (f_1, f_2) denotes the coordinate of the load in WMR. h denotes the displacement from geometric center of the robot to the center point of the drive axis, r denotes the radius of motorized wheels, m_1 denotes the mass of the robot body, m_2 denotes the mass of load, m_w denotes the mass of the derived wheel, I_1 denotes the inertia of the robot body about the vertical axis through point o, I_2 denotes the inertia of the loaded about its center of mass, I_w denotes the inertia of the drived wheel, b denotes the half length of the drive axis, c denotes the constant equals to r/2b, and (c_1, c_2) denotes the coordinate of the CM in WMR.

Consider the WMR with two motorized wheels on an axis driven independently, as illustrated in Fig. 1. Then taking into account the slipping and skidding impact, we have the following three equations:

$$\dot{y}_o \cos \phi - \dot{x}_o \sin \phi = \mu \tag{1}$$

$$\dot{x}_o \cos \phi + \dot{y}_o \sin \phi + b\dot{\phi} = r(\dot{\theta}_1 - \dot{\zeta}_1) \tag{2}$$

$$\dot{x}_o \cos \phi + \dot{y}_o \sin \phi - b\dot{\phi} = r(\dot{\theta}_2 - \dot{\zeta}_2) \tag{3}$$

where (x_o, y_o) is the coordinate of point o in the fixed reference coordinated frame $X - Y$, and ϕ is the heading angle of the mobile robot measured from $X -$ axis, θ_1 and θ_2 are the angular positions of the two driving wheels, respectively. μ indicates the lateral skidding velocity. ζ_1 and ζ_2 denote perturbed angular velocities because of the slipping in two actuated wheels, respectively [13].

By combining (2) and (3), we arrive at

$$2b\dot{\phi} = r(\dot{\theta}_1 - \dot{\theta}_2 - (\dot{\zeta}_1 - \dot{\zeta}_2)) \tag{4}$$

$$\dot{x}_o \cos \phi + \dot{y}_o \sin \phi = cb(\dot{\theta}_1 + \dot{\theta}_2 - (\dot{\zeta}_1 + \dot{\zeta}_2)) \tag{5}$$

From (4) and (5), we get the following compact form of the system model, $A(q_o)\dot{q}_o = \xi$, where $q_o = \begin{bmatrix} x_o & y_o & \theta_1 & \theta_2 \end{bmatrix}^T$, and

$$A(q_o) = \begin{bmatrix} -\sin \phi & \cos \phi & 0 & 0 \\ -\cos \phi & -\sin \phi & cb & cb \end{bmatrix}, \ \xi = \begin{bmatrix} \mu \\ cb(\zeta_1 + \zeta_2) \end{bmatrix}$$

According to (1) and (5), we obtain the kinematics of WMR

$$\dot{q}_o = S(q_o)(\dot{\theta} - \dot{\zeta}) + \varphi(q_o, \mu) \tag{6}$$

in which $\dot{\theta} = \begin{bmatrix} \dot{\theta}_1 & \dot{\theta}_2 \end{bmatrix}^T$, $\dot{\zeta} = \begin{bmatrix} \dot{\zeta}_1 & \dot{\zeta}_2 \end{bmatrix}^T$, and

$$\varphi(q_o, \mu) = [\mu \sin\phi, -\mu \cos\phi, 0, 0]^T, \quad S(q_o) = \begin{bmatrix} cb\cos\phi, cb\sin\phi, 1, 0 \\ cb\cos\phi, cb\sin\phi, 0, 1 \end{bmatrix}^T$$

In addition, we have from (6) that

$$\ddot{q}_o = S(q_o)(\ddot{\theta} - \ddot{\zeta}) + \dot{S}(q_o)(\dot{\theta} - \dot{\zeta}) + \dot{\varphi}(q_o, \mu) \tag{7}$$

2.2 Dynamic Equations of WMR

Because the potential energy of the robot is zero. Therefore, the Lagrangian energy K is given by $K = \frac{1}{2} \sum\limits_{i=1}^{n_i} \left[v_i^T m_i v_i + w_i^T I_i w_i \right]$ [14]. That is

$$\begin{aligned} K = &\tfrac{1}{2}m_1\dot{x}_o^2 + \tfrac{1}{2}m_1\dot{y}_o^2 + \tfrac{1}{2}m_w(\dot{x}_o - b\dot{\phi}\cos\phi)^2 + \tfrac{1}{2}m_w(\dot{y}_o - b\dot{\phi}\sin\phi)^2 \\ &+ \tfrac{1}{2}m_w(\dot{x}_o + b\dot{\phi}\cos\phi)^2 + \tfrac{1}{2}m_w(\dot{y}_o + b\dot{\phi}\sin\phi)^2 + \tfrac{1}{2}I_w(\dot{\theta}_1^2 + \dot{\theta}_2^2) \\ &+ \tfrac{1}{2}m_2(\dot{x}_o - \dot{\phi}f_2 sin\phi + \dot{\phi}f_1\cos\phi + \dot{f}_1\sin\phi + \dot{f}_2\cos\phi)^2 \\ &+ \tfrac{1}{2}m_2(\dot{y}_o + \dot{\phi}f_2\cos\phi + \dot{\phi}f_1\sin\phi - \dot{f}_1\cos\phi + \dot{f}_2\sin\phi)^2 \\ &+ \tfrac{1}{2}I_1\dot{\phi}^2 + \tfrac{1}{2}I_2\dot{\phi}^2 + I_w\dot{\phi}^2 \end{aligned} \tag{8}$$

Furthermore, the dynamical equations of the WMR can be expressed in the matrix form [15]

$$M(q_o)\ddot{q}_o + V_m(q_o, \dot{q}_o)\dot{q}_o + d_1(\cdot) = E(q_o)\tau_o - A(q_o)^T\lambda_{12} \tag{9}$$

where $M(q_o)$ is a positive definite symmetric inertia matrix, $V_m(q_o, \dot{q}_o)$ is the centripetal and coriolis matrix, $d_1(\cdot)$ denotes unknown but bounded disturbances including unstructured unmodeled dynamics, $E(q_o)$ is the input transformation matrix, τ_o is the input vector, and λ_{12} is the vector of constraint forces.

Multiplying (9) by $S(q_o)^T$, we obtain the dynamic model of WMR as

$$S^T(q_o)M(q_o)\ddot{q}_o + S^T(q_o)V_m(q_o, \dot{q}_o) + S^T(q_o)d_1(\cdot) = \tau_o \tag{10}$$

Upon substituting equation (7) into equation (10), it follows from (4) that

$$\bar{M}(\ddot{\theta} - \ddot{\zeta}) + \bar{V}(\dot{\theta} - \dot{\zeta}) + S^T M(q_o)\dot{\varphi}(q_o, \mu) + \varsigma = \tau_o \tag{11}$$

where $\bar{M} = S^T(q_o)M(q_o)S(q_o)$, $\bar{V} = S^T(q_o)M(q_o)\dot{S}(q_o) + S^T(q_o)V_m(q_o, \dot{q}_o)S(q_o)$, $\varsigma = S^T(q_o)d_1(\cdot) + S^T(q_o)V_m(q_o, \dot{q}_o)\varphi(q_o, \mu)$, and

$$\bar{M} = \begin{bmatrix} \bar{m}_{11} & mc^2b^2 - Ic^2 \\ mc^2b^2 - Ic^2 & \bar{m}_{22} \end{bmatrix}, \bar{V} = \begin{bmatrix} 0 & 2m_2c^2bf_2\dot{\phi} \\ -2m_2c^2bf_2\dot{\phi} & 0 \end{bmatrix}$$

in which, $\bar{m}_{11} = mc^2b^2 + 2m_2c^2bf_1 + Ic^2 + I_w$, $\bar{m}_{22} = mc^2b^2 - 2m_2c^2bf_1 + Ic^2 + I_w$. Also, $\bar{M}$ is a positive definite symmetric matrix. Examining the expression of $\bar{m}_{11}$ and $\bar{m}_{22}$ in $\bar{M}$ and $\bar{V}$ reveals that, due to payload loading/unloading and shifting, the skew-symmetric property of the matrix $\dot{\bar{M}} - 2\bar{V}$ disappears, making the control design and stability analysis much more involved.

2.3   State Space Realization

We establish a state space realization of the interest point p [16] with coordinates (x_p, y_p) in the XY plane. According to the relation between point o and p, we have $x_p = x_o + h\cos\phi$, $y_p = y_o + h\sin\phi$. and according to (4) and (6), we get

$$\dot{P} = J(\dot{\theta} - \dot{\zeta}) + \varphi_1 \tag{12}$$

where $P = [x_p, y_p]^T$, $\varphi_1 = [\mu\sin\phi, -\mu\cos\phi]^T$ and

$$J = \begin{bmatrix} cb\cos\phi - ch\sin\phi & cb\cos\phi + ch\sin\phi \\ cb\sin\phi + ch\cos\phi & cb\sin\phi - ch\cos\phi \end{bmatrix}$$

In addition, considering input saturation, τ_o is no longer designed control input $\tau = [\tau_1, \tau_2]^T$, the τ_o is expressed as $B(\tau) = [b_1(\tau_1), b_2(\tau_2)]^T$ here. To facilitate the description, we use $b_i(\tau_i), i = 1, 2$ to denote $b_1(\tau_1)$ and $b_2(\tau_2)$. $b_i(\tau_i)$ denotes the inputs subject to asymmetric non-smooth saturation nonlinearity as follows,

$$b_i(\tau_i) = \begin{cases} \bar{\delta}_i, & \tau_i > \tau_{mb1i} \\ \ell_i(t)\tau_i, & -\tau_{mb2i} \leq \tau_i \leq \tau_{mb1i} \\ -\underline{\delta}_i, & \tau_i < -\tau_{mb2i} \end{cases} \tag{13}$$

where $\ell_i(t)$ is a time-varying function. $\tau_{mb1i} > 0$ and $-\tau_{mb2i} < 0$ represent the break points. $\bar{\delta}_i$ and $\underline{\delta}_i$ are the unknown bounds of input τ_i. A smooth function $w_i(\tau_i) = (\bar{\delta}_i e^{(\varepsilon_{si} + \alpha_{si}\tau_i)} - \underline{\delta}_i e^{-(\varepsilon_{si} + \alpha_{si}\tau_i)})/(e^{(\varepsilon_{si} + \alpha_{si}\tau_i)} + e^{-(\varepsilon_{si} + \alpha_{si}\tau_i)})$ is introduced [17], with $\varepsilon_{si} = 0.5\ln(\underline{\delta}_i/\bar{\delta}_i)$ and $\alpha_{si} > 0$ is a constant, to deal with the non-smooth and asymmetric actuation nonlinearities.

Then $b_i(\tau_i)$ be expressed as $b_i(\tau_i) = w_i(\tau_i) + \delta_i(\tau_i)$ where $\delta_i(\tau_i)$ is the difference between $b_i(\tau_i)$ and $w_i(\tau_i)$. And the function $\delta_i(\tau_i)$ is bounded, i.e. $|\delta_i(\tau_i)| \leq \delta_2$, where δ_2 is a positive and unknown constant. Then we employ the mean value theorem on function $w_i(\tau_i)$ to get $w_i(\tau_i) = w_i(\tau_{i0}) + \frac{\partial w_i(\tau_i)}{\partial\tau_i}\Big|_{\tau_i = \tau_i^\lambda}(\tau_i - \tau_{i0})$, where $\tau_i^\lambda = \lambda\tau_i + (1 - \lambda)\tau_{i0}$ with $0 < \lambda < 1$. By choosing $\tau_{i0} = 0$ and using the fact that $w_i(0) = 0$, we have $w_i(\tau_i) = \frac{\partial w_i(\tau_i)}{\partial\tau_i}\Big|_{\tau_i = \tau_i^\lambda}\tau_i = g_i(\cdot)\tau_i$. Note that $w_i(\tau_i)$ is always bounded in the set $(-\underline{\delta}_i, \bar{\delta}_i)$ and there exists some positive constant g_{max} such that $0 < \partial w_i(\tau_i)/\partial(\tau_i) = g_i \leq g_{max} \leq \infty$, for all $\tau_i \in \Re$ [17,18].

According to the above analysis, from (11) and (12), we obtain

$$\begin{cases} \dot{x}_1 = x_2 \\ \dot{x}_2 = M_2^{-1}(J^T)^{-1}G\tau + L(\cdot) \end{cases} \tag{14}$$

where $x_1 = P$, $M_2 = (J\bar{M}^{-1}J^T)^{-1}$, $G = \text{diag}\{g_i\}$, $\tau = [\tau_1, \tau_2]^T$, and $L(\cdot) = J\dot{\theta} + J\bar{M}^{-1}\delta(\tau) + d(\cdot)$ with $\delta(\tau) = [\delta_1(\tau_1), \delta_2(\tau_2)]^T$ and $d(\cdot) = -J\bar{M}^{-1}\bar{V}\dot{\theta} + J\bar{M}^{-1}\bar{V}\dot{\zeta} - J\bar{M}^{-1}\varsigma - J\bar{M}^{-1}S^T M(q)\dot{\varphi} - \dot{J}\dot{\zeta} + \dot{\varphi}_1$.

Property 1. (1) $\lambda_{min}\|\varrho\|^2 \leq \varrho^T(J^T)^{-1}GJ^{-1}\varrho \leq \lambda_{max}\|\varrho\|^2$, where $\lambda_{max} \geq \lambda_{min} > 0$ denotes the maximum and minimum eigenvalues of $(J^T)^{-1}GJ^{-1}$, respectively. And M_2 is a positive definite symmetric matrix.

The objective in this paper is: designing a neural adaptive controller to make the interest point p of WMR track the desired trajectory.

3   Neural Networks and Functional Approximation

According to the well-known approximation property, any continuous function $\psi(Z)$, there exists an ideal the radial basis function neural networks(RBFNN) capable of approximation it [19], on a compact set $\Omega_z \in \Re^l$, with sufficient accuracy, i.e., $\psi(Z) = W^{*T}\vartheta(Z) + \eta(Z)$, where W^* is the optimal constant weight, $\eta(Z)$ denotes the approximation error and $\vartheta(Z) = [\vartheta_1(Z), \cdots, \vartheta_n(Z)]^T$ is the basis function vector.

One of the typical choices for $\vartheta(Z)$ is $\vartheta_j(Z) = \exp(-(Z - \alpha_j)^T(Z - \alpha_j)/\nu_j^2)$ where $\alpha_j = [\alpha_{j1}, \cdots, \alpha_{jn}]^T, j = 1, \cdots, n$ denote the center of the receptive field, and ν_j is the width of the Gaussian function. Widespread practical application of RBFNNNs show that, if NNs node number n is chosen large enough, then $\eta(Z)$ can be reduced to an arbitrarily small value in a compact set.

4   Control Design and Stability Analysis

In this section, we design the adaptive controller for WMR. In this work, suppose that the desired trajectory is $x^*(t)$, with its 1st and 2nd derivatives $\dot{x}^*$ and $\ddot{x}^*$, each of which is available and bounded, $\forall t \geq 0$. For the analysis convenience, we define intermediate variable s in terms of the tracking error as

$$s = \dot{e} + \beta e \tag{15}$$

where $e = x_1 - x^* = \left[e_{x_p}, e_{y_p}\right]^T$, and $\beta > 0$ is a positive design parameter [20]. Taking the time derivative of the filtered error s along the system model (14) yields that

$$\dot{s} = \dot{x}_2 - \ddot{x}^* + \beta(\dot{x}_1 - \dot{x}^*) = M_2^{-1}(J^T)^{-1}G\tau + N(\cdot) \tag{16}$$

where $N(\cdot) = \eta_{11} + L(\cdot)$, $\eta_{11} = -\ddot{x}^* + \beta(\dot{x}_1 - \dot{x}^*)$. In addition, we need assumption as follows

Assumption 1. $\left\|\dot{M}_2\right\| \leq \gamma q(\dot{\phi})$, where $q(\dot{\phi})$ is a known and computable positive function and γ is an unknown positive constant. $L(\cdot)$ is bounded by an unknown constant L_m, i.e., $\|L(\cdot)\| \leq L_m < \infty$.

Note that $2\|M_2\|\|N(\cdot)\|$ is continuous and well defined over the compact set, thus it can be approximated by NNs as, $2\|M_2\|\|N(\cdot)\| = W^{*T}\vartheta(Z) + \eta(Z) \leq a_1 F_1$ with $F_1 = \vartheta(Z) + 1$, $a_1 = \max(\|W^*\|, \|\eta(Z)\|)$, $Z = [s^T, x_1^T]$. Therefore, $2\|M_2\|\|N(\cdot)\| + \left\|\dot{M}_2\right\|\|s\| \leq a_1 F_1 + \gamma q(\dot{\phi})\|s\| \leq aF$, where $F = F_1 + q(\dot{\phi})\|s\|$, $a = \max(a_1, \gamma)$. If both x_1 and x_2 are bounded, so is F. Upon using Young's inequality, it is seen that $\|s^T\|(2\|M_2\|\|N(\cdot)\| + \left\|\dot{M}_2\right\|\|s\|) \leq aF\|s\| \leq \frac{1}{\rho\lambda_{\min}} + \rho\lambda_{\min}a^2F^2\|s\|^2$.

Theorem 1. *Consider system (14) with Assumption 1, if it is controlled by*

$$\tau = J^{-1}\left(-ks - \frac{\rho}{2}\hat{w}F^2 s\right) \tag{17}$$

where $k > 0$, *and* $\rho > 0$ *are design parameters, and* $\hat{w}$ *is the estimate of* $w = a^2$, *with the updated law,*

$$\dot{\hat{w}} = \rho F^2 \|s\|^2 - \sigma\hat{w} \tag{18}$$

with $\sigma > 0$ *being a design parameter chosen by designer, then (1) the tracking error e is ensured to be ultimately uniformly bounded; (2) all of the signals in the controlled closed-loop system are bounded and continuous.*

Proof. Choosing the Lyapunov function as $V = s^T M_2 s + \frac{\lambda_{\min}}{2}\tilde{w}^2$, where $\tilde{w} = w - \hat{w}$ is the virtual parameter estimation error, and taking the time derivative of V along (16) and (18), we then arrive at

$$\begin{aligned}
\dot{V} &= 2s^T M_2 \dot{s} + s^T \dot{M_2} s + \lambda_{\min}\tilde{w}(-\dot{\hat{w}}) \\
&= 2s^T (J^T)^{-1} G\tau + 2s^T M_2 N(\cdot) + s^T \dot{M_2} s - \lambda_{\min}\tilde{w}\dot{\hat{w}} \\
&\leq 2s^T (J^T)^{-1} G J^{-1}(-ks - \frac{\rho}{2}\hat{w}F^2 s) + \frac{1}{\rho\lambda_{\min}} + \rho\lambda_{\min}wF^2\|s\|^2 \\
&\quad - \lambda_{\min}\tilde{w}(\rho F^2\|s\|^2 - \sigma\hat{w}) \\
&\leq -2k\lambda_{\min}\|s\|^2 + \rho\lambda_{\min}\tilde{w}F^2\|s\|^2 + \frac{1}{\rho\lambda_{\min}} - \lambda_{\min}\tilde{w}(\rho F^2\|s\|^2 - \sigma\hat{w}) \\
&\leq -2k\lambda_{\min}\|s\|^2 + \frac{1}{\rho\lambda_{\min}} + \frac{\sigma\lambda_{\min}}{2}w^2 + \frac{\sigma\lambda_{\min}}{2}\tilde{w}^2 - \sigma\lambda_{\min}\tilde{w}^2 \\
&\leq -\mu_1 V + \chi
\end{aligned} \tag{19}$$

where $\mu_1 = \min(\frac{2k\lambda_{\min}}{\lambda_{\max 1}}, \sigma)$, with $\lambda_{\max 1}$ being the maximum eigenvalue of M_2, and $\chi = \frac{1}{\rho\lambda_{\min}} + \frac{\sigma\lambda_{\min}}{2}w^2$.

Therefore, we can conclude that $V(t) \in L_\infty$, which then implies that $s \in L_\infty$ and $\hat{w} \in L_\infty$. From the definition of s in (15), it thus follows that $e \in L_\infty$ and $\dot{e} \in L_\infty$, which further imply that $x_1 \in L_\infty$ and $\dot{x}_1 \in L_\infty$. Thus $F \in L_\infty$. Furthermore, it holds that $\tau \in L_\infty$ and $\dot{\hat{w}} \in L_\infty$. That is, all of the signals in the closed-loop system are bounded and continuous. In addition, $\dot{V}$ will become negative as long as $s \notin \Omega_s = \{s | \|s\| \leq \sqrt{\chi/(k\lambda_{\min})}\}$. That is, s is confined in the set Ω_s. This further implies that the uniformly ultimately boundedness (UUB) of s is ensured and thus the UUB of e is also ensured.

5   Numerical Simulations

To verify the result obtained in Theorem 1, the simulation runs under the condition that the desired trajectory is $x^* = [2\sin(t), 3\cos(t)]^T$; the system parameters are chosen as $m_1 = 10$, $m_2 = 6 + 2\sin(t)$, $m_w = 1$, $I_1 = 2.5$, $I_2 = 0.4 + 0.1\sin(t)$, $I_w = 0.02$, $r = 0.05$, $b = 0.5$, $h = 0.3$; the coordinate of the load in WMR is chosen as $f_1 = 0.45\sin(t)$, $f_2 = 0.6\cos(t) + 0.5$; slipping/skidding uncertainties are $\mu = 0.01$, $\zeta = [0.001\sin(t), 0.002\cos(t)]^T$; and the asymmetric input saturation is considered as $\bar{\delta}_1 = 20$, $\ell_1 = 1$, $\underline{\delta}_1 = 15$, $\bar{\delta}_2 = 50$, $\ell_2 = 1$,

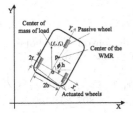

Fig. 1. Nonholonomic wheeled mobile robot

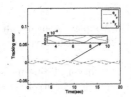

Fig. 2. The system tracking error

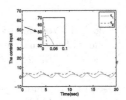

Fig. 3. The designed control input

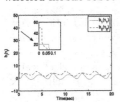

Fig. 4. The saturation control input $b_i(\tau_i)$

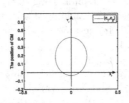

Fig. 5. The variable position of CM in WMR

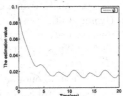

Fig. 6. Evolution of the virtual parameter

$\underline{\delta}_2 = 20$. We take the design parameters as $\beta = 11$, $k = 6$, $\rho = 5$, $\sigma = 0.08$, and initial conditions $x_1 = [x_p, y_p]^T = [0.1, 3.01]^T$, $\theta = [\theta_1, \theta_2]^T = [0, 0]^T$. The simulation results are shown in Figs. 2, 3, 4, 5 and 6. From Fig. 2, we see that position tracking error maintain in a small compact set. It is seen that both the control input signals and the saturation control input signals are continuous and bounded from Figs. 3 and 4; Figs. 5 and 6 show the variable position of CM in WMR and the evolution of the virtual parameter, respectively.

6 Conclusion

While there is a rich collection of technical results on control of WMR, the vast majority of existing works have been focused on the scenario with known and fixed mass center, ignoring either input saturation, or slipping/skidding uncertainties. This paper address these challenging issues simultaneously. First, we establish a model for WMR with the consideration of uncertain mass center due to the impact of uncertain load. Second, by using neural networks, a user-friendly control scheme without the need for precise information on WMR model is developed, which is able to achieve uniformity ultimately bounded trembling, as authenticated theoretically by Lyapunov method and confirmed numerically by simulation.

Acknowledgments. This work is supported in part by technology transformation program of Chongqing higher education university (KJZH17102), the National Natural Science Foundation of China (No. 51374264), and the China Scholarship Council (No. 201508505045).

References

1. Park, B.S., Yoo, S.J., Park, J.B.: Adaptive neural sliding mode control of nonholo-nomic wheeled mobile robots with model uncertainty. IEEE Trans. Control Syst. Tecnol. **17**, 207–214 (2009)
2. Ferrara, A., Rubagotti, M.: Second-order sliding-mode control of a mobile robot based on a harmonic potential field. IET Control Theory Appl. **2**, 807–818 (2008)
3. Wu, J.B., Xu, G.H., Yin, Z.P.: Robust adaptive control for a nonholonomic mobile robot with unknown parameters. J. Control Theory Appl. **7**, 212–218 (2009)
4. Wang, C.C.: A novel variable structure theory applied in design for wheeled mobile robots. IEEE Trans. Control Syst. Tecnol. **16**, 378–382 (2011)
5. Hwang, C.L., Wu, H.M.: Trajectory tracking of a mobile robot with frictions and uncertainties using hierarchical sliding-mode under-actuated control. IET Control Theory Appl. **7**, 952–965 (2013)
6. Hou, Z.G., Zou, A.M., Cheng, L., Tan, M.: Adaptive control of an electrically driven nonholonomic mobile robot via backstepping and fuzzy approach. IEEE Trans. Control Syst. Tecnol. **17**, 803–815 (2009)
7. Xin, L.J., Wang, Q.L., She, J.H.: Robust adaptive tracking control of wheeled robot. Robot. Auto Syst. **78**, 36–48 (2016)
8. Shojaei, K.: Saturated ouput feedback control of uncertain nonholonomic wheeled mobile robots. Robotica **33**, 87–105 (2015)
9. Park, B.S., Yoo, S.J., Park, J.B.: A simple adaptive control approach for tracjectory tracking of electrically driven nonholonomic mobile robots. IEEE Trans. Control Syst. Tecnol. **18**, 1199–1206 (2010)
10. Fierro, R., Lewis, F.L.: Control of a nonholonomic mobile robot using neural networks. IEEE Trans. Neural Netw. **9**, 589–600 (1998)
11. Huang, J.S., Wen, C.Y., Wang, W.: Adaptive stabilization and tracking control of a nonholonomic mobile robot with input saturation and disturbance. Syst. Control Lett. **62**, 234–241 (2013)
12. Chen, H., Wang, C.L., Zhang, B.W.: Saturated tracking control for nonholonomic mobile robots with dynamic feedback. Trans. Inst. Meas. Control **35**, 105–116 (2013)
13. Yoo, S.J.: Adaptive neural tracking and obstacle avoidance of uncertain mobile robots with unknown skidding and slipping. Inf. Sci. **238**, 176–189 (2013)
14. Shojaei, K.A., Shahri, M., Tarakameh, A.: Adaptive trajectory tracking control of a differential drive wheeled mobile robot. Robotica **29**, 391–402 (2011)
15. Yun, X.P., Yamamoto, Y.: Internal dynamics of a wheeled mobile robot. In: Proceedings of 1993 IEEE/RSJ International Conference on Intelligent Robots and Systems, pp. 1288–1294. IEEE Press, Yokohama (1993)
16. Martins, F.N., Celeste, W.C., Carelli, R.: An adaptive dynamic controller for autonomous mobile robot tracjectory tracking. Control Eng. Pract. **16**, 1354–1363 (2008)
17. Zhao, K., Song, Y.D., Wen, C.Y.: Computationally inexpensive fault tolerant control of uncertain non-linear systems with non-smooth asymmetric input saturation and undetectable actuation failures. IET Control Theory Appl. **10**, 1866–1873 (2016)
18. Wen, C.Y., Zhou, J., Liu, Z.T.: Robust adaptive control of uncertain nonlinear systems in the presence of input saturation and external disturbance. IEEE Trans. Autom. Control **56**, 1672–1678 (2011)

19. Sanner, R.M., Slotine, J.J.E.: Gaussian networks for direct adaptive control. IEEE Trans. Neural Netw. **3**, 837–863 (1992)
20. Wang, Y.J., Song, Y.D., Lewis, F.L.: Robust adaptive fault-tolerant control of multiagent systems with uncertain nonidentical dynamics and undetectable actuation failures. IEEE Trans. Ind. Electron. **62**, 3978–3988 (2015)

Adaptive Neural Network Control for Constrained Robot Manipulators

Gang Wang[1], Tairen Sun[1], Yongping Pan[2], and Haoyong Yu[2]([✉])

[1] Jiangsu University, Zhenjiang 212013, China
973196357@qq.com, suntren@gmail.com
[2] National University of Singapore, Singapore 117583, Singapore
{biepany,bieyhy}@nus.edu.sg

Abstract. This paper presents an adaptive neural network (NN) control strategy for robot manipulators with uncertainties and constraints. Position, velocity and control input constraints are considered and tackled by introducing barrier Lyapunov functions in the backstepping procedure. The system uncertainties are estimated and compensated by a locally weighted online NN. The boundedness of the closed-loop control system and the feasibility of the proposed control law are demonstrated by theoretical analysis. The effectiveness of the proposed control strategy has been verified by simulation results on a robot manipulator.

Keywords: Adaptive control · Backstepping · Neural network · System constraint · Barrier Lyapunov function · Robot manipulator

1 Introduction

Control of robot manipulators has gained more and more attention for its applications in industries, agricultures, and teleoperated surgeries. The difficulties in control of robot manipulators mainly include uncertainties and constraints in the position, velocity, and control input. On the one hand, uncertainties always exist in robot manipulator models due to modeling errors and disturbances. On the other hand, control input constraints always exist due to limited control powers, and motion constraints (e.g. position constraints and velocity constraints) are needed to avoid collision or injury to human beings, especially in human-robot interaction. Therefore, the control design for robot manipulators with uncertainties and constraints deserves more research.

Many robust control strategies have been developed for robot manipulators, including sliding mode control [1–3], neural network (NN) control [3–8], fuzzy control [9,10], adaptive control [11], etc. However, sliding mode control usually suffers from chattering and the need of high-frequency bandwidth, adaptive control usually only handles structured uncertainties, and fuzzy control highly depends on the experiences of control engineers. Compared with other control approaches, NN control has its own advantages. NNs can approximate both structured and unstructured uncertainties due to their inherent function

© Springer International Publishing AG 2017
F. Cong et al. (Eds.): ISNN 2017, Part II, LNCS 10262, pp. 118–127, 2017.
DOI: 10.1007/978-3-319-59081-3_15

approximation abilities. The use of NNs estimators in control is possible to obtain desired control performances without high control gains.

Since constraints in robot manipulators need to be considered and ignoring constraints may deteriorate the control performance, some results have been obtained on control of constrained robot manipulators. Set-point regulation control and tracking control laws were designed in [12] and [13] for robot manipulators with velocity constraints, respectively. Quadratic programming-based kinematic control was developed in [14,15] for velocity constrained redundant manipulators. Joint position constraints were considered and optimal control was designed based on adaptive dynamic programming in [16]. Recently, adaptive control was developed for robot manipulators where output or state constraints are tackled by bounding barrier Lyapunov functions (BLFs) in [17,18]. Based on the above analysis, one can see that only position or joint velocity constraints are considered in existing robot manipulators control approaches.

In this paper, an adaptive NN control law is proposed for robot manipulators with uncertainties and constraints, including position, velocity and control constraints. The uncertainties are approximated by locally weighted adaptive NNs and compensated by the NN estimator in the control law. In locally weighted NNs, estimators composed of independently adjusted local models are used to reach the desired approximation accuracy. Thus, fewer neurons are needed to approximate smooth functions in the desired accuracy compared with other NNs. The system constraints are tackled by using BLFs in the backstepping control [19,20] design for robot manipulators, which extends BLFs-based control for output and state constrained systems [17] to state and control constrained systems. It is demonstrated that uniform boundedness of all closed-loop signals is obtained while the constraints are not violated in theory.

2   Problem Statement

Consider a n-link robot manipulator with the following dynamics:

$$M(q)\ddot{q} + V_m(q, \dot{q})\dot{q} + F\dot{q} + G(q) = \tau \tag{1}$$

where $q = q_1 = [q_{11}, q_{12}, \cdots, q_{1n}]^T \in R^n$ is a joint angle, $q_2 = \dot{q}_1 = [q_{21}, q_{22}, \cdots, q_{2n}]^T$ is a joint velocity, $M(q) \in R^{n \times n}$ is an inertia matrix, $V_m(q, \dot{q}) \in R^{n \times n}$ is a centripetal and Coriolis matrix, $F\dot{q} \in R^n$ denotes a viscous friction torque, $G(q) \in R^n$ denotes a gravitation torque, and $\tau \in R^n$ denotes a control torque.

Assumption 1. $M(q)$ satisfies the following inequalities:

$$m_1 ||x||^2 \le x^T M(q)x \le m_2 ||x||^2, \ x \in R^n \tag{2}$$

where m_1, $m_2 \in R$ are positive constants.

Assumption 2. The uncertain function $f(q, \dot{q}) = M^{-1}(q)[V_m(q, \dot{q})\dot{q} + F\dot{q} + G(q)]$ is continuous.

Assumption 3. The reference trajectory is described as $y_d(t) = [y_{d1}, y_{d2}, \cdots,$ $y_{dn}]^T \in R^n$ and satisfies $|y_{di}| \leq A_i$ and $|\dot{y}_{di}| \leq Y_i, i = 1, \cdots, n$.

The objective is to design an adaptive NN control law for the system (refeq1) to track desired trajectory $q_d(t)$ and to satisfy the following constraints:

$$|q_{1i}| \leq b_{1i}, \ |q_{2i}| \leq b_{2i}, \ |\tau_i| \leq \tau_{di}, \ i = 1, 2, \cdots, n. \tag{3}$$

3   BLF-Based Neural Control

3.1   Control Design

Let $e_1 = [e_{11}, e_{12}, e_{13}]^T = q_1 - y_d$ be a tracking error. Consider BLFs as follows:

$$V_1 = \frac{1}{2} \sum_{i=1}^{n} \frac{k_{1i}^2}{k_{1i}^2 - e_{1i}^2} \tag{4}$$

with k_{1i} to k_{1n} being positive design parameters. The time derivative of V_1 is

$$\dot{V}_1 = \sum_{i=1}^{n} \frac{e_{1i}}{k_{1i}^2 - e_{1i}^2}(q_{2i} - \dot{y}_{di}). \tag{5}$$

Design the following virtual control input:

$$\alpha_{1i} = \dot{y}_{di} - \lambda_{1i}e_{1i}, \ i = 1, 2, \cdot, n \tag{6}$$

with $\lambda_{1i}, i = 1, 2, \cdots, n$ being positive parameters.

Let $e_2 = [e_{21}, \cdots, e_{2n}]^T = [q_{21} - \alpha_{11}, \cdots, q_{2n} - \alpha_{1n}]^T$. Then, one has

$$\dot{V}_1 = -\sum_{i=1}^{n} \lambda_{1i} \frac{e_{1i}^2}{k_{1i}^2 - e_{1i}^2} + \sum_{i=1}^{n} \frac{e_{1i}e_{2i}}{k_{1i}^2 - e_{1i}^2}. \tag{7}$$

Consider the following BLFs:

$$V_2 = V_1 + \Lambda_1, \tag{8}$$

$$\Lambda_1 = \sum_{i=1}^{n} \frac{1}{2} \log \frac{k_{2i}^2}{k_{2i}^2 - e_{2i}^2} \tag{9}$$

with k_{2i} to k_{2n} being positive design parameters. The time derivative of Λ_1 is

$$\dot{\Lambda}_1 = \sum_{i=1}^{n} \frac{e_{2i}}{k_{2i}^2 - e_{2i}^2}(\dot{q}_{2i} - \dot{\alpha}_{1i})$$
$$= \xi^T(f(q_1, q_2) + M^{-1}(q_1)\tau - [\dot{\alpha}_{11}, \cdots, \dot{\alpha}_{1n}]^T) \tag{10}$$

where

$$\xi = \left[\frac{e_{21}}{k_{21}^2 - e_{21}^2}, \cdots, \frac{e_{2n}}{k_{2n}^2 - e_{2n}^2} \right]^T. \tag{11}$$

Design the reference signal τ_r for τ as

$$\tau_r = [\tau_{r1}, \cdots, \tau_{rn}]^T = M(q_1)(-\lambda_2 e_2 - \hat{f}(q_1, q_2) - s) \tag{12}$$

where λ_2 is a positive design parameter, $\hat{f}(q_1, q_2)$ is an estimate of $f(q_1, q_2)$, and

$$s = \frac{1}{2}\xi - [\dot{\alpha}_{11}, \cdots, \dot{\alpha}_{1n}]^T + [(k_{21}^2 - e_{21}^2)e_{11}/(k_{11}^2 - e_{11}^2),$$
$$\cdots, (k_{2n}^2 - e_{2n}^2)e_{1n}/(k_{1n}^2 - e_{1n}^2)]^T. \tag{13}$$

Define $e_3 = [e_{31}, \cdots, e_{3n}]^T = \tau - \tau_r$. From (7)–(13), one obtains

$$\dot{V}_2 = -\sum_{i=1}^{n} \lambda_{1i} \frac{e_{1i}^2}{k_{1i}^2 - e_{1i}^2} - \sum_{i=1}^{n} \lambda_2 \frac{e_{2i}^2}{k_{2i}^2 - e_{2i}^2} + \xi^T(\tilde{f} + M^{-1}e_3) - \frac{1}{2}\xi^T\xi \tag{14}$$

Consider the following BLF:

$$V_3 = V_2 + \Lambda_2, \tag{15}$$

$$\Lambda_2 = \sum_{i=1}^{n} \frac{1}{2} \log \frac{k_{3i}^2}{k_{3i}^2 - e_{3i}^2} \tag{16}$$

with k_{3i} to k_{3i} being positive design parameters. Time derivative of Λ_2 is

$$\dot{\Lambda}_2 = \eta^T(\dot{\tau} - \dot{\tau}_r) \tag{17}$$

where

$$\eta = [\frac{e_{31}}{k_{31}^2 - e_{31}^2}, \cdots, \frac{e_{3n}}{k_{3n}^2 - e_{3n}^2}]^T \tag{18}$$

If the control law for the robot manipulator (1) is designed as follows:

$$\tau = -\lambda_3 \int_0^t e_3(\sigma)d\sigma - \int_0^t [\text{diag}\{k_{3i}^2 - e_{3i}^2\}M^{-1}(q_1)\xi](\sigma)d\sigma + \tau_r(t) \tag{19}$$

where λ_3 is a positive parameter, then one gets

$$\dot{V}_3 = -\sum_{i=1}^{n} \lambda_{1i} \frac{e_{1i}^2}{k_{1i}^2 - e_{1i}^2} - \sum_{i=1}^{n} \lambda_2 \frac{e_{2i}^2}{k_{2i}^2 - e_{2i}^2} - \sum_{i=1}^{n} \lambda_3 \frac{e_{3i}^2}{k_{3i}^2 - e_{3i}^2} + \xi^T\tilde{f} - \frac{1}{2}\xi^T\xi.$$
$$\tag{20}$$

3.2 Locally Weighted Online NN Approximation

Let $X = [X_1, \cdots, X_{2n}]^T = [q_1^T, q_2^T]^T$ and $D = \{X : |X_i| \le b_{1i}, |X_{n+i}| \le b_{2i}, i = 1, \cdots, n\}$. The locally weighted NN approximation of $f(X)$ is described by

$$\hat{f}(X) = \frac{\sum_{k=1}^{N} w_k(X)\hat{f}_k(X)}{\sum_{k=1}^{N} w_k(X)} \tag{21}$$

where $w_k(X), k = 1, \cdots, N$ as weighted functions, and the local estimator $\hat{f}_k(X)$ is described as follows:

$$\hat{f}_k(X) = \theta_k^T \phi_k(X), \ \phi_k(X) = [1, (X - c_k)^T]^T. \tag{22}$$

with c_k being the center of the k-th local estimator.

Assume $D \subseteq \cup_{k=1}^N S_k$, where $S_k = \{X : w_k \neq 0\}, k = 1, 2, \cdots, N$ are a series of compact sets. Define $w_k(X)$ as follows:

$$w_k(X) = \begin{cases} (1 - (\|X - c_k\|/\mu_k)^2)^2, & \text{if } \|X - c_k\| \leq \mu_k \\ 0, & \text{otherwise} \end{cases} \tag{23}$$

where μ_k is the radius of S_k. Let $\bar{w}_k(X) = w_k(x)/\sum_k w_k(X)$. Then, (20) can be equivalently expressed as follows:

$$\hat{f}(X) = \sum_{k=1}^N \bar{w}_k \hat{f}_k(X). \tag{24}$$

Define the optimal parameter θ_k^* for $X \in S_k$ as follows:

$$\theta_k^* = \arg\min_{\theta_k} \left(\int_{X \in D} w_k(X) \|f(X) - \hat{f}_k(X)\|^2 dX \right). \tag{25}$$

Also, define the error ϵ_k as follows:

$$\epsilon_k = \begin{cases} f(X) - \hat{f}_k(X), & \text{on } \bar{S}_k \\ 0, & \text{on } D - \bar{S}_k \end{cases} \tag{26}$$

and assume $|\epsilon_k| \leq \epsilon$ with ϵ as a positive constant. Then, $f(x)$ and its locally weighted NN approximation can be expressed to be

$$f = \sum_{k=1}^N \bar{w}_k \theta_k^{*T} \phi_k + \sum_{k=1}^N \bar{w}_k \epsilon_k, \tag{27}$$

$$\hat{f} = \sum_{k=1}^N \bar{w}_k \theta_k^T \phi_k. \tag{28}$$

It is obvious that $|\sum_{k=1}^N \bar{w}_k \epsilon_k| \leq \max(|\epsilon_k|) \sum_{k=1}^{N_i} \bar{w}_k \leq \epsilon$.

Let $\tilde{\theta}_k = \theta_k^* - \theta_k$ and $\Omega_k \triangleq \{\theta_k : \|\theta_k\| \leq c_{\theta k}\}$, and define

$$c = \max_{\theta_k^*, \theta_k \in \Omega_k} \sum_{k=1}^N \tilde{\theta}_k^T \tilde{\theta}_k / \eta$$

with η being a positive design parameter. Design the update law of θ_k to be

$$\dot{\theta}_k = \text{Proj}\left(\eta \bar{w}_k \phi_k \xi^T\right) \tag{29}$$

where Proj(.) is a projection operator given by

$$\text{Proj}(.) = \begin{cases} 0, & \text{if } \theta_k = -c_{\theta k} \text{ and } . < 0 \\ 0, & \text{if } \theta_k = c_{\theta k} \text{ and } . > 0 \\ ., & \text{otherwise} \end{cases} \tag{30}$$

3.3   Stability Analysis

Theorem. Consider the system (1) with constraints (3). Assume Assumptions 1–3 hold and $X(0) \in D, \tau(0) = 0$, and the control law is designed as (19). Let

$$A_{1i} = \max_{(e_{1i}, y_{di}) \in \Omega_{1i}} |\alpha_{1i}(e_{1i}, y_{di})|, \tag{31}$$

$$A_{ri} = \max_{((\bar{e}_{2i}, \bar{y}_{di})) \in \Omega_{ri}} |\tau_{ri}(\bar{e}_{2i}, \bar{y}_{di})|, \tag{32}$$

where $\bar{e}_{2i} = [e_{1i}, e_{2i}]^T$, $\bar{y}_{di} = [y_{di}, \dot{y}_{di}]^T$, and

$$\Omega_{1i} = \{[e_{1i}, y_{di}] : |e_{1i}| \le k_{1i}, |y_{di}| \le A_i\}, \tag{33}$$

$$\Omega_{ri} = \{[\bar{e}_{1i}, \bar{y}_{di}] : |e_{1i}| \le k_{1i}, |e_{2i}| \le k_{2i}, |y_{di}| \le A_i, |\dot{y}_{di}| \le Y_i\}. \tag{34}$$

If there exist $\lambda_{1i}, i = 1, \cdots, n, \lambda_2, \lambda_3$ such that

$$b_{1i} \ge k_{1i} + A_i, \ b_{2i} \ge k_{2i} + A_{1i}, \ \tau_{di} \ge k_{3i} + A_{ri}, \ i = 1, \cdots, n, \tag{35}$$

then the constraints (3) are satisfied and the signals in the closed-loop control system are uniformly ultimately bounded.

Proof. Consider the following Lyapunov function:

$$V = V_3 + \frac{1}{2\eta} tr\{\sum_{k=1}^{N} \tilde{\theta}_k^T \tilde{\theta}_k\} \tag{36}$$

Based on (20) and (36), one obtains

$$\dot{V} = -\sum_{i=1}^{n} \lambda_{1i} \frac{e_{1i}^2}{k_{1i}^2 - e_{1i}^2} - \sum_{i=1}^{n} \lambda_2 \frac{e_{2i}^2}{k_{2i}^2 - e_{2i}^2} - \sum_{i=1}^{n} \lambda_3 \frac{e_{3i}^2}{k_{3i}^2 - e_{3i}^2} - \frac{1}{2} \xi^T \xi$$

$$+ \xi^T (\sum_{i=1}^{N} \bar{w}_k \tilde{\theta}_k^T \phi_k + \sum_{i=1}^{N} \bar{w}_k \epsilon_k) - \frac{1}{\eta} tr\{\sum_{k=1}^{N} \tilde{\theta}_k^T \dot{\hat{\theta}}_k\}$$

$$= -\sum_{i=1}^{n} \lambda_{1i} \frac{e_{1i}^2}{k_{1i}^2 - e_{1i}^2} - \sum_{i=1}^{n} \lambda_2 \frac{e_{2i}^2}{k_{2i}^2 - e_{2i}^2} - \sum_{i=1}^{n} \lambda_3 \frac{e_{3i}^2}{k_{3i}^2 - e_{3i}^2} - \frac{1}{2} \xi^T \xi$$

$$- \frac{1}{\eta} tr\{\sum_{k=1}^{N} \tilde{\theta}_k^T (\dot{\hat{\theta}}_k - \eta \bar{w}_k \phi_k \xi^T)\} + \xi^T \sum_{i=1}^{N} \bar{w}_k \epsilon_k. \tag{37}$$

Substituting (29) into (37), one obtains

$$\dot{V} \le -\sum_{i=1}^{n} \lambda_{1i} \frac{e_{1i}^2}{k_{1i}^2 - e_{1i}^2} - \sum_{i=1}^{n} \lambda_2 \frac{e_{2i}^2}{k_{2i}^2 - e_{2i}^2} - \sum_{i=1}^{n} \lambda_3 \frac{e_{3i}^2}{k_{3i}^2 - e_{3i}^2} + \frac{1}{2} \epsilon^2 \tag{38}$$

As $\log[k_{ji}^2/(k_{ji}^2 - e_{ji}^2)] \le e_{ji}^2/(k_{ji}^2 - e_{ji}^2)$ for $j = 1, 2, 3$. [21], one gets

$$\dot{V} \le -2\lambda V_3 + \frac{1}{2} \epsilon^2 \le -2\lambda V + \beta \tag{39}$$

where $\lambda = \min\{\lambda_{1i}, i = 1, \cdots, n, \lambda_2, \lambda_3\}$ and $\beta = \lambda c + 1/2\epsilon^2$. It is concluded from (39) that V and all closed-loop signals are bounded. Then, based on the forms of BLFs $V_i, i = 1, 2, 3$ and $X(0) \in D$, one gets $|e_{1i}| \leq k_{1i}, |e_{2i}| \leq k_{2i}$ and $|e_{3i}| \leq k_{3i}$. Since (35) holds, one concludes $|q_{1i}| \leq b_{1i}, |q_{2i}| \leq b_{2i}$ and $|\tau_i| \leq \tau_{di}$. According to (39), one also obtains

$$V(t) \leq \exp(-\lambda t)(V(0) - \frac{\beta}{2\lambda}) + \frac{\beta}{2\lambda}. \tag{40}$$

Since $\log \frac{k_{1i}^2}{k_{1i}^2 - e_{1i}^2} \leq 2V(t)$ for $i = 1, \cdots, n$, one gets

$$\log \frac{k_{1i}^2}{k_{1i}^2 - e_{1i}^2} \leq 2\exp(-\lambda t)(V(0) - \frac{\beta}{2\lambda}) + \frac{\beta}{\lambda} \tag{41}$$

from which one obtains

$$\limsup_{t \to \infty} \frac{k_{1i}^2}{k_{1i}^2 - e_{1i}^2} \leq \exp(\beta/\lambda), \tag{42}$$

$$\limsup_{t \to \infty} |e_{1i}| \leq k_{1i}\sqrt{1 - \exp(\beta/\lambda)}. \tag{43}$$

4   Simulation Results

To illustrate the effectiveness of the proposed BLFs-based locally weighted learning control law, simulations are carried out for a one-link robot manipulator with the reference trajectory $y_d = 0.5\cos(0.2t)$. The dynamics of the manipulator is given by

$$ml^2\ddot{q} + d\dot{q} + 0.5mgl\cos(q) = \tau, \tag{44}$$

where $m = 1$ kg, $l = 1$ m, $g = 9.8$ m/s^2, and $d = 1$ kg.m^2/s. The constraints are $|q_1| \leq 1, |q_2| \leq 1$ and $|\tau| \leq 10$ with $q_1 = q, q_2 = \dot{q}$. In the simulation, the initial system states are $q_1(0) = 0.2, q_2(0) = 0.2$.

The control is designed as

$$\tau = -7\int_0^t e_3 d\sigma - \int_0^t \frac{9^2 - e_3^2}{0.6^2 - e_2^2} e_2 d\sigma + \tau_r \tag{45}$$

where $e_1 = q_1 - y_d, e_2 = q_2 - \alpha_1, e_3 = \tau - \tau_r$ and are constrained in $|e_1| \leq 0.5$, $|e_2| \leq 0.6$ and $|e_3| \leq 7$, and the virtual control α_1, τ_r are described by

$$\tau_r = -5e_2 + \dot{\alpha}_1 - \hat{f} - 0.5\frac{e_2}{0.6^2 - z_2^2} - \frac{0.6^2 - e_2^2}{0.5^2 - e_1^2}e_1$$

$$\alpha_1 = -2e_1 + \dot{y}_d$$

where $\hat{f}$ is a localized adaptive NN approximation of $f = -q_2 + 9.8/2\cos(q_1)$. In the NN approximation, the centers location are chosen as $c_1 = [-1, 1]^T$,

$c_2 = [0,1]^T, c_3 = [1,1]^T, c_4 = [-1,0]^T, c_5 = [0,0]^T, c_6 = [1,0]^T, c_7 = [-1,-1]^T, c_8 = [0,-1]^T, c_9 = [1,-1]^T, c_{10} = [-0.5, 0.5]^T, c_{11} = [0.5, 0.5]^T, c_{12} = [-0.5, -0.5]^T, c_{13} = [0.5, -0.5]^T, c_{\theta k} = 0.5, \eta = 100, \mu_k = 1.5$, and the basis functions are chosen as $\phi_i = [1, q_1, q_2]^T - [0; c_i], i = 1, \cdots, 13$.

Simulation results are presented in Fig. 1(a)–(c), where Fig. 1(a) shows the tracking errors e_1, e_2 and e_3, Fig. 1(b) shows the performance of the states q_1, q_2 and the control input τ, and Fig. 1(c) shows the NN approximation error $f - \hat{f}$. From Fig. 1(a), the tracking error is near to 0 after 2 s and the constraints satisfaction $|e_1| \le 0.5, |e_2| \le 0.6, |e_3| \le 7$ and $|q_1| \le 1, |q_2| \le 1, |\tau| \le 10$ is easily seen from Fig. 1(a)–(b). From Fig. 1(c), one sees that the approximation error $f - \hat{f}$ converges to a small neighborhood of zero after 2 s. Therefore, the designed adaptive NN control law makes the system state and control input constraints fulfilled and the tracking error converge to a small neighborhood of 0.

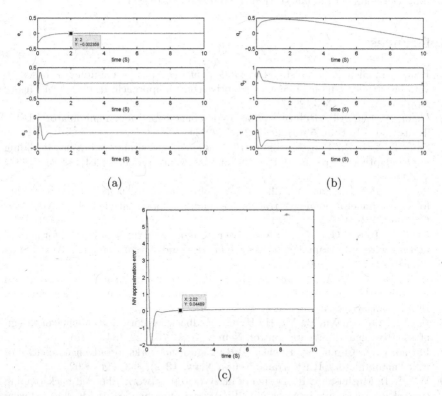

Fig. 1. Control trajectories by the proposed controller. (a) The tracking errors e_1, e_2 and e_3. (b) The states q_1, q_2 and control input τ. (c) The NN approximation error $f - \hat{f}$.

5 Conclusions

A BLFs-based adaptive NN control law was designed for robot manipulators with position, velocity and control constraints. The uncertainties were approximated by locally weighted adaptive NNs and the system constraints were tackled by using BLFs in the backstepping procedure. The control feasibility and uniform boundedness of all closed-loop signals were verified by theoretical analysis. From simulation results, we can see that under the proposed control the system constraints were never violated and absolute value of the tracking error converged to a small neighborhood of zero.

Acknowledgments. This work was supported by the National Natural Science Foundation of China under Grant No. 61503158, the MoE Tier 1 Grant from the Ministry of Education, Singapore, under WBS R-397-000-218-112, and the Priority Academic Program Development of Jiangsu Higher Education Institutions.

References

1. Huh, S.H., Bien, Z.: Robust sliding mode control of a robot manipulator based on variable structure-model reference adaptive control approach. IET Control Theory Appl. **1**(5), 1355–1363 (2007)
2. Islam, S., Liu, X.P.: Robust sliding mode control for robot manipulators. IEEE Trans. Ind. Electron. **58**(6), 2444–2453 (2011)
3. Sun, T., Pei, H., Pan, Y., Zhou, H., Zhang, C.: Neural network-based sliding mode adaptive control for robot manipulators. Neurocomputing **74**(14), 2377–2384 (2011)
4. Sun, T., Pei, H., Pan, Y., Zhang, C.: Robust adaptive neural network control for environmental boundary tracking by mobile robots. Int. J. Robust Nonlinear Control **23**(2), 123–136 (2013)
5. Chen, L., Hou, Z.G., Tan, M.: Adaptive neural network tracking control for manipulators with uncertain kinematics, dynamics and actuator model. Automatica **45**(10), 2312–2318 (2009)
6. Li, T., Duan, S., Liu, J., Wang, L., Huang, T.: A spintronic memristor-based neural network with radial basis function for robotic manipulator control implementation. IEEE Trans. Syst. Man Cybern.: Syst. **46**(4), 582–588 (2016)
7. Pan, Y., Liu, Y., Xu, B., Yu, H.: Hybrid feedback feedforward: an efficient design of adaptive neural network control. Neural Netw. **76**, 122–134 (2016)
8. Patino, H.D., Carelli, R., Kuchen, B.R.: Neural networks for advanced control of robot manipulators. IEEE Trans. Neural Netw. **13**(2), 343–354 (2002)
9. Wai, R.J., Muthusamy, R.: Design of fuzzy-neural-network-inherited backstepping control for robot manipulator including actuator dynamics. IEEE Trans. Fuzzy Syst. **22**(4), 709–722 (2014)
10. Wai, R.J., Chen, P.C.: Intelligent tracking control for robot manipulator including actuator dynamics via TSK-type fuzzy neural network. IEEE Trans. Fuzzy Syst. **12**(4), 552–560 (2004)
11. Seo, D.: Adaptive control for robot manipulator with guaranteed transient performance. In: IEEE Conference on Decision and Control, pp. 2109–2114 (2016)

12. Ngo, K.B., Mahony, R.: Bounded torque control for robot manipulators subject to joint velocity constraints. In: IEEE International Conference on Robotics and Automation, pp. 7–12 (2006)
13. Papageorgiou, X., Kyriakopoulos, K.J.: Motion tasks for robot manipulators subject to joint velocity constraints. In: IEEE/RSJ International Conference on Intelligent Robots and Systems, pp. 2139–2144 (2008)
14. Zhang, Z., Zhang, Y.: Variable joint-velocity limits of redundant robot manipulators handled by quadratic programming. IEEE/ASME Trans. Mechatron. 18(2), 674–686 (2013)
15. Zhang, Y., Ge, S.S., Lee, T.H.: A unified quadratic-programming-based dynamical system approach to joint torque optimization of physically constrained redundant manipulators. IEEE Trans. Syst. Man Cybern. B Cybern. 34(5), 2126–2132 (2004)
16. Subudhi, B., Pradhan, S.K.: Direct adaptive control of a flexible robot using reinforcement learning. In: 2010 International Conference on Industrial Electronics, Control & Robotics, pp. 27–29 (2010)
17. He, W., Chen, Y., Yin, Z.: Adaptive neural network control of an uncertain robot with full-state constraints. IEEE Trans. Cybern. 46, 620–629 (2016)
18. He, W., David, A.O., Yin, Z., Sun, C.: Neural network control of a robotic manipulator with input deadzone and output constraint. IEEE Trans. Syst. Man Cybern. Part A-Syst. 46(6), 759–770 (2016)
19. Kwan, C., Lewis, F.L.: Robust backstepping control of nonlinear systems using neural networks. IEEE Trans. Syst. Man Cybern. Part A-Syst. 30(6), 753–766 (2000)
20. Kuljaca, O., Swamy, N., Lewis, F.L., Kwan, C.: Design and implementation of industrial neural network controller using backstepping. IEEE Trans. Ind. Electron. 50(1), 193–201 (2003)
21. Liu, Y.J., Li, J., Tong, S.C., Philip Chen, C.L.: Neural nework control-based adaptive learning design for nonlinear systems with full-state constraints. IEEE Trans. Neural Netw. Learn. Syst. 27(7), 1562–1570 (2016)

How the Prior Information Shapes Neural Networks for Optimal Multisensory Integration

He Wang[1(✉)], Wen-Hao Zhang[1,2,3], K.Y. Michael Wong[1], and Si Wu[2]

[1] Department of Physics, Hong Kong University of Science and Technology,
Hong Kong, China
{hwangaa,phkywong}@ust.hk, wenhaoz1@andrew.cmu.edu
[2] State Key Laboratory of Cognitive Neuroscience and Learning, and McGovern
Institute for Brain Research, Beijing Normal University, Beijing, China
wusi@bnu.edu.cn
[3] Center for the Neural Basis of Cognition, Carnegie Mellon University,
Pittsburgh, USA

Abstract. Extensive studies suggest that the brain integrates multisensory signals in a Bayesian optimal way. In this work, we consider how the couplings in a neural network model are shaped by the prior information when it performs optimal multisensory integration and encodes the whole profile of the posterior. To process stimuli of two modalities, a biologically plausible neural network model consists of two modules, one for each modality, and crosstalks between the two modules are carried out through feedforward cross-links and reciprocal connections. We found that the reciprocal couplings are crucial to optimal multisensory integration in that their pattern is shaped by the correlation in the joint prior distribution of sensory stimuli. Our results show that a decentralized architecture based on reciprocal connections is able to accommodate complex correlation structures across modalities and utilize this prior information in optimal multisensory integration.

Keywords: Recurrent neural networks · Multisensory processing · Bayesian inference

1   Introduction

Extracting information reliably from ambiguous environments is crucial for the survivorship of organisms. The brain solves this problem by exploiting multiple sensory modalities to gather, from different aspects, as much information as possible about the same entity of interest. It has been reported in a large number of psychophysical and neurobiological studies that the brain can integrate sensory cues in an optimal way, as predicted by Bayesian inference [1–3].

Despite the accumulated behavior evidence, exactly how the brain implements optimal multisensory integration remains largely unknown. In the present

© Springer International Publishing AG 2017
F. Cong et al. (Eds.): ISNN 2017, Part II, LNCS 10262, pp. 128–136, 2017.
DOI: 10.1007/978-3-319-59081-3_16

study, we adopt a theoretical approach to address this challenging issue. We formulate multisensory integration as a mathematical problem of optimizing network structure under the constraint that for a given prior distribution of stimuli, the network's output matches the posterior distribution. This is equivalent to requiring that the network realizes Bayesian inference when the sensory cues are sampled form their prior over many trials. We introduce different prior distributions of the multisensory stimuli and investigate how network structures depend on the choice of priors. We look for evidence to see where information about the prior and that about the likelihood are represented in the network consisting of recurrent and reciprocal connections, cross-links and direct links. These results generate predictions about the structural pre-requisites for multisensory integration. They can be tested in future experiments and shed light on our understanding of how the brain can achieve multisensory integration optimally.

2   Optimal Multisensory Inference with a Composite Prior

Utilizing prior information is important for multisensory information processing. A variety of studies have suggested that the prior distributions are taken into account when animals make perceptual decisions [4–6]. Specifically, multisensory processing relies on the experience about correlations among sensory cues, which usually benefits us in forming a unified and coherent perception of the external world [7], yet sometimes evokes interesting illusions [8,9].

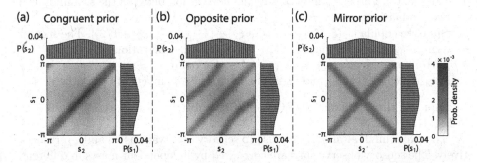

Fig. 1. Three types of the prior. (a) The joint prior distribution constructed from the congruent copula c_1. The marginal priors, which are the same for s_1 and s_2, are plotted to the sides of (a). (b) The joint prior distribution constructed from the opposite copula c_2. (c) The joint prior distribution constructed from the mirror copula c_3. The color code for (a)–(c) are the same and shown to the right of (c). Parameters: for all three cases, $\kappa_s = 0.2, \kappa_p = 11.6, p_c = 0.246$. For the opposite prior in (b) and the mirror prior in (c), $\alpha = 0.5$. (Color figure online)

Different specific forms of the prior distribution have been brought up to characterize different perceptual tasks (see [10] for a review). In general, the joint prior should be composed of an independent part and a correlated part. Suppose s_1 and s_2 are two sensory stimuli in different modalities, whose marginal prior densities are $p(s_1)$ and $p(s_2)$, respectively. The joint prior can be described as $p(s_1, s_2) = (1 - p_c)p(s_1)p(s_2) + p_c q(s_1, s_2)$. Here, $q(s_1, s_2)$ is a correlated distribution and $p_c \in [0, 1]$ describes how often s_1 and s_2 are originated from that distribution. Ideally, $q(s_1, s_2)$ should only affect the correlation between the two underlying stimuli without changing their marginal distributions. This requirement can be satisfied by using a copula, which is a multivariate probability distribution, whose marginal distribution of each variable is uniform [11]. Consider a two-dimensional copula $c(\xi_1, \xi_2)$, satisfying the property that its marginals over ξ_1 or ξ_2 are equal to 1. According to the Sklar's theorem [12], $q(s_1, s_2)$ can be constructed as $q(s_1, s_2) = c(F(s_1), F(s_2))p(s_1)p(s_2)$, where $F(s_i)$ is the cumulative distribution function of $p(s_i)$. It can be verified that the marginal distributions of $q(s_1, s_2)$ are exactly $p(s_1)$ and $p(s_2)$.

In the present work, we consider stimuli such as heading direction residing on a circular space $[-\pi, \pi)$. We use the von Mises distribution as the marginal prior distribution, $p(s_i) \propto e^{\kappa_s \cos s_i}$, $i = 1, 2$, where κ_s is the concentration parameter, and $\propto$ indicates proportionality. For simplicity, we consider the case that the marginal priors are the same for the two modalities, and centered at the origin. In order to observe the dependence of network structure on the prior, three forms of copulas are chosen due to their distinctive profiles:

1. Congruent copula $[c_1(\xi_1, \xi_2) \propto e^{\kappa_p \cos 2\pi(\xi_1 - \xi_2)}]$, which is derived from the von Mises distribution. Similar forms of such prior are widely applied in describing a pair of correlated sensory cues when they are originated from a common cause [8,13]. Larger κ_p indicates higher correlation between the stimuli in the two modalities.
2. Opposite copula $[c_2(\xi_1, \xi_2) = \alpha c_1(\xi_1, \xi_2) + (1 - \alpha)c_1(\xi_1, \xi_2 + 1/2)]$. The second term in c_2 indicated that s_1 and s_2 may come in opposition directions.
3. Mirror copula $[c_3(\xi_1, \xi_2) = \alpha c_1(\xi_1, \xi_2) + (1 - \alpha)c_1(\xi_1, -\xi_2)]$. The second term in c_3 indicates that s_1 and s_2 might be the mirror image of each other.

Examples of three different kinds of joint prior distributions $p(s_1, s_2)$ are shown in Fig. 1(a)–(c).

The two stimuli s_1 and s_2 give rise to sensory observations x_1 and x_2, respectively. The sensory observations are corrupted by independent noises in different sensory pathways. We use the von Mises distribution to represent the likelihood functions, $p(x_i|s_i) \propto e^{\kappa_i \cos(x_i - s_i)}$, $i = 1, 2$, where κ_i is the concentration parameter, which can be understood as the reliability of the sensory input in the corresponding modality.

These uni-sensory observations are supposed to be fed into higher level multisensory regions, where optimal multisensory estimates $\hat{s}_1$ and $\hat{s}_2$ are made. According to the Bayes' theorem, the marginal posterior distribution is given by

$$p(s_1|x_1, x_2) \propto \int p(x_1|s_1)p(x_2|s_2)p(s_1, s_2)\,\mathrm{d}s_2. \tag{1}$$

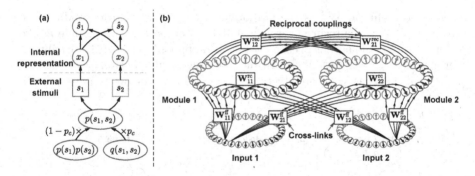

Fig. 2. The multimodal Bayesian inference problem and the recurrent neural network model. (a) A graphical illustration of the Bayesian inference problem. (b) Each small circle portraits one neuron, with the attached arrow indicating the neuron's preferred stimulus. Besides being recurrently connected to each of themselves, the two modules of the network model interact with each other through feedforward cross-links ($\mathbf{W}_{12}^{\mathrm{ff}}$ and $\mathbf{W}_{21}^{\mathrm{ff}}$) and reciprocal couplings ($\mathbf{W}_{12}^{\mathrm{rec}}$ and $\mathbf{W}_{21}^{\mathrm{rec}}$). The inputs of the two modules represent uni-sensory observations, corresponding to x_1 and x_2 in (a). The outputs are multisensory representations, corresponding to $\hat{s}_1$ and $\hat{s}_2$ in (a).

Usually the expected value of s_1 from the posterior distribution is chosen as a Bayesian optimal estimate $\hat{s}_i$ for the underlying stimulus, which minimizes a mean squared error cost function [10]. For circular random variables considered in this work, the Bayesian estimates for the stimuli in two modalities are given by, $\hat{s}_i = \arg\left[\int p(s_i = \phi | x_1, x_2) e^{j\phi}\, d\phi\right]$, for $i = 1, 2$, where $j \equiv \sqrt{-1}$ is the imaginary unit. This Bayesian inference framework for multisensory processing is shown schematically in Fig. 2(a).

3 A Bi-modular Recurrent Neural Network Model

Bi-modular recurrent neural network models have been applied in many studies on the multisensory integration to explain experimental findings and provide insights into the functional roles of connections between brain areas [14,15]. We will explore the capability of such bi-modular recurrent network models in encoding an arbitrary prior distribution and optimally integrating multisensory information based on that prior. Consider a bi-modular recurrent neural network model with its dynamical equation [16],

$$\tau_s \frac{\partial}{\partial t} \begin{bmatrix} \boldsymbol{u_1} \\ \boldsymbol{u_2} \end{bmatrix} = - \begin{bmatrix} \boldsymbol{u_1} \\ \boldsymbol{u_2} \end{bmatrix} + \begin{bmatrix} \mathbf{W}_{11}^{\mathrm{rec}} & \mathbf{W}_{12}^{\mathrm{rec}} \\ \mathbf{W}_{21}^{\mathrm{rec}} & \mathbf{W}_{22}^{\mathrm{rec}} \end{bmatrix} \begin{bmatrix} \boldsymbol{r_1} \\ \boldsymbol{r_2} \end{bmatrix} + \begin{bmatrix} \mathbf{W}_{11}^{\mathrm{ff}} & \mathbf{W}_{12}^{\mathrm{ff}} \\ \mathbf{W}_{21}^{\mathrm{ff}} & \mathbf{W}_{22}^{\mathrm{ff}} \end{bmatrix} \begin{bmatrix} \boldsymbol{I_1} \\ \boldsymbol{I_2} \end{bmatrix}. \quad (2)$$

Here, $\boldsymbol{u_i}$ is a N-element vector, whose m^{th} element $u_{i,m}$ is the synaptic input of the mth neuron in module i. The mth element of the vector $\boldsymbol{r_i}$, $r_{i,m}$, is the firing rates of the mth neuron in module i. The firing rate is related to the synaptic input $\boldsymbol{u_i}$ through an activation function $\boldsymbol{r_i} = f(\boldsymbol{u_i})$. $\boldsymbol{I_i}$ is the external inputs applied on module i. $\mathbf{W}^{ff}_{ij}$ is the feedforward weight matrix from module j to module i. $\mathbf{W}^{rec}_{ij}$ is the recurrent weight matrix from module j to module i. The preferred stimuli of neurons in each module are supposed to be evenly distributed over a circle, $\phi_m = (2\pi m)/N - \pi$. In the following results, both modules consist of $N = 32$ neurons. The architecture of this network model is illustrated in Fig. 2(b).

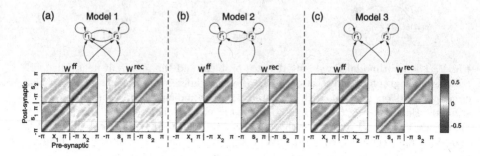

Fig. 3. Comparison of model structures. Network architecture and connection weights for: (a) model 1, a fully connected model; (b) model 2, where feedforward cross-links are cut; and (c) model 3, where reciprocal couplings are cut. All three structures are optimized for congruent copula c_1. Parameters: $\kappa_s = 0.2, \kappa_p = 11.6, p_c = 0.246, \widetilde{\kappa}_1 = \widetilde{\kappa}_2 = 10.7$.

The inputs of the neural network are the neural population representation of the uni-sensory observations of the external stimuli. Due to the uncertain nature of the external world and noisy neuronal firings, the neural population representation is constantly fluctuating, hypothetically sampling the likelihood function. In a similar way, the outputs of the multisensory neural population should sample the posterior distribution. If we consider a time scale that is much longer than this sampling process, the temporal average of the neuronal inputs and outputs should resemble the likelihood function and the network's estimate of the posterior distribution, respectively. Therefore, the external input vector $\boldsymbol{I}_i$ is set to be the same as the likelihood functions $p(x_i|s_i = \phi_m)$, and the stationary firing rates $\boldsymbol{r}_i^*$ will eventually approach the network's estimate of the marginal posterior $\boldsymbol{p}_i$, whose m^{th} element is $p(s_i = \phi_m|x_1, x_2)$, during the network optimization described below.

We use a divisive normalization function as the activation function. Recently, due to its success in accounting for important features of multisensory integration, such as the principle of inverse effectiveness and the spatial principle [17], the divisive normalization model was proposed to be a canonical integration operation [18,19]. Here, we follow the form of divisive normalization in a continuous attractor neural network model [20], $r_{i,m} = [u_{i,m}]_+^2 /\{1 + k_I \sum_n [u_{i,n}]_+^2\}$, for $i = 1, 2$, and $m, n = 1, 2, \ldots, N$. Here, $[x]_+ \equiv \max(x, 0)$, and k_I is the strength of global inhibition. The performance of the network is the best with divisive normalization function, compared with sigmoid or piece-wise linear functions (data not shown here). In the present work, we fix $k_I = 0.1$, while small changes in k_I does not affect the results very much.

3.1 Optimize the Connection Weights Through Stochastic Gradient Descent

We optimize the connection weights in order to minimize the mean squared error L between the stationary network activity r^* and the marginal posterior distribution p_i, $L \equiv \langle \sum_{i=1,2} \|r_i^* - p_i\|^2 \rangle_{p(x_1,x_2)}$. Usually a recurrent neural network is trained using back-propagation through time [21, 22]. Since only the steady state is relevant in this work, we use a simple stochastic gradient descent algorithm to optimize the steady state. Samples of training inputs (I_1, I_2) and training outputs (p_1, p_2) are generated in the following way. Given the prior distribution $p(s_1, s_2)$ and the mean reliabilities of sensory inputs $\tilde{\kappa}_1$ and $\tilde{\kappa}_2$, we first draw the true value of external stimuli s_1 and s_2 from the prior distribution and draw the reliabilities for each sensory input κ_1 and κ_2 independently from log-normal distributions $\ln \mathcal{N}(\tilde{\kappa}_i, \sigma_\kappa^2)$. In this work, we always set σ_κ to be 0.5. Secondly, draw the sensory input x_i from the von Mises distribution $p(x_i|s_i) \propto e^{\kappa_i \cos(s_i - x_i)}$. Then, the training inputs and the training outputs can be calculated according to the Bayes' theorem in Eq. (1).

4   Results

4.1   Model Comparison

Crosstalks between different sensory areas may happen at different levels. In general, we consider two types of communication across modalities: the feedforward cross-links ($\mathbf{W}_{ij}^{ff}$ for $i \neq j$), and the reciprocal couplings ($\mathbf{W}_{ij}^{rec}$ for $i \neq j$). By forcing either of them to be zero, we tested three different model structures (Fig. 3). Model 1 is the fully connected model (Fig. 3(a)). In model 2, the interaction between the two modules are limited to the reciprocal connections, with the feedforward cross-links forced to be zero (Fig. 3(b)). In model 3, the reciprocal connections are set to be zero (Fig. 3(c)). We also tested a purely feedforward network structure, model 4, to see if recurrent connections are essential for optimal multisensory integration. We found that model 1, 2 and 3 are almost indistinguishable in their performances, while the purely feedfoward structure, model 4, is obviously worse than the others (data not shown here). Examples of the connection weights for model 1, 2 and 3 are shown in the lower parts of Fig. 3(a)–(c). In the following part of this work, we will focus on model 2, while general results are similar for model 1 and model 3.

4.2   Coupling Weights for Different Priors

To reveal the impact of the prior information on the recurrent neural network model, we compare the coupling weights of networks optimized with different

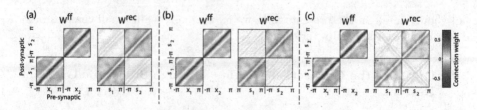

Fig. 4. The optimized coupling weights for three types of the prior. (a) The connection weights of model 2 trained with the congruent prior in Fig. 1(a). (b) The connection weights of model 2 trained with the opposite prior in Fig. 1(b). (c) The connection weights of model 2 trained with the mirror prior in Fig. 1(c). Parameters: for all three cases, $\widetilde{\kappa}_1 = \widetilde{\kappa}_2 = 10.7$.

prior distributions. Three examples are shown in Fig. 4. The prior distributions share the same marginal distributions, and are then constructed using three different types of copulas: the congruent copula c_1 (Fig. 1(a)), the opposite copula c_2 (Fig. 1(b)) and the mirror copula c_3 (Fig. 1(c)). Coupling weights of the networks trained with the three priors are shown in Fig. 4(a)–(c). The same-side connection weights ($\mathbf{W}_{11}^{ff}$, $\mathbf{W}_{22}^{ff}$, $\mathbf{W}_{11}^{rec}$ and $\mathbf{W}_{22}^{rec}$) are nearly identical for the three cases. However, the reciprocal couplings ($\mathbf{W}_{12}^{rec}$ and $\mathbf{W}_{12}^{rec}$) exhibit patterns resembling the corresponding prior distribution. This result strongly suggests that the reciprocal connections, as a bridge between different sensory modules, are able to encode the information of the joint prior distribution, taking the correlation structure between sensory stimuli into account when performing multisensory integration.

5   Conclusion

We have developed a framework to link the network structure of the multisensory processing brain region to the statistical structure of Bayesian inference. We found that a recurrent network structure appears to be necessary for implementing optimal multisensory integration. Furthermore, we have studied the dependence of the network structure for multisensory information processing on the choice of the priors and likelihoods. We found clear evidence that information about the prior is encoded in the indirect couplings (reciprocal connections and cross-links). This can be seen from the correspondence between the profiles of the indirect couplings and the correlation pattern in the joint prior of the stimuli. In the present models, the priors can be encoded in either cross-links or reciprocal connections or both. In the future, we can consider how biological constraints can narrow down these possibilities for realistic architecture exploited by the neural system.

Multisensory integration is not limited to biological systems. In other artificial intelligence applications, such as computer vision and robotics, integrating signals optimally from multiple sensors is also a fundamental technique. The optimal structure we found has implications to the decentralized architecture for multisensory information processing. It demonstrates that composite prior distributions can be encoded in a decentralized fashion in the reciprocal connections.

Acknowledgments. This work is supported by the Research Grants Council of Hong Kong (N_HKUST606/12, 605813 and 16322616) and National Basic Research Program of China (2014CB846101) and the Natural Science Foundation of China (31261160495).

References

1. Alais, D., Burr, D.: No direction-specific bimodal facilitation for audiovisual motion detection. Cogn. Brain Res. **19**(2), 185–194 (2004)
2. Ernst, M.O., Banks, M.S.: Humans integrate visual and haptic information in a statistically optimal fashion. Nature **415**(6870), 429–433 (2002)
3. Gu, Y., Angelaki, D.E., DeAngelis, G.C.: Neural correlates of multisensory cue integration in macaque MSTd. Nat. Neurosci. **11**(10), 1201–1210 (2008)
4. Girshick, A.R., Landy, M.S., Simoncelli, E.P.: Cardinal rules: visual orientation perception reflects knowledge of environmental statistics. Nat. Neurosci. **14**(7), 926–932 (2011)
5. Fischer, B.J., Peña, J.L.: Owl's behavior and neural representation predicted by Bayesian inference. Nat. Neurosci. **14**(8), 1061–1066 (2011)
6. Körding, K.P., Wolpert, D.M.: Bayesian integration in sensorimotor learning. Nature **427**(6971), 244–247 (2004)
7. Ghazanfar, A.A., Schroeder, C.E.: Is neocortex essentially multisensory? Trends Cogn. Sci. **10**(6), 278–285 (2006)
8. Sato, Y., Toyoizumi, T., Aihara, K.: Bayesian inference explains perception of unity and ventriloquism aftereffect: identification of common sources of audiovisual stimuli. Neural Comput. **19**(12), 3335–3355 (2007)
9. Shams, L., Ma, W.J., Beierholm, U.: Sound-induced flash illusion as an optimal percept. NeuroReport **16**(17), 1923–1927 (2005)
10. Shams, L., Beierholm, U.R.: Causal inference in perception. Trends Cogn. Sci. **14**(9), 425–432 (2010)
11. Durante, F., Sempi, C.: Principles of Copula Theory. Taylor & Francis, Boca Raton (2015)
12. Sklar, A.: Random variables, joint distribution functions, and copulas. Kybernetika **9**(6), 449–460 (1973)
13. Körding, K.P., Beierholm, U., Ma, W.J., Quartz, S., Tenenbaum, J.B., Shams, L.: Causal inference in multisensory perception. PLoS One **2**(9), e943 (2007)
14. Zhang, W.H., Chen, A., Rasch, M.J., Wu, S.: Decentralized multisensory information integration in neural systems. J. Neurosci. **36**(2), 532–547 (2016)
15. Magosso, E., Cuppini, C., Ursino, M.: A neural network model of ventriloquism effect and aftereffect. PLoS One **7**(8), e42503 (2012)
16. Amari, S.: Dynamics of pattern formation in lateral-inhibition type neural fields. Biol. Cybern. **27**(2), 77–87 (1977)
17. Ohshiro, T., Angelaki, D.E., DeAngelis, G.C.: A normalization model of multisensory integration. Nat. Neurosci. **14**(6), 775–782 (2011)
18. Carandini, M., Heeger, D.J.: Normalization as a canonical neural computation. Nat. Rev. Neurosci. **13**(1), 51–62 (2012)

19. Van Atteveldt, N., Murray, M.M., Thut, G., Schroeder, C.E.: Multisensory integration: flexible use of general operations. Neuron **81**(6), 1240–1253 (2014)
20. Fung, C.C.A., Wong, K.Y.M., Wu, S.: A moving bump in a continuous manifold: a comprehensive study of the tracking dynamics of continuous attractor neural networks. Neural Comput. **22**(3), 752–792 (2010)
21. Williams, R.J., Zipser, D.: Gradient-based learning algorithms for recurrent networks and their computational complexity. In: Back-Propagation: Theory, Architectures and Applications, pp. 433–486 (1995)
22. Seung, H.S.: Learning continuous attractors in recurrent networks. In: Jordan, M.I., Kearns, M.J., Solla, S.A. (eds.) Advances in Neural Information Processing Systems 10, pp. 654–660. MIT Press, Cambridge (1998)

Fuzzy Uncertainty Observer Based Filtered Sliding Mode Trajectory Tracking Control of the Quadrotor

Yong Wang[1], Ning Wang[1(✉)], Shuailin Lv[1], Jianchuan Yin[2], and Meng Joo Er[1,3]

[1] Marine Engineering College, Dalian Maritime University, Dalian, China
15241192269@163.com, n.wang.dmu.cn@gmail.com, lvshuailin@gmail.com
[2] Navigation College, Dalian Maritime University, Dalian, China
yinjianchuan@gmail.com
[3] School of EEE, Nanyang Technological University, Singapore, Singapore
emjer@ntu.edu.sg

Abstract. In this paper, a filtered sliding mode control (FSMC) scheme based on fuzzy uncertainty observer (FUO) for trajectory tracking control of a quadrotor unmanned aerial vehicle (QUAV) is proposed. To be specific, the dynamics model of QUAV is decomposed into three subsystems. By virtue of the cascaded structure, sliding-mode-based virtual control laws can be recursively designed. In order to remove the smoothness requirements on intermediate signals, a series of first-order filters are employed to reconstruct sliding mode control signals together with their first derivatives. Moreover, fuzzy uncertainty observers are employed to indirectly estimate lumped unknown nonlinearities including system uncertainties and external disturbances and make compensation for the QUAV system. Stability analysis and uniformly ultimately bounded tracking errors and states can be guaranteed by the Lyapunov approach. Simulation studies demonstrate the effectiveness and superiority of the proposed tracking control scheme.

Keywords: Quadrotor unmanned aerial vehicle · Trajectory tracking control · Sliding mode control · Fuzzy uncertainty observer

1 Introduction

Compared with traditional single rotor UAV, the most significant advantage of the quadrotor unmanned aerial vehicles (QUAV) is that the latter has better

N. Wang—This work is supported by the National Natural Science Foundation of P.R. China (under Grants 51009017 and 51379002), Applied Basic Research Funds from Ministry of Transport of P.R. China (under Grant 2012-329-225-060), China Postdoctoral Science Foundation (under Grant 2012M520629), the Fund for Dalian Distinguished Young Scholars (under Grant 2016RJ10), the Innovation Support Plan for Dalian High-level Talents (under Grant 2015R065), and the Fundamental Research Funds for the Central Universities (under Grant 3132016314).

© Springer International Publishing AG 2017
F. Cong et al. (Eds.): ISNN 2017, Part II, LNCS 10262, pp. 137–147, 2017.
DOI: 10.1007/978-3-319-59081-3_17

stability, more compact structure and larger load, *etc.*. So the QUAV pertains to a wide area of possible applications including patrolling for forest fires, traffic monitoring, surveillance rescue, *etc.*, and as a remarkable platform for the UAV, the QUAV has been attracting numerous research [1–6].

The QUAV is a complex nonlinear strongly coupled system with more than one input and output, and thereby leading to great challenges in controller design and synthesis. In [7], PID control scheme is used to achieve the trajectory tracking control. However, this control method is classical linear control scheme, which only work better when the QUAV is near hovering state. Backstepping control scheme has an extensive application in controlling the QUAV in recent years. In [8], the QUAV dynamic system has been divided into two subsystems, i.e., translational subsystem and rotational subsystem and two subcontrollers have been designed. However, general backstepping control schemes need accurate model parameters and is not robust to model uncertainties and external disturbances, for this reason, adaptive integral backstepping control scheme [9] has been applied in the QUAV, which Only suitable for the model uncertainty and external disturbances are slow-varying or constant. Sliding mode control is a powerful control method with characteristics of simple and robust [10]. Combining with adaptive control strategy or observer [11], this kind control methods have widespread used various systems, however, chattering phenomenon is inevitable for the continuous switching logic. Adaptive fuzzy backstepping control has been used to the trajectory tracking control for the QUAV in [12], in which the fuzzy system is employed to approximate directly a model using backstepping techniques. For the reason of underactuation, the virtual controller is designed in most control schemes for the QUAV, while, the derivative of the virtual controller will be complex. In this context, we focus on a QUAV with the lumped unknown nonlinearity including system uncertainties and external disturbances, and a filtered sliding mode trajectory tracking control scheme based on fuzzy uncertainty observer (FUO) for the QUAV is proposed.

2   QUAV Dynamics and Problem Formulation

As shown in Fig. 1, defining the earth-fixed coordinate $OX_0Y_0Z_0$ and the body-fixed coordinate $O'XYZ$ which are respectively considered with the origin coinciding to the starting point and the gravity center of the QUAV. Vectors (x, y, z) and (ϕ, θ, ψ) are respectively denote the positions of the QUAV in earth-fixed coordinate $OX_0Y_0Z_0$ and the Euler angles in body-fixed coordinate $O'XYZ$, in which ϕ refers as to roll angle, θ refers as to pitch angle and ψ refers as to yaw angle.

The position dynamics can be described as follows:

$$\begin{cases} \dot{\boldsymbol{\eta}}_{11} = \boldsymbol{\eta}_{12} \\ \dot{\boldsymbol{\eta}}_{12} = \boldsymbol{f}_1\left(\boldsymbol{\eta}_{12}\right) + \boldsymbol{u}_1\left(\boldsymbol{\eta}_2, \tau\right) + \boldsymbol{d}_1\left(\boldsymbol{\eta}_{11}, \boldsymbol{\eta}_{12}, t\right) \end{cases} \tag{1}$$

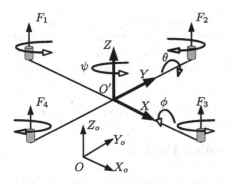

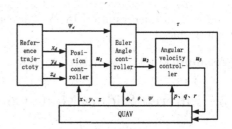

Fig. 1. The configuration of a QUAV.   **Fig. 2.** The overall control diagram.

with the lumped model uncertainties and/or external disturbances $d_1 = [d_{11}, d_{12}, d_{13}]^T$, $f_1 = [D_x \dot{x}^2, D_y \dot{y}^2, D_z \dot{z}^2 - g]^T$ and

$$u_1(\eta_2, \tau) = \frac{\tau}{m} \begin{bmatrix} C_\phi S_\theta C_\psi + S_\phi S_\psi \\ C_\phi S_\theta S_\psi - S_\phi C_\psi \\ C_\phi C_\theta \end{bmatrix} \tag{2}$$

where $\eta_{11} = [x, y, z]^T$ and $\eta_{12} = [\dot{x}, \dot{y}, \dot{z}]^T$ are vectors of the positions and linear velocities in the earth-fixed frame, respectively, m is the mass of the QUAV, g is the acceleration of the gravity, C_* and S_* are the functions $\cos(*)$ and $\sin(*)$, respectively, τ is the total thrust.

The vector of Euler angles $\eta_2 = [\phi, \theta, \psi]^T$ is governed by

$$\dot{\eta}_2 = g_2(\eta_2) u_2(\eta_3) + d_2(\eta_2, t) \tag{3}$$

with the lumped model uncertainties and/or external disturbances $d_2 = [d_{21}, d_{22}, d_{23}]^T$, and

$$g_2(\eta_2) = \begin{bmatrix} 1 & S_\phi T_\theta & C_\phi T_\theta \\ 0 & C_\phi & -S_\phi \\ 0 & \frac{S_\phi}{C_\theta} & \frac{C_\phi}{C_\theta} \end{bmatrix} \tag{4}$$

$$u_2(\eta_3) = \eta_3 \tag{5}$$

where T_* denotes the function $\tan(*)$, $\eta_3 = [p, q, r]^T$ is the angular velocity vector in body-fixed coordinate given by the following dynamics:

$$\dot{\eta}_3 = f_3(\eta_3) + g_3 u_3 + d_3(\eta_3; t) \tag{6}$$

with the diagonal matrix $g_3 = \operatorname{diag}(1/J_x, 1/J_y, 1/J_z)$ where $J_i (i = x, y, z)$ is the moment of inertia with respect to each axis, $d_3 = [d_{31}, d_{32}, d_{33}]^T$ include unmodeled dynamics and/or external disturbances, and $f_3(\eta_3) = [\frac{J_y - J_z}{J_x} qr, \frac{J_z - J_x}{J_y} pr, \frac{J_x - J_y}{J_z} pq]^T$, where $u_3 = [u_{31}, u_{32}, u_{33}]^T$ is the control input and the final control input vector of the QUAV system is $u = [\tau, u_3^T]^T$.

The control objective in this study is to design filtered sliding mode controller of the QUAV with FUO and achieve the trajectory tracking control $(x \to x_d, y \to y_d, z \to z_d, \psi \to \psi_d)$ in presence of the external disturbances and system uncertainties. Before ending this section, the following assumption is introduced:

Assumption 1. *The desired trajectory and its time derivatives are bounded.*

3   Filtered Sliding Mode Controller Design

In this section, three subcontrollers will be designed. The overall control diagram is as shown in Fig. 2.

3.1   Position Controller

Given a reference trajectory $\boldsymbol{\eta}_{11d} := [x_d, y_d, z_d]^T$, combining with position dynamics (1), we design sliding surfaces as follows:

$$s_{11}(t) = e_{11}(t) + k_{11} \int_0^t e_{11}(\tau) d\tau \tag{7}$$

$$s_{12}(t) = e_{12}(t) + k_{12} \int_0^t e_{12}(\tau) d\tau \tag{8}$$

where $\boldsymbol{k}_{11} = \mathrm{diag}(k_{111}, k_{112}, k_{113}) > 0$, $\boldsymbol{k}_{12} = \mathrm{diag}(k_{121}, k_{122}, k_{123}) > 0$, $\boldsymbol{e}_{11} = \boldsymbol{\eta}_{11} - \boldsymbol{\eta}_{11d}$, $\boldsymbol{e}_{12} = \boldsymbol{\eta}_{12} - \bar{\boldsymbol{\eta}}_{12d}$, and $\bar{\boldsymbol{\eta}}_{12d}$ is the filtered output of the virtual control signal $\boldsymbol{\eta}_{12d}$ given by

$$\epsilon_1 \dot{\bar{\boldsymbol{\eta}}}_{12d} + \bar{\boldsymbol{\eta}}_{12d} = \boldsymbol{\eta}_{12d} \tag{9}$$

here, $\epsilon_1 > 0$ is an user-defined filtering time constant and let $\boldsymbol{y}_1 = \bar{\boldsymbol{\eta}}_{12d} - \boldsymbol{\eta}_{12d}$.

In this context, the virtual control signal $\boldsymbol{\eta}_{12d}$ can be selected as follows:

$$\boldsymbol{\eta}_{12d} = -\boldsymbol{p}_{11} s_{11} + \dot{\boldsymbol{\eta}}_{11d} - \boldsymbol{k}_{11} \boldsymbol{e}_{11} - \boldsymbol{e}_{12} \tag{10}$$

where $\boldsymbol{p}_{11} = \mathrm{diag}(p_{111}, p_{112}, p_{113}) > 0$ and a desired position control law for sub-system (1) can be designed as follows:

$$\boldsymbol{u}_1 = -\boldsymbol{p}_{12} s_{12} - \boldsymbol{f}_1(\boldsymbol{\eta}_{12}) + \dot{\bar{\boldsymbol{\eta}}}_{12d} - \boldsymbol{k}_{12} \boldsymbol{e}_{12} - \hat{\boldsymbol{d}}_1 \tag{11}$$

with the FUO given by

$$\hat{\boldsymbol{d}}_1(\boldsymbol{\omega}_1 \mid \hat{\boldsymbol{\vartheta}}_1) = \hat{\boldsymbol{\vartheta}}_1^T \boldsymbol{\xi}_1(\boldsymbol{\omega}_1) \tag{12}$$

Choosing the parameter matrix update rule as

$$\dot{\hat{\boldsymbol{\vartheta}}}_1 = -r_{11} \hat{\boldsymbol{\vartheta}}_1 + r_{12} \boldsymbol{\xi}_1(\boldsymbol{\omega}_1)(s_{12} + \boldsymbol{\varepsilon}_1)^T \tag{13}$$

where $r_{11} > 0$ and $r_{12} > 0$ are user-defined positive definite parameters, $\boldsymbol{\omega}_1 = [\boldsymbol{\eta}_{11}^T, \boldsymbol{\eta}_{12}^T]^T$ is the input vector of the fuzzy system, $\boldsymbol{\varepsilon}_1 = \boldsymbol{\eta}_{12} - \boldsymbol{v}_1$ is the observation error vector with

$$\dot{\boldsymbol{v}}_1 = -r_{13}\boldsymbol{v}_1 + \boldsymbol{f}_1(\boldsymbol{\eta}_{12}) + \boldsymbol{u}_1 + \hat{\boldsymbol{d}}_1(\boldsymbol{\omega}_1|\hat{\boldsymbol{\vartheta}}_1) + r_{13}\boldsymbol{\eta}_{12} \tag{14}$$

where $r_{13} > 0$ is user-defined positive definite parameter.

3.2   Euler Angle Controller

Substituting the control law (11) into the input nonlinearity (2), we can obtain

$$\begin{cases} \tau = m\|\boldsymbol{u}_1\| \\ \phi_d = \arcsin\left(\frac{m}{\tau}(S_{\psi_d}u_{11} - C_{\psi_d}u_{12})\right) \\ \theta_d = \arcsin\left(\frac{\frac{m}{\tau}u_{11} - S_{\psi_d}S_{\phi_d}}{C_{\psi_d}C_{\phi_d}}\right) \end{cases} \tag{15}$$

Let $\boldsymbol{\eta}_{2d} := [\phi_d, \theta_d, \psi_d]^T$ and $\bar{\boldsymbol{\eta}}_{2d} := [\bar{\phi}_d, \bar{\theta}_d, \bar{\psi}_d]^T$ where $\bar{\boldsymbol{\eta}}_{2d}$ is the filtered output of $\boldsymbol{\eta}_{2d}$ given by

$$\epsilon_2\dot{\bar{\boldsymbol{\eta}}}_{2d} + \bar{\boldsymbol{\eta}}_{2d} = \boldsymbol{\eta}_{2d} \tag{16}$$

here, $\epsilon_2 > 0$ is an user-defined filtering time constant and let $\boldsymbol{y}_2 = \bar{\boldsymbol{\eta}}_{2d} - \boldsymbol{\eta}_{2d}$.

Combining with Euler angles dynamics (3), we design a sliding surface as follows:

$$\boldsymbol{s}_2(t) = \boldsymbol{e}_2(t) + \boldsymbol{k}_2 \int_0^t \boldsymbol{e}_2(\tau)d\tau \tag{17}$$

where $\boldsymbol{e}_2 = \boldsymbol{\eta}_2 - \bar{\boldsymbol{\eta}}_{2d}$, $\boldsymbol{k}_2 = \mathrm{diag}(k_{21}, k_{22}, k_{23}) > 0$.

In this context, a desired Euler angles control law for sub-system (3) can be designed as follows:

$$\boldsymbol{u}_2 = \boldsymbol{g}_2^{-1}(\boldsymbol{\eta}_2)[\dot{\bar{\boldsymbol{\eta}}}_{2d} - \boldsymbol{k}_2\boldsymbol{e}_2 - \boldsymbol{p}_2\boldsymbol{s}_2 + \boldsymbol{y}_2 - \hat{\boldsymbol{d}}_2] \tag{18}$$

with the FUO given by

$$\hat{\boldsymbol{d}}_2(\boldsymbol{\omega}_2 \mid \hat{\boldsymbol{\vartheta}}_2) = \hat{\boldsymbol{\vartheta}}_2^T \boldsymbol{\xi}_2(\boldsymbol{\omega}_2) \tag{19}$$

Choosing the parameter matrix update rule as

$$\dot{\hat{\boldsymbol{\vartheta}}}_2 = -r_{21}\hat{\boldsymbol{\vartheta}}_2 + r_{22}\boldsymbol{\xi}_2(\boldsymbol{\omega}_2)(\boldsymbol{s}_2 + \boldsymbol{\varepsilon}_2)^T \tag{20}$$

where $\boldsymbol{p}_2 = \mathrm{diag}(p_{21}, p_{22}, p_{23}) > 0$, $r_{21} > 0$ and $r_{22} > 0$ are user-defined positive definite parameters, $\boldsymbol{\omega}_2 = [\boldsymbol{\eta}_2^T, \dot{\boldsymbol{\eta}}_2^T]^T$ is the input vector of the fuzzy system, $\boldsymbol{\varepsilon}_2 = \boldsymbol{\eta}_2 - \boldsymbol{v}_2$ is the observation error vector with

$$\dot{\boldsymbol{v}}_2 = -r_{23}\boldsymbol{v}_2 + \boldsymbol{g}_2(\boldsymbol{\eta}_2)\boldsymbol{u}_2 + \hat{\boldsymbol{d}}_2(\boldsymbol{\omega}_2|\hat{\boldsymbol{\vartheta}}_2) + r_{23}\boldsymbol{\eta}_2 \tag{21}$$

where $r_{23} > 0$ is user-defined positive definite parameter.

3.3 Angular Velocity Controller

Let $\boldsymbol{\eta}_{3d} := [p_d, q_d, r_d]^T = \boldsymbol{u}_2$, together with angular velocity dynamics (6), we design a sliding surface as follows:

$$s_3(t) = e_3(t) + k_3 \int_0^t e_3(\tau)d\tau \qquad (22)$$

where $\boldsymbol{e}_3 = \boldsymbol{\eta}_3 - \bar{\boldsymbol{\eta}}_{3d}$, $\boldsymbol{k}_3 = \mathrm{diag}(k_{31}, k_{32}, k_{33}) > 0$ and $\bar{\boldsymbol{\eta}}_{3d} := [\bar{p}_d, \bar{q}_d, \bar{r}_d]^T$ is the filtered output of $\boldsymbol{\eta}_{3d}$ given by

$$\epsilon_3 \dot{\bar{\boldsymbol{\eta}}}_{3d} + \bar{\boldsymbol{\eta}}_{3d} = \boldsymbol{\eta}_{3d} \qquad (23)$$

here, $\epsilon_3 > 0$ is an user-defined filtering time constant and let $\boldsymbol{y}_3 = \bar{\boldsymbol{\eta}}_{3d} - \boldsymbol{\eta}_{3d}$.

Accordingly, an nominal angular velocity control law for sub-system (6) can be governed as follows:

$$\boldsymbol{u}_3 = \boldsymbol{g}_3^{-1}[\dot{\bar{\boldsymbol{\eta}}}_{3d} - \boldsymbol{f}_3(\boldsymbol{\eta}_3) - \boldsymbol{k}_3 \boldsymbol{e}_3 - \boldsymbol{p}_3 \boldsymbol{s}_3 + \boldsymbol{y}_3 - \hat{\boldsymbol{d}}_3] \qquad (24)$$

with the FUO given by

$$\hat{\boldsymbol{d}}_3(\boldsymbol{\omega}_3 \mid \hat{\boldsymbol{\vartheta}}_3) = \hat{\boldsymbol{\vartheta}}_3^T \boldsymbol{\xi}_3(\boldsymbol{\omega}_3) \qquad (25)$$

Choosing the parameter matrix update rule as

$$\dot{\hat{\boldsymbol{\vartheta}}}_3 = -r_{31}\hat{\boldsymbol{\vartheta}}_3 + r_{32}\boldsymbol{\xi}_3(\boldsymbol{\omega}_3)(\boldsymbol{s}_3 + \boldsymbol{\varepsilon}_3)^T \qquad (26)$$

where $\boldsymbol{p}_3 = \mathrm{diag}(p_{31}, p_{32}, p_{33}) > 0$, $r_{31} > 0$ and $r_{32} > 0$ are user-defined positive definite parameters, $\boldsymbol{\omega}_3 = [\boldsymbol{\eta}_3^T, \dot{\boldsymbol{\eta}}_3^T]^T$ is the input vector of the fuzzy system, $\boldsymbol{\varepsilon}_3 = \boldsymbol{\eta}_3 - \boldsymbol{v}_3$ is the observation error vector with

$$\dot{\boldsymbol{v}}_3 = -r_{33}\boldsymbol{v}_3 + \boldsymbol{f}_3(\boldsymbol{\eta}_3) + \boldsymbol{g}_3 \boldsymbol{u}_3 + \hat{\boldsymbol{d}}_3(\boldsymbol{\omega}_3|\hat{\boldsymbol{\vartheta}}_3) + r_{33}\boldsymbol{\eta}_3 \qquad (27)$$

where $r_{33} > 0$ is user-defined positive definite parameter.

Then, the final control law is

$$\boldsymbol{u} := \begin{bmatrix} \tau \\ \boldsymbol{u}_3 \end{bmatrix} = \begin{bmatrix} m\|\boldsymbol{u}_1\| \\ \boldsymbol{u}_3 \end{bmatrix} \qquad (28)$$

4 Stability Analysis

Theorem 1. *Consider an uncertain QUAV system (1)–(3)–(6), together with control scheme (11), (18), (24) with FUO given by (12), (19), (25), all system states and signals and all tracking errors are globally uniformly ultimately bounded.*

Proof. Together with system (14), (21) and (27), we have

$$\dot{\varepsilon}_i + r_{i3}\varepsilon_i = d_i - \hat{d}_i(\omega_i|\hat{\vartheta}_i) \tag{29}$$

where $i = 1, 2, 3$.

Define the optimal parameter as

$$\vartheta_i^* = \arg \min_{\hat{\vartheta}_i \in M_{\vartheta_i}} \left(\sup_{\omega_i \in M_{\omega_i}} \|d_i - \hat{d}_i\| \right) \tag{30}$$

where M_{ϑ_i} and M_{ω_i} are bounded sets.

Then we have

$$d_i = \hat{d}_i(\omega_i|\vartheta_i^*) + \zeta_i(\omega_i) \tag{31}$$

where $\zeta_i(\omega_i)$ is reconstruction error vector and $\|\zeta_i(\omega_i)\| < \bar{\zeta}_i$, $\bar{\zeta}_i > 0$. Let $\tilde{\vartheta}_i = \vartheta_i^* - \hat{\vartheta}_i$, together with system (29), we can obtain

$$\dot{\varepsilon}_i + r_{i3}\varepsilon_i = \tilde{\vartheta}_i^T \xi_i(\omega_i) + \zeta_i(\omega_i) \tag{32}$$

Combining with system (12), (19), (25), (31), the following equation holds

$$d_i - \hat{d}_i = \tilde{\vartheta}_i^T \xi_i(\omega_i) + \zeta_i(\omega_i) \quad i = 1, \dots, 3 \tag{33}$$

then together with system (8), (17), (22) and (33), we can obtain

$$\begin{cases} \dot{s}_{12} = -p_{12}s_{12} + \tilde{\vartheta}_1^T \xi_1(\omega_1) + \zeta_1(\omega_1) \\ \dot{s}_i = -p_i s_i + y_i + \tilde{\vartheta}_i^T \xi_i(\omega_i) + \zeta_i(\omega_i) \quad i = 2, 3 \end{cases} \tag{34}$$

Choosing the following Lyapunov function

$$V = \frac{1}{2} \left[\sum_{i=1}^{3} \left(y_i^T y_i + \varepsilon_i^T \varepsilon_i + \frac{\text{tr}(\tilde{\vartheta}_i^T \tilde{\vartheta}_i)}{r_{i2}} \right) + s_{11}^T s_{11} + s_{12}^T s_{12} + s_2^T s_2 + s_3^T s_3 \right] \tag{35}$$

Together with system (32) and (34), the time derivative of (35) can be given as

$$\dot{V} = \sum_{i=2}^{3} \left(-s_i^T p_i s_i + s_i^T y_i - r_{i3}\varepsilon_i^T \varepsilon_i + (s_i^T + \varepsilon_i^T)\zeta_i(\omega_i) + y_i^T \dot{y}_i \right) - s_{11}^T p_{11} s_{11}$$

$$+ s_{11}^T y_1 - s_{12}^T p_{12} s_{12} - r_{13}\varepsilon_1^T \varepsilon_1 + (s_{12}^T + \varepsilon_1^T)\zeta_1(\omega_1) + y_1^T \dot{y}_1$$

$$+ \underbrace{\sum_{i=2}^{3} \left((s_i^T + \varepsilon_i^T)\tilde{\vartheta}_i^T \xi_i(\omega_i) - \frac{\text{tr}(\tilde{\vartheta}_i^T \dot{\hat{\vartheta}}_i)}{r_{i2}} \right) + (s_{12}^T + \varepsilon_1^T)\tilde{\vartheta}_1^T \xi_1(\omega_1) - \frac{\text{tr}(\tilde{\vartheta}_1^T \dot{\hat{\vartheta}}_1)}{r_{12}}}_{M}$$

$$\tag{36}$$

together with systems (13), (20) and (26), we can obtain

$$M = \sum_{i=2}^{3}\left(\sum_{j=1}^{3}(\tilde{\vartheta}_{ij}^{T}((s_{ij}+\varepsilon_{ij})\xi_i(\omega_i) - \frac{\dot{\hat{\vartheta}}_{ij}}{r_{i2}}))\right) + \sum_{j=1}^{3}(\tilde{\vartheta}_{1j}^{T}((s_{12j}+\varepsilon_{1j})\xi_1(\omega_1) - \frac{\dot{\hat{\vartheta}}_{1j}}{r_{12}}))$$

$$= \sum_{i=1}^{3}\frac{r_{i1}}{r_{i2}}\mathrm{tr}(\tilde{\vartheta}_i^{T}\hat{\vartheta}_i) \tag{37}$$

where $s_{12} = [s_{121}, s_{122}, s_{123}]^T$, $s_i = [s_{i1}, s_{i2}, s_{i3}]^T$, $\varepsilon_i = [\varepsilon_{i1}, \varepsilon_{i2}, \varepsilon_{i3}]^T$, $\tilde{\vartheta}_i = [\tilde{\vartheta}_{i1}, \tilde{\vartheta}_{i2}, \tilde{\vartheta}_{i3}]$ and $\hat{\vartheta}_i = [\hat{\vartheta}_{i1}, \hat{\vartheta}_{i2}, \hat{\vartheta}_{i3}]$ with $i = 1, 2, 3$.

Together with systems (9)–(10) and Assumption 1, we can obtain

$$\left\| \dot{\boldsymbol{y}}_1 + \frac{\boldsymbol{y}_1}{\epsilon_1} \right\| \leq z_1(\dot{s}_{11}, \dddot{\chi}_{11d}, \dot{e}_{11}, \dot{e}_{12}) \tag{38}$$

where z_1 is continuous bounded function. Then, we have

$$\boldsymbol{y}_1^T\dot{\boldsymbol{y}}_1 \leq -\frac{\boldsymbol{y}_1^T\boldsymbol{y}_1}{\epsilon_1} + \frac{1}{2}\boldsymbol{y}_1^T\boldsymbol{y}_1 + \frac{1}{2}z_1^2 \tag{39}$$

Similarly, there exists continuous bounded function $z_2(\cdot)$ and $z_3(\cdot)$, such that

$$\boldsymbol{y}_i^T\dot{\boldsymbol{y}}_i \leq -\frac{\boldsymbol{y}_i^T\boldsymbol{y}_i}{\epsilon_i} + \frac{1}{2}\boldsymbol{y}_i^T\boldsymbol{y}_i + \frac{1}{2}z_i^2, \ i = 2, 3 \tag{40}$$

In addition, using the Young's inequality, we have

$$\sum_{i=1}^{3}\left(\frac{r_{i1}}{r_{i2}}\mathrm{tr}(\tilde{\vartheta}_i^{T}\hat{\vartheta}_i)\right) \leq \sum_{i=1}^{3}\left(\frac{r_{i1}}{2r_{i2}}\mathrm{tr}(\vartheta_i^{*T}\vartheta_i^*) - \frac{r_{i1}}{2r_{i2}}\mathrm{tr}(\tilde{\vartheta}_i^{T}\tilde{\vartheta}_i)\right) \tag{41}$$

Substituting system (37), (39), (40) and (41) into system (36), it is easy to obtain

$$\dot{V} \leq \sum_{i=2}^{3}\left(-s_i^{T}(\boldsymbol{p}_i - \boldsymbol{I})s_i - \left(\frac{1}{\epsilon_i} - 1\right)\boldsymbol{y}_i^T\boldsymbol{y}_i - \left(r_{i3} - \frac{1}{2}\right)\varepsilon_i^T\varepsilon_i - \frac{r_{i1}}{2r_{i2}}\mathrm{tr}(\tilde{\vartheta}_i^{T}\tilde{\vartheta}_i)\right)$$

$$- s_{11}^T\left(\boldsymbol{p}_{11} - \frac{\boldsymbol{I}}{2}\right)s_{11} - s_{12}^T\left(\boldsymbol{p}_{12} - \frac{\boldsymbol{I}}{2}\right)s_{12} - \left(r_{13} - \frac{1}{2}\right)\varepsilon_1^T\varepsilon_1 - \left(\frac{1}{\epsilon_1} - 1\right)\boldsymbol{y}_1^T\boldsymbol{y}_1$$

$$- \frac{r_{11}}{2r_{12}}\mathrm{tr}(\tilde{\vartheta}_1^{T}\tilde{\vartheta}_1) + \sum_{i=1}^{3}\left(\bar{z}_i^2 + \frac{r_{i1}}{2r_{i2}}\mathrm{tr}(\vartheta_i^{*T}\vartheta_i^*) + \frac{\bar{z}_i^2}{2}\right) \tag{42}$$

where $\bar{z}_i(t)$ is the upper bound value of $z_i(t)$.

Selecting the following design parameters

$$\boldsymbol{p}_j \geq \frac{2+\alpha}{2}\boldsymbol{I}, \frac{1}{\epsilon_i} \geq 1 + \frac{\alpha}{2}, \ \boldsymbol{p}_{11} \geq \frac{1+\alpha}{2}\boldsymbol{I}, \ \boldsymbol{p}_{12} \geq \frac{1+\alpha}{2}\boldsymbol{I}, \ r_{i3} \geq \frac{1+\alpha}{2}, \frac{r_{i1}}{r_{i2}} \geq \alpha$$

with $j = 2, 3$ and $i = 1, \ldots, 3$, we have

$$\dot{V} \leq -\alpha V + C \qquad (43)$$

with

$$C = \sum_{i=1}^{3} \left(\zeta_i^2 + \frac{r_{i1}}{2r_{i2}} \mathrm{tr}(\boldsymbol{\vartheta}_i^{*T} \boldsymbol{\vartheta}_i^*) + \frac{\bar{z}_i^2}{2} \right) \qquad (44)$$

Together with the system (35) and (42), the following inequality holds

$$0 \leq V(t) \leq V(0)e^{-\alpha t} + (1 - e^{-\alpha t})\frac{C}{\alpha} < \infty \qquad (45)$$

It is obvious that the function $V(t)$ is bounded and together with system (35), we can find that the trajectory error $\boldsymbol{e}_{11}$ and the other error signals are uniformly ultimately bounded.

5 Simulation Studies

In this section, the effectiveness of the proposed control scheme for the QUAV is evaluated. The lumped uncertainties and/or external disturbances are given by $\boldsymbol{d}_i(t) = 3[\sin t, \cos t, \sin t]^T + 0.1\boldsymbol{\eta}_i$, where $\boldsymbol{\eta}_1 = \boldsymbol{\eta}_{11} + \boldsymbol{\eta}_{12}$, $i = 1, 2, 3$.

The reference tracking trajectory is given as $[x_d, y_d, z_d, \psi_d] = [-2\sin t/2, 2\cos t/2, 2\sin t+3, \sin t]$ and the initial conditions of the QUAV are set as follows: $x(0) = 2$, $y(0) = -0.5$, $z(0) = 2$, $\phi(0) = 1$.

Figure 3 shows the tracking of three positions and the yaw, where FSMC denotes the proposed filtered sliding mode control scheme and SMC denotes the traditional sliding mode control scheme. From Fig. 3 we can find that both the proposed control scheme and the SMC scheme are able to robustly stabilize the

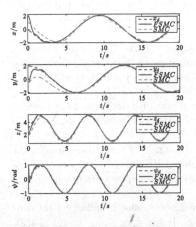

Fig. 3. States of x, y, z and ψ.

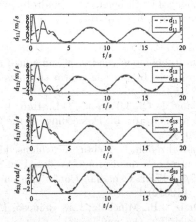

Fig. 4. Unknown nonlinearities.

QUAV and make it track the desired trajectory, while it is obvious that the proposed control scheme has faster response and higher accuracy. Figure 4 shows the estimate state of the FUO for the lumped unknown nonlinearities including system uncertainties and external disturbances on the trajectory, i.e., x, y, z and the yaw ψ, from which we can see, although the unknown lumped nonlinearities continuous change along with the time, FUO can estimate the unknown nonlinearities well. In summary, we can conclude that the proposed tracking control approach can achieve remarkable performance in terms of tracking accuracy and disturbance rejection.

6   Conclusion

In this paper, a filtered sliding mode control scheme based on FUO for trajectory tracking of a QUAV has been proposed. To be specific, three cascaded sub-controllers are designed by incorporating underactuation constraints. First-order filters are employed to reconstruct sliding mode control signals together with their first derivatives, and thereby decoupling the iterative design within the QUAV tracking control scheme. Furthermore, FUOs have been designed to estimate the lumped unknown nonlinearities. By the Lyapunov approach, we have proven that all system states and signals and tracking errors are globally uniformly ultimately bounded. Simulation studies have demonstrated the effectiveness and superiority of the proposed tracking control scheme.

References

1. Cabecinhas, D., Naldi, R., Silvestre, C., Cunha, R., Marconi, L.: Robust landing and sliding maneuver hybrid controller for a quadrotor vehicle. IEEE Trans. Control Syst. Technol. **24**(2), 400–412 (2016)
2. Driessens, S., Pounds, P.: The triangular quadrotor: a more efficient quadrotor configuration. IEEE Trans. Robot. Autom. **31**(6), 1517–1526 (2015)
3. Elfeky, M., Elshafei, M.: Quadrotor with tiltable totors for manned applications. In: Proceedings of the 11th International Systems, Signals and Devices Multi-Conference, pp. 1–5 (2004)
4. Mohd, B., Mohd, A., Husain, A.R., Danapalasingam, K.A.: Nonlinear control of an autonomous quadrotor unmanned aerial vehicle using backstepping controller optimized by particle swarm optimization. J. Eng. Sci. Technol. Rev. **8**(3), 39–45 (2015)
5. Jiang, J., Qi, J.T., Song, D.L., Han, J.D.: Control platform design and experiment of a quadrotor. In: Proceedings of the Chinese Control Conference, pp. 2974–2979 (2013)
6. Sadeghzadeh, I., Mehta, A., Chamseddine, A., Zhang, Y.M.: Active fault tolerant control of a quadrotor UAV based on gainscheduled PID control. In: Proceedings of the IEEE Canadian Conference on Electrical and Computer Engineering, pp. 1–4 (2012)
7. Ortiz, J.P., Minchala, L.I., Reinoso, M.J.: Nonlinear robust H-Infinity PID controller for the multivariable system quadrotor. IEEE Lat. Am. Trans. **14**(3), 1176–1183 (2016)

8. Rashad, R., Aboudonia, A., Ayman, E.B.: Backstepping trajectory tracking control of a quadrotor with disturbance rejection. In: Proceedings of the XXV International Information, Communication and Automation Technologies (ICAT) Conference, pp. 1–7 (2015)

9. Zheng, F., Gao, W.N.: Adaptive integral backstepping control of a Micro-Quadrotor. In: Proceedings of the International Intelligent Control and Information Conferencce, pp. 910–915 (2011)

10. Runcharoon, K., Srichatrapimuk, V.: Sliding Mode Control of quadrotor. In: Proceedings of the International Conference on Technological Advances in Electrical, Electronics and Computer Engineering (TAEECE), pp. 552–557 (2013)

11. Besnard, L., Shtesscl, Y.B., Landrum, B.: Quadrotor vehicle control via sliding mode control driven by sliding mode disturbance observer. J. Franklin Inst. **349**(2), 658–684 (2012)

12. Yacef, F., Bouhali, O., Hamerlain, M.: Adaptive fuzzy backstepping control for trajectory tracking of unmanned aerial quadrotor. In: Proceedings of the International Conference on Unmanned Aircraft Systems (ICUAS), pp. 920–927 (2014)

Local Policy Iteration Adaptive Dynamic Programming for Discrete-Time Nonlinear Systems

Qinglai Wei[1(✉)], Yancai Xu[1], Qiao Lin[1], Derong Liu[2], and Ruizhuo Song[2]

[1] The State Key Laboratory of Management and Control for Complex Systems, Institute of Automation, Chinese Academy of Sciences, University of Chinese Academy of Sciences, Beijing 100190, China
{qinglai.wei,yancai.xu,linqiao2014}@ia.ac.cn
[2] The School of Automation and Electrical Engineering, University of Science and Technology Beijing, Beijing 100083, China
{derong,ruizhuosong}@ustb.edu.cn

Abstract. Adaptive dynamic programming is a hot research topic nowadays. Therefore, the paper concerns a new local policy adaptive iterative dynamic programming (ADP) algorithm. Moreover, this algorithm is designed for the discrete-time nonlinear systems, which are used to solve problems concerning infinite horizon optimal control. The new local policy iteration ADP algorithm has the characteristics of updating the iterative control law and value function within one subset of the state space. Morevover, detailed iteration process of the local policy iteration is presented thereafter. The simulation example is listed to show the good performance of the newly developed algorithm.

Keywords: Nonlinear systems · Approximate dynamic programming · Local policy iteration · Optimal control · Discrete time

1 Introduction

Adaptive dynamic programming (ADP) is always a hot research area since proposed by Werbos [1]. ADP is a very useful and significant intelligent way to solve nonlinear system problems. With the aim of getting optimal control law, the corresponding iterative learning methods are applied to analyze the convergence and optimality characteristics of ADP [2–7].

It has to be admitted that the iterative control laws and the iterative value functions usually have to be updated in the whole state space [8–18], which are also as "global policy iteration algorithms". Moreover, the global policy iteration algorithms have the disadvantages of low efficiency during applications. Most of time, the algorithm has to pause to wait for the accomplishment of a search of the whole state area. Correspondingly, the computation efficiency goes down in the global policy iteration algorithm. The constraint has hindered the development of this research area. Therefore, useful policy iteration algorithms need to be proposed to increase computation efficiency.

© Springer International Publishing AG 2017
F. Cong et al. (Eds.): ISNN 2017, Part II, LNCS 10262, pp. 148–153, 2017.
DOI: 10.1007/978-3-319-59081-3_18

This paper has proposed a new "local policy iteration algorithm" concerning the discrete nonlinear systems. It proves its usage to iterative in a small area. The algorithm has the ability to update the iterative control laws and also the iterative value functions within the given area of the state space. Despite the fact of iterative control laws updating within a preset state space, the system still has the ability to keep stable under any kind of iterative control law. At the end, the simulation part shows the good performance of this newly developed method.

2 Problem Statement

We assume a deterministic discrete-time nonlinear system here

$$s_{k+1} = F(s_k, c_k), \ k = 0, 1, 2, \ldots, \tag{1}$$

where $s_k \in \mathbb{R}^n$ is the state vector. Besides, $c_k \in \mathbb{R}^m$ is the control vector. Assume s_0 as the initial state and $F(s_k, c_k)$ as the system function. Assume $\underline{c}_k = (c_k, c_{k+1}, \ldots)$ as an arbitrary sequence of controls. The performance index function can be defined as

$$J(s_0, \underline{c}_0) = \sum_{k=0}^{\infty} U(s_k, c_k), \tag{2}$$

for state s_0 under the control sequence $\underline{c}_0 = (c_0, c_1, \ldots)$. The utility function $U(x_k, c_k)$ is a positive definite function for s_k and c_k. It is noted that $\underline{c}_k$ changes from k to ∞.

We aim to find an optimal scheme. The scheme has the ability to minimize performance index function (2) while stabilizing system (1).

Assume the control sequence set as $\underline{\mathfrak{U}}_k = \{\underline{c}_k \colon \underline{c}_k = (c_k, c_{k+1}, \ldots), \forall c_{k+i} \in \mathbb{R}^m, i = 0, 1, 2, \ldots\}$.

Then, for an arbitrary control sequence $\underline{c}_k \in \underline{\mathfrak{U}}_k$, the optimal performance index function is

$$J^*(s_k) = \inf_{\underline{c}_k} \left\{ J(s_k, \underline{c}_k) \colon \underline{c}_k \in \underline{\mathfrak{U}}_k \right\}. \tag{3}$$

Based on Bellman principle of optimality, $J^*(s_k)$ meet the requirement of the discrete-time HJB formula

$$J^*(s_k) = \inf_{c_k} \left\{ U(s_k, c_k) + J^*(F(s_k, c_k)) \right\}. \tag{4}$$

Define the law of optimal control as

$$c^*(s_k) = \arg\inf_{c_k} \left\{ U(s_k, c_k) + J^*(F(s_k, c_k)) \right\}. \tag{5}$$

Therefore, the HJB Eq. (4) is

$$J^*(s_k) = U(s_k, c^*(s_k)) + J^*(F(s_k, c^*(s_k))). \tag{6}$$

Overall, there exists the curse of dimensionality. So it is very difficult to obtain the numerical results for the traditional dynamic programming algorithms. Considering this situation, we have proposed a new ADP algorithm thereafter.

3   Descriptions of This New Local Iterative ADP Algorithm

We have designed a new local iterative ADP algorithm. This section gives a detailed description of the algorithm. It is designed to have the ability to get the optimal control law for system (1) correspondingly. Assume $\{\Theta_s^i\}$ as the state sets, $\Theta_s^i \subseteq \Omega_s$, $\forall i$. The value iteration functions and the control laws of the newly developed algorithm have to be updated iteratively.

For all $s_k \in \Omega_s$, assume $v_0(s_k)$ as an admissible control law. Besides, assume $V_0(x_s)$ as the initial iterative value function for all $s_k \in \Omega_s$. The function satisfies the generalized HJB (GHJB) equation

$$\dot{V}_0(s_k) = U(s_k, v_0(s_k)) + V_0(s_{k+1}), \tag{7}$$

where $s_{k+1} = F(s_k, v_0(s_k))$. Then, for all $s_k \in \Theta_s^0$, the local iterative control law $v_1(s_k)$ is computed as

$$v_1(s_k) = \arg\min_{c_k} \{U(s_k, c_k) + V_0(s_{k+1})\} \tag{8}$$

and let $v_1(s_k) = v_0(s_k)$, for all $s_k \in \Omega_s \backslash \Theta_s^0$.

For all $s_k \in \Omega_s$, assume $V_1(s_k)$ as the iterative value function. Therefore, $V_1(s_k)$ satisfies the GHJB equation

$$V_1(s_k) = U(s_k, v_1(s_k)) + V_1(F(s_k, v_1(s_k))). \tag{9}$$

For $i = 1, 2, \ldots$, assume $V_i(s_k)$ as the iterative value function. So $V_i(s_k)$ can satisfy the following GHJB equation

$$V_i(s_k) = U(s_k, v_i(s_k)) + V_i(F(s_k, v_i(s_k))). \tag{10}$$

For all $s_k \in \Theta_x^i$, the iterative control law $v_{i+1}(s_k)$ should be computed as

$$v_{i+1}(s_k) = \arg\min_{c_k} \{U(s_k, c_k) + V_i(s_{k+1})\}$$
$$= \arg\min_{c_k} \{U(s_k, c_k) + V_i(F(s_k, c_k))\}, \tag{11}$$

and for all $s_k \in \Omega_s \backslash \Theta_s^i$, let $v_{i+1}(s_k) = v_i(s_k)$.

The local policy iteration algorithm will be updated within the preset subset of state space according to Eqs. (7) and (11). The given subset is part of whole state space. Therefore, during iterations, once local data of state space is got, the newly developed algorithm can be performed immediately. The advantage is that the algorithm can save lots of time while competing all the data of the whole space in traditional algorithms. Therefore, the computation efficiency can be improved greatly and save a lot of trouble. Besides, if the preset subset of state space is enlarged to all, local policy iteration algorithms equal to the global policy iteration ones.

4   Simulation Examples

First, we have chosen a discretized nonaffine nonlinear system as follows

$$s_{1(k+1)} = (1 - \Delta T)s_{1k} + \Delta T s_{2k}c_k,$$
$$s_{2(k+1)} = (1 - \Delta T)x_{2k} + \Delta T(1 + s_{1k}^2)c_k + \Delta T c_k^3. \tag{12}$$

We choose the utility function as $Q = I_1$ and $R = I_2$. Thereafter, We choose the state space as Ω_s. While I_1 and I_2, are denoted as the identity matrices with suitable dimensions. Let the initial state be $s_0 = [1, -1]^\mathsf{T}$. Based on Algorithm 1 in [16].

The iterative value functions and iterative control laws should be updated accordingly. After 30 iterations, the algorithm has reached corresponding computing precision of $\varepsilon = 0.001$. Figure 1(a) shows that the iterative value function

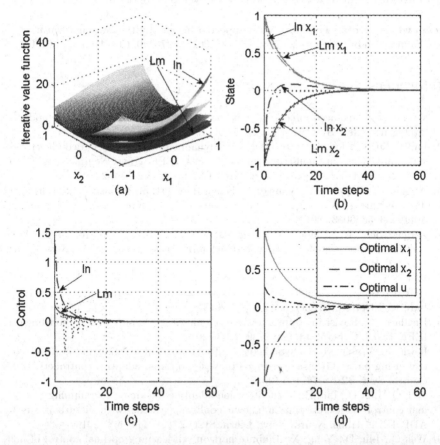

Fig. 1. Simulation results of the new local policy iteration algorithm. (a) Corresponding iterative value function. (b) Corresponding state trajectories. (c) Corresponding control trajectories. (d) Corresponding optimal state and control trajectories.

is monotonically nonincreasing. More importantly, the value function converges to the optimum. Figure 1(b) illustrates the trajectories of simulation states while Fig. 1(c) shows the simulation functions. In Fig. 1(d), we have shown the optimal trajectories of control and also states correspondingly.

5   Conclusion

We proposed a new local policy iteration ADP algorithm in this paper. The algorithm has the ability to greatly improve the computation efficiency of traditional ADP algorithm concerning discrete time nonlinear systems. Therefore, it can reduce computation time greatly which contrast to traditional global policy iteration algorithms. The characteristic concerning this newly developed algorithm is that the iteration control laws and iterative iteration control laws are updated within a preset area of the state space. Besides, the simulation results have proven its effectiveness of the newly developed algorithm.

Acknowledgments. This work was supported in part by the National Natural Science Foundation of China under Grants 61233001, 61273140, 61374105, and 61304079.

References

1. Werbos, P.: Advanced forecasting methods for global crisis warning and models of intelligence. Gen, Syst. Yearb. **22**, 25–38 (1977)
2. Fu, Y., Fu, J., Chai, T.: Robust adaptive dynamic programming of two-player zero-sum games for continuous-time linear systems. IEEE Trans. Neural Netw. Learn. Syst. **26**, 3314–3319 (2015). doi:10.1109/TNNLS.2015.2461452
3. Abouheaf, M., Lewis, F., Vamvoudakis, K., Haesaert, S., Babuska, R.: Multi-agent discrete-time graphical games and reinforcement learning solutions. Automatica **50**(12), 3038–3053 (2014)
4. Zargarzadeh, H., Dierks, T., Jagannathan, S.: Optimal control of nonlinear continuous-time systems in strict-feedback form. IEEE Trans. Neural Netw. Learn. Syst. **26**(10), 2535–2549 (2015)
5. Wei, Q., Liu, D.: Data-driven neuro-optimal temperature control of water gas shift reaction using stable iterative adaptive dynamic programming. IEEE Trans. Industr. Electron. **61**(11), 6399–6408 (2014)
6. Heydari, A.: Revisiting approximate dynamic programming and its convergence. IEEE Trans. Cybern. **44**(12), 2733–2743 (2014)
7. Lewis, F., Vrabie, D., Vamvoudakis, K.: Reinforcement learning and feedback control: using natural decision methods to design optimal adaptive controllers. IEEE Control Syst. **32**(6), 76–105 (2012)
8. Wei, Q., Liu, D., Lin, H.: Value iteration adaptive dynamic programming for optimal control of discrete-time unknown nonlinear systems with disturbance using ADP. IEEE Trans. Neural Netw. Learn. Syst. **27**(2), 444–458 (2016)
9. Wei, Q., Liu, D., Yang, X.: Inifinite horizon self-learning optimal control of nonaffine discrete-time nonlinear systems. IEEE Trans. Neural Netw. Learn. Syst. **26**(4), 879–886 (2015)

10. Wei, Q., Song, R., Yan, P.: Data-driven zero-sum neuro-optimal control for a class of continuous-time unknow nonlinear systems with disturbance using ADP. IEEE Trans. Neural Netw. Learn. Syst. **27**(2), 444–458 (2016)
11. Wei, Q., Wang, F., Liu, D., Yang, X.: Finite-approximation-error based discrete-time iterative adaptive dynamic programming. IEEE Trans. Cybern. **44**(12), 2820–2833 (2014)
12. Wei, Q., Liu, D., Shi, G., Liu, Y.: Optimal multi-battery coordination control for home energy management systems via distributed iterative adaptive dynamic programming. IEEE Trans. Ind. Electron. **42**(7), 4203–4214 (2015)
13. Wei, Q., Liu, D., Shi, G.: A novel dual iterative Q-learning method for optimal battery management in smart residential environments. IEEE Trans. Ind. Electron. **62**(4), 2509–2518 (2015)
14. Wei, Q., Liu, D.: A novel iterative θ-adaptive dynamic programming for discrete-time nonlinear systems. IEEE Trans. Autom. Sci. Eng. **11**(4), 1176–1190 (2014)
15. Wei, Q., Liu, D.: Adaptive dynamic programming for optimal tracking control of unknown nonlinear systems with application to coal gasification. IEEE Trans. Autom. Sci. Eng. **11**(4), 1020–1036 (2014)
16. Liu, D., Wei, Q.: Policy iteration adaptive dynamic programming algorithm for discrete-time nonlinear systems. IEEE Trans. Neural Netw. Learn. Syst. **25**(3), 621–634 (2014)
17. Xu, X., Hou, Z., Lian, C., He, H.: Online learning control using adaptive critic designs with sparse kernel machines. IEEE Trans. Neural Netw. Learn. Syst. **24**(5), 762–775 (2013)
18. Liu, D., Yang, X., Wang, D., Wei, Q.: Reinforcement-learning-based robust controller design for continuous-time uncertain nonlinear systems subject to input constraints. IEEE Trans. Cybern. **45**(7), 1372–1385 (2015)

A Method Using the Lempel-Ziv Complexity to Detect Ventricular Tachycardia and Fibrillation

Deling Xia[1](✉), Yuetian Li[1], Qingfang Meng[2], and Jie He[3]

[1] The Information Office, Liaocheng Vocational and Technical College,
Liaocheng 252000, China
xiadeling@yeah.net
[2] The School of Information Science and Engineering,
University of Jinan, Jinan 250022, China
[3] The Basic Department, Liaocheng Vocational and Technical College,
Liaocheng 252000, China

Abstract. This paper use the Lempel-Ziv complexity to automatically detect ventricular fibrillation (VF) and ventricular tachycardia (VT) based on Wavelet transform (WT) and empirical mode decomposition (EMD). We respectively select WT and EMD to decompose original signals into different sub-bands. Electrocardiogram (ECG) signals were first decomposed into five sub-bands based on Wavelet transform and EMD. Then the complexity of each sub-band was used as a feature to detect VF and VT. A public dataset was utilized. Experimental results show the new method can distinguish VT from VF with the accuracy up to 99.50%.

Keywords: Ventricular fibrillation · Ventricular tachycardia · The L-Z complexity · Wavelet transform · EMD

1 Introduction

The incidence of cardiovascular disease (CVD) is increasing. In the study of cardiovascular disease, the function of cardiac functional decline and lesions hold a large proportion of heart disease. SCD means the end of life if not treatment timely; because of that, many national medical departments of health and biomedical research centers are conducting the research. The researches and experiments show the vast majority of cases of SCD are due largely to ventricular fibrillation (VF) and ventricular tachycardia (VT). Therefore, a kind of efficient automatic detection of VT and VF algorithm is very pressing for the researchers.

Various VF and VT detection methods have been proposed for Electrocardiogram (ECG) arrhythmia recognition in the literature, like the combination of ECG parameters in different domains and nonlinear analysis method [1–4, 8, 19] and so on. Nonlinear analysis methods haven put the nonlinear of VT and VF into consideration. A lots of nonlinear analysis methods for classification of VT and VF have been proposed in recent decade, such as Lyapunov exponent method [8], the Lempel-Ziv complexity

© Springer International Publishing AG 2017
F. Cong et al. (Eds.): ISNN 2017, Part II, LNCS 10262, pp. 154–160, 2017.
DOI: 10.1007/978-3-319-59081-3_19

algorithm [7, 9–11], correlation dimension method [6], approximate entropy and modified approximate entropy method [12], sample entropy and modified sample entropy method [14, 18], empirical mode decomposition method [15].

In 1976, Lempel and Ziv suggested the Lempel-Ziv complexity algorithm. After that, the Lempel-Ziv complexity algorithm has been widely applied to study brain function [7, 10], schizophrenia [9] and mechanomyography [11]. However, this method appeared with some limitations [7].

Decomposition of a time series into a set of components is particularly useful for obtaining the most essential information considering the evolution behaviors of a dynamic system. Over the past few decades, to solve several problems in nonlinear dynamics, many time-series decomposition methods have been reported and success-fully applied, including wavelet transform, empirical mode decomposition (EMD) and so on.

The wavelet transform (WT) is a time-scale representative [14, 16]. The signal of interest is stationary in the Fourier transform. But for the WT, all these disadvantages are overcome. EMD is an intuitive and adaptive signal-dependent decomposition, which was first proposed in 1998 [10, 16, 20].

The complete paper encompasses three principal stages: First we introduce the method of the Lempel-Ziv complexity, wavelet transform, EMD and our proposed method to discriminate VF from VT. Then we report the results by using the Lempel-Ziv complexity and our proposed methods. Finally, discuss the advantages of the new method and other methods, and then give the conclusions. The proposed method suits short data length recording in physiological signals.

2 Data and Method

2.1 Data Selection

The data were selected from MIT-BIH Database and CU Database. A total of 35 single-channel records are contained in CU Database, the length of each records is about 8 min. In the MIT-BIH Database, it contains 35 records, and each record is about 35 min and is saved to three files. In addition, both of the databases are the same in the data format. The sampling rate is 250 Hz and the resolution is 12 bit. 100 VF episodes and 100 VT episodes are extracted. The length of the data is four-second times.

2.2 New Method Proposed in This Paper

In this paper, we used two time-series decomposition: the wavelet transforms (WT) and EMD. In this paper, the WT is performed according to Mallat's, and the method is restricted to binary analysis [17]. The LZ complexity analysis is based on a coarse-graining of the measurements [7, 13, 16].

We first decompose a time series into a set of components containing the essential information reflecting the evolution behaviors of a dynamic system. In this process, we selected two time-series decomposition methods: wavelet transform and EMD. After time-series decomposition, we get five sub-band signals which reflect different frequency band's information. For instance, the first sub-band is the one connected with the locally highest frequency, while the fifth sub-band contains the lowest frequency. Each sub-band contains the information of original signals, so the complexity of each sub-band was used as a feature to detect VF and VT.

The proposed algorithm is defined as follows:

(1) Given a discrete-time signal $X, X = (X_1, X_2 \ldots X_N)$, where N represents the samples of X. Define $X_M = \{X_{(N-1)*m+1}, X_{(N-1)*m+2}, \cdots X_{(N-1)*m+m}\} (M = 1 \sim N)$, m represents the length of each sample.

(2) According to the following formula, make these samples be normalized.

$$X'_M = (X_M - \bar{x})/\sigma \tag{1}$$

Where $\bar{x}$ and σ represent the mean and the standard deviation of the sample X_M.

(3) Repeat step (2) until all samples is normalized. Calculate the LZ complexity of VT and VF using those samples first, and then get the accuracy for detection of VT and VF according to the assessment formula of the algorithm performance. In this paper, we used 0–1 sequence conversion method. The average value is estimated as a threshold Td. By comparing with the threshold, the signal data are converted into a 0–1 sequence $S = u(1), u(2) \ldots, u(r)$, and $x(i) < Td, u(i) = 0$; else $u(i) = 1$.

(4) Get sub-bands of ECG signals. Mallat's algorithm [5] is applied to decompose and reconstruct ECG signals into five wavelets and five Scales respectively. The first five Scales were finally used in this paper. According to the same principle, we used EMD technique to decompose ECG signals into the first five intrinsic mode functions (IMFs), compared to the wavelet transforms. In this paper, we decomposed original signals into five sub-bands signals.

(5) Calculate the complexity of the first Scales of VT and VF, and directly based the feature to classify VT and VF. Repeat this step until other scales are calculated. Similarly, calculate the complexity of the IMFs. After decomposition, the coefficients of variation and fluctuation indexes of VT and VF are different, so we can distinguish VF from VT.

3 Classifications on VF and VT

The LZ complexity analysis is performed first. The results are presented in Fig. 1. Because the clinical signals are very complex, so the results are not good. According to the two time-series decomposition methods, Electrocardiogram (ECG) signals were decomposed into five sub-bands and the complexity of each sub-band is computed. The performances are shown in Tables 1, 2, 3. By using the LZ complexity-Mallat algorithm methods, the sensitivity and specificity can reach 99.00% and 100%, respectively.

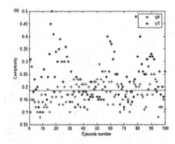

Fig. 1. The complexity is calculated for different VT and VF episodes.

From Tables 1, 2, 3, it can be found that the accuracy of each sub-band for distinguish of VT from VF is higher than only using the LZ complexity. According to the relevant literature, the frequency of VT is 150–200 bmp, while the frequency of VF is the 200–500 bmp. After time-series decomposition, ECG signal is decomposed into five sub-bands. The same scale of VT and VF are used as a feature in order to discriminate between VF and VT. Each sub-band contains the important information of the original signals, so the results are better than only using the LZ complexity. Furthermore, from

Table 1. The detection of VF and VT using the Lempel-Ziv complexity

Method	Component	VF		VT		ACC(%)
		Sensitivity (%)	Specificity (%)	Sensitivity (%)	Specificity (%)	
L-Z complexity		62.09	64.11	64.11	62.09	63.10

Table 2. The Lempel-Ziv complexity and the proposed method for CU and MIT-BIH data(—)

Method	Component	VF		VT		ACC(%)
		Sensitivity (%)	Specificity (%)	Sensitivity (%)	Specificity (%)	
L-Z complexity and Wavelet transform	Scale1	86.08	100.00	100.00	86.08	**93.04**
	Scale2	73.23	89.14	89.14	73.23	81.19
	Scale3	83.17	96.08	96.08	83.17	89.63
	Scale4	63.12	64.07	64.07	63.12	63.60
	Scale5	69.25	84.23	84.23	69.25	76.74
	Scale1-2	92.87	97.65	97.65	92.87	**95.26**
	Scale1-3	100.00	99.00	99.00	100.00	**99.50**
	Scale1-4	98.55	96.00	96.00	98.55	**97.28**
	Scale2-3	90.13	96.21	96.21	90.13	93.17
	Scale2-4	93.15	90.21	90.21	93.15	91.68
	Scale3-4	90.16	92.18	92.18	90.16	91.17

Table 3. The Lempel-Ziv complexity and the proposed method for CU and MIT-BIH data(二)

Method	Component	VF		VT		ACC (%)
		Sensitivity (%)	Specificity (%)	Sensitivity (%)	Specificity (%)	
L-Z complexity and EMD	IMF1	87.12	100.00	100	87.12	**93.56**
	IMF2	82.04	68.16	68.16	82.04	75.10
	IMF3	73.67	70.18	70.18	73.67	71.93
	IMF4	88.13	90.28	90.28	88.13	89.21
	IMF5	98.15	96.01	96.02	98.35	**97.08**
	IMF1-2	84.97	86.85	86.85	84.97	85.91
	IMF1-3	85.75	85.5	85.5	85.75	85.63
	IMF1-4	92.06	94.17	94.17	92.06	93.12
	IMF2-3	94.00	75.00	75.00	94.00	84.50
	IMF2-4	96.00	90.00	90.00	96.00	93.00
	IMF3-4	97.00	84.00	84.00	97.00	90.50

the Tables 2 and 3, we can conclude that using the middle and high frequency of VT and VF to detect VF and VT can get good results. This is fit with the theory that the frequency of VT and VF are in the middle and high frequency band. The proposed method has greatly improved the effect of the classification. In the clinic application, we can use the middle and high frequency of VT and VF to distinguish VF from VT.

For a purpose of comparison, we respectively use the approximate entropy and the sample entropy to detect VF and VT. In the experiment, the number of the sample and the length of the data is the same; all samples are normalized before being used. The results can be found in the Table 4. From the Table 4, the accuracy of the proposed method is far higher than that using the LZ complexity, approximate entropy and sample entropy. This also fully proves that the new method was suit for the classification of VF and VT.

Table 4. The comparison of different methods for CU and MIT-BIH data

Method	TH	VF		VT		ACC (%)
		Sensitivity (%)	Specificity (%)	Sensitivity (%)	Specificity (%)	
L-Z complexity	0.1823	62.09	64.11	64.11	62.09	63.10
Approximate entropy	0.2142	96.00	91.00	91.00	96.00	93.50
Sample entropy	0.1823	91.00	90.00	90.00	91.00	90.50
L-Z complexity and EMD	0.1356	98.15	96.01	96.02	98.35	**97.08**
L-Z complexity and Wavelet transform	0.8685	100.00	99.00	99.00	100.00	**99.50**

4 Discussion and Conclusions

In this paper, a new method that is being used the Lempel-Ziv complexity and the time-series decomposition method for detection of VF and VT is proposed. We selected two time-series decomposition methods: wavelet transform and EMD. The selection of the length of VT and VF is four seconds (1000 points) in the analysis of our proposed method, which is based on the previous work [15]. The accuracy of each sub-band in distinguishing of VT from VF is much higher than only using the LZ complexity. In the clinic application, we can use the middle and high frequency of VT and VF to distinguish VT from VF. This method has greatly improved the classification rate. For a validation using this method, other dataset(s) would be required.

Acknowledgements. This work was supported by the National Natural Science Foundation of China (Grant No. 61201428, 61302090), the Natural Science Foundation of Shandong Province, China (Grant No. ZR2010FQ020, ZR2013FL002).

References

1. Small, M., Simonotto, J., et al.: Uncovering non-linear structure in human ECG recordings. Chaos. Solitons. Fract. **13**, 1755–1762 (2002)
2. Chen, S., et al.: Ventricular fibrillation detection by a regression test on the autocorrelation function. Med. Biol. Eng. Comput. **25**(3), 241–249 (1987)
3. Thakor, N.V., et al.: Ventricular tachycardia and fibrillation detection by a sequential hypothesis testing algorithm. IEEE Trans. Biomed. Eng. **37**, 837–843 (1990)
4. Alonso-Atienza, F., Rojo-álvarez, J.L., Rosado-Munoz, A., Vinagre, J.J., García-Alberola, A., Camps-Valls, G.: Feature selection using support vector machines and bootstrap methods for ventricular fibrillation detection. Expert Syst. Appl. **39**, 1956–1967 (2012)
5. Li, S., Zhou, S., et al.: Feature extraction and recognition of ictal EEG using EMD and SVM. Comput. Biol. Med. **43**, 807–816 (2013)
6. Richman, J.S., Moorman, J.R.: Physiological time-series analysis using approximate and sample entropy. Am. J. Phys. – Heart Circ. Physiol. **278**, H2039–H2049 (2000)
7. Gómeza, C., Hornero, R., Abásolo, D., Fernández, A., López, M.: Complexity analysis of the magnetoencephalogram background activity in Alzheimer's disease patients. Med. Eng. Phys. **28**, 851–859 (2006)
8. Owis, M.I., et al.: Study of features based on nonlinear dynamical modeling in ECG arrhythmia detection and classification. IEEE Trans. Biomed. Eng. **49**, 733–736 (2002)
9. Fernández, A., Gómez, C., Hornero, R., López-Ibor, J.J.: Complexity and schizophrenia. Prog. Neuro-Psychopharmacol. **45**, 267–276 (2013)
10. Abáasolo, D., Hornero, R., Gómez, C., García, M., López, M.: Analysis of EEG background activity in Alzheimer's disease patients with Lempel-Ziv complexity and central tendency measure. Med. Eng. Phys. **28**, 315–322 (2006)
11. Sarlabous, L., Torres, A., Fiz, J.A., Morera, J., Jané, R.: Index for estimation of muscle force from mechanomyography based on the Lempel-Ziv algorithm. J. Electromyogr. Kinesiol. **23**, 548–557 (2013)
12. Small, M., et al.: Deterministic nonlinearity in ventricular fibrillation. Chaos **10**, 268–277 (2000)

13. Aboy, M., Hornero, R., Abásolo, D., Álvarez, D.: Interpretation of the Lempel-Ziv complexity measure in the context of biomedical signal analysis. IEEE Trans. Biomed. Eng. **53**(11), 2282–2288 (2006)
14. Khan, Y.U., Gotman, J.: Wavelet based automatic seizure detection in intracerebral electroencephalogram. Clin. Neurophysiol. **114**(5), 898–908 (2003)
15. Xie, H., Gao, Z., et al.: Classification of ventricular tachycardia and fibrillation using fuzzy similarity-based approximate entropy. Expert Syst. Appl. **38**, 3973–3981 (2011)
16. Pachori, B.R., et al.: Analysis of normal and epileptic seizure EEG signals using empirical mode decomposition. Comput. Methods Programs Biomed. **104**, 373–381 (2011)
17. Burrus, C.S., Gopinath, R.A., et al.: Introduction to Wavelets and Wavelet Transforms: A Primer. Prentice-Hall, Upper Saddle River (1998)
18. Kong, D., Xie, H.: Use of modified sample entropy measurement to classify ventricular tachycardia and fibrillation. Measurement **44**, 653–662 (2011)
19. Wang, J., Chiang, W., et al.: ECG arrhythmia classification using a probabilistic neural network with a feature reduction method. Neurocomputing **116**, 38–45 (2013)
20. Huang, N.E., et al.: The empirical mode decomposition and the Hilbert spectrum for nonlinear and non-stationary time series analysis. Proc. R. Soc. Lond. **454**, 903–995 (1998)

Finite-Time Synchronization of Uncertain Complex Networks with Nonidentical Nodes Based on a Special Unilateral Coupling Control

Meng Zhang and Min Han[(✉)]

Faculty of Electronic Information and Electrical Engineering,
Dalian University of Technology, Dalian 116023, China
minhan@dlut.edu.cn

Abstract. This note investigates finite-time synchronization (FTS) between two uncertain complex networks based on a special unilateral coupling control method. The two networks contain nonidentical nodes, time-varying coupling delayed, unknown parameters and uncertain topological structure. According to the finite-time stability theory and LaSalle's principle, an effective unilateral coupling control scheme and corresponding adaptive laws are proposed to guarantee the FTS. Simultaneously, the unknown parameters are estimated successfully and the weight values of uncertain topology can automatically adaptive to the suitable value. Finally, simulation results are shown the correctness of the theoretical method.

Keywords: Complex networks · Finite-time synchronization (FTS) · Uncertain · Unilateral coupling

1 Introduction

Network synchronization [1] has got much more attention, because it has theoretically and practically value in many fields including sociology, telecommunications, and engineering [2]. Although several kinds of synchronization among complex networks have been studied, most of existing studies concern with basic synchronization types such as global exponential or asymptotic synchronization [3, 4] which synchronization time is infinity. However in reality, people expect a faster convergence rate and to calculation convergence time [5]. Compared to basic types, the FTS can not only compute the maximum of synchronization time but also have a stronger robustness.

The complexity of the complex network is reflects in the complex topology, the diversity and dynamics of the node, other factors interference [6]. Moreover, the network synchronization is unavoidably affected by all these factors. However, in most of the existing researches, the papers [7, 8] suppose all of network nodes have identical dynamics. And some papers [9–11] consider the networks which contain different nodes can achieve basic asymptotic synchronization that the synchronization time is infinity. In addition, most of the topology and node parameters of real-world networks can't be exactly known in advance. And other interference factors such as time delay are also a common phenomenon in practical engineering. People want to find some

© Springer International Publishing AG 2017
F. Cong et al. (Eds.): ISNN 2017, Part II, LNCS 10262, pp. 161–168, 2017.
DOI: 10.1007/978-3-319-59081-3_20

effective methods to overcome the influence of these factors and implementing the network synchronization simultaneously. Some new advances have been reported. In [12, 13], node parameters are unknown, but the topological structure is assumed certain. In [14, 15], the synchronous control is realized and the unknown topological structure and parameters are identified successfully. While, both the two articles are of general asymptotic synchronization and the node dynamic systems are identical.

As is well known, many existing researches are using external input controller for synchronization, and a few other researches take the bidirectional coupling control method [7, 17, 18]. In [17, 18], the two networks are of asymptotic synchronization with identical node and known topology. In [7], one of the method is adopted bidirectional coupling control approach, comparative the [18], to realize the FTS. However, it assumed node dynamical identical as well. Compared with the existing results, in this paper, we investigate the FTS based on a new unilateral coupling control scheme and consider the two networks are of nonidentical nodes, unknown parameters, uncertain topology and time-varying delayed. The proposed method reduces the amount of the coupling and the problem into account is more comprehensive and practical. Finally, simulation results are shown that the method can effective implement the FTS control.

2 Model Description

We consider two uncertain complex networks and take them as drive-response networks respectively.

$$
\begin{cases}
\dot{z}_i(t) = Az_i(t) + F_i(z_i(t)) + S_i(z_i(t))\vartheta_i + \sum_{j=1}^{N} \hat{c}_{ij}z_j(t - l(t)), i = 1, 2 \ldots, N' \\
\dot{z}_i(t) = Bz_i(t) + G_i(z_i(t)) + T_i(z_i(t))\beta_i + \sum_{j=1}^{N} \hat{c}_{ij}z_j(t - l(t)), i = N' + 1, N' + 2, \ldots, N
\end{cases}
$$

$$(1)$$

$$
\begin{cases}
\dot{w}_i(t) = Cw_i(t) + H_i(w_i(t)) + M_i(w_i(t))\xi_i + \sum_{j=1}^{N} \hat{c}_{ij}w_j(t - l(t)) + r_i w_i(t) - z_i(t)), \\
\quad i = 1, 2, \ldots, N^* \\
\dot{w}_i(t) = Dw_i(t) + P_i(w_i(t)) + L_i(w_i(t))v_i + \sum_{j=1}^{N} \hat{c}_{ij}w_j(t - l(t)) + r_i w_i(t) - z_i(t)), \\
\quad i = N^* + 1, N^* + 2, \ldots, N
\end{cases}
$$

$$(2)$$

where $z_i(t) \in R^n$ and $w_i(t) \in R^n$ are the state vectors of the i-th node of drive-response networks respectively. $\iota(t) \geq 0$ is time-varying coupling delay. $F, G, H, P : R^n \to R^n$ are continuous vector functions. Note that A, B, C, D are constant matrixes. $S : R^n \to R^{n \times m_1}$, $T : R^n \to R^{n \times m_2}$, $M : R^n \to R^{n \times m_3}$, $L : R^n \to R^{n \times m_4}$ are continuous function matrixes. $\vartheta_i \in R^{m_1}$, $\beta_i \in R^{m_2}$, $\xi \in R^{m_3}$, $v_i \in R^{m_4}$ are the unknown parameter vectors. $\hat{\vartheta}_i \in R^{m_1}$, $\hat{\beta}_i \in R^{m_2}$, $\hat{\xi} \in R^{m_3}$ and $\hat{v}_i \in R^{m_4}$ are the estimator of unknown parameters respectively. $r_i(i = 1, 2, \ldots, N)$ is unilateral external coupling. We assume $N' > N^*$ and both the external and internal nodes are different. $\hat{C} = [\hat{c}_{ij}]_{N \times N}$ is uncertain coupling configuration matrix, which is defined as: $\hat{c}_{ij} > 0$ it there is a connection from node j to node $i(i \neq j)$ otherwise, $\hat{c}_{ij} = 0$.

Define error system as $e_i(t) = w_i(t) - z_i(t)(i = 1, 2, \ldots, N)$ and the corresponding parameters error as $\tilde{\vartheta}_i = \hat{\vartheta}_i - \vartheta_i, \tilde{\beta}_i = \hat{\beta}_i - \beta_i, \tilde{\xi}_i = \hat{\xi}_i - \xi_i, \tilde{v}_i = \hat{v}_i - v_i$ The networks (1) and (2) can achieve FTS if $e_i(t) \to 0$ as $t \to t_1$.

3 Main Results

By using unilateral coupling control method, the FTS of networks (1) and (2) is studied and the uncertain topology and parameters are estimated.

The unilateral external coupling r_i and updating laws of uncertain topological structure and parameters are shown below:

$$\begin{cases} \dot{r}_i = \lambda_i[r_ie_i(t)/r_i^2(-(C-A)z_i(t) - H_i(w_i(t)) + F_i(z_i(t)) - M_i(w_i(t))\hat{\xi}_i \\ \quad + S_i(z_i(t))\hat{\vartheta}_i) + \Lambda], \ i = 1, 2, \ldots, N^* \\ \dot{r}_i = \lambda_i[r_ie_i(t)/r_i^2(-(D-A)z_i(t) - P_i(w_i(t)) + F_i(z_i(t)) - L_i(w_i(t))\hat{v}_i \\ \quad + S_i(z_i(t))\hat{\vartheta}_i) + \Lambda], \ i = N^*+1, N^*+2, \ldots, N^* \\ \dot{r}_i = \lambda_i[r_ie_i(t)/r_i^2(-(D-B)z_i(t) - P_i(w_i(t)) + G_i(z_i(t)) - L_i(w_i(t))\hat{v}_i \\ \quad + T_i(z_i(t))\hat{\beta}_i) + \Lambda], \ i = N'+1, N'+2, \ldots, N \end{cases} \quad (3)$$

$$\dot{\hat{c}}_{ij} = -\delta_{ij}(e_i^T(t)e_j(t - t(t)) - k|\hat{c}_{ij}|/\hat{c}_{ij}\sqrt{\delta_{ij}}) \quad i = 1, 2, \ldots, N \quad (4)$$

$$\begin{cases} \dot{\hat{\vartheta}}_i = -\mu_{1i}(S_i^T(z_i(t))e_i(t) + ksign(\tilde{\vartheta}_i)/\sqrt{\mu_{1i}}) & i = 1, 2, \ldots, N' \\ \dot{\hat{\beta}}_i = \mu_{2i}(T_i^T(z_i(t))e_i(t) + ksign(\tilde{\beta}_i)/\sqrt{\mu_{2i}}) & i = N'+1, N'+2, \ldots, N \\ \dot{\hat{\xi}}_i = \mu_{3i}(M_i^T(w_i(t))e_i(t) + ksign(\tilde{\xi}_i)/\sqrt{\mu_{3i}}) & i = 1, 2, \ldots, N^* \\ \dot{\hat{v}}_i = \mu_{4i}(L_i^T(w_i(t))e_i(t) + ksign(\tilde{v}_i)/\sqrt{\mu_{4i}}) & i = N^*+1, N^*+2, \ldots, N \end{cases} \quad (5)$$

where $\lambda_i, \delta_{ij}, \mu_{1i}, \mu_{2i}, \mu_{3i}, \mu_{4i}$ are any positive constant.

$$\Lambda = - (1 + \xi)e_i^T(t)e_i(t) - ksign(r_i)/\sqrt{\lambda_i} \\ + r_ie_i(t)(-ksign(e_i(t)) - k/2(1 - \gamma)(\int_{t-i(t)}^t e_i^T(\varphi)e_i(\varphi)d\varphi)^{1/2})/r_i^2$$

Thus, the proposed unilateral external coupling (3), the adaptive laws of uncertain topological structure and unknown parameters (4), (5) will guarantee the realization of networks synchronization and estimation the uncertain parameters and topology in a finite time.

Theorem 1. Suppose the error system is controlled with unilateral external coupling controller (3) and adaptive laws (4), (5). Within the time t_1, The convergence of synchronization error can be realized. That means the two networks can achieve FTS, and t_1 determined by $t_1 \geq 2V^{1/2}(t_0)/\sqrt{2k}$ (Based on the Lemma 1in [7]).

Proof. Consider Lyapunov-Krasovskii function candidate below:

$$V = \frac{1}{2}\sum_{i=1}^{N} e_i^T(t)e_i(t) + \frac{1}{2}\sum_{i=1}^{N'} \tilde{\vartheta}_i^T \tilde{\vartheta}_i + \frac{1}{2}\sum_{i=N'+1}^{N} \frac{1}{\mu_{2i}}\tilde{\beta}_i^T \tilde{\beta}_i + \frac{1}{2}\sum_{i=1}^{N^*} \frac{1}{\mu_{3i}}\tilde{\xi}_i^T \tilde{\xi}_i$$

$$+ \frac{1}{2}\sum_{i=N^*+1}^{N} \frac{1}{\mu_{4i}}\tilde{v}_i^T \tilde{v}_i + \frac{1}{2}\sum_{i=1}^{N}\sum_{j=1}^{N} \frac{1}{\delta_{ij}}\tilde{c}_{ij}^2 + \frac{1}{2}\sum_{i=1}^{N} \frac{1}{\lambda_i}r_i^2 + \frac{1}{2(1-\gamma)}\int_{t-t(t)}^{t} \sum_{i=1}^{N} e_i^T(\varphi)e_i(\varphi)d\varphi$$

(6)

Take the derivative of (6) along the trajectories of $e_i(t)$, and substituting the networks (1) and (2), adaptive laws (3), (4), (5), we have

$$\dot{V} = \sum_{i=1}^{N^*} e_i^T(t)Ce_i(t) + \sum_{i=N^*+1}^{N} e_i^T(t)De_i(t) - \zeta\sum_{i=1}^{N} e_i^T(t)e_i(t) - k\sum_{i=1}^{N} e_i^T(t)sign(e_i(t))$$

$$- k\sum_{i=1}^{N'} \frac{1}{\sqrt{\mu_{1i}}}\tilde{\vartheta}_i^T sign(\tilde{\vartheta}_i) - k\sum_{i=N'+1}^{N} \frac{1}{\sqrt{\mu_{2i}}}\tilde{\beta}_i^T sign(\tilde{\beta}_i^T) - k\sum_{i=1}^{N^*} \frac{1}{\sqrt{\mu_{3i}}}\tilde{\xi}_i^T sign(\tilde{\xi}_i^T)$$

$$- k\sum_{i=N^*+1}^{N} \frac{1}{\sqrt{\mu_{4i}}}\tilde{v}_i^T sign(\tilde{v}_i^T) - k\sum_{i=1}^{N}\sum_{j=1}^{N} \frac{1}{\sqrt{\delta_{ij}}}|\tilde{c}_{ij}| - k\sum_{i=1}^{N} \frac{1}{\sqrt{\lambda_i}}r_i sign(r_i) + \frac{1}{2(1-\gamma)}\sum_{i=1}^{N} e_i^T(t)$$

$$e_i(t) - k\sum_{i=1}^{N} \frac{1}{2(1-\gamma)}\left(\int_{t-t(t)}^{t} e_i^T(\varphi)e_i(\varphi)d\varphi\right)^{1/2} - \frac{1-i(t)}{2(1-\gamma)}\sum_{i=1}^{N} e_i^T(t-t(t))e_i(t-t(t))$$

(7)

That, suppose satisfied Lemma 1 in [17], we have.

$$\dot{V} \leq \sum_{i=1}^{N^*} e_i^T(t)\chi_C e_i(t) + \sum_{i=N^*+1}^{N} e_i^T(t)\chi_D e_i(t) - \zeta\sum_{i=1}^{N} e_i^T(t)e_i(t) - k\sum_{i=1}^{N} \|e_i(t)\|$$

$$- k\sum_{i=1}^{N'} \frac{1}{\sqrt{\mu_{1i}}}|\tilde{\theta}_i| - k\sum_{i=N'+1}^{N} \frac{1}{\sqrt{\mu_{2i}}}|\tilde{\beta}_i| - k\sum_{i=1}^{N^*} \frac{1}{\sqrt{\mu_{3i}}}|\tilde{\xi}_i| - k\sum_{i=N^*+1}^{N} \frac{1}{\sqrt{\mu_{4i}}}|\tilde{v}_i|$$

$$- k\sum_{i=1}^{N}\sum_{j=1}^{N} \frac{1}{\sqrt{\delta_{ij}}}|\tilde{c}_{ij}| - k\sum_{i=1}^{N} \frac{1}{\sqrt{\lambda_i}}|r_i| - k\sum_{i=1}^{N} \frac{1}{2(1-\gamma)}\left(\int_{t-t(t)}^{t} e_i^T(\varphi)e_i(\varphi)d\varphi\right)^{1/2}$$

$$+ \frac{1}{2(1-\gamma)}\sum_{i=1}^{N} e_i^T(t)e_i(t) - \frac{1-i(t)}{2(1-\gamma)}\sum_{i=1}^{N} e_i^T(t-\imath(t))e_i(t-\imath(t))$$

(8)

where χ_C, χ_D are largest eigenvalue of symmetric matrices $(C+C^T)/2$, $(D+D^T)/2$. By employing Lemma 2 and Lemma 3 in [7], we set $e(t) = (e_1^T(t), e_2^T(t), \ldots, e_N^T(t))^T \in R^{n \times N}$, it yields

$$\dot{V} \le e^T(t)\chi e(t) - \zeta e^T(t)e(t) - k \sum_{i=1}^{N} |e_i(t)| - k \sum_{i=1}^{N'} \frac{1}{\sqrt{\mu_{1i}}} |\tilde{\vartheta}_i| - k \sum_{i=N'+1}^{N} \frac{1}{\sqrt{\mu_{2i}}} |\tilde{\beta}_i| - k$$

$$\sum_{i=1}^{N^*} \frac{1}{\sqrt{\mu_{3i}}} |\tilde{\xi}_i| - k \sum_{i=N^*+1}^{N} \frac{1}{\sqrt{\mu_{4i}}} |\tilde{v}_i| - k \sum_{i=1}^{N} \sum_{j=1}^{N} \frac{1}{\sqrt{\delta_{ij}}} |\hat{c}_{ij}| - k \sum_{i=1}^{N} \frac{1}{\sqrt{\lambda_i}} |r_i|$$

$$- k \sum_{i=1}^{N} \frac{1}{2(1-\gamma)} (\int_{t-l(t)}^{t} e_i^T(\varphi)e_i(\varphi)d\varphi)^{1/2} + \frac{1}{2(1-\gamma)} e^T(t)e(t) - \frac{1-i(t)}{2(1-\gamma)} e^T(t-\imath(t))e_i(t-\imath(t))$$

$$\le (\chi + \frac{1}{2(1-\gamma)} - \zeta) e^T(t)e(t) - k(\sum_{i=1}^{N} \|e_i(t)\| - \sum_{i=1}^{N'} \frac{1}{\sqrt{\mu_{1i}}} |\tilde{\vartheta}_i| - \sum_{i=N'+1}^{N} \frac{1}{\sqrt{\mu_{2i}}} |\tilde{\beta}_i| - \sum_{i=1}^{N^*} \frac{1}{\sqrt{\mu_{3i}}}$$

$$|\tilde{\xi}_i| - \sum_{i=N^*+1}^{N} \frac{1}{\sqrt{\mu_{4i}}} |\tilde{v}_i| - \sum_{i=1}^{N} \sum_{j=1}^{N} \frac{1}{\sqrt{\delta_{ij}}} |\hat{c}_{ij}| - \sum_{i=1}^{N} \frac{1}{\sqrt{\lambda_i}} |r_i| - \sum_{i=1}^{N} \frac{1}{2(1-\gamma)} (\int_{t-l(t)}^{t} e_i^T(\varphi)e_i(\varphi)d\varphi)^{1/2})$$

$$\tag{9}$$

where $\chi = \begin{pmatrix} \chi_C I_n N^* & 0 \\ 0 & \chi_D I_n (N - N^*) \end{pmatrix}$. Take $\zeta > \chi + \frac{1}{2(1-\gamma)}$, from Lemma 1 in [7], thus we can further get

$$\dot{V} \le - \sqrt{2}k[\frac{1}{2} \sum_{i=1}^{N} e_i^T(t)e_i(t) + \frac{1}{2} \sum_{i=1}^{N'} \frac{1}{\mu_{1i}} \tilde{\vartheta}_i^T \tilde{\vartheta} + \frac{1}{2} \sum_{i=N'+1}^{N'} \frac{1}{\mu_{2i}} \tilde{\beta}_i^T \tilde{\beta}_i + \frac{1}{2} \sum_{i=1}^{N^*} \frac{1}{\mu_{3i}} \tilde{\xi}_i^T \tilde{\xi}_i +$$

$$\frac{1}{2} \sum_{i=N^*+1}^{N} \frac{1}{\mu_{4i}} \tilde{v}_i^T \tilde{v}_i + \frac{1}{2} \sum_{i=1}^{N} \sum_{j=1}^{N} \frac{1}{\delta_{ij}} \hat{c}_{ij}^2 + \frac{1}{2} \sum_{i=1}^{N} \frac{1}{\lambda_i} r_i^2 + \frac{1}{2(1-\gamma)} (\int_{t-\imath(t)}^{t} \sum_{i=1}^{N} e_i^T(\varphi)e_i(\varphi)d\varphi)]$$

$$\le - \sqrt{2}kV^{1/2}$$

$$\tag{10}$$

According to Lemma 1 in [7], the error e_i can converge to zero in $t_1 \ge 2V^{1/2}(t_0)/\sqrt{2}k$. Obviously, there exists function of unilateral coupling $r_i(i = 1, 2, \ldots, N)$ making the differential of V negative definite. So according to the LaSalle's invariance principle [16], $e_i(t) \to 0$ and $\tilde{\vartheta}_i \to \vartheta_i, \hat{\beta}_i \to \beta_i, \hat{\xi}_i \to \xi_i, \hat{v}_i \to v_i, \hat{C} \to C$ as $t \to t_1$. That means all of the uncertain parameters can be successful estimated, and the two networks can achieve FTS. This completes the proof.

Remark 1. The driver-response complex networks contain two types of different node systems respectively. However, if the networks own more types of different nodes, similar work is easy to be generalized.

4   Illustrative Examples

For further confirm the correctness of the theoretical analysis, a simulation experiment has been carried out. We construct the drive and response networks with well-known Lorenz system, Chen system, Liu system and Rossler system. The system parameters value of Lorenz system are $a_1 = 10, a_2 = 28, a_3 = 8/3$. Chen system parameters are

$d_1 = 35, d_2 = 3, d_3 = 28$. Liu system parameters are $b_1 = 10, b_2 = 40, b_3 = 1$, $b_4 = 2.5, b_5 = 4$. Rossler system parameters are $m_1 = 0.2, m_2 = 0.2, m_3 = 5.7$.

For simplicity, we choose the drive-response networks of size $N' = 6$ and $N^* = 4$. The drive network is consisted of six Lorenz systems $(i = 1, 2, \ldots, 6)$ and four Liu systems $(i = 7, 8, \ldots, 10)$. And the response complex network is consisted of four Rossler systems $(i = 1, 2, \ldots, 4)$ and six Chen systems $(i = 5, 6, \ldots, 10)$. $a_{1i}(i = 1, 2, \ldots, 6), b_{2i}(i = 7, 8, \ldots, 10)$ are the unknown parameters of drive network and $m_{1i}(i = 1, 2, \ldots, 4), d_{2i}(i = 5, 6, \ldots, 10)$ are the unknown parameters of response network. That the real value of the unknown parameters are $a_{1i} = 10$, $b_{2i} = 40, m_{1i} = 0.2, d_{2i} = 3$,, respectively. Taking initial value of the estimator $\hat{a}_{1i}(0) = (12, 14, 6, 8, 4, 5)^T$, $\hat{b}_{2i}(0) = (35, 38, 42, 43)^T$, $\hat{m}_{1i}(0) = (0.3, 0.2, 0.5, 0.4)^T$, $\hat{d}_{2i}(0) = (2, 5, 4, 1, 5, 3)^T$. And $\mu_{1i} = \mu_{2i} = \mu_{3i} = \mu_{4i} = 1$, $k = 2, \lambda_i = 2, \delta_{ij} = 1$, $\iota(t) = 2$.

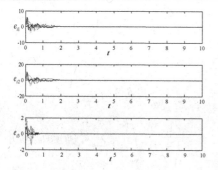

Fig. 1. Synchronization errors $(i = 1, 2, \ldots, 10)$

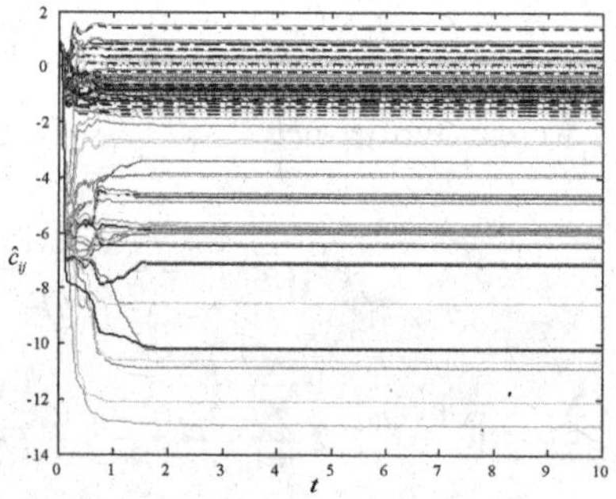

Fig. 2. Weight values of topological structure

Based on Theorem 1, applying the unilateral coupling (3) and adaptive laws (4) and (5), the FTS can be achieved and the unknown parameters and the uncertain coupling configuration matrix are also obtained. In Fig. 1, the synchronous errors $e_{i1}(t), e_{i2}(t), e_{i3}(t)$ converge to zero after $t > 2.5$. Under the action of coupling between the two networks, we can see that the three dimension synchronous error curves starting from the different initial values and after a period of fluctuation, they are stabilized at the origin. Moreover, the error curves are smooth and without shaking behavior. In Fig. 2, the weight values of uncertain topological structure $\hat{c}_{ij}$ adapt to the appropriate constants rapidly. We find that the network topology changes can be well track and the weight topology structure values can be rapidly adaptive to the appropriate value. Figures 3 and 4 show the estimation of the unknown parameters a_{1i}, b_{2i}, d_{1i} and m_{1i}. It is shown that all the parameters can be estimated successfully.

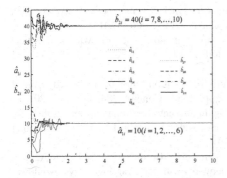

Fig. 3. Estimation of unknown parameters

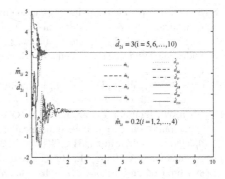

Fig. 4. Estimation of unknown parameters

5 Conclusions

In this paper, the FTS between delayed networks with nonidentical nodes based on estimation of uncertain topological structure and parameters has been investigated. A new unilateral coupling control method is proposed to guarantee the realization of FTS and the corresponding updating laws are also obtained successfully. The simulation results are shown our scheme is feasible.

Acknowledgments. This work was supported by the National Natural Science Foundation of China (No. 61374154).

References

1. Arenas, A., Díaz-Guilera, A., Kurths, J., Moreno, Y., Zhou, C.: Synchronization in complex networks. Phys. Rep. **469**, 93–153 (2008)
2. Suykens, J.A.K., Osipov, G.V.: Introduction to focus issue: synchronization in complex networks. Chaos **18**, 037101-4 (2008)
3. Yang, X., Cao, J.: Exponential synchronization of delayed neural networks with discontinuous activations. IEEE Trans. Circ. Syst. I **60**, 2431–2439 (2013)
4. Guo, Z., Yang, S., Wang, J.: Global exponential synchronization of multiple memristive neural networks with time delay via nonlinear coupling. IEEE Trans. Neural Netw. Learn. Syst. **26**, 1300–1311 (2015)
5. Liu, H., Cheng, L., Tan, M., Hou, Z.G., Wang, Y.P.: Distributed exponential finite-time coordination of multi-agent systems: containment control and consensus. Int. J. Control **88**, 237–247 (2015)
6. Boccaletti, S., Latora, V., Moreno, Y., Chavezf, M., Hwanga, D.U.: Complex networks: Structure and dynamics. Phys. Rep. **424**, 175–308 (2006)
7. Mei, J., Jiang, M., Wang, J.: Finite-time structure identification and synchronization of drive-response systems with uncertain parameter. Commun. Nonlinear Sci. Numer. Simul. **18**, 999–1015 (2013)

8. Sun, Y., Li, W., Zhao, D.: Finite-time stochastic outer synchronization between two complex dynamical networks with different topologies. Chaos **22**, 023152-7 (2012)
9. Yang, X., Cao, J., Lu, J.: Stochastic synchronization of complex networks with nonidentical nodes via hybrid adaptive and impulsive control. IEEE Trans. Circ. Syst. I **59**, 371–384 (2012)
10. Lee, D., Yoo, W., Ji, D., Park, J.: Integral control for synchronization of complex dynamical networks with unknown non-identical nodes. Appl. Math. Comput. **224**, 140–149 (2013)
11. Chen, W., Jiang, Z., Zhong, J., Lu, X.: On designing decentralized impulsive controllers for synchronization of complex dynamical networks with nonidentical nodes and coupling delays. J. Franklin Inst. **351**, 4084–4110 (2014)
12. Abdurahman, A., Jiang, H., Hu, C., Teng, Z.: Parameter identification based on finite-time synchronization for Cohen-Grossberg neural networks with time-varying delays. Nonlinear Anal. **20**, 348–366 (2015)
13. Jing, T., Chen, F., Li, Q.: Finite-time mixed outer synchronization of complex networks with time-varying delay and unknown parameters. Appl. Math. Model. **39**, 23–24 (2015)
14. Liu, H., Lu, J., Lü, J., Hill, D.: Structure identification of uncertain general complex dynamical networks with time delay. Automatica **45**, 1799–1807 (2009)
15. Che, Y., Li, R., Han, C., Cui, S., Wang, J., Wei, X., Deng, B.: Topology identification of uncertain nonlinearly coupled complex networks with delays based on anticipatory synchronization. Chaos **23**, 013127-7 (2013)
16. Khalil, H., Grizzle, J.: Nonlinear Systems. Prentice Hall, Upper Saddle River (2002)
17. Han, M., Zhang, M., Zhang, Y.: Projective synchronization between two delayed networks of different sizes with nonidentical nodes and unknown parameters. Neurocomputing **171**, 605–614 (2016)
18. Xu, Y., Zhou, W., Fang, J., Lu, H.: Structure identification and adaptive synchronization of uncertain general complex dynamical networks. Phys. Lett. A **374**, 272–278 (2009)

A Multiple-objective Neurodynamic Optimization to Electric Load Management Under Demand-Response Program

Xinyi Le[✉], Sijie Chen, Yu Zheng, and Juntong Xi

Shanghai Jiao Tong University, Shanghai 200240, China
{lexinyi,Sijie.chen,yuzheng,jtxi}@sjtu.edu.cn

Abstract. In a power system, a price-based demand-response program offers end electricity users time-varying prices, incentivizing them to shift demand from high-price hours to low-price hours during a day. Heating/cooling (H/C) loads are typical flexible loads to be shifted. Specifically, end users optimize the hourly H/C load to balance electricity costs and comfort. In this paper, a two-time-scale neurodynamic optimization approach is applied for this multi-objective optimization problem. As a result, optimal use of H/C loads is derived that yields significant savings and acceptable comfort. A case study of the Houston City is presented to show the effectiveness of the proposed neurodynamic approach.

Keywords: Demand response · Power system economics · Multi-objective optimization · Two-time-scales · Neurodynamics

1 Introduction

Under the traditional regulated utility structure, most customers are metered monthly and billed a flat average rate for the total amount of electricity used. Nowadays, some demand-response programs have launched in the unbundled electric power system, whereby household and small industrial consumers are organized into coalitions [10] who have opted to limit consumption during energy peaks. In exchange, they also enjoy lower energy costs. The price signals, i.e., variable hourly energy tariffs, provide incentives to the end-use customers to reduce load at peak times of the day when prices are high. Such load control scheme have been supported by FERC in the USA and the European Commission in Europe [18] to vary consumers' consumption habits. It is shown that load control has a dramatic effect on both consumer utility and system-wide peak load.

In the end customers side, heating/cooling (H/C) loads are typical flexible loads which could be rearranged. The human comfort zone is not a point but

J. Xi—The work described in the paper was supported by National Natural Science Foundation of China (Grant No. 51505286), Scientific Research Project of Shanghai Science and Technology Committee (15111107902) and National Key Technology Research and Development of the Ministry of Science and Technology of China (04 project 2014ZX04015021).

© Springer International Publishing AG 2017
F. Cong et al. (Eds.): ISNN 2017, Part II, LNCS 10262, pp. 169–177, 2017.
DOI: 10.1007/978-3-319-59081-3_21

a range. In addition, heating or cooling a house can benefit subsequent hours because thermal insulation enables houses to store heat. As H/C loads represent a considerable portion of residential load, electricity consumers are willing to optimize conditioning costs and comfort by increasing air conditioning use within those hours when prices are off-peak so to maintain indoor temperatures at a desired level, and decreasing air conditioning use within hours characterized by peak prices. Considering the conditioning costs and comfort level, this optimal strategy would require the definition and solution of a multi-objective optimization problem.

Remarkable research has been done with respect to the optimal strategy. In [10], air conditioning loads are scheduled to minimize energy purchase costs. For example, in [18], air conditioning loads are allowed to be self-scheduled by individual consumers by estimating consumers tradeoff between electricity bill savings and living comfort. [2] proposes a simple flexible household utility function that can be calibrated with minimal data to describe diverse household behaviors and reveal household responses to different prices. A two-stage bidding strategies and compensation policies are introduced in [3] considering uncertainties in electricity prices, weather, non-H/C load, and thermal-related house characteristics. According to the changes in electric usage by end-use customers, some state-of-the-art transactive energy approaches have been discussed in [1].

For multi-objective optimization problems, various methods have been developed based on the preference related objectives [17]. Solution techniques with a prior articulation of preferences require a criterion to aggregate the different objectives into a single one before starting the optimization, and then a single solution is obtained. Several objectives are combined using weighted sum methods, value function methods and goal programming methods [6]. Such kind of approaches could only capture one single solution. The whole Pareto front, which is a posteriori articulation of preferences is missing during the combination. Evolutionary algorithms have been widely investigated to find the Pareto front by repeatedly running the weighted method with a varying weights [7]. However, for high-dimension and large-scale problems, the aforementioned methods may require a long time and computing burden to depict the whole optimal region.

Inspired by their biological counterparts, recurrent neural networks have shown great promises in many applications such as classification and regression, time series analysis, and automatic control. Neurodynamic optimization is a promising continuous-time optimization approach, which enjoys the inherent nature of parallel computing and the potential of electronic implementation. In addition, recent advances in neurodynamic engineering offer availability of hardware implementation of neural networks [5], which shows a great potential to run neurodynamic models much faster than iterative methods by orders of magnitude.

Past thirty years witnessed the great developments of neurodynamic optimization since the pioneer work of [21]. From then on, Kennedy and Chua proposed a dynamical canonical nonlinear programming circuit to approximate optimal solutions to nonlinear optimization problems [11]. Wang proposed a deterministic annealing network for convex programming [22]. Forti et al. proposed a generalized neural network for nonsmooth nonlinear programming based on differential inclusions [9]. In recent years, a few neurodynamic optimization models have been developed for nonsmooth or generalized convex optimization with wide applications [4,12,13,16]. Some approaches have been also proposed for more complex optimization problems,such as bilevel programming [20],biconvex problems [14], minimax problems [15] and multi-objective problems [24].

Inspired by above discussions, this paper aims to apply the neurodynamic approach for multi-objective optimization in order to achieve strategic bidding of electricity market. According to the neurodynamic model in [24], a weight vector is varying continuously under the guidance of a dynamical system, where the weights are set to evolve at a relative quite slow rate in comparison with the real-time convergence speed of neural networks. To obtain a description of the whole Pareto-front, the whole neural network has two time scales, which consisting of a "fast-varying" subsystem of neural network and a "slow-varying" subsystem of weight dynamics. This model is proved to be convergent to the Pareto-optimal front. The optimization model is then applied in the strategic bidding bi-objective problem successfully. As a result, the Pareto front has shown the posteriori articulation of preferences for the electricity consumers to achieve economic benefits and enjoy the comparable comforts.

The remain part of this paper is organized as follows. In Sect. 2, preliminary concepts and notations of strategic bidding of electricity market are introduced. In Sect. 3, the neurodynamic models for solving multi-objective functions are applied. In Sect. 4, a case study is presented to illustrate the proposed approach. Section 5 concludes this paper.

2   Preliminaries

Electricity consumers can optimize conditioning costs and comfort level by combining two strategies: increasing the desired indoor temperature but nonetheless taking advantage of the time-varying electricity prices. Indeed, they increase air conditioning use within those hours when prices are off-peak to maintain indoor temperatures at a desired level, and then they decrease air conditioning use within hours characterized by peak prices. $dcomf_i$ is then introduced to show the degradation of living comfort by the deviation from the ideal indoor temperature T_i^{ideal}.

$$dcomf_i = (T_i - T_i^{ideal})^2, \tag{1}$$

where T_i and T_i^{ideal} are, respectively, the actual and ideal residence indoor temperatures at the ith time period when the end users need the air conditioning. The degradation of living comfort is a penalty for the consumers welfare and has been transformed into a cost.

The bi-objective optimization problem can be modeled as follows [18]:

$$\min \ F = [\sum_{i=0}^{N-1} P_i q_i, \sum_{i \in S} dcomf_i] \tag{2}$$

$$\text{s.t. } T_0 - T_N = 0, q_0 - q_N = 0 \tag{3}$$

$$T_i = \alpha T_{i-1} + \beta q_{i-1} + (1-\alpha)T_{i-1}^{\text{out}}, i = 1, \ldots, N \tag{4}$$

$$T^{\min} \leq T_i \leq T^{\max}, i \in S \tag{5}$$

$$0 \leq q_i \leq q^{\max}, i = 0, 1, \ldots, N-1 \tag{6}$$

As is shown in the objective function (2), there are two objectives for minimization. One objective $(\sum_{i=0}^{N-1} P_i q_i)$ is used for saving energy and the other one
$(\sum_{i \in S} \text{dcomf}_i)$ is used for measuring comfort level. S is a time region that the
end users need the air conditioning. The above optimization problem is then a
bi-objective optimization problem. P_i is the electricity price at the ith period, q_i
is the energy consumption at the ith period. α is the residence thermal dispersion
coefficient, β (in $kWh/^\circ C$) represents the reciprocal of total thermal capacity
of the residence, T_i^{out}, $T^{\min}$ and $T^{\max}$ are, the outdoor temperature, lower and
upper bounds of the indoor temperature. The decision variables of the problem
are the amounts of conditioning energy consumption $q = [q_0, q_1, \ldots, q_{N-1}]^T$ and
the indoor temperatures $T = [T_0, T_1, \ldots, T_{N-1}]^T$. Constraints (3) indicate that
initial and final indoor temperatures and energy consumptions are assumed the
same. Constraint (4) is derived from the thermal model. Constraint (5) defines
the maximum temperature variation of the consumers' tolerance. Constraint (6)
shows the limitation of air conditioning capacity.

3   Neurodynamic Approaches

In order to solve the optimization problem (2), a neurodynamic optimization
approach is then introduced. For some general cases, a multi-objective optimization problem can be described as follows:

$$\min F(x) = [F_1(x), F_2(x), \ldots, F_k(x)]^T$$
$$\text{s.t. } Ax = b, x \in \Omega \tag{7}$$

where $x \in \Re^n$ is the vector of decision variables, $F(x) \in \Re^k$ is a vector of
objective functions or criteria with $F_i(x)(i = 1, 2, \ldots, k)$. The state matrix $A \in
\Re^{p \times n}$ has a full row rank. $\Omega \subset \Re^n$ is a closed convex set. Then the feasible region
is defined as the set $\{x \in \Re^n : Ax = b, x \in \Omega\}$.

There always exist tradeoff among all the objectives for multi-objective optimization. If a decision vector $x^* \in X$ is called Pareto optimal if there does
not exist $x \in X$ such that $F_i(x) \leq F_i(x^*)$ for all $(i = 1, 2, \ldots, k)$. Weighted
sum method is a common approach for solving multi-objective optimization as
follows:

$$\min \sum_{i=1}^{k} = w_i F_i(x)$$
$$\text{s.t. } Ax = b, \ x \in \Omega, \ 0 \leq w_i \leq 1, \ w \in \Re^k \tag{8}$$

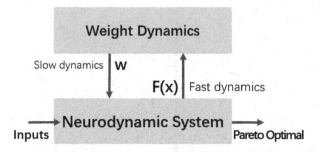

Fig. 1. Dynamic interaction of variables w and x.

where w_i is the weight of the ith objective function. It has been proved in [19] that if (7) is strictly convex, then the optimal solution for (8) is equivalent to the optimal solution to (7). The neurodynamic model has been proposed for solving (8) in [24], which consists of two cooperative parts: neurodynamic system and weights updating dynamics, as depicted in Fig. 1. When the number of objective $k = 2$, the neurodynamic model can be present as follows.

$$\epsilon_1 \dot{x} = -QP_\Omega(x) - (I - Q)(x - P_\Omega(x) + \alpha\nabla F((I - Q)P_\Omega(x) + q)w) - q$$
$$\epsilon_2 \dot{w} = v, \tag{9}$$

where $Q = A^T(AA^T)^{-1}A$, $q = A^T(AA^T)^{-1}b$, $v = [1, -1]^T$ and let the initial value $w_0 = [0, 1]^T$, and ϵ_1 and ϵ_2 are two scaling values which satisfy $\epsilon_1 \ll \epsilon_2$. Set $\epsilon_1 = 10^{-5}$ and $\epsilon_2 = 10^{-2}$. As t goes, the weight vector w will move from $[0, 1]^T$ to $[1, 0]^T$ and covers all the nodes in S^2. P_Ω in (9) is a projection operator. If $\Omega = \{x_i \in \Re^n : l_i \le x_i \le h_i, i = 1, 2, \ldots, n\}$, P_Ω is defined as

$$P_\Omega(x_i) = \begin{cases} l_i, & x_i < l_i \\ x_i, & l_i \le x_i \le h_i \\ h_i, & h_i < x_i \end{cases} \tag{10}$$

4   Simulation Results

A case study is presented in this section, where the coalition in Houston submits bids and purchases electricity for its clients. According to [3], the typical values of α and β are chosen as 0.97 and 0.315 °C/kW/15 min, respectively. The maximum power output q^{max} is set to 8 kW. Desired household temperature T^{ideal} is 22 °C. Highest acceptable household temperature T^{max} is 24 °C and lowest acceptable household temperature T^{min} is 20 °C. The energy delivery day of interest is Dec. 8th, 2016. Figure 2 described the reference outdoor temperature in Dec. 8th 2016 in Houston [23]. The market prices of electricity in Houston is time-varying. Figure 3 demonstrated the transient behavior of electricity price in

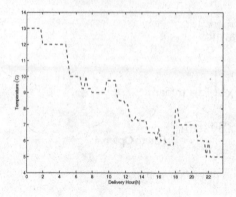

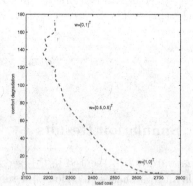

Fig. 2. Reference outdoor temperature in Dec. 8th 2016 in Houston.

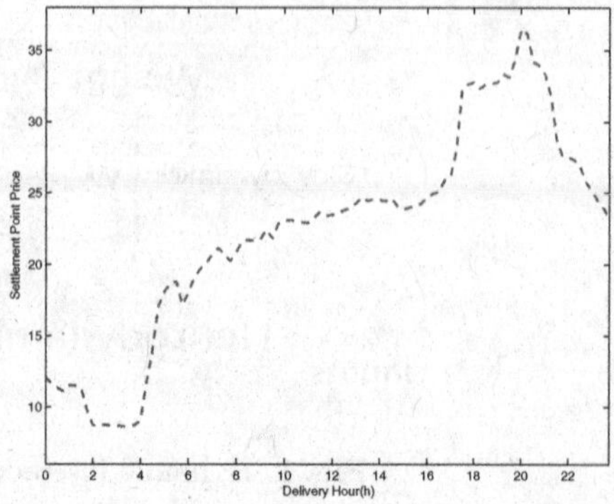

Fig. 3. Transient behavior of electricity price in Dec. 8th 2016 in Houston.

Dec. 8th in Houston [8]. The market prices and temperature data are collected every 15 min. As a result, the length of our case study is $N = 96$. Assume the end-users stay at home from (0:00–8:00) and (17:00–24:00), then S of our interest is $\{N = 1, 2, \ldots, 32, 69, 70, \ldots, 96\}$.

Figure 4 demonstrated transient behaviors of state variables $\{T_i\}$ with $w = [0.5, 0.5]^T$ on integral hours (1:00 am, 2:00 am,...) using two-time-scales neurodynamic model (9). The state variables are convergent in less then 2 ms, which shows the real-time capability of the neural network. Figure 5 shows Pareto front generated from neurodynamic optimization. When $w = [0, 1]^T$, it shows that the comfort level is maximized. Similarly, $w = [1, 0]^T$ means the electricity cost is mostly saved.

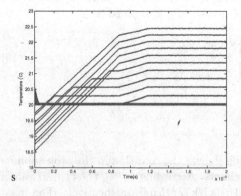

Fig. 4. The transient behaviors of $\{T_i\}$ with $w = [0.5, 0.5]^T$ on integral hours (1:am, 2:am,...)

Fig. 5. Pareto front generated from neurodynamic model.

Figure 6 shows the comparison results of load profiles on Dec. 8th before and after implementing the multi-objective optimization. The optimization shifts load from peak hours (6:00 pm–8:00 pm) to off-peak hours. By shifting H/C

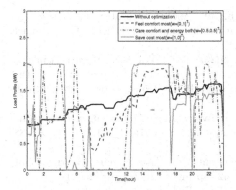
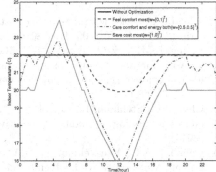

Fig. 6. Comparison results of load profiles before and after implementing optimization in different situations.

Fig. 7. Comparison results of temperature profiles before and after implementing optimization in different situations.

loads from high-price hours to adjoining low-price hours, the electricity cost has reduced 29.3% for mostly saved cases. Figure 7 shows the comparison results temperature profiles before and after implementing optimization in different situations ($w = [0, 1]^T, [0.5, 0.5]^T, [1, 0]^T$). This strategy could be adopted in all situations to avoid heating the house using expensive electricity while maintaining an acceptable temperature for the household.

5 Conclusions

This paper shows the application of two-time-scales neurodynamic optimization approach for solving bi-objective optimization problems. Two objectives, comfort level and the electricity costs are considered in the optimization problem. A case study of Houston in one day has been simulated for demonstrating the effectiveness of the proposed model. Compared to the previous approaches, a Pareto optimal region other than several optimal points has been denoted. In addition, as the electricity market prices and temperature information vary in real time, the proposed method is able to optimize the objectives in real time. After implementing the optimization strategy, "smart" use of air conditioning could be achieved. Future study will consider the unpredictable market prices and temperature changes, and different human habits.

References

1. Chen, S., Liu, C.: From demand response to transactive energy: state of the art. J. Mod. Power Syst. Clean Energy (2017). doi:10.1007/s40565-016-0256-x
2. Chen, S., Love, H.A., Liu, C.C.: Optimal opt-in residential time-of-use contract based on principal-agent theory. IEEE Trans. Power Syst. **31**(6), 4415–4426 (2016)

3. Chen, S., Chen, Q., Xu, Y.: Strategic bidding and compensation mechanism for a load aggregator with direct thermostat control capabilities. IEEE Trans. Smart Grid (2017). doi:10.1109/TSG.2016.2611611

4. Cheng, L., Hou, Z.G., Lin, Y., Tan, M., Zhang, W.C., Wu, F.X.: Recurrent neural network for non-smooth convex optimization problems with application to the identification of genetic regulatory networks. IEEE Trans. Neural Netw. $22(5)$, 714–726 (2011)

5. Chicca, E., Stefanini, F., Bartolozzi, C., Indiveri, G.: Neuromorphic electronic circuits for building autonomous cognitive systems. Proc. IEEE $102(9)$, 1367–1388 (2014)

6. Coello, C.A.C., Van Veldhuizen, D.A., Lamont, G.B.: Evolutionary Algorithms for Solving Multi-objective Problems, vol. 242. Springer, Heidelberg (2002)

7. Deb, K.: Multi-objective Optimization Using Evolutionary Algorithms, vol. 16. Wiley, Hoboken (2001)

8. ERCOT. http://www.ercot.com/mktinfo/prices

9. Forti, M., Nistri, P., Quincampoix, M.: Generalized neural network for nonsmooth nonlinear programming problems. IEEE Trans. Circ. Syst. I: Regul. Pap. $51(9)$, 1741–1754 (2004)

10. Ilic, M., Black, J.W., Watz, J.L.: Potential benefits of implementing load control. In: IEEE Power Engineering Society Winter Meeting, vol. 1, pp. 177–182. IEEE (2002)

11. Kennedy, M.P., Chua, L.O.: Neural networks for nonlinear programming. IEEE Trans. Circ. Syst. $35(5)$, 554–562 (1988)

12. Le, X., Wang, J.: Robust pole assignment for synthesizing feedback control systems using recurrent neural networks. IEEE Trans. Neural Netw. Learn. Syst. $25(2)$, 383–393 (2014)

13. Le, X., Wang, J.: Neurodynamics-based robust pole assignment for high-order descriptor systems. IEEE Trans. Neural Netw. Learn. Syst. $26(11)$, 2962–2971 (2015)

14. Le, X., Wang, J.: A two-time-scale neurodynamic approach to robust pole assignment. In: Eighth International Conference on Advanced Computational Intelligence (ICACI), pp. 60–67. IEEE (2016)

15. Le, X., Wang, J.: A two-time-scale neurodynamic approach to constrained minimax optimization. IEEE Trans. Neural Netw. Learn. Syst. $28(3)$, 620–629 (2017). doi:10.1109/TNNLS.2016.2538288

16. Liu, Q., Guo, Z., Wang, J.: A one-layer recurrent neural network for constrained pseudoconvex optimization and its application for dynamic portfolio optimization. Neural Netw. $26(1)$, 99–109 (2012)

17. Marler, R.T., Arora, J.S.: Survey of multi-objective optimization methods for engineering. Struct. Multidiscip. Optim. $26(6)$, 369–395 (2004)

18. Menniti, D., Costanzo, F., Scordino, N., Sorrentino, N.: Purchase-bidding strategies of an energy coalition with demand-response capabilities. IEEE Trans. Power Syst. $24(3)$, 1241–1255 (2009)

19. Miettinen, K.: Nonlinear Multiobjective Optimization, vol. 12. Springer Science & Business Media, Heidelberg (2012)

20. Qin, S., Le, X., Wang, J.: A neurodynamic optimization approach to bilevel quadratic programming. IEEE Trans. Neural Netw. Learn. Syst. (2016)

21. Tank, D., Hopfield, J.: Simple neural optimization networks: an A/D converter, signal decision circuit, and a linear programming circuit. IEEE Trans. Circ. Syst. $33(5)$, 533–541 (1986)

22. Wang, J.: A deterministic annealing neural network for convex programming. Neural Netw. **7**(4), 629–641 (1994)
23. Wunderground. https://www.wunderground.com/history/
24. Yang, S., Wang, J., Liu, Q.: Multiple-objective optimization based on a two-time-scale neurodynamic system. In: Eighth International Conference on Advanced Computational Intelligence (ICACI), pp. 193–199. IEEE (2016)

Neuro-Adaptive Containment Seeking of Multiple Networking Agents with Unknown Dynamics

Guanghui Wen[1]([⊠]), Peijun Wang[1], Tingwen Huang[2], Long Cheng[3], and Junyong Sun[4]

[1] School of Mathematics, Southeast University, Nanjing 211189, China
ghwen@seu.edu.cn
[2] Texas A&M University at Qatar, Doha 5825, Qatar
tingwen.huang@qatar.tamu.edu
[3] Institute of Automation, Chinese Academy of Sciences, Beijing 100190, China
[4] College of Engineering, Peking University, Beijing 100871, China

Abstract. This paper studies the quasi-containment and asymptotic containment problems of networking uncertain agents with multiple leaders over a fixing communication graph. A kind of containment controller consisting of a linear feedback term, a neuro-adaptive approximation term as well as a non-smooth feedback term is designed to complete the goal of quasi-containment. Under the assumption that the subgraph depicting the underlying communication configuration among multiple followers is detail-balanced and each follower can be at least indirectly influenced by one leader, it is proven that quasi-containment can be realized if the containment controller are appropriately designed. The results are then extended to asymptotic containment of networking uncertain agents with multiple leaders.

Keywords: Neural network · Quasi-containment · Networking agents

1 Introduction

One key scientific problem in the field of distributed multi-agent systems (MASs) is to understand how macroscopic coordination dynamic behavior can emerge from local interactions among neighboring agents [1]. Addressing this question is not only theoretically interesting but also practically significant [2]. A great quantity of attention has been recently paid to the study of emergence for coordination behaviors of MASs [1].

This research was supported by the National Nature Science Foundation of China under Grant No. 61673104 and the National Priority Research Project NPRP 7-1482-1-278 funded by Qatar National Research Fund.

F. Cong et al. (Eds.): ISNN 2017, Part II, LNCS 10262, pp. 178–186, 2017.
DOI: 10.1007/978-3-319-59081-3_22

As one of the most interesting coordination behaviors of MASs, containment of MASs has recently received particular attention [3–8]. Within the field of MASs, the coordination objective of containment is to make the states of all followers in the networking agent systems converge eventually onto a convex hull spanned by those of the multiple dynamic leaders. For MASs with a single leader, the containment problem will reduce to leader-following cooperative tracking problem. In [3], an interesting stop-go strategy was proposed to realize containment in first-order integrator-type MASs with hierarchical structure and multiple leaders. Then, containment problems of first-order integrator-type MASs with directed switching topologies was studied in [4]. Containment control for a class of networking Lagrangian systems in the presence of parametric uncertainties under a directed fixing graph was studied in [5]. Robust containment of uncertain linear MASs with a undirected topology was addressed in [6] via designing a non-smooth containment controller. In [7], distributed containment of linear MASs with multiple dynamic leaders subject to possibly nonzero control inputs was studied. Distributed observer-type containment protocol was designed in [8] to solve containment problem for a class of general linear MASs under directed fixing topology.

This paper is mainly concerned with the quasi-containment and asymptotic containment problems of general linear MASs with multiple leaders subject to nonzero control inputs. Unlike most existing related references where the followers have only nominal dynamics, the dynamic evolution of followers in the considered MASs are allowed to be effected by unknown nonlinear dynamics and external disturbances. A class of containment controllers consisting of a linear feedback term, a neuro-adaptive approximation term as well as a non-smooth feedback term are first designed to make the containment error vector of the closed-loop MASs being uniformly bounded. Under the assumption that the subgraph depicting the coupling configuration among the multiple followers is detail-balanced and each follower can be at least indirectly influenced by one leader, it is proven that quasi-containment can be realized if the structure of the feedback gain matrix and the control gain in the proposed containment controller are appropriately designed. At last, a new kind of containment controllers are designed to achieve asymptotic containment.

Notations. $\mathbb{R}^{n \times m}$ represents the set of $n \times m$ real matrices. Suppose that all the eigenvalues of P are real, $\lambda_{min}(P)$ and $\lambda_{max}(P)$ denotes respectively its smallest and largest eigenvalues. Notation diag$\{A_1, \cdots, A_n\}$ denotes a block-diagonal matrix with A_i as its i-th $(i = 1, \cdots, n)$ diagonal element. Symbol $\otimes$ denotes the Kronecker product. Symbols $\| \cdot \|$ and $\| \cdot \|_F$ denote the Euclidian norm of a vector and the Frobenius norm of a matrix, respectively. The ∞-norm of a vector $x = (x_1, \cdots, x_n)^T \in \mathbb{R}^n$ is denoted by $\|x\|_\infty$. Notation $tr(A)$ indicates the trace of matrix A, for any given $A \in \mathbb{R}^{N \times n}$.

2 Preliminaries on Graph Theory, Matrix Theory and Problem Statement

2.1 Preliminaries

Let $\mathcal{G}(\mathcal{A})$ be a digraph associated with a set of nodes $\mathcal{V} = \{v_1, \cdots, v_N\}$, a set of edges $\mathcal{E} \subseteq \mathcal{V} \times \mathcal{V}$, and an adjacency matrix $\mathcal{A} = [a_{ij}]_{N \times N}$ with non-negative elements a_{ij}. An edge e_{ij} in $\mathcal{G}(\mathcal{A})$ is represented by ordered pair of nodes (v_j, v_i), where v_j and v_i are respectively called the parent and child nodes. And, $e_{ij} \in \mathcal{E}$ if and only if $a_{ij} > 0$. A *directed path* from node v_i to v_j on $\mathcal{G}(\mathcal{A})$ is a finite ordered sequence of edges, $(v_i, v_{k_1}), (v_{k_1}, v_{k_2}), \cdots, (v_{k_l}, v_j)$, with distinct nodes v_{k_m}, $m = 1, \cdots, l$. A directed graph $\mathcal{G}(\mathcal{A})$ is said to be *detail-balanced* if there exist some scalars $\phi_k > 0$, $k = 1, \cdots, N$, to make the following hold: $\phi_i a_{ij} = \phi_j a_{ji}$, for all $i, j = 1, \cdots, N$ [9]. Let $L = [l_{ij}]_{N \times N}$ be the Laplacian matrix of $\mathcal{G}(\mathcal{A})$, defined as follows $l_{ij} = -a_{ij}$ for $i \neq j$, and $l_{ii} = \sum_{k=1, k \neq i}^{N} a_{ik}$.

Lemma 1 [10]. For any given $A \in \mathbb{R}^{n \times m}$, and $B \in \mathbb{R}^{m \times n}$, one has $|tr(AB)| \leq \|A\|_F \|B\|_F$, of which $\|A\|_F = \sqrt{tr(A^T A)}$ and $\|B\|_F = \sqrt{tr(B^T B)}$.

2.2 Problem Statement

The underlying interaction topology among N agents is determined by a digraph $\mathcal{G}(\mathcal{A})$ associated with $\mathcal{V} = \{1, \cdots, N\}$ as the set of nodes where each node represents an agent in the considered MAS. It is assumed that there are M ($M < N$) leaders and $N - M$ followers in the considered MAS. Assume further that the agents labeled from 1 to M ($M > 1$) are the leaders, and the agents labeled from $M + 1$ to N are the followers. Usually, the leaders take the role of exosystems or command generators generating desired trajectories to be tracked by followers. Based on the above statements, one has that it is reasonable to further assume that the leaders have no neighbor, indicating that the evolution of each leader will not be affected by those of the other leaders or the followers. Based on the aforementioned analysis, one obtains that the nodes labeled by $1, \cdots, M$ in graph $\mathcal{G}(\mathcal{A})$ have no neighbor. For convenience, use $\mathbb{L} = \{1, \cdots, M\}$ and $\mathbb{F} = \{M + 1, \cdots, N\}$ to represent, respectively, the leader and the follower sets.

Since the nodes labeled by $1, \cdots, M$ in graph $\mathcal{G}(\mathcal{A})$ have no neighbor, the Laplacian matrix L of digraph $\mathcal{G}(\mathcal{A})$ can be written as

$$L = \begin{bmatrix} 0_M & 0_{M \times (N-M)} \\ L_1 & L_2 \end{bmatrix} \in \mathbb{R}^{N \times N}, \tag{1}$$

where $L_1 \in \mathbb{R}^{(N-M) \times M}$, 0_M and $0_{M \times (N-M)}$ are respectively the $M \times M$ and $M \times (N - M)$ zero matrices, and $L_2 \in \mathbb{R}^{(N-M) \times (N-M)}$.

Assumption 1. For each follower $i \in \mathbb{F}$, there exists at least one leader $j \in \mathbb{L}$ from which there is a directed path to follower i.

Assumption 2. The induced subgraph with vertex set $\mathbb{F}$ in $\mathcal{G}(\mathcal{A})$ is detailed balanced, that is, there exists a vector $\phi = (\phi_{M+1}, \cdots, \phi_N)^T$ with $\phi_i > 0$ for all $i = M + 1, \cdots, N$, such that $\Phi\mathcal{A}_2 = \mathcal{A}_2^T\Phi$, where $\mathcal{A}_2$ is the adjacency matrix of the induced subgraph with vertex set $\mathbb{F}$ in $\mathcal{G}(\mathcal{A})$, and $\Phi = \text{diag}\{\phi_{M+1}, \cdots, \phi_N\}$.

Lemma 2 [5]. Under Assumption 1, one gets that all the eigenvalues of L_2 (defined in (1)) have positive real parts, each row of $-L_2^{-1}L_1$ has sum equal to 1, and each entry of $-L_2^{-1}L_1$ is nonnegative.

The evolution equations of leaders and followers are, respectively, given as

$$\dot{x}_i(t) = Ax_i(t) + Bu_i(t), \quad i \in \mathbb{L}, \tag{2}$$

and

$$\dot{x}_i(t) = Ax_i(t) + B[f_i(x_i(t)) + g_i(t) + u_i(t)], \quad i \in \mathbb{F}, \tag{3}$$

where $A \in \mathbb{R}^{n \times n}$, $x_i(t) \in \mathbb{R}^n$ is the state vector of the i-th agent, $B \in \mathbb{R}^{n \times m}$, $u_i^l(t)$ is the control input of leader i, $u_i(t) \in \mathbb{R}^m$ is the control input acting on follower i, $f_i(x_i(t))$ represents the unknown input nonlinearity which is assumed to be smooth, $g_i(t) \in \mathbb{R}^m$ describes the bounded matching disturbances such that

$$\|g_i(t)\|_\infty \leq \kappa_0, \tag{4}$$

for some given scalar $\kappa_0 > 0$. According to Stone-Weierstrass approximation theorem [11] and the fact that the nonlinear functions $f_i(x_i(t))$, $i \in \mathbb{F}$, in (3) are smooth, $f_i(x_i(t))$ can be thus approximated on a compact set $\Omega \subset \mathbb{R}^m$ by

$$f_i(x_i(t)) = W_i^T \varphi_i(x_i(t)) + \epsilon_i, \quad \forall x_i(t) \in \Omega, \tag{5}$$

where $\varphi_i(\cdot) : \mathbb{R}^n \to \mathbb{R}^s$ is a known basis function, $W_i \in \mathbb{R}^{s \times m}$ represents ideal neural network (NN) weight matrix which is a constant real matrix, $\epsilon_i \in \mathbb{R}^m$ is the NN approximation error vector such that $\|\epsilon_i\|_\infty \leq \epsilon_M$ for all $i \in \mathbb{F}$. For convenience, let $W = \text{diag}\{W_{M+1}, \cdots, W_N\}$. One then gets that there exists a positive scalar W_M to make the following holds: $\|W\|_F \leq W_M$.

Definition 1. *Quasi-containment of MASs with leaders given by (2) and followers given by (3) is said to be achieved, if there exist some nonnegative scalars $p_{ij} \geq 0$ with $\sum_{j=1}^M p_{ij} = 1$, such that*

$$\lim_{t \to \infty} \left\| x_i(t) - \Sigma_{j=1}^M p_{ij} x_j(t) \right\| \leq \varpi, \quad i \in \mathbb{F}, \tag{6}$$

for some given positive scalar ϖ. Asymptotic containment of MASs with leaders given by (2) and followers given by (3) is said to be achieved if $\varpi = 0$ in (6).

3   Main Theoretical Results

Generally, W_i in (5) is unknown. To compensate for the unknown nonlinearities and motivated by the consensus protocols given in [12], a new kind of neuro-adaptive based containment controller is designed in this paper:

$$u_i(t) = -\alpha K\delta_i(t) - \beta \text{sign}(K\delta_i(t)) - \hat{W}_i^T(t)\varphi_i(x_i(t)), \quad i \in \mathbb{F}, \tag{7}$$

of which $\delta_i(t) = \sum_{j=1}^{N} a_{ij}(x_i(t) - x_j(t))$; α and β are the coupling strength to be chosen, $K \in \mathbb{R}^{m \times n}$ is the feedback gain matrix to be designed; $sign(\cdot)$ denotes the element-wise signum operation; $\hat{W}_i(t)$ is the current estimation of the ideal weights for follower i.

3.1 Quasi-Containment of Uncertain MASs

Quasi-containment problem of the considered MASs is studied in this subsection. To complete the goal of quasi-containment, the following neuro-adaptive evolution law for $\hat{W}_i(t)$ in (7) is proposed:

$$\dot{\hat{W}}_i(t) = \nu_i[\phi_i \varphi_i(t)\delta_i^T(t)(P^{-1}B) - c_i\hat{W}_i(t)], \quad i \in \mathbb{F}, \qquad (8)$$

of which ν_i and c_i are two positive scalars, ϕ_i is provided in Assumption 2, P is a positive definite matrix to be designed later. In this paper, the trajectory solutions of all differential systems with non-smooth right-hands should be considered as those in the sense of *Filippov*.

Take $\delta(t) = (\delta_{M+1}^T(t), \cdots, \delta_N^T(t))^T$, $x_f(t) = (x_{M+1}^T(t), \cdots, x_N^T(t))^T$, and $x_l(t) = (x_1^T(t), \cdots, x_M^T(t))^T$. Obviously, $\delta(t) = (L_1 \otimes I_n)x_l(t) + (L_2 \otimes I_n)x_f(t)$. Set $e(t) = x_f(t) - (-L_2^{-1}L_1 \otimes I_n)x_l(t)$ as the containment error vector of the considered networking agents. It can be derived from the above analysis that $e(t) = (L_2^{-1} \otimes I_n)\delta(t)$. This means that

$$\|e(t)\| \le \varrho\|\delta(t)\|, \qquad (9)$$

with ϱ being the largest singular value of L_2^{-1}, i.e., $\varrho = \sqrt{\lambda_{max}\left((L_2^{-1})^T L_2^{-1}\right)}$. Then, combing (7) together with (2)–(5), we have

$$\begin{aligned}
\dot{\delta}(t) = &[(I_{N-M} \otimes A) - \alpha(L_2 \otimes BK)]\delta(t) + (L_1 \otimes B)u_l(t) + (L_2 \otimes B)g(t) \\
&- (L_2 \otimes B)(\tilde{W}^T\Psi - \epsilon) - \beta(L_2 \otimes B)sign((I_{N-M} \otimes K)\delta(t)), \qquad (10)
\end{aligned}$$

where $u_l(t) = (u_1^T(t), \cdots, u_N^T(t))^T$, $g(t) = (g_{M+1}^T(t), \cdots, g_N^T(t))^T$, $\tilde{W} = \text{diag} \{\hat{W}_{M+1}(t) - W_{M+1}, \cdots, \hat{W}_N(t) - W_N\}$, $\epsilon = (\epsilon_{M+1}^T(t), \cdots, \epsilon_N^T(t))^T \in \mathbb{R}^{(N-M)m}$, $\Psi = (\varphi_{M+1}^T(t), \cdots, \varphi_N^T(t))^T \in \mathbb{R}^{(N-M)s}$. To facilitate the analysis in the next section, the following assumption is made.

Assumption 3. For any given $x_l(0) \in \mathbb{R}^{Mn}$, there exist two positive constants $\eta(x_l(0))$ and $\hat{\eta}(x_l(0))$ such that

$$\|x_l(t)\|_\infty \le \eta(x_l(0)), \quad \|u_l(t)\|_\infty \le \hat{\eta}(x_l(0)), \quad \text{for all } t \ge 0. \qquad (11)$$

For notational brevity, let $\phi_{\min} = \min_{i=M+1,M+2,\cdots,N}\phi_i$.

Theorem 1. *Suppose that Assumptions 1–3 hold and the matrix pair (A, B) is stabilizable. Then, quasi-containment for MASs with leaders given by (2) and followers given by (3) under protocol (7) with $\beta > \epsilon_M + \hat{\eta}(x_l(0)) + \kappa_0$,*

$\alpha > \frac{\chi_0 \lambda_{max}(\Phi L_2^{-1})}{2\phi_{min}}$ for some given $\chi_0 > 0$ and $K = B^T P^{-1}$, where $P > 0$ is a solution of the following linear matrix inequality (LMI):

$$AP + PA^T - \chi_0 BB^T + \theta_1 P < 0, \tag{12}$$

of which κ_0 is defined in (4), θ_1 is a given positive scalar.

Proof. Under Assumption 2, one knows that there exists a positive vector $\phi = (\phi_{M+1}, \phi_{M+2}, \cdots, \phi_N)^T > 0$ such that $\Phi L_2 = L_2^T \Phi$ of which $\Phi = \text{diag}\{\phi_{M+1}, \phi_{M+2}, \cdots, \phi_N\} > 0$. On the other hand, it can be obtained from Assumption 1 and Lemma 2 that L_2 is nonsingular. Since $\Phi > 0$, we obtain that ΦL_2 is also a nonsingular matrix. The above analysis indicates that ΦL_2 is a nonsingular and symmetric real matrix. Thus, ΦL_2 is positive definite. As $L_2 \Phi^{-1} = \Phi^{-1}(\Phi L_2)\Phi^{-1}$, one knows $L_2 \Phi^{-1}$ is also positive definite. Based upon the above analysis, we may constructive the following Lyapunov function for system (10):

$$V(t) = \delta^T(t)(\Phi L_2^{-1} \otimes P^{-1})\delta(t) + \sum_{i=M+1}^{N} tr\left(\frac{1}{\nu_i}\tilde{W}_i^T(t)\tilde{W}_i(t)\right), \tag{13}$$

where $\tilde{W}_i(t) = \hat{W}_i(t) - W_i, i = M+1, \cdots, N, P > 0$ is a solution of LMI (12). Calculating the time derivative of $V(t)$ along the solution of (10) and invoking $K = B^T P^{-1}$ give that

$$\dot{V}(t) = \delta^T(t)\left[\Phi L_2^{-1} \otimes (P^{-1}A + A^T P^{-1}) - 2\alpha\Phi \otimes (P^{-1}BB^T P^{-1})\right]\delta(t)$$
$$+ 2\delta^T(t)(\Phi L_2^{-1}L_1 \otimes P^{-1}B)u_l(t) + 2\delta^T(t)(\Phi \otimes P^{-1}B)g(t)$$
$$- 2\delta^T(t)(\Phi \otimes P^{-1}B)(\tilde{W}^T\Psi - \epsilon) - 2\beta\|(\Phi \otimes B^T P^{-1})\delta(t)\|_1 \tag{14}$$
$$+ 2\sum_{i=M+1}^{N} tr\left(\tilde{W}_i^T(t)\phi_i\varphi_i(t)\delta_i^T(t)(P^{-1}B)\right) - 2\sum_{i=M+1}^{N} tr\left(c_i\tilde{W}_i^T(t)\hat{W}_i(t)\right).$$

By Hölders inequality, one obtains

$$2\delta^T(t)(\Phi L_2^{-1}L_1 \otimes P^{-1}B)u_l(t)$$
$$\leq 2\|(L_2^{-1}L_1 \otimes I_n)u_l(t)\|_\infty \cdot \|(\Phi \otimes B^T P^{-1})\delta(t)\|_1. \tag{15}$$

Based on the fact $\|(L_2^{-1}L_1 \otimes I_n)u_l(t)\|_\infty \leq \|(L_2^{-1}L_1 \otimes I_n)\|_\infty \cdot \|u_l(t)\|_\infty \leq \hat{\eta}(x_l(0))$, it can be got from (15) that

$$2\delta^T(t)(\Phi L_2^{-1}L_1 \otimes P^{-1}B)u_l(t) \leq 2\hat{\eta}(x_l(0))\|(\Phi \otimes B^T P^{-1})\delta(t)\|_1. \tag{16}$$

Similarly, one gets $2\delta^T(t)(\Phi \otimes P^{-1}B)g(t) \leq 2\kappa_0\|(\Phi \otimes B^T P^{-1})\delta(t)\|_1$. According to the above analysis, we may further get

$$\dot{V}(t) \leq -\theta_1\delta^T(t)\left(\Phi L_2^{-1} \otimes P^{-1}\right)\delta(t) - 2\sum_{i=M+1}^{N} tr\left(c_i\tilde{W}_i^T(t)\tilde{W}_i(t)\right)$$
$$+ 2\sum_{i=M+1}^{N} tr\left(c_i\tilde{W}_i^T(t)W_i\right), \tag{17}$$

where the last inequality is derived by using LMI (12), the facts $\beta > \epsilon_M + \widehat{\eta}(x_l(0)) + \kappa_0$ and $-2\sum_{i=M+1}^N tr\left(c_i \tilde{W}_i^T(t)\tilde{W}_i(t)\right) \le 0$. By using Lemma 1, it can be got from (17) that

$$\dot{V}(t) \le -c_0 V(t) + 2c_M W_M \|\tilde{W}(t)\|_F, \tag{18}$$

of which $c_0 = \min\{\theta_1, 2c_m\nu_m\}$. Some mathematical calculations give that

$$2c_M W_M \|\tilde{W}(t)\|_F \le 2c_M W_M \sqrt{\nu_M}\sqrt{V(t)} \tag{19}$$

where $\nu_M = \max_{i\in\{M+1,M+2,\cdots,N\}}\nu_i$. Based upon the above analysis, we get

$$\frac{d}{dt}\left(\sqrt{V(t)}\right) \le -(c_0/2)\sqrt{V(t)} + \bar{c}_0/2, \tag{20}$$

where $\bar{c}_0 = 2c_M W_M \sqrt{\nu_M}$. Integrating both sides of (20) from 0 to t gives

$$\sqrt{V(t)} \le \sqrt{V(0)}e^{-\frac{c_0}{2}t} + \frac{\bar{c}_0}{c_0}(1 - e^{-\frac{c_0}{2}t}) \le \sqrt{V(0)} + \frac{\bar{c}_0}{c_0}. \tag{21}$$

The above inequalities indicate that $\|\delta(t)\|$ is uniformly bounded for any given $V(0)$. Recalling $\lambda_{min}(\Phi L_2^{-1} \otimes P^{-1})\|\delta(t)\|^2 \le V(t)$, we may get

$$\|\delta(t)\| \le (\sqrt{V(0)} + \bar{c}_0/c_0)/\sqrt{\lambda_{min}(\Phi L_2^{-1} \otimes P^{-1})}. \tag{22}$$

The proof can be thus completed by combing (9) and (22).   ∎

3.2   Asymptotic Containment of Uncertain MASs

Asymptotic containment problem is addressed in this subsection. Based on the analysis given in the Subsect. 3.1 and motivated by the structures of distributed adaptive controllers given in [12,13], the following neuro-adaptive evolution law for $\hat{W}_i(t)$ in (7) is proposed:

$$\begin{aligned}
\dot{\hat{W}}_i(t) &= \nu_i[\phi_i\varphi_i(t)\delta_i^T(t)(P^{-1}B) - c_i(\hat{W}_i(t) - \overline{W}_i(t))], \\
\dot{\overline{W}}_i(t) &= c_i(\hat{W}_i(t) - \overline{W}_i(t)), \quad i \in \mathbb{F},
\end{aligned} \tag{23}$$

of which ν_i and c_i are two positive scalars, P is a positive definite matrix to be specified later, ϕ_i is provided in Assumption 2, $\overline{W}_i(t)$ is an auxiliary weight matrix.

Theorem 2. *Suppose that Assumptions 1–3 hold and the matrix pair (A, B) is stabilizable. Then, asymptotic containment for MASs with leaders given by (2) and followers given by (3) under protocol (7) associated with adaptive law (23) will be achieved if the control parameters are appropriately designed such that $\alpha > \frac{\chi_1\lambda_{max}(\Phi L_2^{-1})}{2\phi_{min}}$ for some given $\chi_1 > 0, \beta > \epsilon_M + \widehat{\eta}(x_l(0)) + \kappa_0$ and $K = B^T P^{-1}$, where $P > 0$ is a solution of the following LMI:*

$$AP + PA^T - \chi_1 BB^T + \theta_2 P < 0, \tag{24}$$

of which κ_0 is defined in (4), θ_2 is a given positive scalar.

Proof. Consider the following Lyapunov function for system (10):

$$V_1(t) = \delta^T(t)\big(\Phi L_2^{-1} \otimes P^{-1}\big)\delta(t) + \sum_{i=M+1}^{N} tr\Big(\frac{1}{\nu_i}\tilde{W}_i^T(t)\tilde{W}_i(t)\Big)$$

$$+ \sum_{i=M+1}^{N} tr\Big(\overline{\tilde{W}}_i^T(t)\overline{\tilde{W}}_i(t)\Big), \tag{25}$$

where $\tilde{W}_i(t) = \hat{W}_i(t) - W_i$ and $\overline{\tilde{W}}_i(t) = \overline{W}_i(t) - W_i$, $P > 0$ is a solution of LMI (24). This theorem can be then proven by performing some similar steps as those in the proof of Theorem 1.　∎

4　Conclusions

Quasi-containment and asymptotic containment problems have been investigated in this paper for a class of networking linear uncertain MASs with multiple dynamic leaders. The dynamic evolution of leaders in the considered MASs model may be subjected to nonzero control inputs and the evolution of followers may be effected by unknown dynamics. A class of containment controllers consisting of a linear feedback term, a neuro-adaptive approximation term and a non-smooth feedback term have been constructed to ensure quasi-containment. The results are extended to asymptotic containment by designing a new kind of weighting matrix update law.

References

1. Yu, W., Wen, G., Chen, G., Cao, J.: Distributed Cooperative Control of Multiagent Systems. Wiley Press, Singapore (2016)
2. Wen, G., Hu, G., Hu, J., Shi, X., Chen, G.: Frequency regulation of source-grid-load systems: a compound control strategy. IEEE Trans. Ind. Informat. **12**, 69–78 (2016)
3. Ji, M., Egerstedt, M., Ferrari-Trecate, G., Buffa, A.: Hierarchical containment control in heterogeneous mobile networks. In: Porceedings of the 17th International Symposium Mathematical Theory of Networks and Systems, Kyoto, Japan, pp. 2227–2231 (2006)
4. Cao, Y., Ren, W., Egerstedt, M.: Distributed containment control with multiple stationary or dynamic leaders in fixed and switching directed networks. Automatica **48**, 1586–1597 (2012)
5. Mei, J., Ren, W., Ma, G.: Distributed containment control for Lagrangian networks with parametric uncertainties under a directed graph. Automatica **48**, 653–659 (2012)
6. Wen, G., Duan, Z., Zhao, Y., Yu, W., Cao, J.: Robust containment tracking of uncertain linear multi-agent systems: a non-smooth control approach. Int. J. Control **87**, 2522–2534 (2014)
7. Li, Z., Duan, Z., Ren, W., Feng, G.: Containment control of linear multi-agent systems with multiple leaders of bounded inputs using distributed continuous controllers. Int. J. Robust Nonlinear Control **25**, 2101–2121 (2015)

8. Wen, G., Zhao, Y., Duan, Z., Yu, W., Chen, G.: Containment of higher-order multi-leader multi-agent systems: a dynamic output approach. IEEE Trans. Autom. Control **61**, 1135–1140 (2016)
9. Chu, T., Wang, L., Chen, T., Mu, S.: Complex emergent dynamics of anisotropic swarms: convergence vs oscillation. Chaos, Solitons Fractals **30**, 875–885 (2006)
10. Horn, R.A., Johnson, C.R.: Matrix Analysis. Cambridge University Press, Cambridge (1985)
11. Stone, M.: The generalized Weierstrass approximation theorem. Math. Mag. **21**, 237–254 (1948)
12. Sun, J., Geng, Z.: Adaptive consensus tracking for linear multi-agent systems with heterogeneous unknown nonlinear dynamics. Int. J. Robust Nonlinear Control **26**, 154–173 (2016)
13. Mei, J., Ren, W., Chen, J.: Distributed consensus of second-order multi-agent systems with heterogeneous unknown inertias and control gains under a directed graph. IEEE Trans. Autom. Control **61**, 2019–2034 (2016)

Signal, Image and Video Processing

Abnormal Event Detection in Videos Using Spatiotemporal Autoencoder

Yong Shean Chong[✉] and Yong Haur Tay

Lee Kong Chian Faculty of Engineering Science,
Universiti Tunku Abdul Rahman, 43000 Kajang, Malaysia
yshean@1utar.my, tayyh@utar.edu.my

Abstract. We present an efficient method for detecting anomalies in videos. Recent applications of convolutional neural networks have shown promises of convolutional layers for object detection and recognition, especially in images. However, convolutional neural networks are supervised and require labels as learning signals. We propose a spatiotemporal architecture for anomaly detection in videos including crowded scenes. Our architecture includes two main components, one for spatial feature representation, and one for learning the temporal evolution of the spatial features. Experimental results on Avenue, Subway and UCSD benchmarks confirm that the detection accuracy of our method is comparable to state-of-the-art methods at a considerable speed of up to 140 fps.

Keywords: Anomaly detection · Feature learning · Regularity · Autoencoder

1 Introduction

With the rapid growth of video data, there is an increasing need not only for recognition of objects and their behaviour, but in particular for detecting the rare, interesting occurrences of unusual objects or suspicious behaviour in the large body of ordinary data. Finding such abnormalities in videos is crucial for applications ranging from automatic quality control to visual surveillance.

Meaningful events that are of interest in long video sequences, such as surveillance footage, often have an extremely low probability of occurring. As such, manually detecting such events, or anomalies, is a very meticulous job that often requires more manpower than is generally available. This has prompted the need for automated detection and segmentation of sequences of interest. However, present technology requires an enormous amount of configuration efforts on each video stream prior to the deployment of the video analysis process, even with that, those events are based on some predefined heuristics, which makes the detection model difficult to generalize to different surveillance scenes.

Recent effort on detecting anomalies by treating the task as a binary classification problem (normal and abnormal) [12] proved it being effective and accurate, but the practicality of such method is limited since footages of abnormal

© Springer International Publishing AG 2017
F. Cong et al. (Eds.): ISNN 2017, Part II, LNCS 10262, pp. 189–196, 2017.
DOI: 10.1007/978-3-319-59081-3_23

events are difficult to obtain due to its rarity. Therefore, many researchers have turned to models that can be trained using little to no supervision, including spatiotemporal features [3,11], dictionary learning [10] and autoencoders [7]. Unlike supervised methods, these methods only require unlabelled video footages which contain little or no abnormal event, which are easy to obtain in real-world applications.

This paper presents a novel framework to represent video data by a set of general features, which are inferred automatically from a long video footage through a deep learning approach. Specifically, a deep neural network composed of a stack of convolutional autoencoders was used to process video frames in an unsupervised manner that captured spatial structures in the data, which, grouped together, compose the video representation. Then, this representation is fed into a stack of convolutional temporal autoencoders to learn the regular temporal patterns.

Our proposed method is domain free (i.e., not related to any specific task, no domain expert required), does not require any additional human effort, and can be easily applied to different scenes. To prove the effectiveness of the proposed method we apply the method to real-world datasets and show that our method consistently outperforms similar methods while maintaining a short running time.

2   Methodology

The method described here is based on the principle that when an abnormal event occurs, the most recent frames of video will be significantly different than the older frames. Inspired by [2], we train an end-to-end model that consists of a spatial feature extractor and a temporal encoder-decoder which together learns the temporal patterns of the input volume of frames. The model is trained with video volumes consists of only normal scenes, with the objective to minimize the reconstruction error between the input video volume and the output video volume reconstructed by the learned model. After the model is properly trained, normal video volume is expected to have low reconstruction error, whereas video volume consisting of abnormal scenes is expected to have high reconstruction error. By thresholding on the error produced by each testing input volumes, our system will be able to detect when an abnormal event occurs.

2.1   Feature Learning

We propose a convolutional spatiotemporal autoencoder to learn the regular patterns in the training videos. Our proposed architecture consists of two parts — spatial autoencoder for learning spatial structures of each video frame, and temporal encoder-decoder for learning temporal patterns of the encoded spatial structures. As illustrated in Fig. 1, the spatial encoder and decoder have two convolutional and deconvolutional layers respectively, while the temporal encoder is a three-layer convolutional long short term memory (LSTM) model.

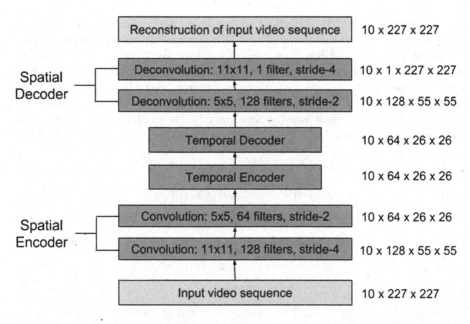

Fig. 1. Our proposed network architecture. It takes a sequence of length T as input, and output a reconstruction of the input sequence. The numbers at the rightmost denote the output size of each layer. The spatial encoder takes one frame at a time as input, after which $T = 10$ frames have been processed, the encoded features of 10 frames are concatenated and fed into temporal encoder for motion encoding. The decoders mirror the encoders to reconstruct the video volume.

Convolutional layers are well-known for its superb performance in object recognition, while LSTM model is widely used for sequence learning and time-series modelling and has proved its performance in applications such as speech translation and handwriting recognition.

Autoencoder. Autoencoders, as the name suggests, consist of two stages: encoding and decoding. It was first used to reduce dimensionality by setting the number of encoder output units less than the input. The model is usually trained using back-propagation in an unsupervised manner, by minimizing the reconstruction error of the decoding results from the original inputs. With the activation function chosen to be nonlinear, an autoencoder can extract more useful features than some common linear transformation methods such as PCA.

Spatial Convolution. The primary purpose of convolution in case of a convolutional network is to extract features from the input image. Convolution preserves the spatial relationship between pixels by learning image features using small squares of input data. Suppose that we have some $n \times n$ square input layer which is followed by the convolutional layer. If we use an $m \times m$ filter W, the convolutional layer output will be of size $(n - m + 1) \times (n - m + 1)$.

Convolutional LSTM. A variant of the long short term memory (LSTM) architecture, namely Convolutional LSTM (ConvLSTM) model was introduced by Shi et al. in [8] and has been recently utilized by Patraucean et al. in [6] for video frame prediction. Compared to the usual fully connected LSTM (FC-LSTM), ConvLSTM has its matrix operations replaced with convolutions. By using convolution for both input-to-hidden and hidden-to-hidden connections, ConvLSTM requires fewer weights and yield better spatial feature maps. The formulation of the ConvLSTM unit can be summarized with (7) through (12).

$$f_t = \sigma(W_f * [h_{t-1}, x_t, C_{t-1}] + b_f) \tag{1}$$

$$i_t = \sigma(W_i * [h_{t-1}, x_t, C_{t-1}] + b_i) \tag{2}$$

$$\hat{C}_t = tanh(W_C * [h_{t-1}, x_t] + b_C) \tag{3}$$

$$C_t = f_t \otimes C_{t-1} + i_t \otimes \hat{C}_t \tag{4}$$

$$o_t = \sigma(W_o * [h_{t-1}, x_t, C_{t-1}] + b_o) \tag{5}$$

$$h_t = o_t \otimes tanh(C_t) \tag{6}$$

In contrast to the FC-LSTM, the input is fed in as images, while the set of weights for every connection is replaced by convolutional filters (the symbol $*$ denotes a convolution operation). This allows ConvLSTM work better with images than the FC-LSTM due to its ability to propagate spatial characteristics temporally through each ConvLSTM state. Note that this convolutional variant also adds an optional 'peephole' connections to allow the unit to derive past information better.

2.2 Regularity Score

Once the model is trained, we can evaluate our models performance by feeding in testing data and check whether it is capable of detecting abnormal events while keeping false alarm rate low. To better compare with [2], we used the same formula to calculate the regularity score for all frames, the only difference being the learned model is of a different kind. The reconstruction error e of all pixel values in frame t of the video sequence is taken as the Euclidean distance between the input frame $x(t)$ and the reconstructed frame $f_W(x(t))$:

$$e(t) = ||x(t) - f_W(x(t))||_2 \tag{7}$$

where f_W is the learned weights by the spatiotemporal model. We then compute the abnormality score $s_a(t)$ by scaling between 0 and 1. Subsequently, regularity score $s_r(t)$ can be simply derived by subtracting abnormality score from 1:

$$s_a(t) = \frac{e(t) - e(t)_{min}}{e(t)_{max}} \tag{8}$$

$$s_r(t) = 1 - s_a(t) \tag{9}$$

3   Experiments

3.1   Datasets

We train our model on five most commonly used benchmarking datasets: Avenue [3], UCSD Ped1 and Ped2 [4], Subway entrance and exit datasets [1]. All videos are taken from a fixed position for each dataset. All training videos contain only normal events. Testing videos have both normal and abnormal events.

3.2   Results and Analysis

Quantitative Analysis: ROC and AUC. Table 1 shows the frame-level AUC and EER of our and of other methods on all five datasets. We outperform all other considered methods in respect to frame-level EER.

Table 1. Comparison of area under ROC curve (AUC) and Equal Error Rate (EER) of various methods. Higher AUC and lower EER are better. Most papers did not publish their AUC/EER for avenue, subway entrance and exit dataset.

Method	AUC/EER (%)				
	Ped1	Ped2	Avenue	Subway entrance	Subway exit
Adam [1]	77.1/38.0	−/42.0			
SF [5]	67.5/31.0	55.6/42.0			
MPPCA [4]	66.8/40.0	69.3/30.0		N/A	
MPPCA+SF [4]	74.2/32.0	61.3/36.0			
HOFME [9]	72.7/33.1	87.5/20.0	N/A	81.6/**22.8**	84.9/17.8
ConvAE [2]	81.0/27.9	**90.0**/21.7	70.2/25.1	**94.3**/26.0	80.7/9.9
Ours	**89.9/12.5**	87.4/**12.0**	**80.3/20.7**	84.7/23.7	**94.0/9.5**

We also present a run-time analysis on our proposed abnormal event detection system, on CPU (Intel Xeon E5-2620) and GPU (NVIDIA Maxwell Titan X) respectively, in Table 2. The total time taken per frame is well less than a quarter second per frame for both CPU and GPU configuration.

Qualitative Analysis: Visualising Frame Regularity. Figures 2, 3, and 4 illustrate the output of the proposed system on samples of the Avenue dataset, Subway entrance and exit scenes respectively; our method detects anomalies correctly in these cases even in crowded scenes.

From Fig. 5, it is easy to see that our method has detected more abnormal events with fewer false alarms compared to [2]. Also, as observed in Fig. 6, our method is able to produce higher regularity score during normal activities and lower scores when there are abnormalities.

Table 2. Details of run-time during testing (second/frame).

	Time (in sec)			
	Preprocessing	Representation	Classifying	Total
CPU	0.0010	0.2015	0.0002	0.2027 (~5fps)
GPU	0.0010	0.0058	0.0002	0.0070 (~143fps)

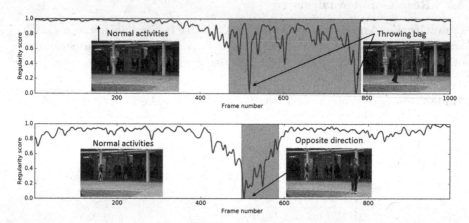

Fig. 2. Regularity score of video #5 (top) and #15 (bottom) from the Avenue dataset.

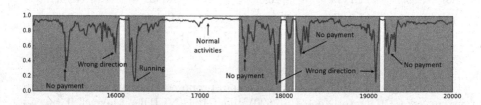

Fig. 3. Regularity score of frames 115000-120000 from the Subway Entrance video.

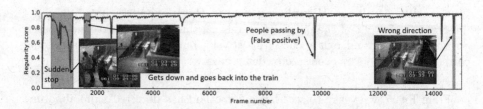

Fig. 4. Regularity score of frames 22500-37500 from the Subway Entrance video.

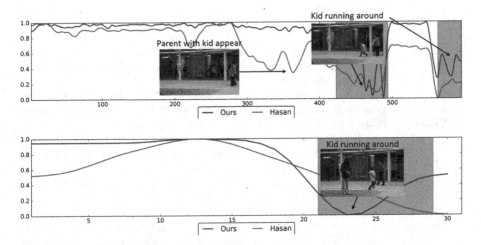

Fig. 5. Comparing our method with ConvAE [2] on Avenue dataset video #7 (top) and #8 (bottom). Best viewed in colour.

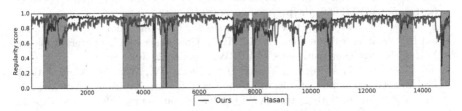

Fig. 6. Comparing our method with ConvAE [2] on Subway Entrance video frames 120000-144000. Best viewed in colour.

4   Conclusion

In this research, we have successfully applied deep learning to the challenging video anomaly detection problem. We formulate anomaly detection as a spatiotemporal sequence outlier detection problem and applied a combination of spatial feature extractor and temporal sequencer ConvLSTM to tackle the problem. The ConvLSTM layer not only preserves the advantages of FC-LSTM but is also suitable for spatiotemporal data due to its inherent convolutional structure. By incorporating convolutional feature extractor in both spatial and temporal space into the encoding-decoding structure, we build an end-to-end trainable model for video anomaly detection. The advantage of our model is that it is semi-supervised – the only ingredient required is a long video segment containing only normal events in a fixed view. Despite the models ability to detect abnormal events and its robustness to noise, depending on the activity complexity in the scene, it may produce more false alarms compared to other methods.

References

1. Adam, A., Rivlin, E., Shimshoni, I., Reinitz, D.: Robust real-time unusual event detection using multiple fixed-location monitors. IEEE Trans. Pattern Anal. Mach. Intell. **30**(3), 555–560 (2008)
2. Hasan, M., Choi, J., Neumann, J., Roy-Chowdhury, A.K., Davis, L.S.: Learning temporal regularity in video sequences. In: 2016 IEEE Conference on Computer Vision and Pattern Recognition (CVPR), pp. 733–742, June 2016
3. Lu, C., Shi, J., Jia, J.: Abnormal event detection at 150 fps in matlab. In: 2013 IEEE International Conference on Computer Vision, pp. 2720–2727, December 2013
4. Mahadevan, V., Li, W., Bhalodia, V., Vasconcelos, N.: Anomaly detection in crowded scenes. In: Proceedings of the IEEE Conference on Computer Vision and Pattern Recognition (CVPR), pp. 1975–1981 (2010)
5. Mehran, R., Oyama, A., Shah, M.: Abnormal crowd behavior detection using social force model. In: 2009 IEEE Computer Society Conference on Computer Vision and Pattern Recognition Workshops, CVPR Workshops 2009, pp. 935–942 (2009)
6. Patraucean, V., Handa, A., Cipolla, R.: Spatio-temporal video autoencoder with differentiable memory. In: International Conference on Learning Representations (2015), pp. 1–10 (2016). http://arxiv.org/abs/1511.06309
7. Sabokrou, M., Fathy, M., Hoseini, M., Klette, R.: Real-time anomaly detection and localization in crowded scenes. In: 2015 IEEE Conference on Computer Vision and Pattern Recognition Workshops (CVPRW), pp. 56–62, June 2015
8. Shi, X., Chen, Z., Wang, H., Yeung, D.Y., Wong, W., Woo, W.: Convolutional LSTM network: a machine learning approach for precipitation nowcasting. In: Proceedings of the 28th International Conference on Neural Information Processing Systems, pp. 802–810. NIPS 2015. MIT Press, Cambridge, MA, USA (2015). http://dl.acm.org/citation.cfm?id=2969239.2969329
9. Wang, T., Snoussi, H.: Histograms of optical flow orientation for abnormal events detection. In: IEEE International Workshop on Performance Evaluation of Tracking and Surveillance, PETS, pp. 45–52 (2013)
10. Yen, S.H., Wang, C.H.: Abnormal event detection using HOSF. In: 2013 International Conference on IT Convergence and Security, ICITCS 2013 (2013)
11. Zhao, B., Fei-Fei, L., Xing, E.P.: Online detection of unusual events in videos via dynamic sparse coding. In: Proceedings of the IEEE Computer Society Conference on Computer Vision and Pattern Recognition, pp. 3313–3320 (2011)
12. Zhou, S., Shen, W., Zeng, D., Fang, M., Wei, Y., Zhang, Z.: Spatial-temporal convolutional neural networks for anomaly detection and localization in crowded scenes. Sig. Process. Image Commun. **47**, 358–368 (2016)

Enhancing Mastcam Images for Mars Rover Mission

Minh Dao[1], Chiman Kwan[1(✉)], Bulent Ayhan[1], and James F. Bell[2]

[1] Applied Research LLC, Rockville, MD, USA
ducminh174@gmail.com, bulentayhan@gmail.com,
chiman.kwan@signalpro.net
[2] Arizona State University, Tempe, AZ, USA
Jim.Bell@asu.edu

Abstract. This paper summarizes some new results in improving the left Mastcam images of the Mars Science Laboratory (MSL) onboard the Mars rover Curiosity. There are two multispectral Mastcam imagers, having 9 bands in each. The left imager has wide field of view, but low resolution whereas the right imager is just the opposite. Our goal is to investigate the possibility of fusing the left and right images to form high spatial resolution and high spectral resolution data cube so that stereo images and data clustering performance can be improved. Many pansharpening algorithms have been investigated. Actual Mastcam images were used in our experiments. Preliminary results indicate that the pansharpened images can indeed enhance the data clustering performance using both objective and subjective evaluations.

Keywords: Mastcam · Curiosity rover · Pansharpening · Image fusion

1 Introduction

Curiosity rover landed on Mars in 2012. Onboard the Curiosity, there is the Mars Science Laboratory (MSL), which has a few instruments [1, 24, 25] for characterizing the Mars surface. There are two Mastcam multispectral imagers in MSL, separated by 24.2 cm [1]. Specifically, the left Mastcam (34 mm focal length) has three times the field of view of the right Mastcam (100 mm focal length). That is, the right imager has 3 times higher resolution than that of the left. Each camera has 9 bands with 6 overlapping bands. For stereo image formation (combining the left and right images to create stereo pairs) and image fusion (merging of the left and right bands to form a 12-band image cube), the current practice is to downsample the right images to the resolution of the left, avoiding artifacts due to Bayer pattern and also lossy JPEG compression. This practice is certainly practical, but may limit the full potential of Mastcams. First, although down-sampling of the right images can preserve the spectral integrity and avoid certain artifacts of the image data, the process will throw away very informative high spatial resolution pixels in the right bands. Second, the current stereo images have lower resolution, which may degrade the augmented reality or virtual reality experience of science fans.

In recent years, significant advances have been made in image super-resolution and new and high performance pansharpening algorithms have been proposed regularly. In

© Springer International Publishing AG 2017
F. Cong et al. (Eds.): ISNN 2017, Part II, LNCS 10262, pp. 197–206, 2017.
DOI: 10.1007/978-3-319-59081-3_24

light of the above development, a natural research question is: can we keep the 9 high resolution bands in the right imager and improve the resolution of all the left bands in the left imager while maintaining the spectral integrity of the left bands? Answering the above question will enhance both the user experience of using high resolution stereo images, as well as the surface characterization performance using 12 bands of high resolution images.

In this research, we investigated the applicability of state-of-the-art pansharpening algorithms to enhance the resolution of the left Mastcam images. Some algorithms were found to be not applicable. For the applicable ones, we performed extensive studies by using over 100 pairs of left and right Mastcam images, selected from over 500,000 images in the NASA database. Both objective and subjective metrics have been used to evaluate the different algorithms. Our studies show that it is indeed possible to enhance the spatial resolution of left bands while maintaining the spectral integrity of the left bands. Here, spectral integrity means that the pansharpened and fused data cube will not perform worse than that of using the images in the original resolution of the left imager in applications such as data clustering [26]. Moreover, our experiments also showed the pansharpened images can be merged with the right bands to form a high spatial resolution 12-band cube, which can improve the clustering performance. Both objective (5 metrics) and subjective evaluations have been used.

This paper is organized as follows. Section 2 briefly reviewed the Mastcam. Section 3 briefly describes those key algorithms used in this study. Section 4 summarizes the data used and all the experimental results. Finally, we conclude the paper and point out some research directions in Sect. 5.

2   Mastcam

Mastcam imager information is shown in Fig. 1. There are 6 overlapping bands and 3 non-overlapping bands (L3, L4 and L5 from the left camera and R3, R4, and R5 from the right camera). More details about Mastcam can be found in [1] and [2].

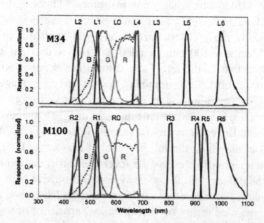

Fig. 1. Normalized MSL/Mastcam system-level spectral response profiles for the left eye M-34 camera (top panel) and the right eye M-100 camera (bottom panel) [1].

3 Key Algorithms

3.1 Image Registration

A two-step image alignment approach was developed [2] and the signal flow is shown in Fig. 2. The first step of the two-step image alignment approach is using RANSAC (Random Sample Consensus) technique [4] for an initial image alignment. In this first step, we use the two RGB stereo images from the left and right Mastcams. First, SURF features [5] and SIFT features [7] are extracted from the two stereo images. These features are then matched within the image pair. This is followed by applying RANSAC to estimate the geometric transformation. Assuming the right camera image is the reference image, the left camera image content is then projected into a new image that is aligned with the reference image using the geometric transformation.

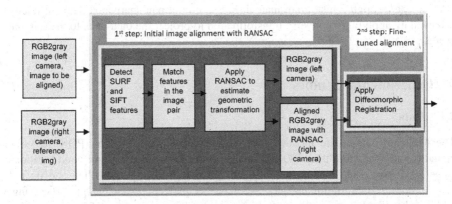

Fig. 2. Block diagram of the two-step image alignment approach.

The second step of the two-step alignment approach uses this aligned image with RANSAC and the left camera image as inputs and applies the Diffeomorphic Registration [6] technique. Diffeomorphic Registration is formulated as a constrained optimization problem, which is solved with a step-then-correct strategy [6]. This second step reduces the registration errors to subpixel levels so that pansharpening can be performed.

3.2 Pansharpening Algorithms

Pansharpening has found wide spread usage in many applications [3, 8–12, 19, 21]. The goal of pan-sharpening is to fuse a low-spatial resolution but high-spectral resolution multispectral (MS) data with a high-spatial resolution panchromatic image (PAN) of a different spectral band, resulting in a data cube with the spectral resolution of the former and the spatial resolution of the latter [8]. In our case, after the two-step registration, the aligned images from the left camera can be considered as blurred versions of the right ones. Therefore, we tentatively apply the pan-sharpening concept to sharpen the aligned versions from the left camera using high spatial resolution images in different filter bands from the right camera as the panchromatic reference image.

Pan-sharpening techniques can be classified into two main categories: (1) the component substitution (CS) approach and (2) the multiresolution analysis (MRA) approach [8]. The former is based on the substitution of a component with the PAN image and the latter relies on the injection of spatial details that are obtained through a multiresolution decomposition of the PAN image into the resampled MS bands. In this paper, we focus on CS-based approach which generally relies upon the projection of the higher spectral resolution image into another space that is capable to separate the spectral information from the spatial structure in different components. Under the assumption that the components containing the spatial structure of multispectral images at all spectral bands are highly correlated, the transformed low-resolution MS images can be enhanced by substituting the components containing the spatial structure of the PAN image through a histogram matching. The output pan-sharpened data are finally achieved by applying the inverse transformation to project the data back to the original space.

4   Experimental Studies

4.1   Data

The Mastcam dataset downloaded from the Planetary Data System (PDS) resource contains a total of more than 500,000 images collected at different times and locations since 2012. Since left and right Mastcams are independently controlled and do not always collect data simultaneously, we have to perform extensive pre-processing, which exhaustively screens through all images to only select pairs of image sets that consist of images of all available spectral bands in both left and right Mastcam cameras. After preprocessing and cleaning up the image sets, we can construct a total of 133 LR-pairs.

4.2   Experimental Results

We extensively compared the performance of enhancing the resolutions of aligned left Mastcam images to the same scale of right Mastcam images using various pan-sharpening algorithms including three conventional CS-based techniques, namely the fast intensity-hue-saturation (indicated as fast IHS or simply IHS) [9], Brovey transform (Brovey) [8], and Gram Schmidt (GS) [10]; and two more advanced CS-based algorithms which are band dependent spatial detail (BDSD) [11], and partial replacement adaptive CS (PRACS) [12]. In each case, the reference PAN image for each of the three blurred left images in non-overlapped bands was selected as the original right image of closest filter band (i.e. the selected PAN images for L4, L3 and L5 bands are R0R, R3 and R4 bands, respectively).

Furthermore, we used 6 overlapped bands to evaluate the performance of pan-sharpening by examining the correspondence between the super-resolved versions of aligned images and the original right images at the same filter bands. Figure 3 shows one example that illustrates the pan-sharpened outputs of the 5 pan-sharpening methods mentioned above for LR-pair that use R0 Gb filter band as PAN reference images to sharpen aligned image from L1 filter band. In each figure, the bigger red rectangles in the top-left corners display the magnified versions of the smaller ones. We can clearly observe the sharpening effects of all methods in comparison with the blurriness in the aligned images (sub-figure (b)). In this example, the peak signal-to-noise ratio (PSNR) of the pan-sharpened image to the original high-resolution right image of PRACS is highest, followed by BDSD algorithm. Lastly, Table 1 summarizes the comparison of PSNRs of the pan-sharpening results, averaged over all tested LR-pairs for the 6 overlapped bands. For each of the overlapped bands, we used the right band as the pan band to pansharpen the corresponding left band.

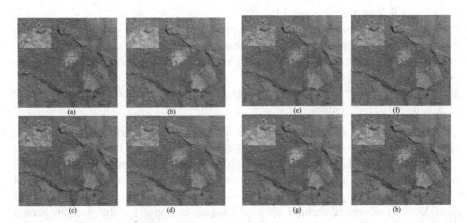

Fig. 3. Pan-sharpening performance of an LR-pair on sol 0726 taken on 08-22-2014 for L1 filter band aligned image using R0 Gb filter image as pan-sharpening reference. (a) Original R1 filter band image; (b) low-resolution aligned image from L1 filter band - PSNR = 27.33 dB; (c) pan-sharpening reference R0 Gb filter image; (d) pan-sharpened image of (b) via IHS - PSNR = 29.95 dB; (e) pan-sharpened image of (b) via Brovey - PSNR = 30.02 dB; (f) pan-sharpened image of (b) via GS - PSNR = 30.05 dB; (g) pan-sharpened image of (b) via BDSD - PSNR = 30.54 dB; and (h) pan-sharpened image of (b) via PRACS - PSNR = 31.45 dB. (Color figure online)

Table 1. Comparison of pan-sharpening performance averaged over all LR-pairs.

Methods	Averaged PSNRs (dB)
Aligned image (bicubic)	27.93
IHS	30.28
Brovey	30.32
GS	30.31
BDSD	31.04
PRACS	**31.52**

4.3   Influence of Image Fusion on Data Clustering Performance

In this section, we will evaluate the impact of Mastcam image fusion on some discrimination applications such as data clustering by comparing the performance of the combined 12-band image data with that of the original 9-band right image data. The clustering results were validated only for the registered results of the 2-step registration with right camera high resolution (i.e., the aligned left images are interpolated to have the same resolution scale of the right camera using bicubic interpolation). In this paper, we perform more thorough evaluation by comparing the outcome results of the original 9-band data with those of the 12-band fused versions after 2nd registration steps at right-camera resolutions, as well as the enhanced data via pan-sharpening algorithms. We only select BDSD and PRACS for comparison since they are the two methods that perform the best for Mastcam data among the techniques studied in the previous section. Specifically, we compare the 4 multispectral data versions:

1. Original 9-band right camera MS cube.
2. 12-band MS cube after 2nd registration step with M-100 resolution (higher-resolution option) using bicubic interpolation for the 3 non-overlapping bands.
3. 12-band MS cube with M-100 resolution using BDSD pan-sharpening for the 3 non-overlapping bands.
4. 12-band MS cube with M-100 resolution using PRACS pan-sharpening for the 3 non-overlapping bands.

It is further noted that all registration results used in this section use SIFT features, as it is observed to perform better than SURF features.

We verify the effectiveness of the registration-based Mastcam image fusion on data clustering using the benchmarked K-means [13] and Gaussian Mixture Model (GMM) [14] clustering models. Figure 4 visualizes the clustering results by color at pixel level of K-means method on one LR-pair example for the 4 multispectral Mastcam data versions listed above. The number of clusters are set to be six (suggested by the Gap statistic method [17] for estimating the best number of clusters in Mastcam data). In each figure, one clustering region are enlarged to show the performance in details. By visual inspection, we can have several observations:

(a) The clustering results of the two-step registration using 12 bands (9 bands from the right and 3 bicubic interpolated bands from the left) are better than that of using 9 bands. This implies that more spectral information will certainly enhance the clustering performance;
(b) With fewer randomly clustered pixels, pan-sharpened data offer the best clustering results, both in clustering performance as well as spatial detail. This is probably due to the fact that, unlike the bicubic interpolation, pansharpening algorithms do not amplify those debayering and JPEG artifacts.

Due to the lack of ground-truth information in the dataset (i.e., the true class labels), the accuracy of clustering results cannot be truly evaluated via quantitative comparisons. Therefore, we propose to use some internal validity indices to measure the clustering quality by using only features and information inherent in the dataset. Particularly, the data clustering results are evaluated using widely used clustering

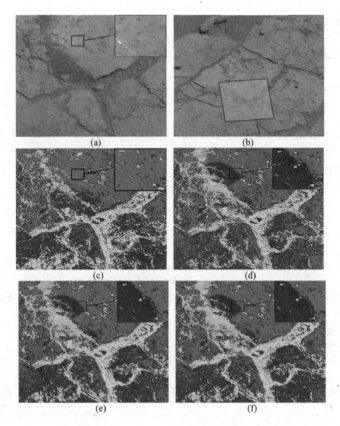

Fig. 4. K-means clustering performance with 6 classes of an LR-pair on sol 0812 taken on 11-18-2014. (a) original RGB right image; (b) original RGB left image with the aligned area from the right image; (c) clustering via 9-band right camera MS cube; (d) clustering via 12-band MS cube after 2nd registration step with higher (M-100) resolution; (e) clustering via pan-sharpened images by BDSD; and (f) clustering via pan-sharpened images by PRACS.

performance indicators including Silhouette index [15], Calinski-Harabasz (CH) index [16], Gap statistic criteria [17], Davies-Bouldin (DB) index [18], and the averaged clustering variance of all classes. Among these indicators, a larger value of Silhouette, CH or Gap index indicates a better quality of a clustering result, while a low value of DB index or clustering variance implies the presence of more compact and well-separated clusters. Tables 2 and 3 demonstrate these cluster validity analysis of the 4 MS data versions using the 5 indicators for K-means and GMM methods. The values are averaged over all LR-pair test sets. These quantitative clustering evaluations are consistent with the visual inspection and they further reinforce the effectiveness of using the combination of our proposed two-step image registration with pan-sharpening techniques in data clustering application.

Table 2. Comparison of averaged K-means clustering results via different cluster validity indices over all LR-pairs.

MS data versions	CH index	Silhouette index	Gap index	DB index	Variance
Original 9-band right camera	1.293×10^5	0.398	1.389	1.684	2.67×10^{-4}
2nd registration step with M-100 resolution	1.518×10^5	0.491	1.614	1.475	1.89×10^{-4}
Pan-sharpening using BDSD	1.614×10^5	0.508	1.651	**1.422**	**1.80 $\times 10^{-4}$**
Pan-sharpening using PRACS	**1.623 $\times 10^5$**	**0.512**	**1.653**	1.424	1.81×10^{-4}

Table 3. Comparison of averaged GMM clustering results via different cluster validity indices over all LR-pairs.

MS data versions	CH index	Silhouette index	Gap index	DB index	Variance
Original 9-band right camera	1.355×10^5	0.377	1.898	1.229	2.48×10^{-4}
2nd registration step with M-100 resolution	1.578×10^5	0.598	2.574	0.948	1.73×10^{-4}
Pan-sharpening using BDSD	1.607×10^5	**0.605**	2.618	0.826	1.65×10^{-4}
Pan-sharpening using PRACS	**1.611 $\times 10^5$**	0.604	**2.624**	**0.822**	**1.62 $\times 10^{-4}$**

5 Conclusions

Through extensive experiments, we thoroughly evaluated a number of pansharpening algorithms in the literature for enhancing the left Mastcam images. First, we observed that BDSD and PRACS performed better than others in terms of both PSNR and subjective visualization. Second, the parsharpened left images were fused with the right images to form a 12-band data cube. It was found that the fused image data cube can yield better data clustering performance than that of using left images only.

Some potential research directions include the following. First, our current implementation of the pansharpening is simple and efficient. Instead of using a single band from the right bands to fuse a corresponding left band, we plan to use the average of all bands from the right to create the pan band. Second, since some of the more sophisticated algorithms require point spread function (PSF) of the imager in the pansharpening process, we plan to estimate the PSF of the left imager and use it for pansharpening. Third, we are also exploring the possibility of using deep learning based approaches to enhancing the Mastcam images. Fourth, we plan to investigate the application of anomaly detection algorithms [20–23] to the pansharpened images.

Acknowledgements. This research was supported by NASA under contract NNX16CP38P.

References

1. Bell, J.F., Godber, A., McNair, S., Caplinger, M.A., Maki, J.N., Lemmon, M.T., Van Beek, J., Malin, M.C., Wellington, D., Kinch, K.M., Madsen, M.B., Hardgrove, C., Ravine, M.A., Jensen, E., Harker, D., Anderson, R.B., Herkenhoff, K.E., Morris, R.V., Cisneros, E.: The mars science laboratory curiosity rover mast camera (Mastcam) instruments: pre-flight and in-flight calibration, validation, and data archiving. AGU J. Earth Space Sci. (2016, submitted)
2. Ayhan, B., Dao, M., Kwan, C., Chen, H., Bell, J.F., Kidd, R.: A novel image alignment approach for Mastcam imagers of mars rover with application to rover guidance. IEEE Trans. Autom. Sci. Eng. (2017, to be submitted)
3. Zhou, J., Kwan, C., Budavari, B.: Hyperspectral image super-resolution: a hybrid color mapping approach. SPIE J. Appl. Remote Sens. **10**(3), 035024 (2016)
4. Hartley, R., Zisserman, A.: Multiple View Geometry in Computer Vision. Cambridge University Press, Cambridge (2003)
5. Bay, H., Ess, A., Tuytelaars, T., Van Gool, L.: SURF: speeded up robust features. Comput. Vis. Image Underst. (CVIU) **110**(3), 346–359 (2008)
6. Chen, H.M., Goela, A., Garvin, G.J., Li, S.: A parameterization of deformation fields for diffeomorphic image registration and its application to myocardial delineation. In: Jiang, T., Navab, N., Pluim, J.P.W., Viergever, Max A. (eds.) MICCAI 2010. LNCS, vol. 6361, pp. 340–348. Springer, Heidelberg (2010). doi:10.1007/978-3-642-15705-9_42
7. Lowe, D.G.: Object recognition from local scale-invariant features. IEEE Int. Conf. Comput. Vis. **2**, 1150–1157 (1999)
8. Vivone, G., Alparone, L., Chanussot, J., Mura, M.D., Garzelli, A., Licciardi, G.: A critical comparison of pansharpening algorithms. IEEE International Conference on Geoscience and Remote Sensing (IGARSS), pp. 191–194 (2014)
9. Tu, T., Huang, P., Hung, C., Chang, C.: A fast intensity-hue-saturation fusion technique with spectral adjustment for IKONOS imagery. IEEE Geosci. Remote Sens. Lett. **1**(4), 309–312 (2004)
10. Aiazzi, A., Baronti, S., Selva, M.: Improving component substitution pansharpening through multivariate regression of MS + Pan data. IEEE Trans. Geosci. Remote Sens. **45**(10), 3230–3239 (2007)
11. Garzelli, A., Nencini, F., Capobianco, L.: Optimal MMSE pan sharpening of very high resolution multispectral images. IEEE Trans. Geosci. Remote Sens. **46**(1), 228–236 (2008)
12. Choi, J., Yu, K., Kim, Y.: A new adaptive component-substitution-based satellite image fusion by using partial replacement. IEEE Trans. Geosci. Remote Sens. **49**(1), 295–309 (2011)
13. Kanungo, T., Mount, D.M., Netanyahu, N.S., Piatko, C.D., Silverman, R., Wu, A.Y.: An efficient k-means clustering algorithm: analysis and implementation. IEEE Trans. Pattern Anal. Mach. Intell. **24**(7), 881–892 (2002)
14. Zeng, H., Cheung, Y.-M.: A new feature selection method for Gaussian mixture clustering. Pattern Recognit. **42**(2), 243–250 (2009)
15. Chen, G., Jaradat, S.A., Banerjee, N., Tanaka, T.S., Ko, M.S., Zhang, M.Q.: Evaluation and comparison of clustering algorithms in analyzing ES cell gene expression data. Stat. Sin. 241–262 (2002)
16. Zhao, Y., Karypis, G.: Data clustering in life sciences. Mol. Biotechnol. **31**(1), 55–80 (2005)

17. Tibshirani, R., Walther, G., Hastie, T.: Estimating the number of data clusters via the gap statistic. J. R. Stat. Soc. B **63**, 411–423 (2001)
18. Kasturi, J., Acharya, R., Ramanathan, M.: An information theoretic approach for analyzing temporal patterns of gene expression. Bioinformatics **19**(4), 449–458 (2003)
19. Kwan, C., Choi, J.H., Chan, S., Zhou, J., Budavari, B.: Resolution enhancement for hyperspectral images: a super-resolution and fusion approach. In: IEEE International Conference on Acoustics, Speech, and Signal Processing, New Orleans (2017)
20. Zhou, J., Kwan, C., Ayhan, B., Eismann, M.: A novel cluster kernel RX algorithm for anomaly and change detection using hyperspectral images. IEEE Trans. Geosci. Remote Sens. **54**(11), 6497–6504 (2016)
21. Wang, W., Li, S., Qi, H., Ayhan, B., Kwan, C., Vance, S.: Identify anomaly component by sparsity and low rank. In: IEEE Workshop on Hyperspectral Image and Signal Processing: Evolution in Remote Sensor (WHISPERS), Tokyo, Japan (2015)
22. Li, S., Wang, W., Qi, H., Ayhan, B., Kwan, C., Vance, S.: Low-rank tensor decomposition based anomaly detection for hyperspectral imagery. In: IEEE International Conference on Image Processing (ICIP), Quebec City, Canada (2015)
23. Qu, Y., Guo, R., Wang, W., Qi, H., Ayhan, B., Kwan, C., Vance, S.: Anomaly detection in hyperspectral images through spectral unmixing and low rank decomposition. In: International Geoscience and Remote Sensing Symposium (IGARSS), Beijing (2016)
24. Ayhan, B., Kwan, C., Vance, S.: On the use of a linear spectral unmixing technique for concentration estimation of APXS spectrum. J. Multidiscip. Eng. Sci. Technol. **2**(9), 2469–2474 (2015)
25. Wang, W., Li, S., Qi, H., Ayhan, B., Kwan, C., Vance, S.: Revisiting the preprocessing procedures for elemental concentration estimation based on CHEMCAM LIBS on MARS rover. In: 6th Workshop on Hyperspectral Image and Signal Processing: Evolution in Remote Sensing (WHISPERS), Lausanne, Switzerland (2014)
26. Kwan, C., Xu, R., Haynes, L.: A new data clustering technique and its applications. data mining and knowledge discovery: theory, tools, and technology III. Proc. SPIE **4384**, 1–5 (2001)

A Collective Neurodynamic Optimization Approach to Nonnegative Tensor Decomposition

Jianchao Fan[1,2(✉)] and Jun Wang[2,3]

[1] Department of Ocean Remote Sensing,
National Marine Environmental Monitoring Center, Dalian 116023, China
jcfan@nmemc.org.cn
[2] School of Control Science and Engineering,
Dalian University of Technology, Dalian 116024, Liaoning, China
[3] Department of Computer Science,
City University of Hong Kong, Kowloon Tong, Hong Kong
jwang.cs@cityu.edu.hk

Abstract. In this paper, a collective neurodynamic optimization approach is proposed to nonnegative tensor factorization. Tensor decompositions are often applied in the data analysis. However, it is often a nonconvex optimization problem, which would cost much time and usually trap into the local minima. To solve this problem, a novel collective neurodynamic optimization approach is proposed by combining recurrent neural networks (RNN) and particle swarm optimization (PSO) algorithm. Each RNN still carries out local search. And then the best solution of each RNN improves through PSO framework. In the end, the global optimal solutions of nonnegative tensor factorization are obtained. Experiments results demonstrate the effectiveness for the nonconvex optimization with constraints.

Keywords: Neurodynamic optimization · Particle swarm optimization · Nonnegative tensor factorization

1 Introduction

Tensor decompositions are appearing as new approaches for data processing, which capture inner structures in multi-dimension datasets and extract latent components hidden in the complex relationships [1,2]. It includes two categories, one is the Canonical Decomposition (CANDECOMP)/Parallel Factor Model (PARAFAC) [3,4], the other is Tucker decomposition [5]. The accuracy of tensor decompositions directly impacts on the effect of feature extractions etc. Nonnegative features are important for the data analysis.

J. Fan—The work described in the paper was supported by the National Key Research and Development Program of China under project 2016YFC1401007, the Foundation of High Resolution Special Research under 41-Y30B12-9001-14/16, and the National Natural Science Foundation of China under project 61273307.

© Springer International Publishing AG 2017
F. Cong et al. (Eds.): ISNN 2017, Part II, LNCS 10262, pp. 207–213, 2017.
DOI: 10.1007/978-3-319-59081-3_25

However, the nonnegative tensor factorization is a nonconvex optimization problem. The multiplicative update (MU) algorithm is a classic method with simple calculation process. However, it often traps into local optimal [6]. Another common method, alternating least squares (ALS) algorithm, has been employed to tensor decomposition. Due to gradient and Hessian matrix are used in the ALS algorithm, it takes high time calculation cost [7]. Furthermore, a hierarchical alternating least squares (NTD-HALS) is modified for nonnegative Tucker decomposition with some constrained cost functions [8]. But these algorithm could not guarantee the nonconvex global optimization and converge slowly.

Neurodynamic optimization, as a global optimization method, is a powerful alternative way to these matrix factorization optimization problems. For convex optimization with bound constraints, recurrent neural networks are available with global convergence to the optimal solution. Classic LU matrix decomposition and Cholesky factorization are analyzed through recurrent neural networks [9]. Neurodynamic optimization is a promising matrix factorization approach to real-time optimization [10]. However, a single RNN could not deal with nonconvex optimization problems [11]. Swarm intelligent, such as Particle Swarm Optimization (PSO) etc. [12], obtain global optimal through stochastic improvement [13]. However, it is deficient in constraint optimization. The advantages of neurodynamic optimization and particle swarm optimization may combine effectively.

To address the above mentioned drawbacks, a collective neurodynamic optimization (CNO) approach is proposed for the optimization problem of nonnegative tensor factorization (NTF) in this paper. Under the PSO framework, a serials of RNNs model as particles in swarm, are combined to deal with NTF problems. Each RNN carries out its own local research. Then, after the information share, the RNN population are improved based on PSO framework. By iteratively calculating and initialing each RNN model to find the global optimal solutions of nonnegative tensor factorization.

The remainder of this paper is organized as follows. The problem formulation of nonnegative tensor decomposition is presented in Sect. 2. The collective neurodynamic optimization for NTF problem is described in Sect. 3. Experimental results are demonstrated in Sect. 4. And conclusion is obtained in the end.

2   Nonnegative Tensor Decomposition

The source signal based on canonical model [14] is represented as

$$\bar{Q}_{cdef} \approx \sum_{r=1}^{n} k_r a_{cr} a_{dr}^* a_{er}^* a_{fr} \tag{1}$$

where $a_{cr}, a_{dr}^*, a_{er}^*, a_{fr}$ are factors of the canonical model, respectively. And k_r is the kurtosis of the r-th source signal. $\bar{Q} = cum(X, X^*, X^*, X) \in R^{m \times m \times m \times m}$ is the fourth-order cumulant tensor of the observed signals, X^* is conjugate of X. $\bar{Q}$ can be expressed as

$$\bar{Q}_{cdef} = \sum_{r=1}^{n} k_r (E_r)_{cd} (E_r)_{ef}^* \tag{2}$$

where E_r are Hermitean matrices and mutually orthonormal. k_r are considered as eigenvalues of the cumulant tensor $\bar{Q}_{cdef}$, which can be represented as

$$Q = (A \odot A^*) K (A \odot A^*)^H \tag{3}$$

here $K = diag(k_1, \cdots, k_n)$ and the diagonal elements of K are eigenvalues of the matrix Q; $\odot$ is the Khatri-Rao product. Eq. (1) is expressed as

$$\begin{aligned}
\bar{Q} &= \bar{K} \times_1 A^{(1)} \times_2 A^{(2)} \times_3 A^{(3)} \times_4 A^{(4)} + \bar{E} \\
&= \sum_{r=1}^{n} k_r (a_r^{(1)} \circ a_r^{(2)} \circ a_r^{(3)} \circ a_r^{(4)}) + \bar{E}
\end{aligned} \tag{4}$$

where $A^{(o)} = [a_1^{(o)}, a_2^{(o)}, \cdots, a_n^{(o)}]$ is all factor of tensor, and the symbol "$\circ$" represents the vector outer product $o = 1, 2, 3, 4$, $\bar{E}$ is residual tensor.

According to (1), the objective function with *Frobenius* norm is expressed as,

$$\begin{aligned}
&D_F(a_r^{(1)}, a_r^{(2)}, a_r^{(3)}, a_r^{(4)}) \\
&= \frac{1}{2} \left\| \bar{Q} - \sum_{r=1}^{n} k_r (a_r^{(1)} \circ a_r^{(2)} \circ a_r^{(3)} \circ a_r^{(4)}) \right\|_F^2
\end{aligned} \tag{5}$$

In order to further simplify the formulation, (5) is rewritten as

$$D_F(a_r^{(1)}, a_r^{(2)}, a_r^{(3)}, a_r^{(4)}) = \frac{1}{2} \left\| \bar{Q}_r^{(o)} - k_r a_r^{(o)} \{a_r\}^{\odot - oT} \right\|_F^2 \tag{6}$$

where

$$\{a_r\}^{\odot - oT} = [a_r^{(N)}]^T \odot \cdots \odot [a_r^{(o+1)}]^T \odot [a_r^{(o-1)}]^T \odot \cdots \odot [a_r^{(1)}]^T \tag{7}$$

and

$$\begin{aligned}
\bar{Q}_r &= \bar{Q} - \sum_{p \neq r}^{n} k_p (a_p^{(1)} \circ a_p^{(2)} \circ a_p^{(3)} \circ a_p^{(4)}) \\
&= \bar{Q} - \sum_{p=1}^{n} k_p (a_p^{(1)} \circ a_p^{(2)} \circ a_p^{(3)} \circ a_p^{(4)}) \\
&\quad + k_r (a_r^{(1)} \circ a_r^{(2)} \circ a_r^{(3)} \circ a_r^{(4)}) \\
&= \bar{Q} - \hat{\bar{Q}} + k_r (a_r^{(1)} \circ a_r^{(2)} \circ a_r^{(3)} \circ a_r^{(4)}) \\
&= \bar{E} + k_r (a_r^{(1)} \circ a_r^{(2)} \circ a_r^{(3)} \circ a_r^{(4)})
\end{aligned} \tag{8}$$

The gradient of the objective function (6) is given by

$$\begin{aligned}
\frac{\partial D_F}{\partial a_r^{(o)}} &= -k_r \bar{Q}_r^{(o)} \{a_r\}^{\odot - o} + k_r^2 a_r^{(o)} \{a_r\}^{\odot - oT} \{a_r\}^{\odot - o} \\
&= -k_r \bar{Q}_r^{(o)} \{a_r\}^{\odot - o} + k_r^2 a_r^{(o)} \gamma_r^{(o)}
\end{aligned} \tag{9}$$

here, the symbol "Θ" represents the Hadamard product, and the scaling coefficients $\gamma_r^{(o)}$ can be calculated as follows,

$$
\begin{aligned}
\gamma_r^{(o)} &= \{a_r\}^{\odot-o\,T}\{a_r\}^{\odot-o} \\
&= \{a_r^T a_r\}^{\Theta-o} \\
&= \{a_r^T a_r\}^{\Theta}/a_r^{(o)T} a_r^{(o)} \\
&= (a_r^{(N)T} a_r^N)/a_r^{(o)T} a_r^{(o)} \\
&= \begin{cases} a_r^{(N)T} a_r^N, o \neq N \\ \quad 1, \qquad o = N \end{cases}
\end{aligned}
\tag{10}
$$

Through taking into account $||a_r^{(l)}||_2 = 1, l = 1, 2, \cdots, N-1, \forall r, \gamma_r^{(o)}$, a updating principle of $a_r^{(o)}$ can be used by making $\nabla(9)$ to zero,

$$
a_r^{(o)} \leftarrow \frac{1}{k_r} \bar{Q}_r^{(o)} \{a_r\}^{\odot-o}.
\tag{11}
$$

3   Collective Neurodynamic Optimization for NTF

From (5), the fourth-order tensor factorization is a nonconvex optimization problem. A serial of recurrent neural networks are exploited collectively for tensor decomposition. It can be also seen as a integrated model between PSO in the swarm intelligent field and neurodynamic optimization in the neural networks field.

3.1   Precise Local Search

Considering the objective function (5) is a nonconvex optimization problem, a one-layer projection neural network [15] is adopted for NTF nonconvex problem expressed as

$$
\dot{x}(t) = -x(t) + P_\Omega(x(t) - \nabla f(x(t)))
\tag{12}
$$

where x is the state of RNN, which are corresponding to the independent variable of NTF algorithm. The gradient of f is defined as,

$$
\nabla f = \frac{1}{k_r} \bar{Q}_r^{(o)} \{a_r\}^{\odot-o}
\tag{13}
$$

Projection operator P_Ω is represented as

$$
P_\Omega(u) = \arg \min_{v \in \Omega} \|u - v\|
\tag{14}
$$

In the NTF optimization problem, Ω is a box set. And $\Omega = \{u \in \Re^{4mn} : l_i \leq u_i \leq h_i\}$, P_Ω is defined specifically as

$$
P_\Omega(u_i) = \begin{cases} l_i, & u_i < l_i \\ u_i, & l_i \leq u_i \leq h_i \\ h_i, & u_i > h_i \end{cases}
\tag{15}
$$

here $l_i \geq -\infty$ and $h_i \leq +\infty$, which can control the resolution space.

3.2 Information Exchange

The basic principle of particle swarm optimization [12] fully considers population intelligence. The location of the RNN is represented as $u_i = \{u_{i1},$ $u_{i2}, \cdots, \cdots, u_{iD}\}$. The best previous position P_{best} of the ith RNN is denoted as $p_i = (p_{i1}, p_{i2}, \cdots, p_{iD})$. Let $v_i = (v_{i1}, v_{i2}, \cdots, v_{iD})$ be the current velocity. The position P_{Gbest} of the best one among all the RNNs represented as $p_g = (p_{g1}, p_{g2}, \cdots, p_{gD})$. So the update rule is expressed as

$$\begin{cases} v_{id}(k+1) = wv_{id}(k) + c_1 rand_{1d}[p_{id}(k) - u_{id}(k)] \\ \qquad\qquad + c_2 rand_{2d}[p_{gd}(k) - u_{id}(k)] \\ u_{id}(k+1) = u_{id}(k) + v_{id}(k+1) \end{cases} \tag{16}$$

where $w \in [0,1]$ is the inertia weight, $rand_1$, $rand_2 \in [0,1]$ denote random numbers, and c_1, c_2 are acceleration constants, respectively.

The better search information are exchanged among these RNNs according to (16). The better initial positions of each RNN are refreshed to begin new round optimization process. Hence, the advantages of neurodynamic optimization and swarm intelligent optimization are combined appropriately.

4 Experimental Results for Blind Identification

In this section, the ICALAB benchmark dataset for signal processing is used for qualitative performance evaluations of the proposed collective neurodynamic optimization for nonnegative tensor factorization (NTF-CNO). Three public benchmark datasets are considered in the experiment, which are Speech4. Let $\varepsilon = 10^{-6}$, the number of recurrent neural networks is five in Speech4 experiment. 100 times tests are run with different random initialization. The number of the observed signals m is referenced to [14], so $m = 3$.

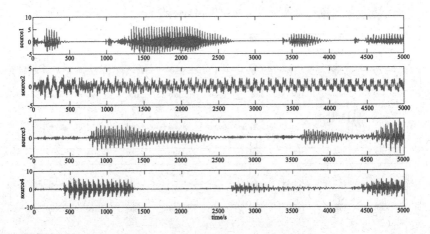

Fig. 1. The original source signals

The original source signals are deficted in Fig. 1 to demonstrate the similarity intuitively. Let observed signal dimension be 3. A is randomly generated by MATLAB shown as follows,

$$A = \begin{bmatrix} 0.2165 & 0.4787 & 0.0787 & 0.8905 \\ 0.7245 & 0.1826 & 0.8434 & 0.4099 \\ 0.8671 & 0.3817 & 0.4260 & 0.1802 \end{bmatrix} \tag{17}$$

Figure 2 describes the estimated source signals obtained by NTD-HALS approach. Figure 3 shows the estimated source signals obtained of Speech4 dataset by proposed NTF-CNO method. It is seen that NTF-CNO approach can obtain the most similar estimated source signals to the original ones.

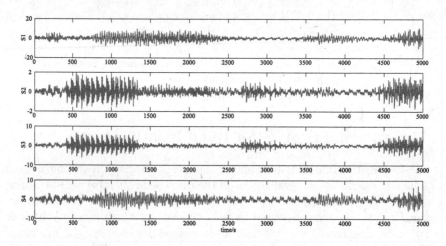

Fig. 2. The estimated source signals by NTD-HALS approach

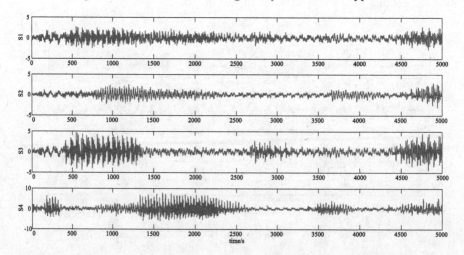

Fig. 3. The estimated source signals by NTF-CNO approach

5   Conclusion

A CNO approach is proposed for nonnegative tensor factorization. A serial of RNNs are integrated in framework of PSO, which has superior capabilities of handling global optimization. NTF-CNO can sufficiently guarantee the nonconvex global optimization and performance effectiveness.

References

1. Chen, Y., Han, D., Qi, L.: New ALS methods with extrapolating search directions and optimal step size for complex-valued tensor decompositions. IEEE Trans. Sig. Process. **59**, 5888–5898 (2011)
2. Cichocki, A., Mandic, D., Lathauwer, L., Zhou, G., Zhao, Q., Caiafa, C.: Tensor decompositions for signal processing applications: from two-way to multiway component analysis. IEEE Sig. Process. Mag. **32**, 145–163 (2015)
3. Dauwels, J., Srinivasan, K., Reddy, M., Cichocki, A.: Near-lossless multichannel EEG compression based on matrix and tensor decompositionss. IEEE J. Biomed. Health Inform. **17**, 708–714 (2013)
4. Kolda, T., Bader, B.: Tensor decompositions and applications. SIAM Rev. **51**, 455–500 (2009)
5. Chen, Y., Hsu, C., Hsu, H., Liao, H.: Simultaneous tensor decomposition and completion using factor priors. IEEE Trans. Pattern Anal. Mach. Intell. **36**, 577–591 (2014)
6. Lee, D., Seung, H.: Algorithms for nonnegative matrix factorization. In: Advances in Neural Information Processing Systems, pp. 556–562 (2000)
7. Cichocki, A., Zdunek, R., Phan, A., Amari, S.: Nonegative Matrix and Tensor Factorizations: Application to Exploratory Multi-way Data Analysis and Blind Source Separation. Wiley, Chichester (2009)
8. Phan, A., Cichocki, A.: Extended HALS algorithm for nonnegative Tucker decomposition and its applications for multiway analysis and classification. Neurocomputing **74**, 1956–1969 (2011)
9. Wang, J., Wu, G.: Recurrent neural networks for LU decomposition and Cholesky factorization. Math. Comput. Model. **18**, 1–8 (1993)
10. Xia, Y.S., Wang, J.: A recurrent neural network for nonlinear convex optimization subject to nonlinear inequality constraints. IEEE Trans. Circ. Syst. I Regul. Pap. **51**, 1385–1394 (2004)
11. Yan, Z., Wang, J., Li, G.C.: A collective neurodynamic optimization approach to bound-constrained nonconvex optimization. Neural Netw. **55**, 20–29 (2014)
12. Kennedy, J., Eberhart, R.: Particle swarm optimization. In: IEEE International Conference on Neural Networks, pp. 1942–1948 (1995)
13. Han, M., Fan, J.C., Wang, J.: A dynamic feedforward neural network based on gaussian particle swarm optimization and its application for predictive control. IEEE Trans. Neural Netw. **22**, 1457–1468 (2011)
14. Lathauwer, L.D., Castaing, J., Cardoso, J.F.: Fourth-order cumulantbased blind identification of underdetermined mixtures. IEEE Trans. Sig. Process. **55**, 2965–2973 (2007)
15. Liu, Q.S., Wang, J.: A one-layer recurrent neural network with a discontinuous hard-limiting activation function for quadratic programming. IEEE Trans. Neural Netw. **19**, 558–570 (2008)

A Novel Spatial Information Model for Saliency Detection

Hang Gao, Bo Li$^{(\boxtimes)}$, and Han Liu

School of Electronic and Information Engineering,
South China University of Technology, Guangzhou, China
{g.hang,eeliuhan}@mail.scut.edu.cn, leebo@scut.edu.cn

Abstract. Saliency detection is one of the critical issues in computer vision. The location of saliency object is a widely used cue in the procedure of detecting saliency map, which is named as spatial information. Center prior and Harris-points are two popular cues of spatial information. In this paper, we propose a novel spatial information model which is different from center prior and Harris-points. In the proposed method, a cost function and a rectangle are used to seek the saliency object accurately, and to discriminate the saliency object from background effectively. The model can be used to optimize previous saliency detection approaches. In the experiment, the model is used to optimize five state-of-the-art approaches on two popular datasets ASD and ECCSD. The experiment results demonstrate the feasibility and validity of our method. And the performance of our method is better than center prior and Harris-points.

Keywords: Saliency detection · Spatial information · Image processing

1 Introduction

Saliency detection is an efficient way to understand and analyze image. Its goal is to discriminate the salient region from background in an image, which can be applied in image editing [1], segmentation [2], image retargeting [3], object detection and recognition [5].

In addition to color contrast [4,6,7] and background prior [8,9] which are widely used in saliency detection, spatial information is another popular cue. Spatial information represents the geometric information of a saliency map, on which the location of the salient region can be determined, and then the undetermined regions away from the salient region can be suppressed. Up to now, there are two methods to obtain the spatial information:

1. Center prior [1,6]. It is assumed that the salient region is placed near the center of an image. It is obvious that this cue is not accurate enough for detecting saliency map, because the salient region can be anywhere in an image. In other words, the salient region may be placed at the corner or the boundary in a natural image.

© Springer International Publishing AG 2017
F. Cong et al. (Eds.): ISNN 2017, Part II, LNCS 10262, pp. 214–221, 2017.
DOI: 10.1007/978-3-319-59081-3_26

2. Harris-points [10]. Harris-points are those points which have great rate of change on vertical and horizontal directions. Combined with convex hull, Harris-points can determine the salient region [11,12]. For an image with simple and smooth background, this cue performs better. However, if the scene of the image is complex, Harris-points has a greater probability of missing the salient region.

Considering the aforementioned issues, we propose a novel spatial information which can accurately explore the position of salient region and can be used to optimize the saliency map generated by most of present detection approaches. The experiment results demonstrate that our method performs better than center prior and Harris-points on datasets. The remainder of this paper is organized as: in Sect. 2, our method is presented detailedly. In Sect. 3, we evaluate and compare our method with center prior and Harris-points on 2 benchmark dataset, and the conclusion follows in Sect. 4.

2 Refined Spatial Information

Our method can optimize the saliency map which is generated by commonly using saliency detection approaches. In this section, for convenience, we employ RBD [9] approach as an instance to generate an initial saliency map. The initial saliency map is denoted as $SM(i)$ indicating the salient score of the ith pixel in the map. The working flow of our method is shown in Fig. 1.

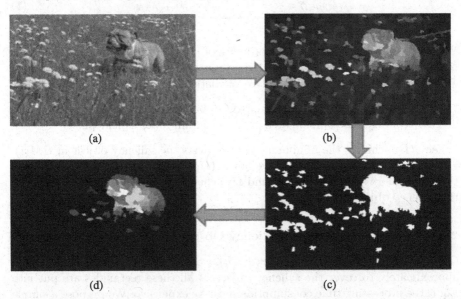

Fig. 1. Working flow of our method. (a) Source image. (b) The saliency map generated by RBD [9]. (c) Binarization using OTSU [13]. (d) Optimization with our method and the final saliency map.

2.1 The Model of Our Method

In order to reduce the cost of floating-point calculation, we use OTSU [13] algorithm to process $SM(i)$, and obtain a binary map $BM(i)$ (as shown in Figs. 1(c) and 2) whose size is assumed as $w \times h$ in pixel (where w and h donate width and height respectively). In $BM(i)$, there are some regions including the saliency object and noisy regions.

Fig. 2. Seeking the saliency object with Ω and $\xi(\Omega)$ in $BM(i)$.

For seeking the saliency object in $BM(i)$, we design a rectangle Ω to cover it according to the cost function:

$$\xi(\Omega) = \arg \min_{\Omega} \{ \underbrace{\sum_{BM(x,y) \in \Omega} (BM(x,y) - 1)^2}_{E1} + \underbrace{\sum_{BM(x,y) \notin \Omega} (BM(x,y) - 0)^2}_{E2} \} \quad (1)$$

Where (x, y) and $BM(x, y)$ are coordinate of pixel and its value in BM respectively. Obviously, $BM(x, y)$ is 0 or 1. The cost function $\xi(\Omega)$ contains two terms, $E1$ and $E2$, which impose different constraints on Ω:

1. $E1$ encourages Ω to cover fewer pixels of background.
2. $E2$ encourages the outside area of Ω to contain fewer salient pixels.

When $\xi(\Omega)$ achieves the minimum, Ω just covers the saliency object in $BM(i)$. For example, as shown in Fig. 2, we have $\xi(\Omega_1) > \xi(\Omega_2) > \xi(\Omega_3)$. It is obvious that Ω_3 is more suitable than Ω_1 and Ω_2. Then we can use Ω and Eq. 1 to seek the saliency object in $BM(i)$ map.

2.2 Procedure of Seeking Saliency Object

Owing to the size of $BM(i)$ map is $(w \times h)$, we will generate $(w \times h)^2$ rectangles approximately to cover the saliency object. If all these rectangles are put into Eq. 1, the processing time consumption must be expensive. We propose a simple and effective way to solve this problem in two steps.

Step 1. As shown in Fig. 3, we initialize a rectangle Ω_0 with fixed size of $(w/N \times h/N)$ at the top-left corner in $BM(i)$. Then we move Ω_0 toward the right and

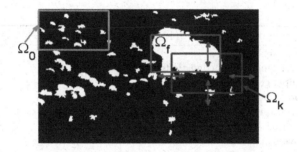

Fig. 3. Adjusting the edges of Ω_k to cover the saliency object.

down direction at regular step w/M and h/M respectively, by which r rectangles must be generated. These rectangles are expressed as $\Omega = \{\Omega_0, \Omega_1, \Omega_2, \cdots, \Omega_r\}$, in which we choose the rectangle whose $\xi(\cdot)$ is minimum. Let us assume that Ω_k is the chosen one, which expresses the coarse position of the saliency object. Ω_k possibly covers part of the saliency object or all of the saliency object, so it needs to be adjusted. We will present how to adjust Ω_k in next step. The specific procedure of Step 1 is presented in Algorithm 1.

There are two parameters N and M, which control the size of Ω_0 and its moving step size. For N, we set it as 4, in order to ensure that the Ω_0 is larger than noisy region. M is set as 8, which ensures that Ω_0 moves faster and does not miss the salient region.

Step 2. The four edges of Ω_k are adjusted respectively until Ω_k just covers the salient region. As showed in Fig. 3, each edge of Ω_k can be adjusted on two directions which will change $\xi(\Omega_k)$. So we adjust the four edges in turn on the direction of making $\xi(\Omega_k)$ decrease until $\xi(\Omega_k)$ achieves minimum. As shown in Fig. 3, Ω_k is adjusted to the position where Ω_f is. Then the rectangle just covers the salient region. The specific procedure of this step is presented in Algorithm 2.

2.3   Application in Saliency Detection

By using Algorithms 1 and 2, the rectangle Ω_f can be identified accurately, which is named as the refined spatial information (RSI) and can be used to optimize the initial saliency map $SM(i)$. Ω_f covers the salient region in $BM(i)$, and then its centroid can be viewed as the centroid of the salient region, which is denoted as (X_c, Y_c). We introduce it into $SM(i)$, and obtain the final saliency map $SM\_SP(i)$:

$$SM\_SP(i) = SM(i) \times exp\{-\frac{(X_i - X_c)^2 + (Y_i - Y_c)^2}{2\sigma^2}\} \qquad (2)$$

Where X_i and Y_i denotes the normalized coordinates of pixel i, and σ controls the effect of spatial information. In this paper, we set σ as 0.4 with pixel coordinate normalized to $[0, 1]$.

Algorithm 1. Seeking Ω_k whose $\xi(\Omega_k)$ is minimum.

Require: BM, w, h, N, M;

Ensure: Ω_k=(upedge$_k$,bottomedge$_k$,leftedge$_k$,rightedge$_k$);

Initialize Ω_0: upedge$_0$ = leftedge$_0$=0,

                bottomedge$_0$=h/N, rightedge$_0$=w/N;

     % *Place Ω_0 at the top-left corner in BM.*

Ω_k=Ω_0;

while bottomedge$_0 \leqslant$h **do**

  if $\xi(\Omega_0)<\xi(\Omega_k)$ **then**

    Ω_k=Ω_0;

  end if

  for rightedge$_0 \leqslant$w **do**

    Moving Ω_0 toward the right direction at step size w/M. If Ω_0 beyond the border of BM at step size w/M, just moving Ω_0 to the right border of BM;

    if $\xi(\Omega_0)<\xi(\Omega_k)$ **then**

      Ω_k=Ω_0;

    end if

  end for

  Moving Ω_0 toward the down direction at step h/M. If Ω_0 beyond the bottom of BM at step size w/M, just moving Ω_0 to the bottom border of BM;

end while

Return Ω_k;

Algorithm 2. Adjusting Ω_k to obtain Ω_f which just covers the saliency object.

Require: Ω_k, $BM(i)$, ε=0.01

  % ε *is used to determine whether $\xi(\cdot)$ achieves minimum or not.*

Ensure: Ω_f=(upedge$_f$,bottomedge$_f$,leftedge$_f$,rightedge$_f$)

$\Omega_f = \Omega_k$;

while 1 **do**

  upedge$_f$=upedge$_f$+1;

  if $\xi(\Omega_k)<\xi(\Omega_f)$ **then**

    upedge$_f$=upedge$_f$-2;

    if $\xi(\Omega_k)<\xi(\Omega_f)$ **then**

      upedge$_f$=upedge$_f$+1;

    end if

  end if

  bottomedge$_f$, leftedge$_f$, rightedge$_f$ are adjusted as same as upedge$_f$;

  if $|\xi(\Omega_k)-\xi(\Omega_f)|<\varepsilon$ **then**

    Break;

  else

    $\Omega_k = \Omega_f$;

  end if

end while

Return Ω_f;

---

3   Experiments

In this section, beside RBD, other four state-of-the-art approaches (MDF [15], DRFI [7], DSR [17], MC [18]) are employed as optimized object. We use the five approaches to generate the initial saliency maps, on which we make a comparison among our method, center prior and Harris-points. The experiments are tested on two publicly available datasets: ASD [16] and ECSSD [14]. ASD contains 1000 images, whose background is simple and smooth. Those images in ECSSD have complex scenes, and the saliency objects are harder to be detected than the images in ASD. Similar with Achanta [16], we also use the precision-recall (PR) rate to demonstrate the accuracy of the saliency map. The experiments are made at a PC with Inter Core i5-4460 CPU@3.20 GHz and 8 GB RAM.

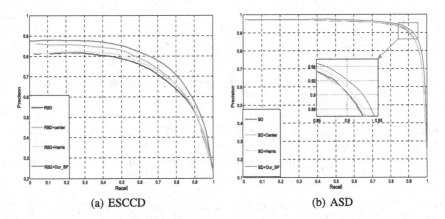

(a) ESCCD   (b) ASD

Fig. 4. Experiment on ASD and ECCSD datasets. Our model, center prior and Harris-points are used respectively to optimize the saliency map generated by RBD. In the figure, "BD" means RBD approach; "BD+Center", "BD+Harris" and "BD+Our_SP" mean the optimization with center prior, Harris-points and our model respectively.

Experiment on RBD: The results are shown in Fig. 4, in which the PR curves demonstrate that our method can improve the performance of RBD, and performs better than center prior and Harris-points. Our method can identify the saliency object and suppress the noisy regions effectively, and the performance is improved significantly compared with RBD, center prior and Harris-points (as shown in Fig. 4(a)). However, the performance improved on ASD is not as good as that on ECCSD (as shown in Fig. 4(b)). There are two reasons: (1) The background of the images in ASD are simple and smooth, so there are few noisy regions in the initial saliency maps; (2) The main purpose of RSI is to suppress the noisy regions. Thus the improvement is not significant on ASD.

Experiment on MDF, DRFI, DSR, MC: The results are shown in Fig. 5, from which we can see that our method performs much better than original approach, center prior and Harris-points. Compared with traditional approaches that only use low level cues, the performance of DRFI is the best one [19].

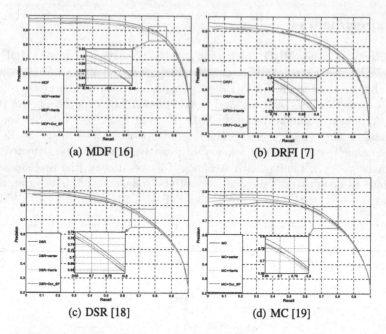

Fig. 5. Evaluation for optimization of MDF [15], DRFI [7], DSR [17] and MC [18] on ESCCD dataset. Same with the first experiment, our model, center prior and Harris-points are introduced to optimize the saliency maps generated by the four approaches.

MDF is a new approach based on deep learning, and its performance is much better than almost all traditional approaches [15]. When our method is introduced into DRFI and MDF, the performance can be further improved (as shown in Fig. 5(a) and (b)).

4   Conclusion

In the paper, we propose a novel model as a new spatial information cue to detect the saliency object. In the model, we construct a cost function to produce a rectangle, by which the saliency object can be sought out accurately. Then the saliency object can be effectively discriminated from the background in an image. Our model can be introduced into saliency detection approaches to improve their performances. The results of experiments on ASD and ECCSD datasets demonstrate that the performance of our method is better than center prior and Harris-points. Actually, our work has a limitation that the saliency map with only one saliency object can be detected accurately. If there are more than two detached saliency objects in an image, the proposed method could not work effectively. Then, the main task of our future work will focus on how to solve this problem.

Acknowledgement. This research was supported by the National Natural Science Foundation of China (Grant Nos. 11627802, 51678249), by the Science and Technology Projects of Guangdong (2013A011403003), and by the Science and Technology Projects of Guangzhou (201508010023).

References

1. Goferman, S., Zelnik-Manor, L., Tal, A.: Context-aware saliency detection. IEEE Trans. Pattern Anal. Mach. Intell. **34**(10), 915–1926 (2012)
2. Ko, B.C., Nam, J.Y.: Object-of-interest image segmentation based on human attention and semantic region clustering. J. Opt. Soc. Am. **23**(10), 2462–2470 (2006)
3. Ding, Y., Xiao, J., Yu, J.: Importance filtering for image retargeting. In: Proceedings of the CVPR (2011)
4. Li, J., Levine, M.D., An, X., Xu, X., He, H.: Visual saliency based on scale-space analysis in the frequency domain. IEEE Trans. Pattern Anal. Mach. Intell. **35**(4), 996–1010 (2013)
5. Rutishauser, U., Walther, D., Koch, C., Perona, P.: Is bottom-up attention useful for object recognition. In: Proceedings of the CVPR (2004)
6. Kim, J., Han, D., Tai, Y.: Salient region detection via high-dimensional color transform. In: Proceedings of the CVPR (2014)
7. Jiang, H., Wang, J., Yuan, Z., Wu, Y., Zheng, N., Li, S.: Salient object detection: a discriminative regional feature integration approach. In: Proceedings of the CVPR (2013)
8. Yang, C., Zhang, L., Lu, H., Ruan, X., Yang, M.H.: Saliency detection via graph-based manifold ranking. In: Proceedings of the CVPR (2013)
9. Zhu, W., Liang, S., Wei, Y., Sun, J.: Saliency optimization from robust background detection. In: Proceedings of the CVPR (2014)
10. Harris, C., Stephens, M.: A combined corner and edge detector. In: Proceedings of the Fourth Alvey Vision Conference (1988)
11. Tong, N., Lu, H., Zhang, L., Ruan, X.: Saliency detection with multi-scale superpixels. IEEE Sig. Process. Lett. **21**(9), 1035–1039 (2014)
12. Yang, C., Zhang, L., Lu, H.: Graph-regularized saliency detection with convex-hull-based center prior. IEEE Sig. Process. Lett. **20**(7), 637–640 (2013)
13. Otsu, N.: A threshold selection method from gray-level histograms. IEEE Trans. Sys. Man Cyber. **9**(1), 62–66 (1979)
14. Yan, Q., Xu, L., Shi, J., Jia, J.: Hierarchical saliency detection. In: Proceedings of the CVPR (2013)
15. Li, G., Yu, Y.: Visual saliency based on multiscale deep features. In: Proceedings of the CVPR (2015)
16. Achanta, R., Hemami, S., Estrada, F., Susstrunk, S.: Frequency-tuned salient region detection. In: Proceedings of the CVPR (2009)
17. Li, X., Lu, H., Zhang, L., Ruan, X., Yang, M.: Saliency detection via dense and sparse reconstruction. In: Proceedings of the ICCV (2013)
18. Jiang, B., Zhang, L., Lu, H., Yang, C., Yang, M.: Saliency detection via absorbing Markov chain. In: Proceedings of the ICCV (2013)
19. Borji, A., Cheng, M., Jiang, H., Li, J.: Salient object detection: a benchmark. IEEE Trans. Image Process. **24**(12), 5706–5722 (2015)

Sparse Representation with Global and Nonlocal Self-similarity Prior for Single Image Super-Resolution

Weiguo Gong[✉], Xi Chen, Jinming Li, Yongliang Tang, and Weihong Li

Key Lab of Optoelectronic Technology and Systems of Education Ministry, Chongqing University, Chongqing 400044, China
wggong@cqu.edu.cn

Abstract. Nonlocal self-similarity sparse representation models exhibit good performance in single image super-resolution (SR) application. However, due to the independent coding process of each image patch, the global similarity information among all similar image patches in whole image is lost. Consequently, the similar image patches may be encoded as the totally different code coefficients. In this paper, considering that low-rank constraint is better at capturing the global similarity information, a new sparse representation model combining the global low-rank prior and the nonlocal self-similarity prior simultaneously is proposed for single image super-resolution. The weighted nuclear norm minimization (WNNM) method is then introduced to effectively solve the proposed model. Extensive experimental results validate that the presented model achieves convincing improvement over many state-of-the-art SR models both quantitatively and perceptually.

Keywords: Single image super-resolution · Sparse coding · Global and nonlocal self-similarity · Low-rank constraint · Weighted nuclear norm minimization

1 Introduction

Image super-resolution, one of the active research topics in the image processing community, is dedicated to restore the high resolution (HR) image from its degraded version. To handle the ill-posed nature of SR problems, a variety of image super-resolution methods have been proposed including interpolation methods, reconstruction based methods and example learning methods. Recent studies put more emphasis on example learning methods for its strong image super-resolution capability.

Example learning method plays a significant role in SR problems. It presumes that the high-frequency details lost in low-resolution (LR) image can be predicted by learning the co-occurrence relationship between LR- HR training patches. Freeman *et al.* [1] first proposed a Markov network model to estimate the co-occurrence relationship. Yang *et al.* [2] proposed a SR method based on sparse coding in consideration of the assumption that LR-HR image patch pairs share the same sparse codes with respect to their own dictionaries. Several works aiming at obtaining more image priors to improve the accuracy of sparse codes have been proposed. The so-called nonlocal self-similarity (NSS) prior

© Springer International Publishing AG 2017
F. Cong et al. (Eds.): ISNN 2017, Part II, LNCS 10262, pp. 222–230, 2017.
DOI: 10.1007/978-3-319-59081-3_27

which means that a local patch can find many similar patches across the whole image has demonstrated its great success in image restoration. Dong *et al.* [3] presented the non-locally centralized sparse representation (NCSR) framework by unifying the NSS prior and the sparsity prior to improve the robustness of SR results. Exploring both the column and row NSS priors among the sparse code matrix, Li *et al.* [4] presented a dual-sparsity regularized sparse representation (DSRSR) model. However, due to the independent coding process of each image patch via these NSS sparse representation models, the global similarity in whole image is lost despite the NSS prior is fully utilized.

Recent researches have demonstrated that the low-rank constraint is better at capturing the global similarity information [5]. Inspired by [5], we propose a new sparse representation model, which concerns the global low-rank prior and nonlocal self-similarity prior simultaneously to preserve such global and nonlocal self-similarity information. The weighted nuclear norm minimization [6] is introduced to solve the proposed model. Experimental results show that our method outperforms many state-of-the-art SR methods in terms of subjective and objective qualities. The rest of the paper is organized as follows: Sect. 2 describes our formulation and solution of the presented model for SR. Section 3 presents the extensive experimental results together with relevant discussions and Sect. 4 concludes the paper.

2   Proposed Method

The idea of sparse representation for SR is to approximate each input patch x_i with a weighted linear combination of a few elementary atoms chosen from a redundancy dictionary ϕ. First for an image $x \in \mathbb{R}^N$, let $x_i = R_i x$ denote an image patch of size $p \times p$ extracted at location i. Then, x_i can be approximated as $x_i = \phi s_i$ by a straight-forward least-square solution [7]:

$$x_i \approx \phi \circ s_i = \left(\sum\nolimits_{i=1}^{N} R_i^T R_i \right)^{-1} \sum\nolimits_{i=1}^{N} \left(R_i^T \phi s_{x,i} \right) \tag{1}$$

we can reformulate Eq. (1) as the following minimization problem [2]

$$s_i = \underset{s}{argmin} \|x_i - \phi s_i\|_2^2 + \lambda \|s_i\|_1 \tag{2}$$

Because of the degradation of the observed image, using only the local l_1-norm sparsity prior ignores image structures and leads to an inaccurate SR results. The SR model incorporating NSS prior has been boosted studied. In the NCSR framwork [3], the general NSS sparse representation model is modeled as

$$s_i = \underset{s}{argmin} \|y_i - H\phi s_i\|_2^2 + \lambda_1 \|s_i\|_1 + \lambda_2 \|s_i - \beta_i\|_2^2 \tag{3}$$

where $\beta_i = \sum\limits_{j=1}^{q} w_j s_j$ is a good estimation of s_i, s_j is the coefficient corresponding to the j-th similar image patch of x_i, and the regularization parameters λ_1, λ_2 quantify the tradeoff between the sparsity and the NSS prior. The iterative shrinkage algorithm [8] can be used to solve Eq. (3), therefore, in each interation, s_i can be solved as

$$s_i^{l+1}(j) = T_\tau(v_{i,j}^{(l)} - \beta_i(j)) + \beta_i(j) \tag{4}$$

where $v^{(l)} = \phi^T \circ H^T(Y - H\phi \circ \alpha^{(l)})/c + \alpha^{(l)}$, τ is a parameter to guarantee the convexity of Eq. (4) and $T_\tau()$ is the classic soft-thresholding operator.

2.1   WNNM for Global Low-Rank Constraint

Clearly, it is expected that similar image patches should be encoded as totally similar code coefficients in whole image. However, due to the independence of sparse coding for each image patch in traditional NSS methods, similar image patches may be encoded totally differently. Therefore, it is necessary to impose a constraint on all similar code coefficients for preserving the global similarity information.

Recent researches have demonstrated that the latent structure underlying image similar patches forms a low-dimensional subspace [9]. The general matrix rank minimization problem can be expressed as

$$\begin{aligned} &\min rank(S) \\ &\text{subject to } S \in \Gamma \end{aligned} \tag{5}$$

where $S \in \mathbb{R}^{m \times n}$ is a matrix and Γ is a convex set. In view of that Eq. (5) is hard to solve, the nuclear norm minimization (NNM) was presented as a convex relaxation of matrix rank minimization problem. The nuclear norm minimization of a matrix S, denoted by $\|S\|_*$, is defined as the sum of its singular values, i.e.,

$$\|S\|_* = \sum_i |\sigma_i(S)|_1 \tag{6}$$

where $\sigma_i(S)$ denote the i-th singular values of the matrix S. It is proposed by Cai et al. [10] that soft-thresholding operation on the singular values of matrix S can easily solves the NNM based low rank matrix approximation. In other words, the solution of

$$\hat{S} = \underset{s}{argmin} \|S - S_0\|_F^2 + \tau \|S\|_* \tag{7}$$

where τ is a positive constant, can be obtained by

$$\hat{S} = UT_\tau(\Sigma)V^T \tag{8}$$

where $S = U\Sigma V^T$ is the SVD of S_0 and $T_\tau(\Sigma)$ is the soft-thresholding operator on Σ with parameter τ. However, the standard NNM treats each singuar value equally and ignores the prior knowledge on the matrix singular values. In [6], the NNM has been extended to weighted nuclear norm minimization and achieved excellent results in image denoising. The WNNM problem can be expressed as

$$\|S\|_{w,*} = \sum_i |w_i \sigma_i(S)|_1 \tag{9}$$

where $w_i = [w_1 \ldots w_n]$ and $w_1 \geq 0$ is the weight assigned to $\sigma_i(S)$.

2.2 Global and Nonlocal Self-similarity Prior Guided SR Model

In [3], Dong *et al.* proposed the NCSR framework to estimate the sparse code by nonlocal similar patches, which ignores the global self-similarity in whole image. To overcome this defect, we introduce low-rank constraint to preserve the global similarity information. In our model, we cluster all image patches into M categories by K-means method, then the columns of the global similar coefficient matrix S are composed of similar code coefficients whose corresponding image patches belong to the same category. The low-rank constraint of similar code coefficient matrix S can be defined as:

$$R(S) = \|S\|_* \tag{10}$$

To incorporate Eq. (10) into the nonlocal self-similarity sparse representation model, we rewrite Eq. (3) as

$$S = \underset{S}{argmin} \|Y - H\phi S\|_F^2 + \lambda_1 \|S\|_1 + \lambda_2 \|S - SW\|_F^2 \tag{11}$$

where the matrix Y is composed of all similar image patches y_i to be encoded in whole image. The matrix S is the code coefficient matrix corresponding to the matrix Y.

By combining Eqs. (10) and (11), the proposed sparse representation model can be defined as follows:

$$S = \underset{S}{argmin} \|Y - H\phi S\|_F^2 + \lambda_1 \|S\|_1 + \lambda_2 \|S - SW\|_F^2 + \lambda_3 \|S\|_* \tag{12}$$

The first term in Eq. (12) is the data error term that measures the error between sparse linear combination and real values. The second term is the sparsity prior and the third term is the NSS prior that assumes code coefficients to be predicted by weighted average of neighborhoods. The last term is low-rank constraint to ensure global similar image patches can be encoded as similar code coefficients.

2.3 Algorithm of the Proposed Method for SR

The proposed SR method is composed of two parts: the learning phase and the image reconstruction phase. In learning phase, we utilize a set of compact Principal Components Analysis (PCA) sub-dictionaries which are trained by using the adaptive sparse domain selection (ASDS) strategy proposed in [7]. We first extract patches across different image scales from the training images to obtain training examples and then cluster the image patches into K clusters $\{L_1, L_2, \ldots L_k\}$ with the certain iteration loop by K-means clustering method. Finally, K sub-dictionaries $\{\phi_1, \phi_2, \ldots \phi_k\}$ are trained with the PCA method. For each image patch, we adaptively select one of PCA sub-dictionaries to code it. It enforces the code coefficients of the coded patch over other sub-dictionaries to be 0, which ensures the sparsity of the code coefficients. Thus, the sparsity regularization term can be removed, we can rewrite Eq. (12) as

---

Algorithm 1: the algorithm of the proposed method for SR

Input: the test LR image y, scaling factor s

Output: the final HR image x^*

1.Initialization:

   (a) set the size of LR patch $p \times p$, initial regularization and step parameters: $\lambda_2, \lambda_2, \delta$.

   (b) Upscale y to y^* using Bicubic interpolation with a factor of s.

2. Outer loop (dictionary learning and clustering): for each iteration $i = 0$ to I do

   (a) Update the sub-dictionaries $\{\phi_k\}$ via K-means clustering and PCA methods;

   (b) Inner loop: for each iteration $j = 0$ to J do

   1) $\hat{x}^{(j+1/2)} = \hat{x}^{(j)} + \delta H^T(Y - H\hat{x}^{(j)})$, where δ is the pre-determined constant.

   2) Compute $v^{(j)} = [\phi_{k1}^T \hat{x}_1^{(j+1/2)}, ..., \phi_{kN}^T \hat{x}_N^{(j+1/2)}]$, where ϕ_{kl} is the dictionary assigned to patch $\hat{x}_l$, and N is the total number of image patches;

   3) Compute $\hat{s}^{(j+1)}$ using the shrinkage operator given in Eqn. (4);

   4) If $mod(j, J_0) = 0$, partition $\hat{x}^{(j+1/2)}$ into a sequence of image patches, then cluster these image patches into M clusters via K-means clustering. Let $X_n^{(j+1/2)}$ be n-th cluster and $S_n^{(j+1/2)}$ be the code coefficients matrix corresponding to $X_n^{(j+1/2)}$, compute $S_n^{(j+1/2)}$ using Eqn. (14);

   5) Image estimate update: $\hat{x}^{(j+1)} = \phi \circ S^{j+1}$ using Eqn. (1).

$$S = \underset{S}{argmin} \|Y - H\phi S\|_F^2 + \lambda_2 \|S - SW\|_F^2 + \lambda_3 \|S\|_* \qquad (13)$$

In image reconstruction phase, we first select a PCA sub-dictionary to code the patch by Eq. (4), and then we cluster these image patches into M clusters via K-means clustering. For all patches belong to the same category, the corresponding coefficient matrix S can be solved by using low-rank approximation. Compared Eq. (8) with Eq. (9), we have

$$\hat{S}_i = UT_{w_i}(\Sigma)V^T \qquad (14)$$

The main procedure of SR algorithm is summarized in Algorithm 1.

3 Experimental Results

In this section, numerous experimental studies on SR were carried out to validate the performance of our SR model. We compare the proposed method with several state-of-the-art image super-resolution methods including NE [1], SCSR [2], ASDS [7], NCSR [3], A+ [11], SRCNN [12] and evaluate the quality by Peak Signal to Noise Ratio (PSNR) and Structural Similarity (SSIM). The source codes of all competing algorithm are obtained from their original authors and we use the default parameter settings. In the experiments, the basic parameters are set as follows: two testing benchmarks Set5 and Set14 containing 5 and 14 commonly used images respectively are adopted and they are down-sampled by a decimated factor of s, blurred by 7×7 Gaussian filter with standard deviation 1.6 to generate LR images. By experience, we set the scaling factors s to be 2, 3. The patch size is 5×5 and the number of cluster used to train dictionary K is set to 70. For image super-resolution, $\delta = 3.5$, $\lambda_2 = 7$, $N = 40$, $I = 6$ and $J = 160$ respectively. For low-rank approximation of code coefficients, combining the noise knowledge, we make the weight w_i of ith category to be $w_i = c * 2\sqrt{2}\sigma_i^2/(\lambda_3 + \epsilon)$, where c and σ_i represent the number of coefficients and the mean value of patch noise respectively, and $\lambda_3 = 1$.

Table 1 lists some quantitative comparison results in terms of PSNR index. It can be seen that our reported PSNR scores are higher than the aforementioned methods. In Table 2, we show the average PSNR and SSIM values of reconstructed images for

Table 1. PSNR(dB) results for scaling factors $3\times$ for reconstructed images

Images	NE	SCSR	ASDS	NCSR	A+	SRCNN	Proposed
Bird	31.18	33.04	35.53	35.73	34.65	33.04	**35.96**
Butterfly	22.49	23.38	27.35	28.10	26.32	23.37	**28.69**
Woman	27.35	28.69	31.43	31.84	30.36	29.14	**32.12**
Pepper	22.52	30.78	32.43	33.10	32.24	30.37	**33.51**
ppt3	24.20	22.88	25.57	25.94	25.46	22.81	**26.44**
Zebra	23.32	24.81	28.97	29.32	28.16	24.64	**29.47**
Average	25.18	27.26	30.21	30.67	29.53	27.23	**31.03**

Table 2. Aveage PSNR(dB)/SSIM results for scaling factors $2\times$, $3\times$ for reconstructed images

Images	Ratios	NE	SCSR	ASDS	NCSR	A+	SRCNN	Proposed
Set5	2	29.85	28.59	33.77	35.10	33.57	28.74	**35.22**
		0.8658	0.8467	0.9506	0.9349	0.9220	0.8447	**0.9356**
	3	28.62	29.73	32.64	32.97	31.86	29.72	**33.20**
		0.8263	0.8647	0.9053	0.9104	0.8958	0.8590	**0.9133**
Set14	2	26.07	26.68	31.09	31.18	29.94	26.53	**31.33**
		0.6947	0.7526	0.8731	0.8756	0.8418	0.7473	**0.8768**
	3	26.21	26.93	28.91	29.01	28.63	26.81	**29.15**
		0.7286	0.7640	0.8168	0.8205	0.7993	0.7688	**0.8223**

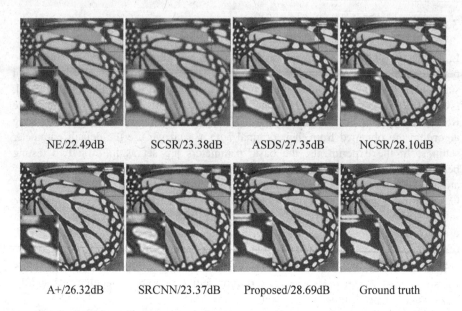

<table>
NE/22.49dB   SCSR/23.38dB   ASDS/27.35dB   NCSR/28.10dB

A+/26.32dB   SRCNN/23.37dB   Proposed/28.69dB   Ground truth
</table>

Fig. 1. Comparison of super resolution results by 3× on "butterfly" image (Color figure online)

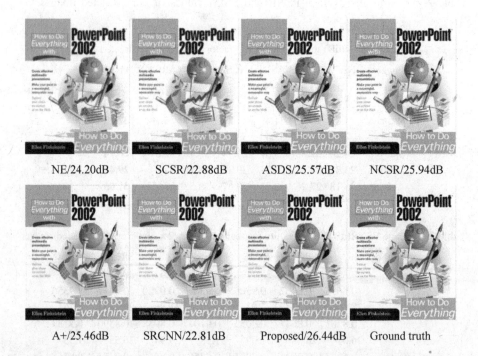

NE/24.20dB   SCSR/22.88dB   ASDS/25.57dB   NCSR/25.94dB

A+/25.46dB   SRCNN/22.81dB   Proposed/26.44dB   Ground truth

Fig. 2. Comparison of super resolution results by 3× on "ppt3" image (Color figure online)

scaling factors of 2 and 3. It indicates that when compared with these competing methods, the proposed method achieves the highest PSNR and SSIM measures on test images regradless of the scaling factor. For $3\times$ scaling factor, the average PSNR gains of the proposed method over the second best method (i.e., the NCSR method) and the third best method can be up to 0.23 dB and 0.56 dB respectively on Set5, 0.14 dB and 0.24 dB respectively on Set14. Figures 1 and 2 present some subjective results for $3\times$ scaling factor to further validate the superiority of our method. The red block with its corresponding magnification on the left-bottom corner of each image shows the reconstruction details. Because of introducing the global and nonlocal self-similarity, the proposed method can recover fewer artifacts, sharper edges and fine details.

4   Conclusion

In this paper, we developed a SR model containing the global low-rank prior and the nonlocal self-similarity prior simultaneously for solving the ill-posed SR problem. The global low-rank prior is utilized to capture global similarity information in whole image. A WNNM algorithm is introduced to effectively solve the global low-rank approximation of code coefficients. Extensive experiments demonstrate that the presented model achieves convincing improvement over many state-of-the-art SR methods in PSNR, SSIM and visual quality assessment. Moreover, we can recover sharper edges and richer textures, which indicates the effectiveness of our model in preserving the global similarity information.

References

1. Freeman, W., Jones, T., Pasztor, E.: Example-based super resolution. IEEE Comput. Graph. Appl. **22**(2), 56–65 (2002)
2. Yang, J., Wright, J., Huang, T., Ma, Y.: Image super-resolution via sparse representation. IEEE Trans. Image Process. **19**(11), 2861–2873 (2010)
3. Dong, W., Zhang, L., Shi, G., Li, X.: Nonlocally centralized sparse representation for image restoration. IEEE Trans. Image Process. **22**(4), 620–1630 (2013)
4. Li, J., Gong, W., Li, W.: Dual-sparsity regularized sparse representation for single image super-resolution. Inform. Sci. **298**, 257–273 (2015)
5. Liu, G., Lin, Z., Yan, S., Sun, J., Yu, Y., Ma, Y.: Robust recovery of subspace structures by low-rank representation. IEEE Trans. Pattern Anal. Mach. Intell. **35**(1), 171–184 (2013)
6. Gu, S., Zhang, L., Zuo, W., Feng, X.: Weighted nuclear norm minimization with application to image denoising. In: CVPR, pp. 2862–2869 (2014)
7. Dong, W., Zhang, L., Shi, G., Wu, X.: Image deblurring and super resolution by adaptive sparse domain selection and adaptive regularization. IEEE Trans. Image Process. **20**(7), 1838–1857 (2011)
8. Daubechies, I., Defriese, M., DeMol, C.: An iterative thresholding algorithm for linear inverse problems with a sparsity constraint. Commun. Pure Appl. Math. **57**(11), 1413–1457 (2004)
9. Chen, F., Zhang, L., Yu, H.: External patch prior guided internal clustering for image denoising. In: IEEE International Conference on Computer Vision, pp. 603–611 (2015)

10. Cai, J.-F., Candès, E.J., Shen, Z.: A singular value thresholding algorithm for matrix completion. SIAM J. Optim. **20**(4), 1956–1982 (2010)
11. Timofte, R., De Smet, V., Van Gool, L.: A+: adjusted anchored neighborhood regression for fast super-resolution. In: Cremers, D., Reid, I., Saito, H., Yang, M.-H. (eds.) ACCV 2014. LNCS, vol. 9006, pp. 111–126. Springer, Cham (2015). doi:10.1007/978-3-319-16817-3_8
12. Dong, C., Chen, C.L., He, K., et al.: Image super-resolution using deep convolutional networks. IEEE Trans. Pattern Anal. Mach. Intell. **38**(2), 295–307 (2016)

EMR: Extended Manifold Ranking
for Saliency Detection

Bo Li[⊠], Hang Gao, and Han Liu

School of Electronic and Information Engineering,
South China University of Technology, Guangzhou, China
leebo@scut.edu.cn, {g.hang,eeliuhan}@mail.scut.edu.cn

Abstract. A novel and outstanding saliency detection approach based
on color features and background prior is proposed in this paper. Specifically, background prior is used in saliency detection widely, which considers the image boundaries as part of background. Then we propose
an extended manifold ranking (EMR) algorithm to propagate the background prior to other image regions. Compared with GMR, EMR eliminates the negative effect of the initial assumption that non-boundary
areas are all saliency regions. Furthermore, gradient boosting decision
tree (GBDT) is introduced to refine the saliency map generated by EMR.
The experimental results on three benchmark datasets demonstrate that
our algorithm outperforms 10 state-of-the-art methods based on low-level
features.

Keywords: Background prior · Manifold ranking · EMR · GBDT

1 Introduction

Saliency detection is an efficient way to understand and analyze images. Its goal
is to distinguish the saliency object of an image. Generally, saliency models can
be roughly categorized into top-down [1] and bottom-up [2–4] approaches. The
former one is by task-driven and the later one is by data-driven. In this paper,
we focus on bottom-up approach to detect saliency objects.

The color contrast [3] is widely used to detect saliency object. Most of the
previous works use LAB mean color feature to describe images. In this paper,
it is found that the color contrast computed in different color spaces could show
different performances. Moreover, color histogram features can describe images
more specifically. Hereby, the mean color features and color histogram features
extracted from Lab, HSV, RGB color spaces are employed in the paper.

On the other hand, background prior [2,4,5] is a widely used cue, which
chooses the image boundaries as part of background. Yang et al. [2] (GMR) utilize the manifold ranking to propagate background prior to the whole image and
obtain a satisfied result. Specifically, in addition to choose the image boundaries
as background, GMR assumes the non-boundary areas are all saliency regions,
then applies manifold ranking to revise the saliency regions based on background

© Springer International Publishing AG 2017
F. Cong et al. (Eds.): ISNN 2017, Part II, LNCS 10262, pp. 231–238, 2017.
DOI: 10.1007/978-3-319-59081-3_28

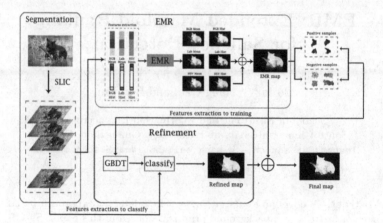

Fig. 1. The pipeline of our method. Firstly, the image is segmented into multi-scale superpixels and six different color features are extracted. Based on each color feature, the background prior is propagated to get the saliency map via EMR. Finally, GBDT is applied to refine the saliency map, the positive and negative samples are extracted based on EMR map. The progress is repeated at multiple scales and results are fused.

prior. However, the assumption is inappropriate, since it can bring the negative effect to detect the background regions. To address the above problem, an extended manifold ranking (EMR) algorithm is proposed in this paper. By using the background prior alone and making no any assumption on saliency regions, EMR could perform much better than GMR.

For the refinement part, the most believable saliency and background regions are extracted from the saliency map generated by EMR, and are combined into a feature vector. Gradient boosting decision tree (GBDT) [6] is applied to learn the relationship among these features automatically. The pipeline of the proposed approach is showed in Fig. 1.

In summary, contributions of this paper can be summarized as follows: (1) The mean color features and color histogram features are both extracted. (2) An extended manifold ranking method is proposed to diffuse the background prior. (3) GBDT is introduced to seek the relationship of different features automatically.

2   Image Pre-processing

2.1   Multi-scale Segmentation

As in [4,5], SLIC [8] is used to segment the whole image into N superpixels with l different scales. $l = 5$ and $N = 100, 150, 200, 400$ are considered in this paper. For each scale, a saliency map will be generated based on our approach, and the averaged saliency maps can achieve a better result. Next, we will take one of these scales as an instance. The superpixel set is defined as $V = \{P_1, P_2, \ldots, P_n\}$, where n is the number of superpixels and P_i is one superpixel of them.

2.2   Graph Construction

A single layer graph $G = (V, E)$ is used to represent the relationship of different superpixels, where V denotes the set of nodes and E denotes the set of undirected edges. In this paper, each node is defined as a superpixel, and the edges E are weighted by an affinity matrix $W = [w_{ij}]_{n \times n}$. Here, w_{ij} denotes the similarity between any two adjacent superpixels.

As discussed before that the mean color features and color histogram features extracted from three different color spaces can describe the images from different perspectives. Considering this fact, for the mean color feature, the color contrast CC_{ij} between P_i and P_j is computed as follows: $CC_{ij}^k = \|CS_i^k - CS_j^k\|_2, k = 1, 2, 3$. where CS_i^k denotes the mean color of P_i in three color spaces respectively. For the histogram feature, The contrast between two histograms is computed by chi-squared distance as follows: $CC_{ij}^k = \sum_{m=1}^{K} \frac{2(h_{ik}(m)-h_{jk}(m))^2}{h_{ik}(m)+h_{jk}(m)}, k = 4, 5, 6$ Where K denotes the number of bins($K = 8$ in this paper).

Based on the above features, in the paper, the similarity matrix W is defined as follows:

$$w_{ij} = \begin{cases} \frac{1}{CC_{ij}+\varepsilon} & P_j \in N(i) \\ 0 & i = j \text{ or otherwise} \end{cases} \tag{1}$$

where $N(i)$ denotes the neighbor superpixels of P_i, and $\varepsilon > 0$ is a constant to avoid $CC_{ij} = 0$. In this paper, ε is set to 10^{-4}.

We can obtain six similarity matrixes respectively. Based on the similarity matrixes, six saliency maps will be obtained using the proposed propagation method and the average of them could reach a better result.

3   Propagation

In the following section, the graph-based manifold ranking [2] (GMR) for saliency detection will be briefly introduced, and then the proposed approach (EMR) will be presented in detail.

3.1   Manifold Ranking

Background prior is widely used in saliency detection. In the superpixel set V, the superpixels which locate at the image boundaries are labelled as queries and the rest need to be ranked according to their relevances. GMR aims to assign a ranking value f_i to each superpixel P_i, and the ranking values can be viewed as a vector $\mathbf{f} = [f_1, f_2, \ldots, f_n]^T$. Let $\mathbf{y} = [y_1, y_2, \ldots, y_n]^T$ be an indication vector, in which $y_i = 1$ if P_i is a query, and $y_i = 0$ otherwise. Given the affinity matrix W, the degree matrix is denoted as $D = diag\{d_{11}, d_{22}, \ldots, d_{nn}\}$, where $d_{ii} = \sum_j w_{ij}$. Then the ranking value of $\mathbf{f}$ can be solved as follows:

$$\mathbf{f}^* = \arg\min_f \frac{1}{2}(\sum_{i,j=1}^{n} w_{ij}\|\frac{f_i}{\sqrt{d_{ii}}} - \frac{f_j}{\sqrt{d_{jj}}}\|^2 + \mu \sum_{i=1}^{n} \|f_i - y_i\|^2) \tag{2}$$

In Eq. (2), the first term ensures that the ranking values of two adjacent super-pixels which have similar color features should not change to much(smoothness constraint), and the second term ensures the optimal result $\mathbf{f}^*$ should not differ too much from the initial query assignment(fitting constraint). μ acts as a balance parameter(in [2], μ is set to 0.99). The optimal solution of Eq. (2) can be obtained as: $\mathbf{f}^* = (D - \alpha W)^{-1}\mathbf{y}$. Then, the saliency of superpixels are measured using the normalized ranking score $1 - \mathbf{f}^*$.

3.2   Extended Manifold Ranking

In Eq. (2), the superpixels labelled as queries ($y_i = 1$) are part of the background. For those background superpixels labelled as non-queries ($y_i = 0$), the optimal ranking values restrained by the second term in Eq. (2) (fitting constraint) should not differ too much from 0. However, to be judged as background, the ranking values of background pixels are desired to be 1. So the second term in Eq. (2) has a negative effect on those background pixels labelled as non-queries.

To solve the problem, we remove the second term in Eq. (2), and introduce an extern term to highlight the saliency regions. Let $\mathcal{L}$ be a superpixel set, which consists of all the boundary superpixels, and $\mathbf{s} = [s_1, s_2, \ldots, s_n]^T$ be the saliency value of each superpixel. s_i is fixed to a small value $\epsilon(\epsilon = 0.01)$ if $s_i \in \mathcal{L}$. In this paper, the proposed extended manifold ranking algorithm can be obtained by minimizing the following energy function:

$$\arg\min_{\mathbf{s}} \frac{1}{2} \sum_{i,j}^{n} w_{i,j}(s_i - s_j)^2 - \sum_{i=1}^{n} s_i \tag{3}$$

$$\text{s.t.   } s_i = \epsilon, \forall s_i \in \mathcal{L}$$

Compared with Eq. (2), in Eq. (3), the first term is simplified for better comprehending and solution, and the second term is used to highlight the saliency regions. To gain a more comprehensive interpretation, we minimize Eq. (3) with respect to $\mathbf{s}$ by setting its derivative to zero and obtain the following solution:

$$s_i = \frac{1}{\sum_j w_{i,j}} \sum_j w_{i,j}s_j + \frac{1/2}{\sum_j w_{i,j}} \tag{4}$$

As the similarity matrix W defined in Sect. 2.2, $w_{i,j} > 0$ only if P_i is adjacent to P_j. According to Eq. (4), the saliency value of P_i is only affected by its surrounding superpixels. The more similar is P_j to P_i, the larger impact P_j has on P_i. If P_i is similar to one or more of its surrounding superpixels, there exists at least one $w_{i,j}$ which is large, then the second term in Eq. (4) is small. Consequently, Eq. (4) can ensure the two adjacent superpixels which show similar color characteristic have similar saliency values.

The key of detecting the saliency regions is the second term in Eq. (3). Since the proposed approach is based on the background prior, it required to propagate the background prior to the whole image. According to Eq. (4), when the saliency values of boundary superpixels are fixed to a small value, then other background superpixels will be also updated to small values. While for the saliency regions, we have not fixed the saliency values to them, and the background has fewer impact on them. So in order to minimize the Eq. (3), the saliency values of them should be large, consequently.

To solve the energy function in Eq. (3), matrix $L = D - W$ will be used, where D is given in Sect. 3.1. We re-order the saliency value vector $\mathbf{s}$ as $\mathbf{s} = [\mathbf{s}_l^T \quad \mathbf{s}_u^T]^T$ and $L = \begin{bmatrix} L_{ll} & L_{lu} \\ L_{ul} & L_{uu} \end{bmatrix}$ where l indicates the boundary superpixel set and u corresponds to the no-boundary superpixel set. The energy minimization problem can then be re-written as follows:

$$\mathbf{s}^* = \arg\min_{\mathbf{s}} \frac{1}{2}\mathbf{s}^T L \mathbf{s} - \mathbf{s}^T \mathbf{1}_n \tag{5}$$

where $\mathbf{1}_n := [1, \ldots, 1]^T \in \mathbb{R}^n$. By setting the derivative of Eq. (5) to be zero, the final saliency values of no-boundary superpixels are computed as:

$$\mathbf{s}_u = L_{uu}^{-1}[-\frac{1}{2}(L_{ul}\mathbf{s}_l + L_{lu}^T\mathbf{s}_l) + \mathbf{1}_u] \tag{6}$$

4 Refinement

In this paper, we employ gradient boosting decision tree(GBDT) [6] to refine the saliency map. To train GBDT, samples are extracted from the saliency map that generated by EMR. We set a high threshold T_h and a low threshold T_l. If $s_i > T_h$, the superpixel P_i acts as a positive sample, and if $s_i < T_l$, P_i acts as a negative sample. In this paper, we set $T_h = \max[1.5 \times \bar{\mathbf{s}}, 0.1]$, and $T_l = 0.05$. $\bar{\mathbf{s}}$ denotes the mean saliency value of the saliency map.

Following this way, we can consider all the features in an uniform way, and GBDT can seek the relationship of different features. Different from other supervised methods for saliency detection [1], the positive and negative samples are extracted from the saliency map instead of the ground truth labeled by humans.

5 Experiments

We evaluate the proposed approach on three benchmark datasets: MSRA10K [3], ECSSD [7], and DUT-OMRON [2]. These datasets have been widely used to evaluate the performance of saliency detection.

5.1 Quantitative Evaluation

All methods are evaluated by precision-recall curve. Usually, neither precision nor recall can evaluate the quality of a saliency map comprehensively. In most cases, high precision and recall are both required. So we also use F-measure to measure the overall performance:

$$F_\beta = \frac{(1 + \beta^2) \cdot precision \cdot recall}{\beta^2 \cdot precision + recall} \tag{7}$$

As suggested in [9], we set $\beta^2 = 0.3$ to emphasize the precision value.

5.2   Analysis of Our Model

In this section, we will analyze the performance of three components in our approach: low-level features, extended manifold ranking, and refinement.

In this experiment, we generate the saliency maps respectively via EMR without refinement on ECSSD dataset. The P-R curves are shown in Fig. 2(a). From the figure, we can see that the features extracted from HSV color space perform best, and the color histogram features outperform the mean color features by a large margin. The average of these saliency maps can achieve a better result (red line in Fig. 2(a)).

As shown in Fig. 2(b), by applying the GBDT to refine the saliency map, the result can be improved significantly.

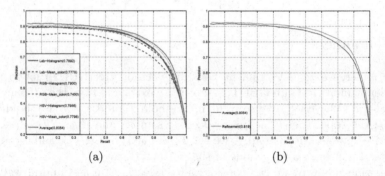

(a)   (b)

Fig. 2. The performance of six features and refinement. (a) PR curves of saliency maps generated by six features, and the values shown in the parenthesis are the max value of F-measure. (b) The performance of refinement. (Color figure online)

We compare the proposed EMR with GMR on the three datasets. In this experiment, we only take the mean color feature to generate the saliency maps in a single layer ($N = 200$). The results are shown in Fig. 3. As it can be found that the proposed EMR outperforms GMR by a large margin.

5.3   Comparison with State-of-the-Art Approaches

We compare the proposed approach with ten of the state-of-the-art approaches, including DRFI [1], EQCUT [5], RBD [4], ST [10], GMR [2], RC [3], HDCT [11], BSCA [12], IDCL [13], BL [14]. The saliency maps of the first seven methods on different datasets are provided by Borji's work et al. [15]. The others are achieved by running available codes.

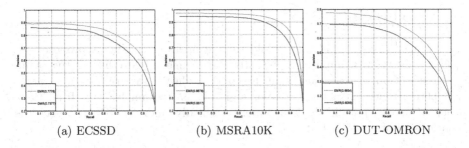

(a) ECSSD   (b) MSRA10K   (c) DUT-OMRON

Fig. 3. The comparison between the proposed EMR and GMR [2] on three datasets.

As shown in Fig. 4, our approach performs best on these datasets, especially on MSRA10K and ECSSD. It is noted that DRFI is a supervised method which is also based on the low-level cues, while it is expensive for data collection. Our approach does not rely on the ground truth labeled by humans and just use six simple features.

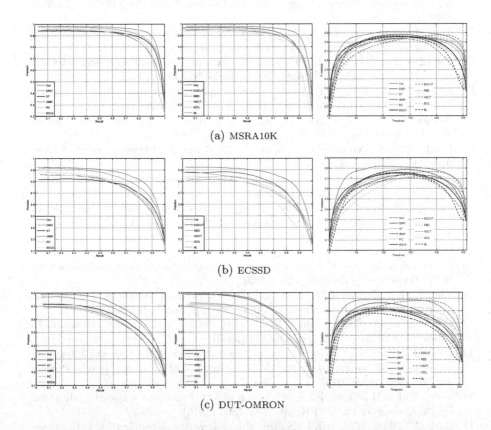

(a) MSRA10K

(b) ECSSD

(c) DUT-OMRON

Fig. 4. Comparison with ten state-of-art methods on three benchmark datasets

6   Conclusion

In this paper, an extended manifold ranking method (EMR) is proposed for saliency detection. Compared with GMR, EMR can eliminate the negative of initial assumption, which assumes all the non-boundary areas are saliency regions. Furthermore, we extract mean color and color histogram features from three color spaces to describe the image more detailedly, and use GBDT to seek the relationship of different features.

Acknowledgement. This research was supported by the National Natural Science Foundation of China (Grant Nos. 11627802, 51678249), by the Science and Technology Projects of Guangdong (2013A011403003), and by the Science and Technology Projects of Guangzhou (201508010023).

References

1. Jiang, H., Wang, J., Yuan, Z., et al.: Salient object detection: a discriminative regional feature integration approach. In: Proceedings of the CVPR, pp. 2083–2090 (2013)
2. Yang, C., Zhang, L., Lu, H., et al.: Saliency detection via graph-based manifold ranking. In: Proceedings of the CVPR, pp. 3166–3173 (2013)
3. Cheng, M.M., Zhang, G.X., et al.: Global contrast based salient region detection. IEEE Trans. Pattern Anal. Mach. Intell. **37**(3), 569–582 (2015)
4. Zhu, W., Liang, S., Wei, Y., Sun, J.: Saliency optimization from robust background detection. In: Proceedings of the CVPR, pp. 2814–2821 (2014)
5. Aytekin, C., Ozan, E.C., Kiranyaz, S., et al.: Visual saliency by extended quantum-cuts. In: Proceedings of the ICIP, pp. 1692–1696 (2015)
6. Friedman, J.H.: Greedy function approximation: a gradient boosting machine. Ann. Stat. **29**(5), 1189–1232 (2000)
7. Yan, Q., Xu, L., Shi, J., Jia, J.: Hierarchical saliency detection. In: Proceedings of the CVPR, pp. 1155–1162 (2013)
8. Achanta, R., Shaji, A., et al.: Slic superpixels compared to state-of-the-art superpixel methods. IEEE Trans. Pattern Anal. Mach. Intell. **34**(11), 2274–2282 (2012)
9. Achanta, R., Hemami, S., Estrada, F., et al.: Frequency-tuned salient region detection. In: Proceedings of the CVPR, pp. 1597–1604 (2009)
10. Liu, Z., Zou, W., Le, M.O.: Saliency tree: a novel saliency detection framework. IEEE Trans. Image Process. **23**(5), 1937–1952 (2014)
11. Kim, J., Han, D., Tai, Y.W., Kim, J.: Salient region detection via high-dimensional color transform. In: Proceedings of the CVPR, pp. 883–890 (2014)
12. Qin, Y., Lu, H., Xu, Y., Wang, H.: Saliency detection via cellular automata. In: Proceedings of the CVPR, pp. 110–119 (2015)
13. Zhou, L., Yang, Z., Yuan, Q., et al.: Salient region detection via integrating diffusion-based compactness and local contrast. IEEE Trans. Image Process. **24**(11), 3308–3320 (2015)
14. Tong, N., Lu, H., Xiang, R., et al.: Salient object detection via bootstrap learning. In: Proceedings of the CVPR, pp. 1884–1892 (2015)
15. Borji, A., Sihite, D.N., Itti, L.: Salient object detection: a benchmark. IEEE Trans. Image Process. **24**(12), 5706–5722 (2015)

Hybrid Order l_0-Regularized Blur Kernel Estimation Model for Image Blind Deblurring

Weihong Li[(✉)], Yangqing Chen, Rui Chen, Weiguo Gong,
and Bingxin Zhao

Key Lab of Optoelectronic Technology and Systems of Education Ministry,
Chongqing University, Chongqing 400044, China
weihongli@cqu.edu.cn

Abstract. Most of blur kernel estimation models may fail when the blurred image contains some complex structures or is contaminated by large blur. In this paper, we propose a hybrid order l_0-regularized blur kernel estimation model for solving the problem. Firstly, we regularize the latent image in a hybrid order case involving both first-order and second-order regularization term, in which l_0 sparse prior is introduced. Secondly, we introduce an improved adaptive adjustment factor into the model for removing detrimental structures and obtaining more useful information. Finally, we develop an efficient optimization algorithm based on the half-quadratic splitting technique to obtain an accurate blur kernel. Extensive experiments results on both synthetic and some challenged real-life images show that proposed model can estimate a more accurate blur kernel and can effectively recover the latent image when it contains complex structures or is contaminated by large blur.

Keywords: Blur kernel estimation · Image blind deblurring · Hybrid order l_0-regularized · Adaptive adjustment factor · Half-quadratic splitting technique

1 Introduction

Blur kernel estimation is a key problem in image blind deblurring. There are many blur kernel estimation methods were proposed [1–12]. Image prior information plays an important role in blur kernel estimation model. Fergus *et al.* [4] adopted a zero-mean mixture of Gaussian to fit the heavy-tailed statistical properties of natural image gradient, and a mixture of exponential distribution to approximate the sparse prior of blur kernel, which can acquire a better estimate result, but also exist some staircase effect. Levin *et al.* [5] used the same prior as Fergus *et al.* [4] but followed a different optimization algorithm by searching for the blur kernel that brings the distribution of the deblurred image closest to that of observation in [6]. Goldstein and Fattal [7] introduced a model and a spectral whitening formula to calculate the power spectrum of the blur kernel, which is efficiency but cannot operate on small images. Shan *et al.* [8] exploited two piecewise continuous functions to match the heavy-tailed natural image prior, and used the l_1 norm of the blur kernel intensities to regularize the sparsity of the blur kernel, which can recover more details. Xu and Jia [9] found sharp edges are very useful in estimating a more accurate blur kernel but the details remarkably damage it,

© Springer International Publishing AG 2017
F. Cong et al. (Eds.): ISNN 2017, Part II, LNCS 10262, pp. 239–247, 2017.
DOI: 10.1007/978-3-319-59081-3_29

thus they performed a small-scale structures removal strategy to adaptively select useful edges for blur kernel estimation, but the deblurred results of this method strongly depends on the image filters. Krishnan *et al.* [10] introduced a normalized sparse prior to estimate the blur kernel, which favored latent images over blurred ones. Xu *et al.* [11] used the l_0 sparse prior for estimating the blur kernel and established a unified framework for both uniform and non-uniform image deblurring, which effectively improved the quality of the deblurred image.

Besides the above analysis, most of the mentioned methods may fail when blurred image contains some complex tiny structures or is contaminated by large blur. So, this paper proposes a hybrid order l_0-regularized blur kernel estimation model for solving the problem. The hybrid order l_0-regularized term can recover the step-edges as well as possible and eliminates the staircase effect at the same time and the l_0 sparse prior can effectively reflect the sparse characteristic of original image. Then we introduce an improved adaptive adjustment factor to choose reliable structures of image. And we develop the half-quadratic splitting technique to obtain an accurate blur kernel. Extensive experiments results demonstrate the efficiency of the proposed model on the blurred image that contains some complex tiny structures or is contaminated by large blur.

2   Proposed Hybrid Order l_0-Regularized Blur Kernel Estimation Model

It is difficult to estimate an accurate blur kernel when image is contaminated by large blur, as shown in Fig. 1 or contains some complex structures, as shown in Fig. 2.

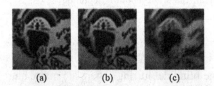

Fig. 1. Images are contaminated by blur kernels in different size. (a) clear image. (b) blur kernel size is 5 × 5. (c) blur kernel size is 27 × 27.

Fig. 2. Real-life images with different structures. (a) image contains simple structure. (b) image contains complex structures.

Xu et al. [11] found the extracted structures of image in the middle procedure often contain more salient edges, which are usually sparse and the l_0 sparse prior is a most natural choice to recover the original sparsity. Hence, they introduced the l_0 sparse prior into a unified model to estimate a better blur kernel for blind deblurring in uniform and non-uniform blurred images. But when image contains some complex tiny structures or is contaminated by large blur, their model can't effectively remove the ringing effects.

2.1   Proposed Blur Kernel Estimation Model

This paper proposes a hybrid order l_0-regularized blur kernel estimation model, and then the estimated blur kernel is used for the non-blind deblurring frame [15] to recover the latent clean image.

At each scale level, our proposed model for spatially uniform blurring is:

$$\min_{\partial L,k}\left\{\|\partial L \otimes k - \partial B\|_2^2 + \gamma\|k\|_2^2 + \omega\lambda\left(\sigma\|\partial L\|_0 + \|\partial^2 L\|_0\right)\right\} \tag{1}$$

where $\partial L = \sqrt{L_x^2 + L_y^2}$, L_x and L_y denote the first order finite difference of sharp image L in x and y direction, respectively. $\partial B = \sqrt{B_x^2 + B_y^2}$, B_x and B_y denote the first order finite difference of sharp image B in x and y direction, respectively. $\partial^2 L$ denotes the second-order finite difference of L, $\partial^2 L = \sqrt{(L_{xx})^2 + (L_{xy})^2 + (L_{yx})^2 + (L_{yy})^2}$. L_{xx} and L_{xy} denote the first order finite difference of L_x in x and y direction, respectively. L_{yx} and L_{yy} denote the first order finite difference of L_y in x and y direction, respectively. $\|\partial L\|_0$ and $\|\partial^2 L\|_0$ are the l_0-norm that count the number of non-zero values of ∂L and $\partial^2 L$, the scalar weights γ, λ and σ control the relative strength of the blur kernel and image regularization terms. ω is the adaptive structure selection factor.

Compared to the model proposed in Ref. [11], our proposed model mainly has the following advantages.

(1) The blur kernel estimation model is done in the image gradient domain, and the data fidelity term is $\|\nabla f - \nabla u \otimes k\|_2^2$ for acquiring a large amount of helpful information in salient edges of the blurred image.

(2) Introduce the hybrid order and l_0 sparse prior into image regularization term $\sigma\|\nabla u\|_0 + \|\nabla^2 u\|_0$ to protect the sparsity of the image edges, and effectively suppress noise in the intermediate image even though the image contains some complex tiny structures or is contaminated by large blur.

(3) Introduce an improved adaptive adjustment factor ω to remove detrimental structures of image, which can adjust the value of λ, defined as:

$$\omega = \exp\left(-|r(p)|^{\frac{2}{3}}\right), \text{ where } r(p) = \frac{\left\|\sum_{q\in N_h(p)} \partial B(q)\right\|_2}{\sum_{q\in N_h(p)} \|\partial B(q)\|_2 + 0.5} \tag{2}$$

in which $N_h(p)$ is an $h \times h$ window centered at pixel p. A large ω implies that the local region is flat where need a strong penalty, and a small ω means that there exist strong image structures in the local window.

2.2   Developed Optimization Algorithm for Proposed Model

For convenience of description, we introduce two auxiliary variables x and y denote ∂L and ∂B, respectively. The proposed model is rewritten as following:

$$\min_{x,k}\|x \otimes k - y\|_2^2 + \gamma\|k\|_2^2 + \omega\lambda(\sigma\|x\|_0 + \|\partial x\|_0) \tag{3}$$

where the term $\|x \otimes k - y\|_2^2$ is the data fidelity to ensure the recovered data consistent with the observation. The term $\gamma\|k\|_2^2$ is the l_2-norm based regularization on k used to stabilize the blur kernel estimation with a fast solver. The term $\omega\lambda(\sigma\|x\|_0 + \partial\|x\|_0)$ is the hybrid l_0 regularization term on x that can preserve the sparsity of natural image gradients.

We first estimate x with k. The x sub-problem is given by:

$$\min_{x}\|x \otimes k - y\|_2^2 + \omega\lambda(\sigma\|x\|_0 + \|\partial x\|_0) \tag{4}$$

This sub-problem is seriously non-convex due to the incorporation of new hybrid l_0 regularization term. Based on the half-quadratic splitting technique [14], we develop an efficient optimization algorithm to solve the problem by alternating minimization. We first introduce two auxiliary variables u and $g = (g_h, g_v)^T$ corresponding to x and ∂x, respectively, and rewrite the objective function as:

$$\min_{x,u,g}\|x \otimes k - y\|_2^2 + \beta\|x - u\|_2^2 + \eta\|\partial x - g\|_2^2 + \omega\lambda(\sigma\|u\|_0 + \|g\|_0) \tag{5}$$

where β and η are two positive penalty parameters which will be varied during the optimization. As β and η are close to ∞, the solution of Eq. (5) converges to that of Eq. (4). With this formulation, we alternate minimizing x, u and g independently by fixing the remaining two variables.

We first update x by giving a fixed u and g produced by previous iterations, and the values of u and g are initialized to be zeros, the objective function Eq. (5) reduced to

$$\min_{x}\|x \otimes k - y\|_2^2 + \beta\|x - u\|_2^2 + \eta\|\partial x - g\|_2^2 \tag{6}$$

The closed-form solution for this quadratic minimizing problem can be easily obtained using Fast Fourier Transform.

And then update u and g. Because u and g are not coupled with each other in Eq. (5), we can optimize them independently by solving two separate objective functions:

$$\min_{u} \beta\|x - u\|_2^2 + \omega\lambda\sigma\|u\|_0 \tag{7}$$

$$\min_{g} \eta\|\partial x - g\|_2^2 + \omega\lambda\|g\|_0 \tag{8}$$

Equations (7) and (8) involve a discrete optimization problem, which is difficult to solve by traditional gradient decent methods. We can obtain the solutions of u and g based on [14].

$$u = \begin{cases} x, & |x|^2 \geq \frac{\omega\lambda\sigma}{\beta} \\ 0, & \text{otherwise} \end{cases}, g = \begin{cases} \partial x, & |\partial x|^2 \geq \frac{\omega\lambda}{\eta} \\ 0, & \text{otherwise} \end{cases} \tag{9}$$

After estimating x, we estimate k with x. The k sub-problem is given by:

$$\min_{k} \|x \otimes k - y\|_2^2 + \gamma\|k\|_2^2 \tag{10}$$

Obviously, Eq. (10) is a least square minimization problem whose closed-form solution can be computed in the frequency domain by FFT. After obtaining k, a dynamic threshold constraint is used to eliminate the noises during the iterative process. And the constraints on kernel k (non-negativity) are following from the physical principles of blur formation.

3 Experimental Results and Evaluations

We execute some experiments to evaluate different models which include state-of-the-art methods [6, 10, 11, 13, 16] on synthetic images and real-life images.

Our proposed model is executed in a multi-scale setting using a coarse-to-fine pyramid of image resolution similar to the method [4] to ensure the model converges to global optimal solution. Initial size of the estimated blur kernel in multi-scale setting should be a square matrix and odd number larger than 3. The sampling ratio between the upper and the lower level are $\sqrt{2}$. The parameters γ, λ and σ in our model are empirically set to 0.001, 0.04 and 1, respectively. $\beta_{\max}$ and $\eta_{\max}$ are set to 2^3 and 1e5, which are calculated by the minimum value of ε according to the reference literature [4]. The number of iteration T is set to 5.

3.1 Experiments with Synthetically Blurred Images

We first implement our model on the public available dataset from [5]. For fairness, we compare the estimated blur kernels by our model with those of, Refs. [6, 10, 11, 13, 16], and perform image non-blind deblurring using the algorithm of [15] with the same parameter settings. We measure the quality of estimated blur kernel using the error metric proposed by Ref. [5].

To facilitate the description process, we name the 4 clear images and 8 blur kernels from [5] in Fig. 3. In addition, the name of the formed blurred images are concatenated by the names of clear image and blur kernel, such as the clear image im01 is contaminated by blur kernel ker01 form a blurred image is named im01_ker01.

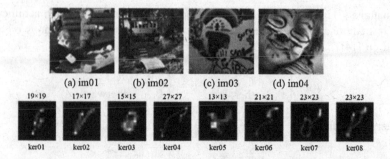

(a) im01   (b) im02   (c) im03   (d) im04

19×19   17×17   15×15   27×27   13×13   21×21   23×23   23×23

ker01   ker02   ker03   ker04   ker05   ker06   ker07   ker08

Fig. 3. The 4 clear images and 8 blur kernels from literature [5]

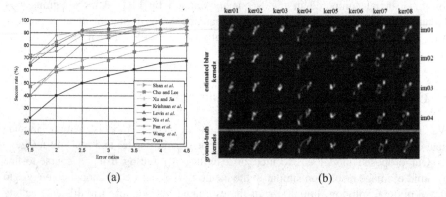

(a)   (b)

Fig. 4. Evaluation of blur kernels estimation results on the dataset from [5]. (a) Cumulative histogram of the estimated blur kernel error ratios across test examples. (b) Estimated blur kernels by our model and the ground-truth kernels

Figure 4 are the experiment results on above-mentioned dataset. Figure 4(a) shows the cumulative error ratios where higher curve denotes a more accurate result, and the horizontal axis is the number of n shows the percentage of test cases whose deblurring error ratios are below n. As indicated in [5], error ratios over 2 will make the result visually implausible. Our model outperforms the other models as a whole, with 83% of the images achieving an error ratio less than 2. The whole 32 estimated blur kernels by our model, show in Fig. 4(b), and are very close to the ground-truth blur kernels.

For a more objective evaluation of the various models, we employ PSNR (peak signal to noise ratio) to evaluate the quality of deblurred images. Table 1 shows the evaluation results, and we can see that our model is better than other methods.

Figure 5 shows one example named im04_ker04 with a blur kernel size 27×27 from the test dataset. As we can see, Refs. [10, 13] fail to provide reasonable blur kernel, and their image deblurring results still contain some obvious blur or ringing artifacts. The deblurring results of Refs. [6, 16] look better, but the estimated kernels still contain some noises. In contrast, our model generates a better blur kernel which is noiseless and sparse. Moreover, the deblurring result of our method can comparable with method of Ref. [11].

3.2   Experiments with Real-Life Blurred Images

We evaluate the performance of our model on some challenged real-life blurred images and comparison with some state-of-the-art deblurring methods.

Table 1. The PSNR of the deblurred images by various methods

Name	[10]	[6]	[11]	[16]	[13]	Ours
im04_ker01	21.05	27.62	28.06	28.11	27.53	28.56
im04_ker02	22.05	27.13	27.23	26.89	26.31	27.52
im04_ker03	22.19	26.39	27.64	27.15·	26.86	28.13
im04_ker04	17.35	22.43	23.12	23.01	22.34	24.29
im04_ker05	23.28	28.67	28.32	28.33	27.86	28.43
im04_ker06	20.19	24.32	25.31	26.05	25.86	26.54
im04_ker07	18.54	23.24	23.76	24.18	23.05	24.73
im04_ker08	18.12	24.67	24.86	24.64	24.33	25.12

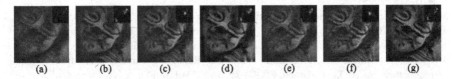

(a)   (b)   (c)   (d)   (e)   (f)   (g)

Fig. 5. Deblurring results on synthetic image. (a) Blurred image and the ground truth blur kernel. (b)–(f) are the deblurring results of Refs. [10], [6], [11], [16], [13], respectively. (j) is our deblurring results.

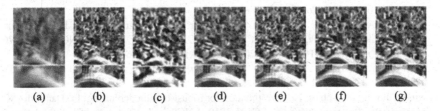

(a)   (b)   (c)   (d)   (e)   (f)   (g)

Fig. 6. Deblurring results on real-life image with complex contour structures. (a) Blurred image, (b)–(f) are the deblurring results of Refs. [10], [6], [11], [16], [13], respectively. (g) is our deblurring result.

Figure 6 shows a challenging example that selected from internet. Due to the complex contour structures of building, most of the existing methods fail to choose the appropriate image structures for blur kernel estimation, which would lead to a bad restored result. The deblurring result of Ref. [10] contain some obvious blur. The results of Ref. [16] seem better but still contain ringing artifacts. Note that our deblurring result is visually comparable with that of Ref. [6, 11, 13] seeming more nature.

4 Conclusion

This paper proposes an effective blur kernel estimation model for blind deblurring from a single image. The proposed model impose hybrid order l_0-regularization term on image, and introduce improved adaptive adjustment factor to increase its robustness. Although the l_0-regularized problem is hard to be optimized, we developed an efficient optimization algorithm based on a half-quadratic splitting technique to get an accurate blur kernel. Both quantitative and qualitative evaluations on synthetic and some challenged real-life images verify the effectiveness of our purposed model.

Acknowledgement. This work was supported by the National Science and Technology Program for Public Wellbeing, China (Grant No. 2013GS500303).

References

1. Chan, T.F., Wong, C.K.: Total variation blind deconvolution. IEEE Trans. Image Proces. **7**, 370–375 (1998)
2. Cho, S., Lee, S.: Fast motion deblurring. ACM Trans. Graph. (TOG) **28**, 145 (2009)
3. Li, W., Chen, R., Xu, S., et al.: Blind motion image deblurring using nonconvex higher-order total variation model. J. Electron. Imaging **25**, 053033 (2016)
4. Fergus, R., Singh, B., Hertzmann, A., et al.: Removing camera shake from a single photograph. ACM Trans. Graph. (TOG) **25**, 787–794 (2006)
5. Levin, A., Weiss, Y., Durand, F., et al.: Understanding and evaluating blind deconvolution algorithms. In: 2013 IEEE Conference on Computer Vision and Pattern Recognition, pp. 964–1971 (2009)
6. Levin, A., Weiss, Y., Durand, F., et al.: Efficient marginal likelihood optimization in blind deconvolution. In: IEEE Conference on Computer Vision and Pattern Recognition. pp. 2657–2664. IEEE Computer Society (2011)
7. Goldstein, A., Fattal, R.: Blur-Kernel estimation from spectral irregularities. In: Fitzgibbon, A., Lazebnik, S., Perona, P., Sato, Y., Schmid, C. (eds.) ECCV 2012. LNCS, vol. 7576, pp. 622–635. Springer, Heidelberg (2012). doi:10.1007/978-3-642-33715-4_45
8. Shan, Q., Jia, J., Agarwala, A.: High-quality motion deblurring from a single image. ACM Trans. Graph. (TOG) **27**, 73 (2008)
9. Xu, L., Jia, J.: Two-phase kernel estimation for robust motion deblurring. In: Daniilidis, K., Maragos, P., Paragios, N. (eds.) ECCV 2010. LNCS, vol. 6311, pp. 157–170. Springer, Heidelberg (2010). doi:10.1007/978-3-642-15549-9_12
10. Krishnan, D., Tay, T., Fergus, R.: Blind deconvolution using a normalized sparsity measure. In: 2011 IEEE Conference on Computer Vision and Pattern Recognition, pp. 233–240 (2011)
11. Xu, L., Zheng, S., Jia, J.: Unnatural l0 sparse representation for natural image deblurring. In: 2013 IEEE Conference on Computer Vision and Pattern Recognition, pp. 1107–1114 (2013)
12. Lefkimmiatis, S., Bourquard, A., Unser, M.: Hessian-based norm regularization for image restoration with biomedical applications. IEEE Trans. Image Process. **21**, 983–995 (2012)
13. Pan, J., Liu, R., Su, Z., et al.: Kernel estimation from salient structure for robust motion deblurring. Signal Process.: Image Commun. **28**, 1156–1170 (2013)

14. Xu, L., Lu, C., Xu, Y., et al.: Image smoothing via L_0 gradient minimization. ACM Trans. Graph. (TOG) **30**, 174 (2011)
15. Krishnan, D., Fergus, R.: Fast image deconvolution using hyper-Laplacian priors. In: Advances in Neural Information Processing Systems, pp. 1033–1041 (2009)
16. Wang, K., Shen, Y., Xiao, L., Wei, Z., Sheng, L.: Blind motion deblurring based on fused ℓ_0-ℓ_1 regularization. In: Zhang, Y.-J. (ed.) ICIG 2015. LNCS, vol. 9218, pp. 1–10. Springer, Cham (2015). doi:10.1007/978-3-319-21963-9_1

Saliency Detection Optimization via Modified Secondary Manifold Ranking and Blurring Depression

Han Liu, Bo Li$^{(\boxtimes)}$, and Hang Gao

School of Electronic and Information Engineering,
South China University of Technology, Guangzhou, China
{eeliuhan,g.hang}@mail.scut.edu.cn, leebo@scut.edu.cn

Abstract. We propose an unsupervised saliency optimization method mainly via modified secondary manifold ranking and blurring depression (SMBD). Generally, saliency object is detected insufficiently by most methods. To solve this problem, a modified manifold ranking is circulated twice to detect saliency object completely. A blurry degree detection approach is introduced to locate blurring regions, which is more likely to be background. As a result, blurring regions are depressed by SMBD to avoid mistaking background as foreground. Our method is performed based on hierarchical luminance for better performance. Extensive experimental results demonstrate that SMBD is able to promote the performances of state-of-the-art saliency detection algorithms significantly.

Keywords: Saliency detection · Manifold ranking · Blurring depression

1 Introduction

Saliency detection aims to predict the position of the most important part of an image and it is increasingly popular with the rapid development of computer vision. Saliency detection has been applied to a great number of vision tasks. For instance, image segmentation [12], image cropping [13] and video summarization.

There exist two categories of saliency detection, top-down and bottom-up methods. Top-down methods [14,15] are task-driven while bottom-up methods [2,4] are data-driven. In this paper, we mainly focus on bottom-up saliency detection methods.

Prior principles are essential in saliency detection. Generally, image boundaries are more likely to be background, which is called boundary prior [4]. In contrast, center prior is proposed by the principle that the nearer to the image center, the more possibility of the object is salient according to photographing habit of people. Besides, contrast prior is considered regularly. Color variation, texture gradient and hue histogram are frequently employed to calculate the contrast between foreground and background [2]. Various methods for saliency

© Springer International Publishing AG 2017
F. Cong et al. (Eds.): ISNN 2017, Part II, LNCS 10262, pp. 248–256, 2017.
DOI: 10.1007/978-3-319-59081-3_30

detection have been proposed during the past few decades. However, saliency object is not detected completely almost by all of these methods. To address this problem, we propose modified secondary manifold ranking (MSMR) to improve the results of existing approaches. A modified secondary circulation is conducted by manifold ranking, whose source image is removed partly based on original saliency map to detect saliency objects completely. People always set the most important object on focal plane when they are taking photos. The regions outside the focal plane are blurry and they are more likely to be background. This principle is introduced into our algorithm to further promote the accuracy of our method. Blurring regions of source image are detected and corresponding locations on saliency map are depressed as background. Contrast between objects will be altered as the change of luminance. Better saliency detection result can be delivered by integrating saliency maps got from multiple luminance.

Contributions in this paper include: (1) MSMR tends to detect more saliency objects than state-of-the-art saliency detection methods. (2) Background can be accurately depressed by detecting blurry regions of an image. (3) Integrating various saliency maps based on multiple luminance is an effective optimization method.

2   Related Work

Itti et al. [1] predict visual fixation by combining multi-scale image features, who firstly propose the principle of saliency detection. Then, numerous unsupervised algorithms are raised by introducing saliency cues contrast such as color, spatial position, hue and texture [2,3]. Certain methods are proposed based on boundary prior, since [4] makes use of boundary superpixels.

Some algorithms [6] make use of the influence of similar neighbors through iterations, which are proved quite effective for further reinforcing foreground and suppressing background. A uniformly high-response saliency map can be obtained by using a hierarchical framework that infers importance values from three image layers in different scales [16]. Recently, deep learning grows rapidly in saliency detection [10,11], which has achieved favorable result. However, deep learning requires expensive training time and hand-craft images are difficult to collect.

3   Proposed Method

Our algorithm optimizes saliency detection via three main processes. MSMR and blurring region depression are complementary to each other. Multi-luminance is an effective optimization hierarchical method, which is firstly proposed by us.

3.1   Modified Secondary Manifold Ranking

Different from previous methods, our MSMR algorithm is carried out on the basis of source image whose saliency object is removed. After obtaining saliency map of the changed source image by manifold ranking in [4], its saliency object is merged with original saliency map. Figure 1 is the procedure chart of MSMR algorithm.

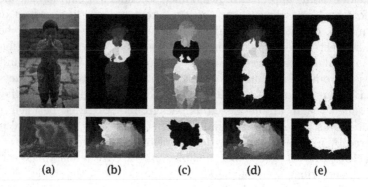

Fig. 1. Procedure of MSMR. (a) Source images, (b) original saliency maps got by MR method, (c) secondary saliency maps, (d) final saliency maps and (e) Ground truth.

Firstly, we segment the source image into N superpixels by the simple linear iterative clustering (SLIC) algorithm. Then, a graph structure $M = (V, E)$ is constructed. Nodes V represent superpixels and edges E are established based on the connections of any two nodes in M.

Mean saliency value for each superpixel i of original saliency map is denoted as S_i and we set $I_i' = I_i$. Then saliency object is found out and removed from corresponding source image as follows:

$$I_i'(S_i > \xi) = \varnothing, (i = 1, 2, ..., N) \tag{1}$$

where $\varnothing$ is empty set. I_i represents source image and I_i' stands for the image whose saliency object has been removed. Next, we take a secondary circulation on I_i' via Eq. 2 [4], then the secondary saliency map S_i^s can be obtained as follows:

$$f^* = (D - \alpha W)^{-1} y \tag{2}$$

where $y_i = 1$ if superpixel i belongs to boundary, and $y_i = 0$ otherwise. $D = diag(d_1, ..., d_n)$ is the degree matrix, in which $d_i = \sum_j w_{ij}$. The weight matrix is denoted as:

$$w_{ij} = exp(-\frac{||c_i - c_j||^2}{\sigma^2}) \tag{3}$$

where c_i and c_j are the mean CIE Lab colors of nodes i and j.

Finally, we set $S_i^m = S_i$ and then the saliency object of S_i^s is merged with S_i. So, the merged saliency map S_i^m can be obtained:

$$S_i^m(S_i^s > \psi) = 255, (i = 1, 2, ..., N) \tag{4}$$

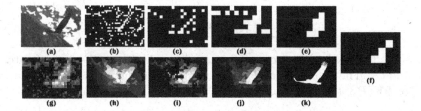

Fig. 2. If the blurring value of a patch is large, then the patch will be background. (a) Source image. (b)–(f) Blurring detection maps with patch size of 10, 20, 30, 40 and 50. (g) Fusion of the five images ahead. (h) Final saliency map obtained by modified secondary manifold ranking. (i) Saliency map with blurring regions being depressed. (j) Saliency map optimized via Single-layer Cellular Automata. (k) Ground truth.

As shown in the second row of Fig. 1, if original saliency map has been fully detected, the boundaries of secondary saliency map will be bright owing to existing no highly contrast objects in it. So, we define boundary brightness (BB) to judge if the MSMR is necessary for different images. Let S_t, S_b, S_l, and S_r denote the sum of saliency values of pixels in top, bottom, left and right boundaries. If BB is higher than threshold T_{BB}, then we will choose the original saliency map as the final saliency map S_i^f:

$$BB = S_t + S_b + S_l + S_r \tag{5}$$

$$S_i^f = \begin{cases} S_i & BB > T_{BB} \\ S_i^m & BB < T_{BB} \end{cases}$$

3.2 Blurring Regions Depression

With the popularization of single-lens reflex camera, there are a large number of pictures with blurring background. As we know, saliency objects should be on focal plane in an image and they are definitely clear, while blurring regions in an image are more likely to be background.

As a pre-processing procedure, an image is divided into numerous little patches P_r with five different sizes. Then blurring value (BV) of every patch is measured in turn based on gray map G. A Gaussian filter is used on G, and G_s can be obtained.

SSIM(x,y) is proposed to measure image quality [19]. In this paper, we use it to measure the blurry degree of images:

$$SSIM(x, y) = \frac{(2\mu_x\mu_y + C_1)(2\sigma_{xy} + C_2)}{(\mu_x^2 + \mu_y^2 + C_1)(\sigma_x^2 + \sigma_y^2 + C_2)} \tag{6}$$

where μ_x and μ_y are mean pixel values of a patch for G and G_s. Standard deviations of the patches in maps G and G_s are denoted as σ_x and σ_y. σ_{xy} is defined as:

$$\sigma_{xy} = \frac{1}{L-1} \sum_{i=1}^{L} (x_i - \mu_x)(y_i - \mu_y) \tag{7}$$

where L is the pixel number of patch r. x_i and y_i denote pixel values of patch r in G and G_s. The clearer patch P_r is, the larger difference exists between it and G_s. So, $BV(P_r)$ is large for blurry patch and small for clear patch as:

$$BV(P_r) = SSIM(G, G_s) \tag{8}$$

Table 1. F-measure values of MR, RBD, MC, ST, DRFI methods and optimized by SMBD and SCA on JuddDB, ECSSD and MSRA-10K datasets. Ori are original F-measure values. Opt are optimized values by SMBD and Pro are the percentage of promotion.

Method	JuddDB			ECSSD					MSRA-10K		
	Ori	Opt	Pro	Ori	Opt	Pro	SCA	SCA-Pro	Ori	Opt	Pro
MR	0.454	0.476	+2.2%	0.743	0.756	+1.3%	0.742	−0.1%	0.846	0.863	+1.7%
RBD	0.457	0.481	+2.4%	0.720	0.759	+3.9%	0.749	+2.9%	0.855	0.874	+1.9%
MC	0.460	0.484	+2.4%	0.739	0.759	+2.0%	0.745	+0.6%	0.848	0.867	+1.9%
ST	0.455	0.491	+3.6%	0.747	0.776	+2.9%	0.759	+1.2%	0.867	0.882	+1.5%
DRFI	0.475	0.504	+2.9%	0.786	0.790	+0.4%	0.769	−1.7%	0.881	0.890	+0.9%

After calculating all the blurring values of patches in the whole image, we make some changes to saliency map S_i^f. If $BV(P_r) > threshold\ T_{BV}$, then the patch r will be depressed as background, as shown in Fig. 2. The final blurring map S_r^b is obtained by fusing the five blurring detection maps with different sizes together. Rough outline of saliency object in Fig. 2(a) has been delineated by Fig. 2(h). If pixel value of blurring map S_r^b is lower than T_{BM}, then the pixel is treated as background in saliency map S_i^f, got by modified secondary manifold ranking. We set $S_r^{bv} = S_i^f$ and denote the new saliency map as:

$$S_r^{bv}(S_r^b < T_{BM}) = 0 \tag{9}$$

Figure 2 shows the advantages of blurring region depression. However, saliency map obtained from blurring regions depression is not smooth enough. So, applying an optimization step proposed by [6], named Single-layer Cellular Automata (SCA), will further improve the saliency map. We denote the optimized saliency map got by SCA as S^{op}, as shown in Fig. 2(j).

3.3 Hierarchical Luminance

Contrast between foreground and background is varied as the change of luminance. As a result, multi-scale contrast information can be detected all sidely by applying hierarchical luminance. In this paper, luminance of source image is changed to $L \in (1, 0.7, 0.5, 0.3, 0.1)$ of initial value respectively. All of the optimization steps aforementioned are applied on source images with the five luminance. Finally, saliency maps obtained from five scale luminance are integrated by Eq. 10.

$$S_l^H = \sum S_l^{op} \ (l = 0.1, 0.3, 0.5, 0.7, 1) \tag{10}$$

4   Experiments

4.1   Datasets

We test our optimization method, SMBD on three publicly available datasets: MSRA-10K [3], ECSSD [16] and JuddDB [17]. MSRA-10K contains 10,000 relative simple images. While ECSSD and JuddDB are made up of 1000 and 900 images severally, including more complex scenes compared with MSRA-10K.

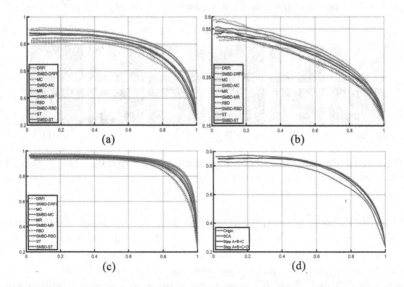

Fig. 3. (a)–(c) P-R curves on the ECSSD, JuddDB and MSRA-10K datasets. Dotted lines represent original saliency map generated by MR, RBD, MC, ST and DRFI methods respectively. Solid lines represent saliency maps optimized by our SMBD method. (d) P-R curves of SMBD for RBD method on ECSSD dataset, in which steps A-D represent modified secondary manifold ranking, blurring background depression, SCA optimization and hierarchical luminance.

4.2   Experimental Setup

We set the number of superpixel nodes N as 200 in our experiments. These parameters are empirically chosen, $\sigma^2 = 0.1$, $\alpha = 0.99$, $C_1 = 0.2$ and $C_2 = 0.6$.

ξ and ψ are set as 150 and 180 according to experimental results. The larger σ and α are, the more saliency objects in S_i and S_i^s will be included. We set threshold T_{BB} as 800000 to avoid mistaking background as saliency object. Blurry degree of source image is measured by threshold T_{BV}, which is set as 0.18, 0.04, 0.025, 0.02 and 0.02 respectively for the five scale patches. Background is depressed efficiently when $T_{BM} = 40$.

4.3 Precision and Recall

To analyze the accuracy of our method, we employ Precision-Recall curve and F-measure value. P-R curve is generated by varying the threshold from 0 to 255 and compared with corresponding ground-truth map. F-measure value fuses precision and recall together:

$$F = \frac{(1 + \beta^2) \cdot precision \cdot recall}{\beta^2 \cdot precision + recall} \tag{11}$$

We set β as 0.3 because precision values more in this experiment [18].

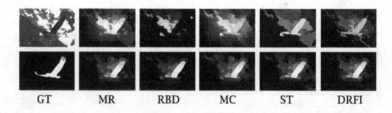

| GT | MR | RBD | MC | ST | DRFI |

Fig. 4. First row are original saliency maps obtained by existing methods. Second row are optimized saliency maps by SMBD.

Figure 3(d) shows the P-R curves of several steps in our SMBD method. We find out that twice circulations for manifold ranking are enough by extensive experimental results. Saliency map optimized by step C is better than just applying SCA for original saliency map owing to A and B steps, which can also be seen from Table 1. The last step further promotes accuracy effectively via changing the luminance of source image.

The proposed algorithm is applied for five existing methods, MR [4], MC [7], ST [8], RBD [5] and DRFI [9]. P-R curves of refined saliency maps are plotted in Fig. 3(a), (b) and (c). The closer curve locates to (1,1), the better performance does the algorithm possess. Obviously, our optimization method can improve the performance of the five methods above significantly no matter in simple or complex datasets. F-measure values of refined saliency maps obtained by our optimization method are summarized in Table 1. No matter in which dataset, SMBD can improve F-measure values of state-of-the-art saliency detection algorithms greatly, demonstrating the universality of our method.

SCA method is conducted directly on original saliency maps and F-measure values of SCA saliency maps for ECSSD are showed in Table 1. Compared with our algorithm, SCA can only promote F-measure value slightly or even make it worse for MR and DRFI methods. Figure 4 illustrates that even though original saliency maps are varied and unsatisfying, they are all optimized by our method to similar and favorable results.

5   Conclusion

In this paper, we propose a novel optimization method (SMBD) for saliency detection algorithms. SMBD takes advantage of modified manifold ranking in order to detect overall saliency objects. Then the background of an image is depressed according to the principle that blurring regions are more likely to be background. Introducing hierarchical luminance further promotes the accuracy of our method. The results of experiments demonstrate that our optimization method performs favorably on the MSRA-10K, ECSSD and JuddDB datasets. In the future, we will further explore different kinds of hierarchical modes for saliency detection.

Acknowledgments. This research was supported by the National Natural Science Foundation of China(Grant Nos. 11627802, 51678249), by the Science and Technology Projects of Guangdong (2013A011403003), and by the Science and Technology Projects of Guangzhou (201508010023).

References

1. Itti, L., Koch, C., Niebur, E.: A model of saliency-based visual attention for rapid scene analysis. IEEE Trans. Pattern Anal. Mach. Intell. **20**, 1254–1259 (1998)
2. Cheng, M.M., Mitra, N.J., Huang, X., Torr, P.H.S., Hu, S.M.: Global contrast based salient region detection. In: Proceedings of Computer Vision and Pattern Recognition, pp. 409–416 (2015)
3. Liu, T., Sun, J., Zheng, N.N., Tang, X., Shum, H.Y.: Learning to detect a salient object. In: Proceedings of Computer Vision and Pattern Recognition, pp. 1–8 (2007)
4. Yang, C., Zhang, L., Lu, H., Ruan, X., Yang, M.H.: Saliency detection via graph-based manifold ranking. In: Proceedings of Computer Vision and Pattern Recognition, pp. 3166–3173 (2013)
5. Zhu, W., Liang, S., Wei, Y., Sun, J.: Saliency optimization from robust background detection. In: Proceedings of Computer Vision and Pattern Recognition, pp. 2814–2821 (2014)
6. Qin, Y., Lu, H., Xu, Y., Wang, H.: Saliency detection via cellular automata. In: Proceedings of Computer Vision and Pattern Recognition, pp. 110–119 (2015)
7. Jiang, B., Zhang, L., Lu, H., Yang, C., Yang, M.H.: Saliency detection via absorbing Markov chain. In: Proceedings of International Conference on Computer Vision, pp. 1665–1672 (2013)
8. Liu, Z., Zou, W., Meur, O.L.: Saliency tree: a novel saliency detection framework. IEEE Trans. Image Process. **23**, 1937–1952 (2014)

9. Jiang, H., Wang, J., Yuan, Z., Wu, Y., Zheng, N., Li, S.: Salient object detection: a discriminative regional feature integration approach. In: Proceedings of Conference on Computer Vision and Pattern Recognition, pp. 2083–2090 (2014)

10. Zhao, R., Ouyang, W., Li, H., Wang, X.: Saliency detection by multi-context deep learning. In: Proceedings of Computer Vision and Pattern Recognition, pp. 1265–1274 (2015)

11. Wang, L., Lu, H., Xiang, R., Yang, M.H.: Deep networks for saliency detection via local estimation and global search. In: Proceedings of Conference on Computer Vision and Pattern Recognition, pp. 3183–3192 (2015)

12. Lempitsky, V., Kohli, P., Rother, C., Sharp, T.: Image segmentation with a bounding box prior. In: Proceedings of International Conference on Computer Vision, pp. 277–284 (2009)

13. Marchesotti, L., Cifarelli, C., Csurka, G.: A framework for visual saliency detection with applications to image thumbnailing. In: Proceedings of International Conference on Computer Vision, pp. 2232–2239 (2009)

14. Kanan, C., Tong, M.H., Zhang, L., Cottrell, G.W.: Sun: top-down saliency using natural statistics. Vis. Cogn. **17**, 979–1003 (2009)

15. Yang, M.H., Yang, J.: Top-down visual saliency via joint CRF and dictionary learning. In: Proceedings of Computer Vision and Pattern Recognition, pp. 2296–2303 (2012)

16. Yan, Q., Xu, L., Shi, J., Jia, J.: Hierarchical saliency detection. In: Proceedings of Computer Vision and Pattern Recognition, pp. 1155–1162 (2013)

17. Borji, A.: What is a salient object? A dataset and a baseline model for salient object detection. Trans. Image Process. **24**, 742–756 (2014)

18. Achanta, R., Estrada, F., Wils, P., Süsstrunk, S.: Salient region detection and segmentation. In: Gasteratos, A., Vincze, M., Tsotsos, J.K. (eds.) ICVS 2008. LNCS, vol. 5008, pp. 66–75. Springer, Heidelberg (2008). doi:10.1007/978-3-540-79547-6_7

19. Wang, Z., Bovik, A.C., Sheikh, H.R., Simoncelli, E.P.: Image quality aassessment: from error visibility to structural similarity. IEEE Trans. Image Process. **13**, 600–612 (2004)

Noise Resistant Training for Extreme Learning Machine

Yik Lam Lui, Hiu Tung Wong, Chi-Sing Leung$^{(\boxtimes)}$, and Sam Kwong

City University of Hong Kong, Kowloon Tong, Hong Kong
eeleungc@cityu.edu.hk

Abstract. The extreme learning machine (ELM) concept provides some effective training algorithms to construct single hidden layer feedforward networks (SHLFNs). However, the conventional ELM algorithms were designed for the noiseless situation only, in which the outputs of the hidden nodes are not contaminated by noise. This paper presents two noise-resistant training algorithms, namely noise-resistant incremental ELM (NRI-ELM) and noise-resistant convex incremental ELM (NRCI-ELM). For NRI-ELM, its noise-resistant ability is better than that of the conventional incremented ELM algorithms. To further enhance the noise resistant ability, the NRCI-ELM algorithm is proposed. The convergent properties of the two proposed noise resistant algorithms are also presented.

Keywords: Node noise · Extreme learning machines · Incremental algorithm

1 Introduction

Single hidden layer feedforward networks (SHLFNs) can act as universal approximators [1]. With the traditional training algorithms, such as backpropagation based algorithms, we need to estimate all the connection weights, including the input weights from the input layer to the hidden layer, and the output weights from the hidden layer to the output node. Training all the connection weights may have some problems, such as local minimum. Huang et al. [2] proposed the extreme learning machine concept, where the hidden nodes are generated randomly. Besides, they showed that SHLFNs with the ELM concept can act as universal approximators too. In [2,3], Huang et al. developed the incremental ELM (I-ELM) [2] algorithm and the convex incremental ELM (CI-ELM) algorithm [3]. The mean square error (MSE) performances of these two algorithms are very well under the noiseless situation, where there is no node noise in the implementation.

In the implementation of neural networks, noise take place unavoidably [4]. When we use the finite precision technology to implement a trained network, multiplicative noise or additive noise would be introduced [5]. Also, when the implementation is at the nano-scale, transient noise may occur [6]. For traditional

© Springer International Publishing AG 2017
F. Cong et al. (Eds.): ISNN 2017, Part II, LNCS 10262, pp. 257–265, 2017.
DOI: 10.1007/978-3-319-59081-3_31

neural network models, some batch mode learning algorithms for trained neural networks under the imperfection situation were reported [8]. To the best of our knowledge, there are not many literatures related to the noise-resistant ELMs.

This paper considers the multiplicative node noise and the additive node noise as the imperfect conditions for the SHLFN model. We first derive the training set error expression of noisy SHLFNs. Afterwards, we develop two noise-resistant incremental ELM algorithms, namely noise-resistant I-ELM (NRI-ELM) and noise-resistant CI-ELM (NRCI-ELM). For the NRI-ELM algorithm, we keep all the previously trained weights unchanged, and we adjust the output weight of the newly inserted node. The noise-resistant performance of the NRI-ELM algorithm is better than that of I-ELM and CI-ELM. For the NRCI-ELM algorithm, we use a simple rule to update all the previously trained weights, and we estimate the output weight of the new node to maximize the reduction in the training set error of noisy SHLFNs. The noise-resistant ability of the NRCI-ELM algorithm is much better than that of I-ELM, CI-ELM, and NRI-ELM. In addition, we prove that in terms of the training set error of noisy SHLFNs, the NRCI-ELM algorithm and the NRCI-ELM algorithm converges.

The rest of this paper is organized as follows. Section 2 presents the background of the ELM concept and the node noise models. Section 3 derives the two proposed noise resistant incremental ELM algorithms. Section 4 presents the simulation result. Section 5 concludes the paper.

2   ELM and Node Noise

The nonlinear regression problem is considered in this paper. The training set is denoted as $\mathbb{D}_t = \{(\boldsymbol{x}_k, o_k) : \boldsymbol{x}_k \in \mathbb{R}^M, o_k \in \mathbb{R}, k = 1, \ldots, N\}$, where $\boldsymbol{x}_k$ and o_k are the input and the target output of the k-th sample, respectively. The test set is denoted as $\mathbb{D}_f = \{(\boldsymbol{x}'_{k'}, o'_{k'}) : \boldsymbol{x}'_{k'} \in \mathbb{R}^M, o'_{k'} \in \Re, k' = 1, \ldots, N'\}$. In a SHLFN with n hidden nodes, the network output is given by $f_n(\boldsymbol{x}) = \sum_{i=1}^{n} \beta_i h_i(\boldsymbol{x})$, where $h_i(\boldsymbol{x})$ is the output of the ith hidden node, and β_i is the output weight of the ith hidden node. In this paper, we use the sigmoid function as the activation function. Hence the output of the ith hidden node is given by

$$h_i(\boldsymbol{x}) = \frac{1}{1 + \exp\{-(\boldsymbol{w}_i^\mathrm{T} \boldsymbol{x} + b_i)\}}, \tag{1}$$

where b_i is the input bias of the ith hidden node, and $\boldsymbol{w}_i$ is the input weight vector of the ith hidden node.

In the ELM approach [2,3], the bias terms b_i's and the input weight vectors $\boldsymbol{w}_i$'s are randomly generated. We only need to estimate the output weights β_i's. For a trained SHLFN, the training set error is given by

$$\mathcal{E} = \sum_{k=1}^{N}(y_k - \sum_{i=1}^{n} \beta_i h_i(\boldsymbol{x}_k))^2 = \left\| \boldsymbol{o} - \sum_{i=1}^{n} \beta_i \boldsymbol{h}_i \right\|_2^2, \tag{2}$$

where $\boldsymbol{o} = [o_1, \ldots, o_N]^\mathrm{T}$, and $\boldsymbol{h}_i = [h_i(\boldsymbol{x}_1), \ldots, h_i(\boldsymbol{x}_N)]^\mathrm{T}$.

In the implementation of a network, node noise may not be avoided. When we use the digital implementation, finite precision can be modelled as multiplicative node noise or additive node noise [5]. When we use the floating point approach, the round-off error can be modelled as multiplicative noise. On the other hand, when we use the fixed point approach, the round-off error can be modelled as additive noise.

Given the kth input vector, when a hidden node is affected by the multiplicative noise and additive noise concurrently, its output can be modelled as

$$\tilde{h}_i(\boldsymbol{x}_k) = (1 + \delta_{ik})h_i(\boldsymbol{x}_k) + \epsilon_{ik}, \forall i = 1, \ldots, n \text{ and } \forall k = 1, \ldots, N, \quad (3)$$

where δ_{ik}'s are the noise factors that describe the deviation due to the multiplicative node noise, and ϵ_{ik}'s are the noise factors that describe the deviation due to the additive node noise. Note that in the multiplicative noise case, the magnitude of the noise component "$\delta_{ik}h_i(\boldsymbol{x})$" is proportional to the magnitude of the output $h_i(\boldsymbol{x})$. This paper assumes that the noise factors δ_{ik}'s and ϵ_{ik}'s are zero-mean identically independently distributed random variables with variances equal to σ_δ^2 and σ_ϵ^2, respectively.

From [3], the CI-ELM algorithm works very well for the noiseless situation. For example, as shown in Fig. 1(a), the network outputs fit the training samples very well. However, when node noise exists, the network outputs contain a lot of noise with large magnitude, as shown in Fig. 1(b). When our proposed NRCI-ELM is used, the noise in the network outputs can be greatly suppressed, as shown in Fig. 1(c) and (d).

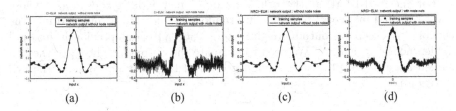

    (a)   (b)   (c)   (d)

Fig. 1. Illustration of the noise resistant ability of CI-ELM and NRCI-ELM. (a) The network output of a noiseless network with CI-ELM. (b) The network output of a noisy network with CI-ELM. (c) The network output of a noiseless network with NRCI-ELM. (d) The network output of a noisy network with NRCI-ELM. In this example, $\sigma_\epsilon^2 = \sigma_\delta^2 = 0.01$.

3   Noise Resistant Incremental Learning

For a SHLFN with a particular noise pattern, the training set error can be expressed as

$$\tilde{\mathcal{E}} = \sum_{k=1}^{N} \left(o_k - \sum_{i=1}^{n} \beta_i \tilde{h}_i(\boldsymbol{x}_k) \right)^2. \quad (4)$$

According to the properties of δ_{ik}'s and ϵ_{ik}'s, the statistics of $\tilde{h}_i(\boldsymbol{x}_k)$'s are given by

$$\langle \tilde{h}_i(\boldsymbol{x}_k) \rangle = h_i(\boldsymbol{x}_k), \tag{5}$$

$$\langle \tilde{h}_i^2(\boldsymbol{x}_k) \rangle = (1 + \sigma_\delta^2)h_i^2(\boldsymbol{x}_k) + \sigma_\epsilon^2, \tag{6}$$

$$\langle \tilde{h}_i(\boldsymbol{x}_k)\tilde{h}_j(\boldsymbol{x}_k) \rangle = h_i(\boldsymbol{x}_k)h_j(\boldsymbol{x}_k), \forall\, i \neq j. \tag{7}$$

Taking the expectation over all possible noise patterns, we obtain the training set error of noisy SHLFNs, given by

$$\langle \tilde{\mathcal{E}} \rangle = \left\langle \sum_{k=1}^{N} \left(o_k - \sum_{i=1}^{n} \beta_i \big((1+\delta_{ik})h_i(\boldsymbol{x}_k) + \epsilon_{ik} \big) \right)^2 \right\rangle. \tag{8}$$

From (5)–(7), Eq. (8) becomes

$$\langle \tilde{\mathcal{E}} \rangle = \left\| \boldsymbol{o} - \sum_{i=1}^{n} \beta_i \boldsymbol{h}_i \right\|_2^2 + \sigma_\delta^2 \sum_{i=1}^{n} \beta_i^2 \|\boldsymbol{h}_i\|_2^2 + \sigma_\epsilon^2 N \sum_{i=1}^{n} \beta_i^2. \tag{9}$$

Similarly, we can obtain the test set error of noisy SHLFNs, given by

$$\langle \tilde{\mathcal{E}}_t \rangle = \left\| \boldsymbol{o}' - \sum_{i=1}^{n} \beta_i \boldsymbol{h}'_i \right\|_2^2 + \sigma_\delta^2 \sum_{i=1}^{n} \beta_i^2 \|\boldsymbol{h}'_i\|_2^2 + \sigma_\epsilon^2 N' \sum_{i=1}^{n} \beta_i^2, \tag{10}$$

where $\boldsymbol{o}' = [o'_1, \ldots, o'_N]^{\mathrm{T}}$, and $\boldsymbol{h}_i = [h_i(\boldsymbol{x}'_1), \ldots, h_i(\boldsymbol{x}'_{N'})]^{\mathrm{T}}$.

For the NRI-ELM, at the nth iteration, a new hidden node $h_n(\cdot)$, whose input bias and input weight vector are randomly generated, is inserted into the network. We keep the output weights $\{\beta_1, \ldots, \beta_{n-1}\}$ of the previously inserted hidden nodes unchanged. We need to estimate the output weight β_n of the nth hidden node. From (9), the training set error of the noisy networks at the nth iteration is

$$L_n = \left\| \boldsymbol{o} - \sum_{i=1}^{n} \beta_i \boldsymbol{h}_i \right\|_2^2 + \sigma_\delta^2 \sum_{i=1}^{n} \beta_i^2 \|\boldsymbol{h}_i\|_2^2 + \sigma_\epsilon^2 N \sum_{i=1}^{n} \beta_i^2. \tag{11}$$

Define

$$\boldsymbol{f} = \sum_{i=1}^{n} \beta_i \boldsymbol{h}_i, \quad \boldsymbol{e}_n = \boldsymbol{o} - \sum_{i=1}^{n} \beta_i \boldsymbol{h}_i, \quad v_n = \sum_{i=1}^{n} \beta_i^2 \|\boldsymbol{h}_i\|_2^2, \quad u_n = N \sum_{i=1}^{n} \beta_i^2. \tag{12}$$

From (12), Eq. (11) can be rewritten as

$$L_n = \|\boldsymbol{e}_n\|_2^2 + \sigma_\delta^2 v_n + \sigma_\epsilon^2 u_n. \tag{13}$$

From (13), the change in the training set error between the nth-iteration and $(n-1)$th-iteration is given by

$$\triangle_n = L_n - L_{n-1} = -2\beta_n \boldsymbol{e}_{n-1}^{\mathrm{T}} \boldsymbol{h}_n + (1+\sigma_\delta^2)\beta_n^2 \|\boldsymbol{h}_n\|_2^2 + \sigma_\epsilon^2 \beta_n^2 N. \tag{14}$$

Since $\triangle_n$ is a quadratic function of β_n with a minimum value equal to a negative value, the optimal value of β_n to maximize the decrease in the training set error is given by

$$\beta_n = \frac{e_{n-1}^{\mathrm{T}} h_n}{(1+\sigma_\delta^2)\|h_n\|_2^2 + N\sigma_\epsilon^2}. \tag{15}$$

With (15), the change in the training set error between two consecutive iterations is

$$\triangle_n = -\frac{\left(e_{n-1}^{\mathrm{T}} h_n\right)^2}{(1+\sigma_\delta^2)\|h_n\|_2^2 + N\sigma_\epsilon^2}. \tag{16}$$

Equation (16) means that when we inserted a new hidden node, the training set error of noisy networks decreases. That means, in terms of the training set MSE of noisy network, the NRI-ELM algorithm converges. Algorithm 1 shows the proposed NRI-ELM algorithm. From Steps (5)–(8) in Algorithm 1, for the NRI-ELM algorithm, the computational complexity is $O(N)$ for each iteration.

Algorithm 1. NRI-ELM

1: Set n equal to zero $(n = 0)$, $e_0 = y$, and $f_0 = 0$.
2: **while** $n \leq n_{\mathrm{max}}$ **do**
3:     $n = n + 1$.
4:     Insert a new hidden node.
5:     Compute the output vector h_n of this hidden node.
6:     Compute the output weight of the newly inserted node: $\beta_n = \frac{e_{n-1}^{\mathrm{T}} h_n}{(1+\sigma_\delta^2)\|h_n\|_2^2 + N\sigma_\epsilon^2}$.
7:     $f_n = f_{n-1} + \beta_n h_n$.
8:     $e_n = y - f_n$.
9: **end while**

In [3], the CI-ELM algorithm was proposed. Under the noiseless situation [3], the training set error of the original CI-ELM algorithm is better than that of I-ELM algorithm. However, as shown in Sect. 4, the original CI-ELM algorithm has a very poor noise resistant ability. Hence it is interesting to develop a noise resistant version of CI-ELM, namely *NRCI-ELM*.

In the NRCI-ELM case, after we estimate the output weight β_n at the nth iteration, we update all the previously trained weights by

$$\beta_i^{new} = (1 - \beta_n)\beta_i, \tag{17}$$

for $i = 1$ to $n - 1$. Hence we have the recursive definitions for f_n, e_n, v_n and u_n

$$f_n = (1 - \beta_n)f_{n-1} + \beta_n h_n, \quad e_n = y - f_n,$$
$$v_n = (1 - \beta_n)^2 v_{n-1} + \beta_n^2 \|h_n\|_2^2, \quad u_n = (1 - \beta_n)^2 u_{n-1} + \beta_n^2 N,$$

where $f_0 = o$, $e_0 = y$, $v_0 = 0$, and $u_0 = 0$. With this new updating scheme for the previously trained output weights, the change in the training set error between the nth-iteration and $(n-1)$th-iteration is given by

$$\triangle_n = L_n - L_{n-1} = -2\beta_n \left(e_{n-1}^{\mathrm{T}} r_n + \sigma_\delta^2 v_{n-1} + \sigma_\epsilon^2 u_{n-1} \right)$$
$$+ \beta_n^2 \left(\|r_n\|_2^2 + \sigma_\delta^2 (v_{n-1} + \|h_n\|_2^2) + \sigma_\epsilon^2 (u_{n-1} + N) \right), \quad (18)$$

where $r_n = h_n - f_{n-1}$.

Similar to the NRI-ELM case, to maximize the decrease in the training set error of noisy networks, β_n should be given by

$$\beta_n = \frac{e_{n-1}^{\mathrm{T}} r_n + \sigma_\delta^2 v_{n-1} + \sigma_\epsilon^2 u_{n-1}}{\|r_n\|_2^2 + \sigma_\delta^2 (v_{n-1} + \|h_n\|_2^2) + \sigma_\epsilon^2 (u_{n-1} + N)}. \quad (19)$$

With (19), the change in the training set error between two consecutive iterations is

$$\triangle_n = -\frac{\left(e_{n-1}^{\mathrm{T}} r_n + \sigma_\delta^2 v_{n-1} + \sigma_\epsilon^2 u_{n-1} \right)^2}{\|r_n\|_2^2 + \sigma_\delta^2 (v_{n-1} + \|h_n\|_2^2) + \sigma_\epsilon^2 (u_{n-1} + N)}. \quad (20)$$

Equation (20) means that when we insert a new hidden node, the training set error of noisy networks decreases. That means, in terms of the training set error of noisy network, the NRCI-ELM algorithm converges too. Algorithm 2 shows the proposed NRCI-ELM algorithm. At each each iteration, the complexity of the NRCI-ELM algorithm is "$O(n) + O(N)$". Compared to the NRI-ELM case whose complexity is equal to $O(N)$, the additional complexity $O(n)$ is due to the update of the previous weights.

Algorithm 2. NRCI-ELM

1: Set $n = 0$, $e_0 = y$, $f_0 = 0$, $v_0 = 0$, $v_0 = 0$ and $r_0 = 0$.
2: **while** $n \leq n_{\mathrm{max}}$ **do**
3:     $n = n + 1$.
4:     Insert a new hidden node whose b_n and w_n are randomly generated.
5:     Compute the output vector h_n for this new hidden node.
6:     Compute $r_n = h_n - f_{n-1}$.
7:     Compute the new weight: $\beta_n = \frac{e_{n-1}^{\mathrm{T}} r_n + \sigma_\delta^2 v_{n-1} + \sigma_\epsilon^2 u_{n-1}}{\|r_n\|_2^2 + \sigma_\delta^2 (v_{n-1} + \|h_n\|_2^2) + \sigma_\epsilon^2 (u_{n-1} + N)}$.
8:     $f_n = (1 - \beta_n) f_{n-1} + \beta_n h_n$.
9:     $e_n = y - f_n$.
10:    $v_n = (1 - \beta_n)^2 v_{n-1} + \beta_n^2 \|h_n\|_2^2$.
11:    $u_n = (1 - \beta_n)^2 u_{n-1} + \beta_n^2 N$.
12:    $\beta_i = (1 - \beta_n)\beta_i$, for all $i = 1, \ldots, n-1$.
13: **end while**

---

4   Simulation

Two real life datasets from the UCI data repository are used. They are Abalone [9] and Housing Price [10]. The Abalone dataset has 4,177 samples. Each sample has eight inputs and one output. Two thousand samples are randomly taken as the training set. The other 2,177 samples are used as the test set. The Housing Price dataset has 506 samples. Each sample has 13 inputs and one output. The training set contains 250 samples, while the test set has 256 samples.

This section considers four incremented algorithms. They are the original I-ELM algorithm, the original CI-ELM algorithm, the proposed NRI-ELM algorithm, and the proposed NRCI-ELM algorithm, respectively. Figure 2 shows the MSE performance versus the number of hidden nodes, where the noise level is equal to $\sigma_\epsilon^2 = \sigma_\delta^2 = 0.09$. It can be seen that the proposed NRI-ELM algorithm is better than the two original incremental algorithms. Also, the MSE performance of the original CI-ELM algorithm is very poor. When we use more hidden nodes, the performance of the CI-ELM algorithm suddenly becomes very poor. To sum up, the proposed NRCI-ELM algorithm is much better than the original I-ELM algorithm, the original CI-ELM algorithm and the proposed NRI-ELM algorithm.

Table 1 shows the average test set MSE values of noisy networks over 100 trials for various node noise levels. In Table 1, the number of hidden nodes is equal to 500. It can be seen that the performance of the CI-ELM algorithm is very poor. The noise resistant ability of the NRI-ELM algorithm is better than that of the I-ELM algorithm. In addition, the NRCI-ELM algorithm is much better than the other three algorithms. For instance, in the Abalone dataset with noise level $\sigma_\epsilon^2 = \sigma_\delta^2 = 0.01$, the test set MSE of I-ELM is equal to 0.01421. When the NRI-ELM is used, the test set error is reduced to 0.01367. The NRCI-ELM algorithm can further reduce the test set error to is 0.00815.

For high noise levels, the improvement of the NRCI-ELM algorithm is more significant. For instance, with the node noise level equal to $\sigma_\epsilon^2 = \sigma_\delta^2 = 0.09$ The test set MSE of the I-ELM algorithm is equal to 0.05855. With the

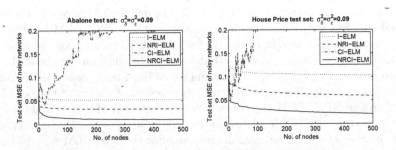

Fig. 2. The performance of the four incremental methods versus the number of additive nodes. The noise level is $\sigma_\epsilon^2 = \sigma_\delta^2 = 0.09$. The Abalone dataset is considered.

Table 1. Average test set MSEs of noisy networks. The average values are taken over 100 trials. There are 500 hidden nodes.

	Node noise level $\sigma_\epsilon^2, \sigma_\delta^2$	I-ELM mean(std)	NRI-ELM mean(std)	CI-ELM mean(std)	NRCI-ELM mean(std)
Abalone	0.01, 0.01	0.01421(0.00180)	0.01367(0.00142)	0.06153(0.03175)	0.00815(0.00007)
	0.09, 0.09	0.05855(0.01561)	0.03386(0.00312)	0.49954(0.28633)	0.01002(0.00009)
	0.25, 0.25	0.14723(0.04334)	0.04648(0.00202)	1.37555(0.79548)	0.01174(0.00015)
Housing	0.01, 0.01	0.02558(0.00649)	0.02425(0.00440)	0.05488(0.02510)	0.01478(0.00029)
	0.09, 0.09	0.11266(0.05269)	0.05921(0.00706)	0.39941(0.22699)	0.02026(0.00044)
	0.25. 0.25	0.28682(0.14524)	0.08081(0.00450)	1.08848(0.63079)	0.02528(0.00050)

NRI-ELM algorithm, the test set MSE is reduced to 0.03386. When the NRCI-ELM algorithm is used, the test MSE is reduced to 0.01002.

Another interesting property of the NRCI-ELM algorithm is that the test set error is insensitive to the node noise level. In the Abalone dataset, when the noise level is $\sigma_\delta^2 = \sigma_\epsilon^2 = 0.01$, the test set error of the NRCI-ELM algorithm is equal to 0.00815. When the noise level is greatly increased to $\sigma_\delta^2 = \sigma_\epsilon^2 = 0.25$, the test set error of the NRCI-ELM algorithm is slightly increased to 0.01174 only.

One may suggest that we should use the NRCI-ELM algorithm only because its test set error of noisy network is the best. The difference between the NRCI-ELM algorithm and the NRI-ELM algorithm is the computation complexity. For the NRI-ELM algorithm, the complexity is $O(N)$. But for the NRCI-ELM algorithm, the computation complexity is "$O(N) + O(n)$".

5   Conclusion

This paper proposed two incremental ELM algorithms, namely NRI-ELM and NRCI-ELM, for handling node noise. They insert the randomly generated hidden nodes into the network in the one-by-one manner. The NRI-ELM algorithm adjusts the output weight of the newly inserted hidden node only. Its noise-resistant ability is better than that of the original I-ELM algorithm and the original CI-ELM algorithm. Besides, we proposed the NRCI-ELM algorithm. It estimates the output weight of the newly additive node, and uses a single rule to modify the previously trained output weights. In addition, we prove that for the two proposed algorithms, the training set MSE of noisy networks converges. Simulation examples illustrate that the noise resistant ability of the NRI-ELM algorithm and NRCI-ELM algorithm is better than that of I-ELM and CI-ELM. In addition, the NRCI-ELM algorithm has the best noise resistant ability, compared to other three incremental algorithms. For the NRI-ELM algorithm, the complexity is $O(N)$. For the NRCI-ELM algorithm, the computation complexity is "$O(N) + O(n)$".

Acknowledgment. The work was supported by a research grant from the Government of the Hong Kong Special Administrative Region (CityU 11259516).

References

1. Hornik, K., Stinchcombe, M., White, H.: Multilayer feedforward networks are universal approximators. Neural Netw. **2**(5), 359–366 (1989)
2. Huang, G.B., Chen, L., Siew, C.K.: Universal approximation using incremental constructive feedforward networks with random hidden nodes. IEEE Trans. Neural Netw. **17**(4), 879–892 (2006)
3. Huang, G.B., Chen, L.: Convex incremental extreme learning machine. Neurocomputing **70**, 3056–3062 (2007)
4. Burr, J.: Digital neural network implementations. In: Neural Networks, Concepts, Applications, and Implementations. Prentice Hall, Englewood Cliffs, NJ (1995)
5. Liu, B., Kaneko, K.: Error analysis of digital filter realized with floating-point arithmetic. Proc. IEEE **57**(10), 1735–1747 (1969)
6. Mahvash, M., Parker, A.C.: Synaptic variability in a cortical neuromorphic circuit. IEEE Trans. Neural Netw. Learn. Syst. **24**(3), 397–409 (2013)
7. Leung, C.-S., Wang, H., Sum, J.: On the selection of weight decay parameter for faulty networks. IEEE Trans. Neural Netw. **21**(8), 1232–1244 (2010)
8. Leung, C.-S., Sum, J.: RBF networks under the concurrent fault situation. IEEE Trans. Neural Netw. Learn. Syst. **23**(7), 1148–1155 (2012)
9. Sugiyama, M., Ogawa, H.: Optimal design of regularization term and regularization parameter by subspace information criterion. Neural Netw. **15**(3), 349–361 (2002)
10. Lichman, M.: UCI machine learning repository (2013). http://archive.ics.uci.edu/ml

Phase Constraint and Deep Neural Network for Speech Separation

Zhuangguo Miao[1], Xiaohong Ma[1(✉)], and Shuxue Ding[2]

[1] School of Information and Communication Engineering,
Dalian University of Technology, Dalian, China
maxh@dlut.edu.cn
[2] School of Computer Science and Engineering, University of Aizu Fukushima,
Aizuwakamatsu, Japan
sding@u-aizu.ac.jp

Abstract. The phase response of speech is an important part in speech separation. In this paper, we apply the complex mask to the speech separation. It both enhances the magnitude and phase of speech. Specifically, we use a deep neural network to estimate the complex mask of two sources. And considering the importance of the phase, we also explore a phase constraint objective function, which can ensure the phase of the sum of estimated sources that is close to the phase of the mixture. We demonstrate the efficiency of the method on the TIMIT speech corpus for single channel speech separation.

Keywords: Speech separation · Deep neural network · Objective function · Complex mask

1 Introduction

Speech separation is an important part of source separation, and it plays an important role in real life. This problem becomes even more challenging in the single channel case [16]. Researchers have devoted to solving this problems from various aspects, which can be divided into two parts from whether to consider the phase response.

Traditional method based non-negative matrix factorization (NMF) processes the mixture in the time-frequency (T-F) domain by enhancing the magnitude response [9]. Recently with the rising of method based on deep neural network (DNN), Kang et al. combine NMF with DNN by using DNN to learn the encoding vectors of NMF [5]. Besides supervised mask estimation based on DNN also has been shown to improve separation result in very noisy conditions [3,6]. This method solves this problem by learning a mask, such as binary or soft mask [7,13,14] for the magnitude of the source. And mask mostly operates in the magnitude domain and uses the noisy phase during signal reconstruction. The above methods all ignore the phase part mainly because of the studies in [12]. In [12] some well-designed experiments have shown that enhancing speech phase cannot improve speech quality significantly. Besides the research by Ephraim

© Springer International Publishing AG 2017
F. Cong et al. (Eds.): ISNN 2017, Part II, LNCS 10262, pp. 266–273, 2017.
DOI: 10.1007/978-3-319-59081-3_32

and Malah [1] reveals that we do not need to enhance the phase response when the minimum-mean square error (MMSE) is applied to enhancing noisy speech.

However recently a study made by Paliwal et al. showed that only enhancing the phase of sources also can lead to the speech quality improvements [8]. This research has led some researchers to consider phase recover in the speech separation. Based on phase enhancement research, Erdogan used the phase-sensitive masking for the speech separation [2]. Wang combined a subnet as the inverse fast Fourier transform to obtain the clean time-domain signal [15], but it still uses the phase of mixture. And Williamson et al. [16] provided a complex masking for both enhancing magnitude and phase response simultaneously with deep neural network.

In this paper, we apply the complex mask to the speech separation of two speakers and obtain the complex spectrum of two sources together. Moreover, a new objective function which has a constraint between the phase of the sum of two sources and mixture is also provided. The organization of this paper is as follows: Sect. 2 introduces the proposed methods, including the complex mask and phase constrained objective function, Sect. 3 presents the experimental setting and results evaluated on the TIMIT corpus. The conclusion is given is Sect. 4.

2 Proposed Method

2.1 Problem Formulation

In this paper, we assume the observed signal is a mixture of source signals of two speakers [10]. Ignored the noise, the problem can be formulated as

$$\mathbf{y}(t) = \mathbf{s}_1(t) + \mathbf{s}_2(t) \tag{1}$$

where $\mathbf{s}_1(t)$ and $\mathbf{s}_2(t)$ represent the two sources.

2.2 System Framework

The proposed framework is showed in Fig. 1. Firstly, the feature of the speech is fed to the network. Then with the network output, we can obtain the complex spectrum of estimated sources using the Eqs. (2) and (6). Next we process the magnitude of estimated sources with ideal ratio mask (IRM) due to the powerful role of time-frequency masking [14]. Finally, an overlap add method is used to synthesize the time domain signal with the processed magnitude and estimated phase.

2.3 Complex Mask

The time-frequency mask is a powerful tool for speech separation. Traditional mask is defined in the magnitude domain, and based on the research in phase enhancement [16], Williamson provided a complex mask in the complex domain,

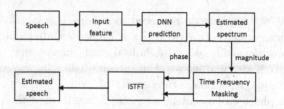

Fig. 1. System framework

which can obtain the complex spectrum of the clean speech from the spectrum of mixture [11]. This part is a brief derivation for the proposed complex mask. They define a mask $M_{t,f}$ to get the clean source signal $S_{t,f}$ directly from the complex spectrum of mixture signal $Y_{t,f}$ as follows:

$$S_{t,f} = M_{t,f} * Y_{t,f} \tag{2}$$

where t, f represents different time-frequency (T-F) unit. The following definitions are operated on each T-F unit and these subscripts do not show for simplicity. '$*$' indicates complex multiplication, and $M_{t,f}$ is the complex mask. The complex mask M can be obtained from the Eq. (2) as followings:

$$M = \frac{S}{Y} = \frac{S_r + iS_i}{Y_r + iY_i} = \frac{Y_r S_r + Y_i S_i}{Y_r^2 + Y_i^2} + i\frac{Y_r S_i - Y_i S_r}{Y_r^2 + Y_i^2} = M_r + iM_i \tag{3}$$

where the subscripts r and i represent the real and imaginary components respectively.

Note that the range of real and imaginary of M is belong to R because of S_r, S_i, Y_r and $Y_i \in R$, and it is not easy to estimate for the DNN. Therefore they deal with this mask as follow:

$$cIRM_x = K\frac{1 - e^{-CM_x}}{1 + e^{-CM_x}} \tag{4}$$

where x can be r or i, indicating the real and imaginary components of the complex mask. This process makes mask values within $[-K, K]$. During testing, the unprocessed mask can be obtained by the following inverse function:

$$\hat{M}_x = -\frac{1}{C}\log(\frac{K - O_x}{K + O_x}) \tag{5}$$

where the O is the output of the DNN.

2.4 New Objective Function

In this paper, DNN is trained to learn the processed complex mask of sources from the mixed magnitude spectrum. The output of DNN, $O = [c\hat{I}RM^1; c\hat{I}RM^2]$, is the processed masking of two sources and is expected to have small error with the target output, where $cI\hat{R}M^1$, $cI\hat{R}M^2$ represent DNN estimation of target output $cIRM^1$, $cIRM^2$, so conventionally one can optimize the neural network parameters by minimizing the mean squared error:

$$J_{MSE} = d(O_r^1, cIRM_r^1) + d(O_i^1, cIRM_i^1) \atop +d(O_r^2, cIRM_r^2) + d(O_i^2, cIRM_i^2) \tag{6}$$

where the O represents the output of the DNN, and the subscript indicates the real and imaginary of the processed complex mask of two sources. And $d(a, b)$ represents the mean square error within a and b.

We present a new objective function with phase constraints. Given $\hat{s}_1$, $\hat{s}_2$ as the complex spectrum of recovered two sources, the sum of them is expected to have small error with the mixture source:

$$\hat{s}_1 + \hat{s}_2 = (\hat{M}_1 + \hat{M}_2) * y = y \tag{7}$$

where the y is the complex spectrum of the mixture. Considering $\hat{s}_1$, $\hat{s}_2$ and y all are complex number, we can divide them into real and imaginary parts.

$$\hat{M}_1 = \hat{M}_{1r} + \hat{M}_{1i}i \tag{8}$$

$$\hat{M}_2 = \hat{M}_{2r} + \hat{M}_{2i}i \tag{9}$$

where the r and i represents the real and imaginary parts of the two sources. Then we can get the following equation from the Eq. (8):

$$\hat{M}_1 + \hat{M}_2 = 1 \tag{10}$$

where $\hat{M}_{1r}$, $\hat{M}_{2r}$, $\hat{M}_{1i}$ and $\hat{M}_{2i}$ are obtained from the output of the DNN, and the relations between the imaginary parts can be obtained from the above three equations:

$$0 = \hat{M}_{1i} + \hat{M}_{2i} \tag{11}$$

$$\hat{M}_{1i} = -\frac{1}{C} \log \left(\frac{K - O_i^1}{K + O_i^1} \right) \tag{12}$$

$$\hat{M}_{2i} = -\frac{1}{C} \log \left(\frac{K - O_i^2}{K + O_i^2} \right) \tag{13}$$

where the Eq. (11) can ensure that the phase of the sum of estimated speech is equal to mixture. And the Eqs. (12) and (13) are obtained from Eq. (5). We can further derive the final result through the above three equations:

$$0 = O_i^1 + O_i^2 \tag{14}$$

Taking into account of the importance of phase, we add this phase constraint to the original objective function in Eq. (6):

$$J = d(O_r^1, cIRM_r^1) + d(O_i^1, cIRM_i^1) \atop +d(O_r^2, cIRM_r^2) + d(O_i^2, cIRM_i^2) + \frac{1}{2}\alpha \sum \left\| O_i^1 + O_i^2 \right\|_2^2 \tag{15}$$

the above equation is the proposed objective function, and the last part represents the phase constraint and α is determined experimentally.

3   Experiments

3.1   Setting

We conduct experiments for single channel speech separation on TIMIT corpus [17] to evaluate the proposed method. Two speakers, same gender or opposite gender, are selected from database. We select eighty percent of the sentences for training, ten percent of the sentences for development set and the remaining for testing from each speaker. The sentences are mixed at 0 dB signal-to-noise ratio (SNR). We also circularly shift the signal from one speaker and mix it with the other source in order to increase the number of training data [4].

In order to compare with DNN [4], the neural network consists of 2 hidden layers which have 160 Rectified Linear Unit (ReLU) units each. And the input feature is a 257-dimensional log energy spectrum, computed using a 32 ms window with a frame shift of 16 ms. Empirically, the function in Eq. (4) the value of parameters K and C is setted 10 and 0.1.

The separation performance is evaluated in terms of three metrics, signal to distortion ratio (SDR), signal to interference ratio (SIR), and signal to artifacts ratio (SAR) [11]. SDR reflects the distortion of original source. It is valid as a global performance measure. SIR shows the ability of rejection of interferences caused by other sources and SAR reports errors caused by extraneous artifacts introduced during the source separation procedure. The higher SDR, SIR and SAR are, the better performance a method achieves [4].

We compare the experimental results with those standard NMF and based on DNN [4]. For standard NMF, the basis matrices $\mathbf{D}_1$ and $\mathbf{D}_2$ are first learned from the training data of two speakers respectively, then we can obtain two coefficients $\mathbf{H}_1$, $\mathbf{H}_2$. Finally a soft mask is applied to obtaining the final results, and the estimated time frequency representations of two sources can be obtain by:

$$\hat{s}_1(t, f) = \frac{\mathbf{D}_1\mathbf{H}_1}{\mathbf{D}_1\mathbf{H}_1 + \mathbf{D}_2\mathbf{H}_2} \odot \mathbf{y}(t, f) \tag{16}$$

$$\hat{s}_2(t, f) = \frac{\mathbf{D}_2\mathbf{H}_2}{\mathbf{D}_1\mathbf{H}_1 + \mathbf{D}_2\mathbf{H}_2} \odot \mathbf{y}(t, f) \tag{17}$$

where $\odot$ denotes element-wise multiplication. $\hat{s}$ represents the spectrum of recovered source and the y is the spectrum of the mixture. As for the method based on DNN, the architecture in [4] is applied here.

3.2   Experimental Results

First, we conduct experiments with opposite gender and the results are displayed in Table 1. DNNcIRM represents the method based DNN with complex mask and objective function in Eq. (6). As for DNNcIRM-dis, the new objective function is added in the above method. It can be seen that the method based neural network achieves better results compared to standard NMF, which confirms that neural network has better generalization and separation capacity. Comparing the last two rows in Table 1, we can find that SIR, SDR and SAR all have improved 0.3–0.8 dB. It verify that it is useful to add the phase constraint in the object function. Besides, we compare the results between [4] and proposed method. We find that the results of only using complex mask are close to the [4], which may because the estimation of phase is not accurate enough. But when it uses new object function, the SIR achieves around 1.3 dB gain compared to [4], and the other two are close. These results show that phase is helpful to the speech separation.

Table 1. Speech separation results of female and male.

Method	Measurement (dB)		
	SDR	SIR	SAR
NMF	6.008	8.722	7.624
DNN [16]	7.70	11.53	8.07
DNNcIRM	7.40	12.00	7.50
DNNcIRM-dis	7.67	12.83	7.89

It is more difficult for the same gender speech separation. Derived from the above results, we only compare the results between the [4] and DNNcIRM-dis method.

Table 2. Speech separation results of two females.

Method	Measurement (dB)		
	SDR	SIR	SAR
DNN [16]	5.71	9.19	6.08
DNNcIRM-dis	6.18	9.79	6.77

The results are displayed in Tables 2 and 3. For the separation between same gender, obviously the result is worse than the opposite gender. And the proposed method achieves a 0.2–0.7 dB gain in SDR, SIR and SAR which is similar with the opposite gender. This shows that proposed method can also play an important role in the same gender separation.

Table 3. Speech separation results of two males.

Method	Measurement (dB)		
	SDR	SIR	SAR
DNN [16]	5.67	8.19	6.38
DNNcIRM-dis	5.81	8.79	6.77

4   Conclusions

In this paper, a framework for jointly separating the magnitude and phase of the sources with a deep neural network has been proposed. And we can obtain two source signals at the same time. Besides, an improvement has been proposed to further enhance the separation performance. The phase constraint on the objective function can ensure the phase of the sum of two source which is close to the mixture. The proposed algorithm achieves better results through a series of experiments on speech separation. But it is still speaker-dependent. The future work will consider extending it to the case of speak-independent.

References

1. Ephraim, Y., Malah, D.: Speech enhancement using a minimum-mean square error short-time spectral amplitude estimator. IEEE Trans. Acoust. Speech Signal Process. **33**(2), 443–445 (1985)
2. Erdogan, H., Hershey, J.R., Watanabe, S., Roux, J.L.: Phase-sensitive and recognition-boosted speech separation using deep recurrent neural networks. In: 2015 IEEE International Conference on Acoustics, Speech and Signal Processing, ICASSP 2015, pp. 708–712 (2015)
3. Healy, E.W., Yoho, S.E., Wang, Y., Wang, D.: An algorithm to improve speech recognition in noise for hearing-impaired listeners. J. Acoust. Soc. Am. **135**(4), 3029–3038 (2014)
4. Huang, P.S., Kim, M., Hasegawa-Johnson, M., Smaragdis, P.: Deep learning for monaural speech separation. In: 2014 IEEE International Conference on Acoustics, Speech and Signal Processing, ICASSP 2014, pp. 1562–1566 (2014)
5. Kang, T.G., Kwon, K., Shin, J.W., Kim, N.S.: NMF-based target source separation using deep neural network. IEEE Signal Process. Lett. **22**(2), 229–233 (2015)
6. Kim, G., Lu, Y., Hu, Y., Loizou, P.C.: An algorithm that improves speech intelligibility in noise for normal-hearing listeners. J. Acoust. Soc. Am. **126**(3), 1486–1494 (2009)
7. Narayanan, A., Wang, D.L.: Ideal ratio mask estimation using deep neural networks for robust speech recognition. In: IEEE International Conference on Acoustics, Speech and Signal Processing, pp. 7092–7096 (2013)
8. Paliwal, K., Wjcicki, K., Shannon, B.: The importance of phase in speech enhancement. Speech Commun. **53**(4), 465–494 (2011)
9. Schmidt, M.N., Olsson, R.K.: Single-channel speech separation using sparse non-negative matrix factorization. In: ICSLP, Ninth International Conference on Spoken Language Processing, INTERSPEECH 2006, Pittsburgh, PA, USA, September 2006
10. Tu, Y., Du, J., Xu, Y., Dai, L.: Speech separation based on improved deep neural networks with dual outputs of speech features for both target and interfering speakers. In: International Symposium on Chinese Spoken Language Processing, pp. 250–254 (2014)
11. Vincent, E., Gribonval, R., Fevotte, C.: Performance measurement in blind audio source separation. IEEE Trans. Audio Speech Lang. Process. **14**(4), 1462–1469 (2006)
12. Wang, D., Lim, J.S.: The unimportance of phase in speech enhancement. IEEE Trans. Acoust. Speech Signal Process. **30**(4), 679–681 (1982)

13. Wang, G.X., Hsu, C.C., Chien, J.T.: Discriminative deep recurrent neural networks for monaural speech separation. In: IEEE International Conference on Acoustics, Speech and Signal Processing (2016)

14. Wang, Y., Narayanan, A., Wang, D.L.: On training targets for supervised speech separation. IEEE/ACM Trans. Audio Speech Lang. Process. **22**(12), 1849–1858 (2014)

15. Wang, Y., Wang, D.L.: A deep neural network for time-domain signal reconstruction. In: 2015 IEEE International Conference on Acoustics, Speech and Signal Processing, ICASSP 2015, pp. 4390–4394 (2015)

16. Williamson, D.S., Wang, Y., Wang, D.: Complex ratio masking for monaural speech separation. IEEE/ACM Trans. Audio Speech Lang. Process. **24**(3), 1–1 (2016)

17. Zue, V., Seneff, S., Glass, J.: Speech database development at MIT: TIMIT and beyond. Speech Commun. **9**(4), 351–356 (1990)

A Neural Autoregressive Framework for Collaborative Filtering

Zhen Ouyang, Chen Sun, and Chunping Li[✉]

Tsinghua University, Haidian District, Beijing, China
oy_zhen@126.com, sunchen_jlu@163.com, cli@mail.tsinghua.edu.cn

Abstract. Restricted Boltzmann Machine (RBM) is a two layer undirected graph model that capable to represent complex distributions. Recent research has shown RBM-based approach has comparable performance with or even better performance than previous models on many collaborative filtering (CF) tasks. However, the intractable inference makes the training of RBM sophisticated, which prevents it from practical application. We present a novel feedforward neural framework for collaborative filtering called NACF, which is extended from the Neural Autoregressive Distribution Estimator (NADE). Because of the autoregressive feed-forward architecture, NACF can be trained with efficient stochastic gradient descent, instead of slow and hard-to-tune truncated Gibbs sampling for RBM. By introducing linear visible units and dual reversed ordering, NACF show faster convergence and better results than Probabilistic Matrix Factorization (PMF) and corresponding RBM models on MovieLens dataset. Besides, by combining NACF results, the rating prediction of efficientsignificantly improved, showing NACF is an effective and efficient model for collaborative filtering.

Keywords: Neural network · Collaborative filtering · RBM · NADE

1 Introduction

In recent years, recommendation system not only gives people easier access to news, movies, music or other products or services on the Internet, but also brings more and more business opportunities for companies. One of the most common recommendation problem is user-item rating prediction, and Collaborative Filtering is a simple but effective approach for this problem. Memory-based models [11] use the user-item ratings to calculate similarity of the users or items, while model-based methods can model the ratings by learning latent factors of users or items. Dimension reduction methods such as Singular Value Decomposition (SVD) [1,7], principle component analysis (PCA) [3,6] can deal with the scalability problem and giving good rating predictions despite the sparsity problem.

Restricted Boltzmann Machine (RBM) is often used as a generative model, to model the distribution of the input vector. Recent research also bring it into the problem of CF tasks [2,10]. It has been shown to have good performance as

© Springer International Publishing AG 2017
F. Cong et al. (Eds.): ISNN 2017, Part II, LNCS 10262, pp. 274–281, 2017.
DOI: 10.1007/978-3-319-59081-3_33

matrix decomposition models like SVD. And combining the result can improve the predicted ratings. Combining the results of RBM and Probabilistic Matrix Factorization (PMF) reduces the RMSE 7% than baseline on Netflix competition dataset [9]. However, the RBM is undirected graphic model which is intractable to inference. So another neural latent factor model, Neural Autoregressive Density Estimator (NADE) [8], is extended from RBM for providing tractable distribution estimation using an autoregressive approach. In this paper, we introduce a novel Neural Autoregressive framework for Collaborative Filtering (NACF) to the task of user-item rating prediction, and evaluate it in different ways.

The rest of the paper is organized as follows: Sect. 2 presents preliminary knowledge for our model, Sect. 3 illustrates the NACF model and how we extend it to adapt to CF tasks, Sect. 4 describes our experiments and analysis and Sect. 5 concludes this paper.

2 Preliminaries

2.1 RBM for Collaborative Filtering

RBM is an undirected graphical model, a bipartite whose distribution is energy based. The energy function of binary hidden and visible units is

$$E(v, h) = -h^T W v - b^T h - c^T v \qquad (1)$$

The marginal distribution of visible units is given by

$$p(v) = \frac{\sum_h exp(-E(v, h))}{\sum_{v', h} exp(-E(v', h)))} \qquad (2)$$

W is the weight matrix, v, h represent the visible input vector and hidden vector respectively. b is the bias for visible units, c is the bias for the hidden units. For parameter learning, usually use the Gibbs sampling to sample between hidden and visible layers. The Contrastive Divergence (CD) algorithm use the truncated instead of full Gibbs sampling, to approximate the gradient, yields good results for RBM training [4].

For CF task, given $N \times M$ user-item matrix, N is the number of users, M is the number of items, each rating entry is a integer value from 1 to K. The corresponding RBM has M visible softmax units to model 1 to K ratings, F binary hidden units, and weight matrix W size is $M \times F \times K$. Each row of the matrix, which is the ratings for all the items of a particular user, is a training case of the RBM. For different users, the missing rating entries tend to be different. In the training procedure, we just ignore the corresponding entries of W, which means the structure of the RBM varies between different training cases, yet sharing the same weight matrix. In the inference procedure, we use non-empty item ratings to get the hidden values and then construct unseen ratings.

2.2   Neural Autoregressive Distribution Estimator

NADE is a generative model extended from RBM by factoring the marginal distribution of the visible units $p(\boldsymbol{v})$ to conditional distribution

$$p(\boldsymbol{v}) = \prod_{i=1}^{D} \frac{p(v_i, \boldsymbol{v}_{<i})}{p(\boldsymbol{v}_{<i})} = \prod_{i=1}^{D} \frac{\sum_{\boldsymbol{v}_{>i}} \sum_h exp(-E(\boldsymbol{v}, \boldsymbol{h}))}{\sum_{\boldsymbol{v}_{\geq i}} \sum_h exp(-E(\boldsymbol{v}, \boldsymbol{h}))} \qquad (3)$$

D is the dimension of the visible input vector. From Eq. (2) we can derive the $p(v_i = 1 | \boldsymbol{v}_{<i})$ is intractable because of its partition function. However, it can be obtained by finding $q(v_i, \boldsymbol{v}_{>i}, \boldsymbol{h} | \boldsymbol{v}_{<i})$ to approximate the true conditional $p(v_i, \boldsymbol{v}_{<i}, \boldsymbol{h} | \boldsymbol{v}_{<i})$ by minimizing the KL-divergence [8]. The conditional distribution is computed in an autoregressive feed forward approach:

$$\boldsymbol{h}^i(\boldsymbol{v}_{<i}) = sigm(\boldsymbol{c} + \boldsymbol{W}_{:<i}^T \boldsymbol{v}_{<i}) \qquad (4)$$

$$p(v_i | \boldsymbol{v}_{<i}) = sigm(\boldsymbol{b} + \boldsymbol{V}_{:<i} \boldsymbol{h}^i(\boldsymbol{v}_{<i})) \qquad (5)$$

Like previous definition for RBM, $\boldsymbol{W}$ is the weight matrix from visible units to hidden units, $\boldsymbol{c}$ and $\boldsymbol{b}$ are the bias of hidden and visible units respectively. $\boldsymbol{V}$ is the weight matrix from hidden units to visible units. This is different from RBM, the untied weight matrix $\boldsymbol{V}$ can lead to better distribution estimator [8]. $\boldsymbol{v}_{<i}$ is the visible units left to v_i, $\boldsymbol{W}_{:<i}^T$ and $\boldsymbol{V}_{:<i}$ is the corresponding weight matrix for $\boldsymbol{v}_{<i}$. $\boldsymbol{h}^i(\boldsymbol{v}_{<i})$ denotes the hidden activation contributed by visible units left to v_i and its superscript i denotes that it is only for calculating the conditional distribution $p(v_i | \boldsymbol{v}_{<i})$. $\hat{v}_i$ is the expectations for $p(v_i = 1)$ which is calculated from the contributions from $\boldsymbol{v}_{<i}$. For the first visible unit, the expectation is given only by the bias $\boldsymbol{c}$. Thus, the neural network can be trained by standard back-propagation algorithm to minimize the cross-entropy error.

3   The Theoretical Model

3.1   The Neural Autoregressive Framework for CF

Now we bring in the NACF model. As the RBM model for CF, we want it to ignore the missing ratings for a particular user in the training process. As shown in Eq. (5) the conditional distribution is directly calculated from $\boldsymbol{h}^i(\boldsymbol{v}_{<i})$, so we can just ignore the contribution of missing rating entries when calculating $\boldsymbol{h}^i(\boldsymbol{v}_{<i})$ using Eq. (4).

As shown in Fig. 1, if there are continuous missing entries between two visible units, the activation of hidden units remains the same. That's to say, if a rating for a visible unit v_i is missing, the corresponding hidden status for calculating $\boldsymbol{h}^i(\boldsymbol{v}_{<i})$ is the same as $\boldsymbol{h}^{i-1}(\boldsymbol{v}_{<i-1})$. This might introduce some extra bias to predict some empty ratings when they are the left most in the specific ordering. We show how to address this effect later in Sect. 3.

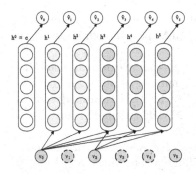

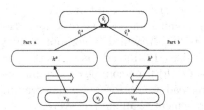

Fig. 1. Figure for NACF model. For a user-based NACF, the visible units are the rating for all items of a user. The visible units with dashed border like v_1, v_3 and v_4 represent the empty entries of item rating. Because there's no contribution from v_1, v_3 and v_4, we have $h^1 = h^2$, $h^3 = h^4 = h^5$.

Fig. 2. Dual reversed ordering NACF (Dual-NACF) is composed by two NACF with reversed orderings denoted as part a and part b. The white hollow arrows represents the autoregressive directions, and the two orderings are exactly reversed.

3.2 Modeling Rating Values

We demonstrate that a basic NACF model can be easily applied to modeling the binary ratings. RBM model use softmax visible units which are overparameterized and the training complexity should times K. However, because of the autoregressive approach to model the conditional distribution, we need to find other approaches to bypass the problem.

We first try to normalize each entry of user-item ratings to zero mean, standard deviation, just like RBM with Gaussian visible units [5]. We denote this as Gaussian-NACF, and its training and inference procedure is the same as basic NACF model.

We also extend the NACF to linear visible units instead of binary stochastic units, and this brings many advantages. First, with the linear units, it is easy to adapt to any possible rating values, keeping the number of parameters to $M \times F$ instead of $M \times F \times K$ for RBM. Second, the linear visible units are also naturally more fit to rating values than 1-of-K representation [2]. If we want to use linear units, we also need to change the error function from cross entropy to mean square error $E(\hat{\boldsymbol{v}}) = \frac{1}{D} \sum_{i=1}^{D} (\hat{v}_i - v_i)^2$.

Another point is that, different user or item have different rating bias [7]. For user-based NACF, the visible unit bias can be seen as item specific bias, however the user bias are not considered. We add a user specific bias term $\hat{r}_{u,i} = \hat{v}_{u,i} + b_u$. Similarly, an item specific bias term are added to item-based NACF.

3.3 Multiple Orderings to Improve Rating Prediction

Because the ordering of conditional distribution chain of NACF brings extra bias of the data, we can improve the rating prediction by reducing this effect.

As explained before in Sect. 3, some visible units may have few of other units left to them, and the corresponding hidden states may have enough information to predict the ratings precisely.

Random orderings of input units have different biases, we can reduce the bias by combining all possible orderings. However, the number of all possible orderings is $O(D!)$, which is computationally impossible for NACF of considerable input size. [12] introduced a method to generate NADE ensembles, but it still takes $O(D!)$ time complexity to inference the result.

Instead we introduce a new ensemble method for NACF. It brings more improvement than simply averaging models of different orderings, and keeping the same training time complexity. As shown in Fig. 2, the output of dualordering NACF model is computed by the two NACF components. The two parts have different hidden units and different weight matrix, but they share the input data, and the final output is calculated by averaging of the two outputs. The two components have reversed orderings. By using reversed orders, each visible unit can 'see' the contribution from exactly all visible units except itself, which might have better predictions than single random ordering model.

The way to calculate the output is defined as:

$$\hat{v}_i = (\hat{v}_i^a + \hat{v}_i^b)/2 \tag{6}$$

The $\hat{v}_i^a$ and $\hat{v}_i^b$ is the output value of visible unit i in part a and part b respectively. When updating the weight matrix in part a and b, we use $\hat{v}_i$ instead of $\hat{v}_i^a$ or $\hat{v}_i^b$ to calculate error. Our experiments show this approach is much better than simply averaging the output of several different orderings.

4   Experiments

4.1   Datasets and Metrics

We evaluate the NACF and its variations on two MovieLens datasets with different size. All the user-item ratings of both two datasets are from 1 to 5. The MovieLens-100k dataset has 100,000 user-item ratings for 1,682 items by 943 users. The MovieLens-1M dataset has 1,000,000 ratings by 6,040 movies for 3,952 items. As previous research usually did, we use 5-fold cross validation training/testing datasets to evaluate the performance. We use *Root Mean Square Error* (RMSE) and *Mean Absolute Error* (MAE) as evaluation metrics for the 1 to 5 score prediction.

$$RMSE = \sqrt{\frac{\sum_{i,j}(r_{ij} - \hat{r}_{ij})^2}{N_t}} \qquad MAE = \frac{\sum_{i,j}|r_{ij} - \hat{r}_{ij}|}{N_t} \tag{7}$$

4.2   Experiment Setup

Without much parameter tuning, we fix the learning rate of NACF model to 0.1, weight decay parameter λ fixed to 0.001. The number of visible units and hidden units varies from 10 to 50 regard to different datasets and NACF implementations.

4.3   Result of Modeling Rating Values

The RMSE on MovieLens-100k of different visible units are as Fig. 3(a) shows. We can see that, the decrease speed of RMSE of NACF with linear visible unit is the fastest. The Gaussian visible units, initially converge faster than linear units, but later its RMSE and MAE becomes higher. The reason is that, the Gaussian units are more sensible to learning rate than binary units or linear units [5]. Note that for experiments after this section, our result all is given by NACF with linear visible units. From Fig. 3(b) we can see that NACF with bias terms, converge faster and get better rating prediction results than raw NACF.

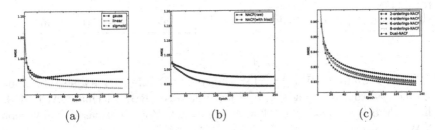

(a)   (b)   (c)

Fig. 3. The results for NACF on MovieLens-100k dataset. (a) Convergence curve for different visible units; (b) convergence curve for NACF with/without bias; (c) convergence curve for NACF ensemble of different orderings and Dual-NACF.

4.4   Result of Multiple Orderings

We test NACF with different orderings, and evaluate the dual reversed ordering NACF (denoted as Dual-NACF) on the MovieLens-100k. As shown in Fig. 3(c), the RMSE and MAE of NACF with 2, 4 and 6 orderings decreases with incorporating more orderings. However, we can see that the RMSE and MAE for 6 orderings and 8 orderings are almost the same. This means simply averaging the results from more NACF with different orderings does not help. Our proposed Dual-NACF outperforms simple averaging approach significantly.

4.5   Result of Comparison with Previous Existing Models

Here we compare the performance with previous collaborative filtering models on MovieLens-100k and MovieLens-1M. We evaluate the Dual-NACF and the single ordering NACF with linear visible units without combining any other information. The result of RBM-based models comes from [2], while the PMF and NACF results are evaluated with common 5-fold cross validation.

For the MovieLens-100k dataset, our NACF models, despite its single random ordering, performs better than corresponding user based or item based RBM model. And the Dual-NACF performs better than corresponding NACF model. For the MovieLens-1M dataset, we can see from Table 1 that, our I-NACF and I-Dual-NACF model outperforms I-RBM model. The U-RBM, I-RBM,

Table 1. MAE results

(a) MAE of CF models on MovieLens-100k dataset.

CF Model	MAE
PMF	0.729
U-RBM	0.779
I-RBM	0.775
U-NACF	0.772
I-NACF	0.727
U-Dual-NACF	0.750
I-Dual-NACF	0.715
I-RBM+INB	0.699
I-Dual-NACF+INB	**0.695**

(b) MAE of CF models on MovieLens-1M dataset.

CF Model	MAE
PMF	0.689
U-RBM	0.711
I-RBM	0.710
U-NACF	0.694
I-NACF	0.693
U-Dual-NACF	0.688
I-Dual-NACF	0.686
I-RBM+INB	0.669
I-Dual-NACF+INB	**0.664**

I-RBM+INB are from RBM based methods [2]. I-RBM+INB is a neighborhood based method that utilizing the results of itembased RBM. I-Dual-NACF+INB is the neighborhood boosted method use the results from item-based dual version of NACF.

4.6 Result of Combining Tests

Ensemble of different CF models can improve the final results. Linear regression is usually used to combine rating predictions of different models. As shown in Table 2, NACF model tend to be a good model to combine. And the coefficients indicate NACF is an effective CF model.

Table 2. Model ensemble results and coefficients. RMSE↓ and MAE↓ indicate the decreasing rate of results corresponding to the first model. COEF1 and COEF2 are the linear regression weights of the first and second model respectively.

CF Model	RMSE	MAE	RMSE↓	MAE↓	COEF1	COEF2
PMF+NACF	0.8534	0.6710	1.49%	2.19%	0.5047	0.5282
RBM+PMF	0.8654	0.6837	3.87%	3.70%	0.1575	0.8474
PMF+NACF	0.8595	0.6755	4.52%	4.86%	0.2464	0.7557

5   Concluding Remarks

In this paper, we present a novel neural autoregressive model for collaborative filtering, and evaluate the model in different ways. By introducing linear visible units and user/item specific bias, our model converges faster and yields better results than RBM, showing its efficiency. By boosting the result using the average outputs of NACF with exactly reversed orderings, the model get significant improvement compared to simply average several models. And experiments show NACF is actually an effective model. In future work, we are going to adapt the extensions of NACF to more datasets to integrate heterogeneous kinds of information and going to implement more efficient CPU or GPU parallel training algorithms for NACF.

Acknowledgements. This work is supported by China NSFC under Grant 61672309 and National Fundamental Research Special Program funded by China MOST under Grant 2017FY201407.

References

1. Billsus, D., Pazzani, M.J.: Learning collaborative information filters. In: ICML, vol. 98, pp. 46–54 (1998)
2. Georgiev, K., Nakov, P.: A non-IID framework for collaborative filtering with restricted Boltzmann machines. In: Proceedings of the 30th International Conference on Machine Learning (ICML 2013), pp. 1148–1156 (2013)
3. Goldberg, K., Roeder, T., Gupta, D., Perkins, C.: Eigentaste: a constant time collaborative filtering algorithm. Inf. Retr. **4**(2), 133–151 (2001)
4. Hinton, G.E., Salakhutdinov, R.R.: Reducing the dimensionality of data with neural networks. Science **313**, 504–507 (2006)
5. Hinton, G.: A practical guide to training restricted Boltzmann machines. Momentum **9**(1), 926 (2010)
6. Kim, D., Yum, B.J.: Collaborative filtering based on iterative principal component analysis. Expert Syst. Appl. **28**(4), 823–830 (2005)
7. Koren, Y., Bell, R., Volinsky, C.: Matrix factorization techniques for recommender systems. Computer **42**(8), 30–37 (2009)
8. Larochelle, H., Murray, I.: The neural autoregressive distribution estimator. J. Mach. Learn. Res. **15**, 29–37 (2011)
9. Salakhutdinov, R., Mnih, A.: Probabilistic matrix factorization. In: NIPS, vol. 1, no. 1, pp. 2–1 (2007)
10. Salakhutdinov, R., Mnih, A., Hinton, G.: Restricted Boltzmann machines for collaborative filtering. In: Proceedings of the 24th International Conference on Machine Learning, pp. 791–798. ACM (2007)
11. Sarwar, B., Karypis, G., Konstan, J., Riedl, J.: Item-based collaborative filtering recommendation algorithms. In: Proceedings of the 10th International Conference on World Wide Web, pp. 285–295. ACM (2001)
12. Uria, B., Murray, I., Larochelle, H.: A deep and tractable density estimator. arXiv preprint arXiv:1310.1757 (2013)

Deep Semantics-Preserving Hashing Based Skin Lesion Image Retrieval

Xiaorong Pu[(✉)], Yan Li, Hang Qiu, and Yinhui Sun

Provincial Key Laboratory of Digital Media,
Health Big Data Science Research Center, Big Data Research Center,
School of Computer Science and Engineering, University of Electronic Science
and Technology of China, Chengdu 611731, People's Republic of China
puxiaor@uestc.edu.cn

Abstract. This study proposes a content-based pigmented skin lesion image retrieval scheme on semantic hash clustering on the output of the deep neural networks. The skin lesion images are acquired with standard digital cameras or mobile phones. To retrieval skin lesion images efficiently online, semi-supervised deep convolutional neural network incorporated with hash functions jointly learn feature representations, for preserving similar semantics between skin lesion images, and mappings to hash codes. The target candidates are clustered by Affinity Propagation (AP) for ranking, which are selected among the outputs of layer F7 based on the Hamming distance of their semantic hash codes. Experiments on 4 disease categories of pigmented skin lesions of a set of 239 images yielded a specificity of 93.4% and a sensitivity of 80.89%.

Keywords: Semantics hash coding · Pigmented skin lesion · Image retrieval · Affinity propagation cluster

1 Introduction

With the deterioration of environmental problems, skin diseases are becoming a common high incidence of common diseases. Pigmented skin lesions, such as a deadly cancer of malignant melanoma, threat to human health seriously. Traditional diagnosis for skin disease is highly dependent on subjective judgment of physicians, but the differentiation of the pigmented skin lesions from others is not trivial even for experienced dermatologists. With emerging development of computer vision and image processing, computer-aided diagnosis (CAD) for skin lesions receives more and more attentions. Based on image processing and artificial intelligence, automatic image-based diagnosis of skin lesions can both benefit patients and dermatologists. Patients can identify roughly the skin lesion through a retrieval system without going to hospital, while physician can be guided by predicting the disease of a particular case and to assist them in diagnosis.

However, the majority of CAD for skin lesions development so far has been mainly focused on dermoscopic images, which have constant illumination, different texture patterns, and characteristics that are not measurable in regular camera images, such as

© Springer International Publishing AG 2017
F. Cong et al. (Eds.): ISNN 2017, Part II, LNCS 10262, pp. 282–289, 2017.
DOI: 10.1007/978-3-319-59081-3_34

lesion area and perimeter. One of the most widely used methods for dermoscopic skin lesion images is the Asymmetry, Border, Color and Diameter, called ABCD rule [1].

The image retrieval in this work is designed for images of skin lesions taken by a non-professional digital camera or mobile phone, which is convenient, no damage to patients as well as easy to operate, while generally suffer from lack of high quality images. As can be seen in Fig. 1, general camera images of pigmented skin lesions typically contain various noises such as hair, uneven illumination and shading areas, which will decrease the efficiency of image analysis. In order to reduce image artifacts, preprocessing techniques should be adopted, such as hair removal [2], image enhancement and segmentation [3].

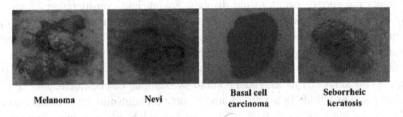

| Melanoma | Nevi | Basal cell carcinoma | Seborrheic keratosis |

Fig. 1. Sample images used in our work

Semantic-preserving feature representations are essential for a content-based image retrieval (CBIR) system. The learned deep features capture rich image representations and provide better performance than the handcrafted features on image retrieval [4]. In our work, for automatically analyzing the actual contents of the skin lesion image based on computer vision and machine learning, deep semantic-preserving features are jointly learned through a semi-supervised deep convolutional neural network [5] incorporated into hash functions, which avoids the limitation of semantic representation power of hand-crafted features.

A typical CBIR system ranks the relevance between the query image and any target image in proportion to a similarity measure calculated from the features. CLUE (cluster-based retrieval of images by unsupervised learning) has been proposed to retrieves image clusters rather than a set of ordered images [6]. In this study, we apply Affinity Propagation, proposed by Frey and Dueck in the journal of SCIENCE [7], to cluster the target candidates which are selected among the outputs of layer F7 based on the Hamming distance of their semantic hash codes.

The rest of this paper is organized as follows. In the next section, we review related work and technologies. Section 3 presents our proposed scheme. In Sect. 4, the experiments and the results are shown. We finally conclude and discuss future work in Sect. 5.

2 Related Works

There exists many semantic structure preserving hash methods. For multi-label images retrieval, a deep semantic ranking based hashing incorporates deep convolutional neural network into hash functions to jointly learn feature representations and mappings from them to hash codes [4]. Semi-supervised hashing minimizes an empirical error over the labeled pairs of points and makes hash codes balanced and uncorrelated to avoid over-fitting [8]. Our method leverages deep learning model to discover deep semantic similarity of skin lesion images.

Deep convolutional neural networks (CNNs) have achieved great success in image classification, retrieval and object detection [9, 10] since it can learn image similarity metric. Krizhevsky et al. [5] achieved a set of feature vectors from the 7th layer of a deep CNN for image retrieval on ImageNet. Yang et al. [11] applied a supervised learning algorithm of semantics-preserving hashing with deep neural networks for large-scale image search.

A CBIR system has the capability of finding out nearest images to the query image and outputs the sorted results to the display. Taking into consideration of relations among retrieved images, the query image and neighboring target images are clustered by Affinity Propagation cluster [7], which can be treated as a procedure on searching for minima of an energy function that depends on a set of K hidden labels corresponding to the K data points. Each label represents the exemplar to which the point belongs, so the similarity of each data point to its exemplar can be calculated. AP views each data point as a node in a network and recursively transmits real-valued messages along edges of the network until a good set of exemplars and corresponding clusters.

3 The Proposed Image Semantics Learning and Retrieval Ranking

3.1 Deep Learning of Binary Hash Codes for Skin Lesion Image Retrieval

Inspired by [11], we apply the semi-supervised semantics-preserving deep hashing for pigmented skin lesion image retrieval, by learning the rich mid-level image descriptors based on deep CNN, as shown in Fig. 2. A set of semantic labels are hypothetically associated with a set of latent attributes for classification.

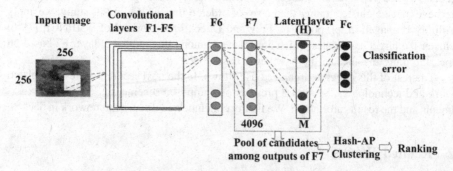

Fig. 2. Framework of the proposed scheme

Hash functions can be treated as a set of mapping of data inputs onto binary codes. Given a dataset of N images $L = \{I_n\}_{n=1}^{N}$ and their associated M classes of label vectors $y = \{y_n\}^{M \times N}$, where y_n is 1 if an image I_n belongs to the corresponding class and 0 otherwise, our goal is to learn a hash function $h : L \rightarrow \{0,1\}^{K \times N}$ which maps images to their k-bit binary codes $B = \{b_n\} \in \{0,1\}^{K \times N}$, while preserving the semantic similarity between image data.

The binary encoding function is given by [10]

$$b_n = sign\left(\sigma(a_n^7 W^H + e^H) - 0.5\right) = sign(a_n^H - 0.5) \tag{1}$$

Where $sign(v) = 1$ if $v > 0$ and 0 otherwise. $W^H \in R^{d \times K}$ is the weight matrix in the latent layer of the network, $a_n^7 \in R^d$ denotes the feature vector of layer F7 of an image I_n, $a_n^H = \sigma(a_n^7 W^H + e^H)$ is the activations of the units in H, where e^H is the bias term and $\sigma(z) = 1/(1 + \exp(-z))$ is the logistic sigmoid function with real value z.

When the data labels are available, binary hash codes can be learned by employing a hidden layer of a CNN model for representing the latent concepts that dominate the class labels. Hash functions are constructed as a latent layer of a deep CNN in which binary hash codes are learned by the optimization of an objective function defined over classification error and other properties of hash codes as:

$$\arg\min_{W}\left(\left(a \sum_{n=1}^{N} L(y_n, \hat{y}_n) + \lambda \|W\|^2\right) - \beta \sum_{n=1}^{N} \left\|a_n^H - 0.5\vec{E}\right\| + \gamma \sum_{n=1}^{N} (mean(a_n^H) - 0.5)^2\right) \tag{2}$$

where $\hat{y}_n$ is the output of the layer F8, $W^C \in R^{K \times M}$ denotes the weights of the network, $L(\bullet)$ is a loss function that minimizes classification error, λ governs the relative importance of the regularization term, α, β and γ are parameters that control the weighting of each term.

With this scheme, classification and retrieval are unified in a single learning model where the image similarity metric can be learned to preserve the semantic similarity among images. In addition, hash functions are constructed through incorporating the CNN model to learn image representations as hash codes in a point wised manner which is naturally scalable to large-scale datasets. Semi-supervised semantics-preserving deep hashing is quite simple and can be easily realized by a slight modification of an existing deep architecture for classification.

3.2   The Hierarchical Retrieval Ranking on Hash-AP

To rank the target candidates according to the semantic similarities among skin lesion images, we apply AP to cluster the target candidates selected among the outputs of layer F7 based on the Hamming distance of their semantic hash codes, output of the latent layer, which is named Hash-AP.

The input of AP is a similarity matrix $SM(i,k) = -\|x_i - x_k\|^2$ representing the log-likelihood of identity point i and its exemplar point k. Let availability matrix be $AM(i,k)$, then the responsibility $RM(i,k)$ can be computed as

$$RM(i,k) \leftarrow SM(i,k) - \max_{k \neq k'}\{AM(i,k') + SM(i,k')\} \tag{3}$$

The availability $AM(i,k)$ is updated by

$$AM(i,k) \leftarrow \min\{0, RM(k,k) + \sum_{i' \notin \{i,k\}} \max\{0, RM(i',k)\}\}, i \neq k$$

$$\text{or} \quad AM(i,k) \leftarrow \sum_{i',s.t,i' \neq k} \max\{0, RM(i',k)\}, i = k \tag{4}$$

The goal of AP is to have the clusters reach high aggregation by actual value interaction of each group of nodes. At any point during affinity propagation, availabilities and responsibilities can be combined to identify exemplars by

$$RM(i,k) + AM(i,k) \leftarrow SM(i,k) + AM(i,k) - \max_{k' \neq k}\{AM(i,k') + SM(i,k')\} \tag{5}$$

The point k is the cluster-head if the value of $RM(i,k) + AM(i,k)$ greater than zero.

We construct the hash functions as a latent layer with $M(M \leq 4096)$ units between the image representation layer and classification outputs in a CNN model, originally comprising 5 convolutional layers, 2 fully connected layers, and an output layer. Hash-AP takes inputs from images and learns image representations, binary codes, and classification through the optimization of an objective function that combines a classification loss with desirable properties of hash codes. Similarity matching is carried out between the query and candidate images based on Hamming distance.

3.3 Performance Measures

Performance measures such as precision and recall quantifies the retrieval efficiency [12]. These measures are calculated by:

$$precision = \frac{number \quad of \quad retrieved \quad relevant \quad images}{number \quad of \quad all \quad retrieve \quad images} \tag{6}$$

$$recall = \frac{number \quad of \quad retrieved \quad relevant \quad images}{number \quad of \quad all \quad images \quad in \quad the \quad database} \tag{7}$$

4 Experiments

We test the proposed scheme on clinical skin lesion image dataset provided by Sichuan Institute of Dermatology and Sichuan Provincial People's Hospital. This dataset contains about 800 skin lesion images of 4 disease categories such as melanoma, basal cell carcinoma, nevi and seborrheic keratosis shown in Fig. 1. All the dermatosis images have been taken by regular digital cameras or mobile phones. All the samples were resized to 256 × 256 prior to training or testing.

Among the skin lesion image database, 189 images of 4 skin lesion categories are randomly selected for training and retrieval, and 50 images of 4 different classes are randomly selected for testing queries.

We implements the proposed methods based on open source Caffe [13] framework. In our experiments, the Hamming threshold is set to 0.3. The ranking performance of top-5 relevant items of Hamming distance to the target image less than 0.3 is evaluated. We have implemented a fusion strategy color, texture and shape feature vectors which are extracted by conventional handcrafted methods for image retrieval. In the CaffeNet deep learning framework, the size of feature vector learned from layer F7 is 4096. After the deep semantic hash codes are translated, the size is reduced to 48. Figure 3 shows the result of iterations of 10000 with loss of 0.5 and accuracy of 75%, where the accuracy is acceptable though nonsignificant due to insufficient dermatologist-labelled image data.

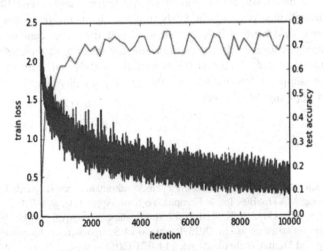

Fig. 3. Results of iterations 10000 times

Some features output from layer F7 are applied to Hash-AP module for clustering and ranking. Table 1 shows that the proposed methods can effectively retrieval skin lesion images. Comparing the proposed methods with the conventional methods based on handcrafted fusion features, it shows that the performance of our method is significantly better than the method based on handcrafted features. By using the CNN model to construct hash functions and clustering by AP, our method have higher learning capability and is able to exploit more semantic information than both the conventional methods based on handcrafted features as well as the deep hash features only, which potentially suffer from some crucial information loss.

Table 1. Results of skin lesion images retrieval

Method	Melnoma	Basal cell carcinoma	Nevi	Seborrheic keratosis	Average precision
Handcrafted features	0.445	0.739	0.81	0.855	0.712
Deep hash	0.389	0.891	0.857	0.892	0.757
Hash-AP	0.556	0.934	0.833	0.916	0.810

However, the precision of Melanoma is relatively low. The possible reason may be that its sample size is too small to applying deep learning framework, which is a kind of rare skin disease in Southwest of china.

5   Conclusions

Based on semantics-preserving hashing via deep neural networks, we have presented a CBIR system as a diagnostic aid for pigmented skin lesion images. Hash functions are constructed through incorporating the CNN model to learn image representations as semantics hash codes, while AP is applied to cluster the target candidates selected among the outputs of layer F7 based on the Hamming distance of their semantic hash codes. Experiments demonstrate that the proposed methods can learn the semantics hash codes on simple CNN model and the ranking quality can be significantly improved after applying AP cluster.

References

1. Nilkamal, S.R., Shweta, V.J.: ABCD rule based automatic computer-aided skin cancer detection using MATLAB®. Int. J. Comput. Technol. Appl. **4**(4), 691–719 (2013)
2. Pu, X.R., Wu, X.J., Ouyang, K.M., Ji, L.P.: Automatic hair removal for skin lesion images from regular digital cameras. In: 2015 International Symposium on Electrical, Electronic Engineering and Digital Technology, pp. 714–721 (2015)
3. Cavalcanti, P., Yari, Y., Scharcanski, J.: Pigmented skin lesion segmentation on macroscopic images. In: The 25th International Conference on Image and Vision Computing, pp. 1–7 (2010)
4. Zhao, F., Huang, Y., Wang, L., Tan, T.N.: Deep semantic ranking based hashing for multi-label image retrieval. In: IEEE Conference on Computer Vision and Pattern Recognition, pp. 1556–1564 (2015)
5. Krizhevsky, A., Sutskever, I., Hinton, G.E.: Imagenet classification with deep convolutional neural networks. In: Advances in Neural Information Processing Systems, pp. 1097–1105 (2012)

6. Chen, Y., Wang, J.Z., Krovetz, R.: CLUE: cluster-based retrieval of images by unsupervised learning. IEEE Trans. Image Process. **14**(8), 1187–1201 (2005)
7. Frey, B.J., Dueck, D.: Clustering by passing messages between data points. Science **315** (5814), 972–976 (2007)
8. Wang, J., Kumar, S., Chang, S.F.: Semi-supervised hashing for scalable image retrieval. In: IEEE Conference on Computer Vision and Pattern Recognition, pp. 3424–3431 (2010)
9. Girshick, R., Donahue, J., Darrell, T., Malik, Z.J.: Rich feature hierarchies for accurate object detection and semantic segmentation. In: IEEE Conference on Computer Vision and Pattern Recognition, pp. 580–587 (2014)
10. Wang, J., Song, Y., Leung, T., Rosenberg, C., Wu, Y.: Learning fine-grained image similarity with deep ranking. In: IEEE Conference on Computer Vision and Pattern Recognition, pp. 1386–1393 (2012)
11. Yang, H.F., Lin, K., Chen, C.S.: Supervised learning of semantics-preserving hashing via deep neural networks for large-scale image search. Computer Science (2015)
12. Henning, M., Nicolas, M., David, B., Antoine, G.: A review of content-based image retrieval systems in medical applications-clinical benefits and future directions. Int. J. Med. Inform. **73**(1), 1–23 (2004)
13. Jia, Y., Shelhamer, E., Donahue, J., et al.: Caffe: convolutional architecture for fast feature embedding. In: The ACM International Conference on Multimedia, pp. 675–678 (2014)

Fast Conceptor Classifier in Pre-trained Neural Networks for Visual Recognition

Guangwu Qian, Lei Zhang[(✉)], and Qianjun Zhang

Machine Intelligence Laboratory, College of Computer Science, Sichuan University,
24# South Section 1, Yihuan Road, Chengdu 610065, People's Republic of China
g.qian@stu.scu.edu.cn, leizhang@scu.edu.cn, zqjblue@163.com

Abstract. Training large neural network models from scratch is not feasible due to over-fitting on small datasets and time-consuming on large
datasets. Hence, to utilize the feature extracting capacity learned by large
models, many investigations have been done on various neural network
models. At the classifying stage of those models, they employ either a linear SVM classifier or a Softmax classifier, which is the only trained part
of the whole model. In this paper, following this line of work, we propose
a classifier based on conceptors called *Fast Conceptor Classifier* (FCC),
which is simple to construct and GPU accelerate. Its evaluations with
pre-trained and no fine-tuning neural networks have been investigated on
Caltech-101 and Caltech-256 datasets, where it achieves state-of-the-art
results with the training time reduced by a factor of 60 on average.

Keywords: Conceptors · Pre-trained neural networks · Convolutional
neural networks

1 Introduction

In the last decade, deep neural networks (DNNs) have risen to the skies
of machine learning [11,12,15,21]. Especially, convolutional neural networks
(CNNs) have recently achieved a great success in the large-scale image and
video recognition [15,18,22]. In the advance of deep visual recognition architectures, the ImageNet Large-Scale Visual Recognition Challenge (ILSVRC) [17]
has played an important role, due to its difficulty brought by a large number
of data (1.2 million images) over a large number of classes (1000 categories).
Although deep CNNs have demonstrated impressive classification performance
on the ImageNet benchmark, they still suffer from the long training time, even
accelerated by modern Graphics Processing Units (GPUs). Besides the high
requirements of the computational capacity and the experimenters of significant professional skills and experience, this issue is another tremendous obstacle
preventing deep learning from practical applications.

Therefore, various approaches have been proposed to address this issue, such
as non-iterative methods [2,3] and transfer learning [9,20,22], the major idea
of which is to simplify the training procedure and improve the training speed.

© Springer International Publishing AG 2017
F. Cong et al. (Eds.): ISNN 2017, Part II, LNCS 10262, pp. 290–298, 2017.
DOI: 10.1007/978-3-319-59081-3_35

Non-iterative methods in neural networks mainly remove iterative training procedures, like backpropagation (BP), and attempt to obtain the analytic solution in a very short time. Such approach settles the problem in some sense, but it usually causes the reduction of the classification performance and brings a tradeoff between the classifying quality and the training speed. As it turns out that deep image representations learned on ImageNet generalize well to other datasets [20], there has been a lot of interest in such a use case [4,5,19,22]. In practice, training an entire CNN with random initialization from scratch is rare, because of the insufficient size of data. Instead, it is common to pre-train a CNN on a very large dataset, like ImageNet, and then use the CNN either as an initialization or a fixed feature extractor for the task of interest. Since the visualization of CNN and a well designed CNN motivated by the visualizing method are proposed in [22], this brief idea becomes prevalent, especially in the industrial practice. A further delicate CNN model presented in [9], introducing spatial pyramid pooling (SPP), achieves the state-of-the-art classifying accuracy 93.42% on Caltech-101 [6]. VGG Nets [20], which spread quickly in the deep learning community, have contributed the guidance of building very deep CNNs by using very small convolution filters. They achieve the state-of-the-art classifying accuracy 86.2% on Celtech-256 [8].

In this paper, we propose a classifier based on conceptors [13], called *Fast Conceptor Classifier* (FCC) and evaluate it with VGG-16 Net [20], Resnet-50 and Resnet-152 [10], all of which are pre-trained by ImageNet datasets and without fine-tuning, on Caltech-101 and Caltech-256 classification benchmarks. FCC is in several aspects inspired by echo state networks [14] and conceptors. It is essentially a quadratic classifier so that it is faster to train such a classifier than Support Vector Machine (SVM) [1] and Softmax classifiers. Besides, it is easy to construct and almost parameterless, namely, only 1 parameter called aperture. According to the experimental results, we demonstrate that FCC boosts the accuracy of a variety of pre-trained CNN architectures despite their different designs. With these advantages, FCC should in general not only improve all pre-trained image classification methods but also accelerate the training step of classifiers.

The rest of this paper is organized as follows. We provide preliminaries in Sect. 2. In Sect. 3 we describe our proposed model in detail. In Sect. 4, experimental results on Caltech-101 and Caltech-256 are reported. Finally, discussion and conclusion are given in Sect. 5.

2 Preliminaries

For most classifiers, it is better to apply them on extracted features instead of raw data, because the representations of features on a higher dimension are likely more separable, i.e. to be better classified. In this paper, several prevalent pre-trained CNNs are adopted for the feature extraction. Then, the extracted features are fed into FCC to complete the classifying task. As CNN is well known, we only introduce the fundamental aspects of conceptors, which are the inspirations of FCC.

2.1 Conceptors

Conceptors, proposed in [13] for recurrent neural networks (RNNs), can be understood as filters which characterize temporal neural activation patterns. Conceptors take the form of soft projection matrices, which achieves a direction-selective damping of high-dimensional network signals. So far conceptors have been applied to temporal neural data, but here we make use of the same constructions for static data that arise in the pre-trained neural networks. In order to accomplish this, several necessary changes have been done to original conceptors to fit static data.

Here is a brief explanation of conceptors. A conceptor is a square matrix C which linearly transforms an n-dimensional input vector x to an n-dimensional vector y,

$$y = Cx. \tag{1}$$

The defining objective of a conceptor matrix is to replicate the input as accurately as possible while having small matrix entries, which leads to the empirical loss function

$$\mathcal{L} = \|X - CX\|_{fro}^2 + \alpha^{-2} \|C\|_{fro}^2, \tag{2}$$

where $X \in \mathbb{R}^{N \times M}$ collects M sample inputs x as columns and $\alpha \in (0, \infty)$ is a balancing parameter named *aperture* in a conceptor context. The matrix norm in (2) is the Frobenius norm. It is easily derived [13] that this loss is minimized by

$$C = R \left(R + \alpha^{-2} I\right)^{-1}, \tag{3}$$

where $R = 1/M\ XX^T$ is the input data correlation matrix. Conceptor matrices are positive semi-definite and have eigenvalues (= singular values for positive semi-definite matrices) of at most unit size. A conceptor matrix computed from empirical data typically has a low numerical rank, that is, most singular values are so close to zero that they can be neglected. This circumstance admits representing conceptor matrices economically by their low-rank Singular Value Decomposition (SVD), a fact that we exploit to trim down computation times in the latter phase.

2.2 Conceptors Based Classifier

When input data X come in different classes X_j, for each class a separate conceptor $C_j = R_j(R_j + \alpha_j^{-2}I)^{-1}$ is computed, where $R_j = 1/M_j\ X_j X_j^T$.

Given $j = 1, \ldots, K$ input pattern classes, and having computed conceptors C_j, they can be used to classify a test pattern $x \in \mathbb{R}^n$ by computing the K *positive evidences* [13]

$$E^+(x, i) = x^T C_i\, x \tag{4}$$

and deciding for class $j = \text{argmax}_i E^+(x, i)$.

3   Classification Framework

An FCC classification framework is made from two components, a pre-trained CNN for feature extraction and an FCC for classification. Compared with other classifiers, the advantages of FCC are summarized as follows:

Simple. FCC is as simple as a Softmax classifier and easy to implement by any programming language. Although the linear SVM classifier is also available as a function library, it's much more complicated inside.

Fast. FCC mainly involves singular value decomposition (SVD) and matrix multiplication, both of which are fundamental linear algebra operations and can be GPU accelerated easily, while only a few GPU implementations exist for SVM. Besides, even the only training targets are the weights of the last fully-connected layer attached to the Softmax classifier, it costs much longer time by using a gradient descent algorithm, which is an iterative optimization method.

High-Performance. According to our experimental results, FCC achieves a better classification performance than linear SVM and Softmax on equal conditions. In addition, FCC with Resnet-152, which is, to our best knowledge, the deepest CNN available publicly, achieves the state-of-the-art classifying performance on Caltech-101 and Caltech-256.

3.1   Feature Extraction

The most straightforward way to extract features is to keep pre-trained CNN weights fixed and feed the input into the CNN to get the generated features. Only minor changes of the network structure have to be done, including removing the last fully-connected layer, which performs 1000-way ImageNet classification, and replacing the Softmax classifier with FCC. This strategy is carried out on FCC with Resnet-50 and FCC with Resnet-152. However, to take advantage of the benefit of the aggregation of multi-scale features, the fully-connected layers have to be converted to convolutional layers, namely a fully convolutional network (FCN) [16]. The conventional CNNs require a fixed input image size, which limits both the aspect ratio and the scale of the input image. When fed with images of arbitrary sizes, those CNNs demand a pre-processing step to generate images with the fixed size. This pre-processing step could be cropping [15,22], warping [5,7] or simply resizing, which prejudices either the aspect ratio or the scale of the original image. This technical issue can be easily handled by an FCN, as it only involves convolving and pooling operations so that it can accept images of arbitrary sizes. VGG Nets are evaluated on Caltech-101 and Caltech-256 in this way and achieves the state-of-the-art performance.

Due to the tolerance of high dimensionality and the low requirement of computational capacity of FCC, it is possible to implement another extracting strategy, i.e. the extraction of intermediate features generated by any layer in a CNN. Such approach is quite rare if an SVM classifier or a Softmax classifier is used. In general, the intermediate layers in neural networks possess more hidden units than the penultimate layer, namely the last layer before the classifier layer, since more hidden units mean better feature extracting ability but the final features should be manageable for the future demand. Moreover, as the feature hierarchies become deeper, they learn increasingly powerful features [22]. With this premise supported by a lot of literature, features generated by the intermediate layers are usually neglected. Thanks to the simplicity of FCC, classifying the intermediate features is affordable.

3.2 Classification

Conceptors were introduced in [13] in a context of recurrent neural networks as filters for recurrent states. Here we adopt the basic definition of conceptors (compare (3)) for CNNs and employ them for the final classification layer, using (4). In order to save the computation times for the SVD, that is implicitly computed in (3), within reasonable bounds, we use the compact SVD.

By applying compact SVD, the correlation matrix can be computed as follows

$$
\begin{cases}
X = U_r \Sigma_r V_r^T \\
R = \dfrac{XX^T}{M} = \dfrac{U_r \Sigma_r^2 U_r^T}{M} = \dfrac{U_r D_r U_r^T}{M}
\end{cases}
\tag{5}
$$

where r is the rank of the matrix $X \in \mathbb{R}^{N \times M}$. In our case, $r = min(N, M)$ is usually decided by the number of samples, because most of time, the matrix X is full rank and the number of samples of each class is far less than the dimension of features. The explicit solution of a conceptor matrix in Eq. (3) can be rewritten as

$$
\begin{aligned}
C(R, \alpha) &= arg\ min_C E\left[\|X - CX\|^2\right] + \alpha^{-2}\|C\|_{fro}^2 \\
&= U_r \left[D_r(D_r + \alpha^{-2}I_r)^{-1}\right] U_r^T \\
&= U_r S_r U_r^T
\end{aligned}
\tag{6}
$$

and the computation of a positive evidence becomes

$$
E^+(x, j) = x^T C_j x = x^T U_{r_j} S_{r_j} U_{r_j}^T x,
\tag{7}
$$

where j stands for the jth class and x is a test sample. In fact, as a classifier, the conceptor matrix is only needed in testing. Moreover, instead of storing a huge conceptor matrix, compact SVD components U_{r_j} and Sr_j are stored to avoid generating a huge matrix in training and used to compute the evidence in testing.

4   Experiments and Results

In this section, we present the image classification results achieved by VGG-16 Net, Resnet-50 and Resnet-152 on Caltech-101 and Caltech-256 with and without FCC. Caltech-101 contains 9 K images labeled into 102 classes (101 object categories and a background class), while Caltech-256 is larger with 31 K images and 257 classes. On Caltech-101, 30 images per category are randomly sampled for training and up to 50 images per category are randomly sampled for testing. On Caltech-256, 60 images per category are randomly sampled for training and the rest is used for testing.

4.1 VGG-16 Net, Resnet-50 and Resnet-152 with FCC

VGG Nets are very successful deep convolutional networks and achieve several state-of-the-art results. Two best-performing CNN models in VGG Nets, VGG-16 Net and VGG-19 Net, are available publicly. As VGG-16 Net and VGG-19 Net possess very close classification performances and the structures of them are also similar, only VGG-16 Net is investigated in our experiments. The only pre-processing we do is subtracting the mean RGB value, computed on the training set, from each pixel. Also, as in [20], the evaluation is carried out over multiple scales and 4096-D activations of the penultimate layer are collected as features. In our case, three scales $Q \in \{256, 384, 512\}$ have been used. In the meanwhile, to further investigate whether FCC can work with a very deep network, we have also implemented FCC with Resnet-50 and Resnet-152, which are deepest networks available publicly so far. Since no obvious performance improvement by involving multiple scales on Resnets, only the scale 224 is employed. The results of the comparison experiments between FCC and the state-of-the-art methods on Caltech-101 and Caltech-256 are listed in Table 1.

Table 1. Classification accuracy on Caltech-101 and Caltech-256

Method	Caltech-101	Caltech-256
Zeiler and Fergus [22]	86.5	74.2
Chatfield et al. [4]	88.4	77.6
He et al. [9]	93.4	-
VGG-16 Net [20]	91.8	84.57
Resnet-50 [10]	92.65	82.43
Resnet-152 [10]	95.23	90.24
FCC (VGG-16 Net)	91.87	84.67
FCC (Resnet-50)	93.08	82.81
FCC (Resnet-152)	**95.55**	**90.87**

As can be seen, on Caltech-256, FCC with VGG-16 Net outperforms [20]. On Caltech-101, FCC (VGG-16 Net) is competitive with the approach of [9] and better than original VGG-16 Net, the pre-trained neural network FCC is based on. It suggests that FCC can improve the classification performance on a pre-trained neural network. Both on Caltech-101 and Caltech-256, Resnet-152 with FCC achieves the state-of-the-art classification accuracy. Besides FCC, we suppose that the good results benefit from the better generalized feature extracting capacity because Resnet-152 is trained on a larger ImageNet dataset over 11 K classes.

4.2 Evaluation on Features of Different Layers

As mentioned before, the intermediate features can be easily utilized by FCC. The features, generated by the intermediate layers before the global average pooling layer, are different in size, so spatial pyramid pooling (SPP) [9] is employed to generate features in the same size (as the same as the feature generated by a 224 × 224 input image). The classification accuracy of features from the last 3 layers (last pooling layer, the penultimate fully-connected layer and the last fully-connected layer before the Softmax layer) in VGG-16 Net and the last two layers (last pooling layer and the activation layer before) in Resnet-50 and Resnet-152 are listed in Table 2.

Table 2. Classification accuracy on features from different layers of VGG-16 Net, Resnet-50 and Resnet-152 with FCC

Layer		Caltech-101	Caltech-256
VGG-16 Net [20]	Pool5	**93.96**	83.96
	FC6	91.87	**84.67**
	FC7	90.92	84.02
Resnet-50 [10]	ReLU	**94.32**	**83.83**
	Pool	93.08	82.81
Resnet-152 [10]	ReLU	**95.59**	**91.27**
	Pool	95.55	90.87

From this evaluation, we can observe that some intermediate features are better than the final features. In our opinions, this is caused by the inevitable information loss of the last few layers, where the dimensional reduction is necessary to guarantee a manageable feature size.

4.3 Runtime Comparisons

Besides the good classifying performance, the runtime of FCC is also impressive. The CPU (Intel Xeon CPU E5-2620 v3 @ 2.40 GHz) runtime of VGG-16 Net, Resnet-50, and Resnet-152 with different classifiers are listed in Table 3.

The classifier compared in our experiments is linear SVM. If a Softmax classifier is used instead, it will take much longer time and perform worse. As indicated in Table 3, the training time of FCC is extremely little. Because the main step SVD in the training stage is applied on a sample collection matrix $X_j \in R^{N \times M_j}$ of class j, where $N >> M_j$, its consumed time is rather little. Compared with linear SVM, the training time of FCC is reduced by a factor of 60 on average. Even for the testing time, FCC only needs at most half of what linear SVM demands.

Table 3. The runtime (second) of VGG-16 Net, Resnet-50, and Resnet-152 with different classifiers

Method	Caltech-101		Caltech-256	
	Training	Testing	Training	Testing
VGG-16 Net	118.31	118.28	2345.07	3114.48
FCC(VGG-16 Net)	1.76	65.2	26.16	1103.25
Resnet-50	16.03	21.59	82.43	554.12
FCC(Resnet-50)	0.33	15.19	2.64	220.99
Resnet-152	13.07	20.24	229.93	497.93
FCC(Resnet-152)	0.32	15.33	2.73	223.21

5 Conclusions

In this paper, a classifier named FCC, which is simple to construct and GPU accelerate, is proposed. Meanwhile, a classification framework embracing various pre-trained networks and FCC is also presented. Experiments have been conducted to test FCC on VGG16-Net, Resnet-50, and Resnet-152. Among them, FCC outperforms linear SVM and Softmax classifier at a significantly reduced runtime. Compared with linear SVM, FCC reduces the training time by a factor of 60 on average. In particular, FCC with Resnet-152 achieves the state-of-the-art classification performance on Caltech-101 and Caltech-256. Given those experimental results, we suppose that it's promising to extend FCC to more neural network models. Under runtime sensitive and pre-trained neural network employed circumstances, FCC is also preferable.

Acknowledgments. This work was supported by Fok Ying Tung Education Foundation (grant 151068); and National Natural Science Foundation of China (grants 61322203, 61332002). The authors would like to appreciate enormous help from Prof. Dr. Herbert Jaeger.

References

1. Boser, B.E., Guyon, I.M., Vapnik, V.N.: A training algorithm for optimal margin classifiers. In: Proceedings of the Fifth Annual Workshop on Computational Learning Theory, pp. 144–152. ACM (1992)
2. Bruna, J., Mallat, S.: Invariant scattering convolution networks. Pattern Anal. Mach. Intell. IEEE Trans. **35**(8), 1872–1886 (2013)
3. Chan, T.H., Jia, K., Gao, S., Lu, J., Zeng, Z., Ma, Y.: PCANet: a simple deep learning baseline for image classification? Image Process. IEEE Trans. **24**(12), 5017–5032 (2015)
4. Chatfield, K., Simonyan, K., Vedaldi, A., Zisserman, A.: Return of the devil in the details: delving deep into convolutional nets. arXiv preprint arXiv:1405.3531 (2014)

5. Donahue, J., Jia, Y., Vinyals, O., Hoffman, J., Zhang, N., Tzeng, E., Darrell, T.: DeCAF: a deep convolutional activation feature for generic visual recognition. In: ICML, pp. 647–655 (2014)
6. Fei-Fei, L., Fergus, R., Perona, P.: Learning generative visual models from few training examples: an incremental bayesian approach tested on 101 object categories. Comput. Vis. Image Underst. **106**(1), 59–70 (2007)
7. Girshick, R., Donahue, J., Darrell, T., Malik, J.: Rich feature hierarchies for accurate object detection and semantic segmentation. In: Proceedings of the IEEE Conference on Computer Vision and Pattern Recognition, pp. 580–587 (2014)
8. Griffin, G., Holub, A., Perona, P.: Caltech-256 object category dataset (2007)
9. He, K., Zhang, X., Ren, S., Sun, J.: Spatial pyramid pooling in deep convolutional networks for visual recognition. In: Fleet, D., Pajdla, T., Schiele, B., Tuytelaars, T. (eds.) ECCV 2014. LNCS, vol. 8691, pp. 346–361. Springer, Cham (2014). doi:10.1007/978-3-319-10578-9_23
10. He, K., Zhang, X., Ren, S., Sun, J.: Deep residual learning for image recognition. arXiv preprint arXiv:1512.03385 (2015)
11. Hinton, G.E., Osindero, S., Teh, Y.W.: A fast learning algorithm for deep belief nets. Neural Comput. **18**(7), 1527–1554 (2006)
12. Hinton, G.E., Salakhutdinov, R.R.: Reducing the dimensionality of data with neural networks. Science **313**(5786), 504–507 (2006)
13. Jaeger, H.: Controlling recurrent neural networks by conceptors. arXiv preprint arXiv:1403.3369 (2014)
14. Jaeger, H., Haas, H.: Harnessing nonlinearity: predicting chaotic systems and saving energy in wireless communication. Science **304**(5667), 78–80 (2004)
15. Krizhevsky, A., Sutskever, I., Hinton, G.E.: Imagenet classification with deep convolutional neural networks. In: Advances in Neural Information Processing Systems, pp. 1097–1105 (2012)
16. Long, J., Shelhamer, E., Darrell, T.: Fully convolutional networks for semantic segmentation. In: Proceedings of the IEEE Conference on Computer Vision and Pattern Recognition, pp. 3431–3440 (2015)
17. Russakovsky, O., Deng, J., Su, H., Krause, J., Satheesh, S., Ma, S., Huang, Z., Karpathy, A., Khosla, A., Bernstein, M., et al.: Imagenet large scale visual recognition challenge. Int. J. Comput. Vis. **115**(3), 211–252 (2015)
18. Sermanet, P., Eigen, D., Zhang, X., Mathieu, M., Fergus, R., LeCun, Y.: Overfeat: integrated recognition, localization and detection using convolutional networks. arXiv preprint arXiv:1312.6229 (2013)
19. Sharif Razavian, A., Azizpour, H., Sullivan, J., Carlsson, S.: CNN features off-the-shelf: an astounding baseline for recognition. In: Proceedings of the IEEE Conference on Computer Vision and Pattern Recognition Workshops, pp. 806–813 (2014)
20. Simonyan, K., Zisserman, A.: Very deep convolutional networks for large-scale image recognition. arXiv preprint arXiv:1409.1556 (2014)
21. Wan, L., Zeiler, M., Zhang, S., Cun, Y.L., Fergus, R.: Regularization of neural networks using dropconnect. In: Proceedings of the 30th International Conference on Machine Learning (ICML-2013), pp. 1058–1066 (2013)
22. Zeiler, M.D., Fergus, R.: Visualizing and understanding convolutional networks. In: Fleet, D., Pajdla, T., Schiele, B., Tuytelaars, T. (eds.) ECCV 2014. LNCS, vol. 8689, pp. 818–833. Springer, Cham (2014). doi:10.1007/978-3-319-10590-1_53

Video-Based Fire Detection with Saliency Detection and Convolutional Neural Networks

Lifeng Shi, Fei Long[✉], ChenHan Lin, and Yihan Zhao

Software School, Xiamen University,
Xiamen 361005, Fujian, People's Republic of China
flong@xmu.edu.cn

Abstract. Much work has been done in fire detection by using color model and hand-designed features. However, these methods are difficult to meet the needs of various fire detection scenarios. In this paper we propose a new method of video-based fire detection by combining image saliency detection and convolutional neural networks. Our method consists of two modules: (1) utilize saliency detection method to extract flame candidate region proposals. (2) extract features from each candidate region by using convolutional neural networks, and then classify these features into fire or non-fire. This method can automatically learn effective features from video sequences. The experimental results show that our method achieves classification results superior to some hand-designed features for fire detection. We also compare color model method and saliency detection method for obtaining flame candidate regions.

Keywords: Video-based fire detection · Saliency detection · CNNs

1 Introduction

The occurrence of fire disaster is usually unpredictable and its consequences are always incalculable. Fire accidents frequently cause economic and ecologic damage as well as endangering people's lives. Thus it's critical to find the fire as early as possible. Heat sensors, smoke sensors, gas detectors, as the typical traditional fire detection technologies, have a number of limitations. For instance, they require a close proximity to fire sources so that they aren't enough suitable for the outdoor scene, e.g., forest, open land, stadiums. In contrast, video-based fire detection can be effectively applied to open or large spaces and offer us much extra information, such as flame spreading trend, growing rate and so on. And in recent years, many video-based fire detection methods have been proposed and achieve better true positive rate and lower false alarm in order to make possible a commercial exploitation. However, previous work on video-based fire detection heavily rely on hand-designed features.

In this paper, we propose a new video-based method for fire detection, which can automatically learn effective features for further classification. In our method, we firstly utilize a general saliency detection algorithm to extract candidate flame regions. Furthermore, on the basis of the saliency detection, we add

© Springer International Publishing AG 2017
F. Cong et al. (Eds.): ISNN 2017, Part II, LNCS 10262, pp. 299–309, 2017.
DOI: 10.1007/978-3-319-59081-3_36

fire-pixels color confidence measure to further acquire accurate flame regions. Secondly, those region proposals are feed into a convolutional neural network for automatically learning effective features and classification. Contrasting to hand-designed features methods, our method achieves a better true positive rate on our fire dataset including indoor, outdoor and forest fire occasions.

2   Related Work

Previous work on fire detection has focused on color extraction and hand-designed features. Both color and motion are very distinct characteristics of flame and generally used to determine candidate flame areas in videos. Chen et al. [1] propose three decision rules based on RGB color model, (1) $R \geq G > B$; (2) $R > R_T$ (3) $S > (255 - R)S_T R_T$, to identify fire pixels from an image, which is one of the earliest flame detection methods for flame candidate region extraction. However, in fact, some high intensity pixels may not satisfy $R > G$. Celik et al. [2] make a set of rules in YCbCr color space. Besides RGB and YCbCr, various color spaces such as HSV, Lab, HIS and so on, are tried to extract flame pixels in [3–9]. In [3], the authors use a large amount of labeled samples to estimate class conditional probability densities of flame and background, and they propose four detection rules with difference of RGB channels and five discriminate models to achieve the extraction of flame pixels. Liu et al. [10] propose a flame detection algorithm that is based on a saliency detection technique and uniform local binary pattern (ULBP), but its computational efficient is lower than the method in [2]. In recent years, various spatiotemporal features, such as flickering, spatiotemporal energy, are proposed to model the fire behavior. For example, in [11], authors divide fire videos into spatiotemporal blocks and use covariance-based features extracted from these blocks to detect fire, and the method can be extented to non-stationary cameras. However, its computational efficient is lower than the method proposed in [10].

However, These features are designed with much expert experiences and problem domain knowledge. Different from traditional hand-designed features, feature learning methods can automatically get effective feature representations based on raw input data. In recent years, deep learning methods, such as convolutional neural networks (CNNs), have been applied with great success to many computer vision tasks, for example, scene classification [12], marine animal classification [13], vehicle detection tasks [14] and so on. In [15], authors propose a new combination method of region proposals and CNNs: Regions with CNN features (R-CNN). This method firstly generates category-independent region proposals by selective search, and then it uses a large convolutional neural network to extract a feature vector from each region. Finally, a set of class-specific linear SVMs are applied to classification. The algorithm gives a 30% relative improvement over the best previous results on PASCAL VOC 2012.

3   Our Method

Inspired by the idea of R-CNN, we propose a new method for fire detection with image saliency detection method and convolutional neural network. Figure 1 shows the basic flowchart of our approach, which consists of two modules. Firstly, we extract flame candidate regions by saliency detection from input video sequences, and apply fire-pixels color confidence measure to further acquire accurate flame regions. The second module is a convolutional neural network that learns effective features from each region and distinguish it into fire or non-fire region. In this section, we will describe the design for each module in detail.

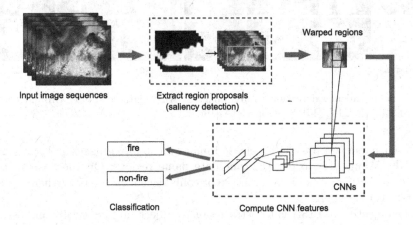

Fig. 1. Framework of proposed fire detection method

3.1   Candidate Region Extraction

In the most real-world scenarios, as shown in Fig. 2, it is obvious that flame pixels are in red-yellow color range. The method, proposed by Chen et al. [1], uses three decision rules based on RGB color model that can obtain decent results. And it has lower computational complexity and spends less running time. However, as we all know, due to the differences in the material of the combustion, the flame presents different colors, such as red-yellow, blue, white, etc., which is shown in Fig. 3. So it's difficult for meeting various circumstances by using color space methods (RGB, HIS, YCbCr, etc.).

Saliency detection, as shown in Fig. 4, which aims to identify the most important and conspicuous object regions in an image, has received increasingly more interest in recent years. Serving as a preprocessing step, it can efficiently extract the interesting image regions related to specific tasks and broadly facilitates computer vision applications such as segmentation, image classification and so on. Generally, the fire or flame is very salient in various scenes, so we can use saliency detection to obtain candidate flame regions.

Fig. 2. An example of common RGB values of flame pixels (Color figure online)

Fig. 3. Different materials presents different colors (Color figure online)

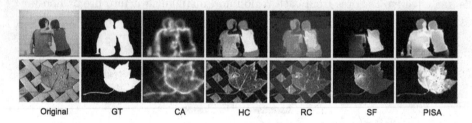

Original   GT   CA   HC   RC   SF   PISA

Fig. 4. Some saliency detection results on PASCAL-S data sets. GT: Ground truth. CA [16], HC [17], RC [17], SF [18], PISA [19].

In our paper, we use the Pixelwise Image Saliency Aggregating (PISA) proposed by Wang. [19] to extract candidate flame regions. On the basis of the saliency detection, we add fire-pixels color confidence measure to further acquire accurate flame regions.

The objective of PISA is to extract salient objects automatically and assign consistently high saliency levels to them. This method introduces two types of features to capture contrast information of salient regions, they are a color-based contrast feature $U^c(p)$ and a structure-based feature $U^g(p)$. This method also considers some spatial priors that salient pixels tent to distribute near the image center and away from the image boundary. Thus they use $D^c(p)$ and $D^g(p)$ to denote the integration of image center spatial distance and image boundary exclusion and to reweight the color and structure contrast measurement. So computing the feature-based saliency confident $\hat{f}$ for each pixel p by aggregating the two contrast measures with the spatial priors, as

$$\hat{f}(p) = U^c(p) \cdot D^c(p) + U^g(p) \cdot D^g(p) \tag{1}$$

and then a sigmoid-like function is used to normalize the saliency confidence to the discrete saliency level set $\{0, 1, \ldots, L-1\}$:

$$f(p) = R\left(\frac{L-1}{1 + exp(-\hat{f}(p))}\right) \tag{2}$$

where R is a round down function.

The goal of PISA is to assign a saliency level S_p for each pixel according the normalized saliency confidence under some constrains. In addition, to suppress spurious and non-uniform saliency assignment, this method requires that the assigned saliency level S_p should be consistent with its neighborhood pixels within its local observation region Ω_p in the image domain. And this coherence constraint $C(S_p)$ is defined as

$$C(S_p) = \sum_{q \in \Omega_p} \omega_{pq} ||S_p - S_q||_2^2 \tag{3}$$

where q represents a neighboring pixel to p in Ω_p, and S_q is the saliency level assigned to q. ω_{pq} encodes the similarities between p and q within Ω_p.

According to the above theory, we can compute the assigned saliency level S_p for each pixel p by minimizing the following energy function

$$E = \sum_{q \in I} A(S_p) + C(S_p) \tag{4}$$

where $A(S_p)$ represents the cost of labeling pixel p with the saliency level S_p, and $A(S_p) = ||S_p - f(p)||_2^2$.

Saliency detection aims at highlighting salient foreground objects automatically from the background. So the results of saliency detection may contain other objects which are considered to be conspicuous, such as pedestrian, traffic signs, light and so on. For the results, we expect that flame pixels can be remained as far as possible, while removing irrelevant information. Because flame regions are further analyzed by the convolutional neural network, we just extract the minimum enclosing rectangle containing flame pixels. Based on the above considerations, we propose two confidence measures.

$$F_c = \frac{\sum f(x, y)}{\sum R(x, y)} \tag{5}$$

$$S_c > Th \tag{6}$$

The fire-confidence F_c denotes the probability of that the extracted minimum enclosing rectangle region is flame or includes flame. Where $f(x, y) = 1$ indicates that the pixel located at (x, y) of rectangle region belongs to flame pixel, and $f(x, y) = 0$ otherwise; $\sum R(x, y)$ represents the number of the pixels in region; The saliency-confidence S_c describes a rough saliency level used to the adjustment of image binarization. Th is a threshold to extract minimum enclosing rectangle. Figure 5 presents the extracted flame bounding box postprocessed with same saliency-confidence and different fire-confidence. It's obvious that F_c could remove some invalid saliency areas. Figure 6 presents the extracted flame bounding box using different saliency-confidence and same fire-confidence. It shows that the appropriate threshold can decrease redundance pixels.

We also compare RGB color model method [1] and saliency detection method. More details are discussed in Sect. 4.

3.2   Feature Learning and Classification

We learn features from each processed flame region using the CNN described by Krizhevsky et al. [20] on Caffe platform. Features are computed by forward

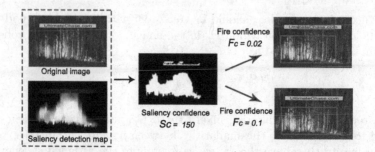

Fig. 5. Flame bounding box with different fire-confidence

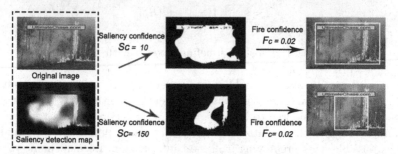

Fig. 6. Flame bounding box with different saliency-confidence

propagating a mean-subtracted 227×227 RGB image through five convolutional layers and three fully connected layers. The details of the architecture are listed in Table 1. Each layer contains learnable parameters and consists of a linear transformation followed by a nonlinear mapping, which is implemented by Rectified Linear Unites (ReLUs) to accelerate the training process. Response-normalized is applied to the all layers to help generalization. And Max pooling is applied to the first, second and firth convolutional layers for translational invariance. The dropout procedure is used after the first and the second fully connected layers to avoid overfitting. Finally, we use a softmax regression model in the output layer to classify regions into fire (1) or non-fire (0).

Table 1. Architecture details of networks. C: convolutional layer; R: ReLUs; N: response-normalized; D: dropout

Networks Architecture								
Layer	1	2	3	4	5	6	7	8
Input size	227*227*3	27*27*96	13*13*256	13*13*348	13*13*348	6*6*256	4096	4096
Type	C+R+N	C+R+N	C+R	C+R	C+R	F+R+D	F+R+D	F+S
Output	96	256	384	384	256	4096	4096	2
Filter size	11*11	5*5	3*3	3*3	3*3	-	-	-
Pooling size	3	3	-	-	3	-	-	-
Pooling stride	2	2	-	-	2	-	-	-

In training stage, for the video clips without label information provided, we manually label fire regions on images in the training set. At test stage, we obtain candidate regions by using the saliency detection method mentioned above. Similarly, convert these candidate regions into 227 × 227. And then these wrapped candidate regions are feed into the CNN to extract features for determining whether regions are fire regions or non-fire regions.

4   Experiments

4.1   Datasets

There are no public and standard data sets for fire detection available on Internet. Therefore, we build our own data sets which use some small public data and other fire videos from the Internet and the real life. This dataset is composed of 102 fire and non-fire video clips. There are 16 forest fire clips, 23 indoor fire clips, 22 outdoor fire clips and 41 non-fire clips among them. About 1/4 data is used for testing, the rest for training. So the test set contains 4 forest fire clips, 5 indoor fire clips, 5 out fire clips and 10 non-fire clips. Each video contains 200∼300 frames. For negative samples, we capture some videos containing similar flame, such as including car light, red-moving flag, neon lamp and so on. Figure 7 shows some example frames of the dataset.

Fig. 7. Sample images of different scenarios from the collected data set. Top is positive samples including forest, indoor, outdoor scenarios. Below is negative samples including similar-flame factor. (Color figure online)

4.2   Experimental Results and Discussion

We compare our method with other previous fire detection methods. We also carry out experiments to compare different methods for flame candidate region extraction. We evaluate the methods using the true detection rate and the false alarm rate. The true detection rate (TDR) is defined as:

$$TDR = \frac{TF}{F} \tag{7}$$

where TF is the number of correctly classified frames, which contain fire, in test video; F is number of frames which contain fire in test video.

The false alarm rate(FAR) is defined as:

$$FAR = \frac{FF}{N} \tag{8}$$

where FF is the number of mis-classified frames, which do not contain fire, in test video; N is number of frames which do not contain fire in test video.

In Table 2, we present the results of the true detection rate using RGB color model method, YCbCr color model method and saliency detection method (with and without confidence measures) for 14 test videos including 4 forest fire, 5 indoor fire and 5 outdoor fire clips. We notice that the RGB color rules can achieve better performance on some forest and outdoor fire test data because of containing red-yellow flame. And YCbCr color rules are effective for the bright light of the outdoors or indoors. But on the whole, our method performs better on different fire scenarios.

Table 2. Comparison of color-model methods and saliency detection for the candidate flame regions. TDR indicates true detection rates.

	TDR: true detection rates			
	RGB	YCbCr	Saliency	Saliency+Conf
forest01	163/195(83.5%)	148/195(76.1%)	166/195(85.3%)	172/195(88.2%)
forest02	164/203(81.0%)	153/203(75.6%)	166/203(81.8%)	178/203(87.9%)
forest03	246/295(85.3%)	217/295(73.4%)	254/295(86.1%)	261/295(88.5%)
forest04	189/231(86.2%)	177/231(76.7%)	200/231(86.4%)	203/231(87.7%)
indoor01	178/211(60.5%)	148/211(70.1%)	158/211(74.9%)	171/211(81.2%)
indoor02	150/281(53.5%)	183/281(65.1%)	209/281(74.4%)	206/281(73.3%)
indoor03	125/197(63.7%)	141/197(71.6%)	149/197(75.6%)	152/197(77.2%)
indoor04	114/223(51.1%)	171/223(76.7%)	162/223(72.6%)	165/223(74.2%)
indoor05	96/213(45.1%)	157/213(73.7%)	160/213(75.0%)	178/213(83.4%)
outdoor01	211/241(87.5%)	145/241(60.2%)	212/241(88.1%)	212/241(88.1%)
outdoor02	211/263(85.1%)	187/263(71.1%)	227/263(85.3%)	238/263(90.6%)
outdoor03	168/201(83.6%)	138/201(68.7%)	167/201(83.2%)	175/201(87.1%)
outdoor04	85/181(47.3%)	146/181(80.7%)	145/181(80.2%)	149/181(88.2%)
outdoor05	128/253(50.5%)	199/253(78.8%)	201/253(79.3%)	203/253(80.1%)

We compare our method with covariance matrix-based method [11] which is popular in video-based fire detection. Table 3 shows the true detection rates of the two methods, and Table 4 shows the false alarm rates. The results show that our method performs better than covariance matrix-based method for fire detection.

Table 3. Comparison of our method with covariance matrix-based method (Cov). TDR indicates true detection rates.

	TDR/#of frames			TDR/#of frames	
	Cov	Our method		Cov	Our method
forest01	73.3%(195)	88.2%(195)	indoor04	65.9%(223)	74.2%(223)
forest02	73.7%(203)	87.9%(203)	indoor05	71.8%(213)	83.4%(213)
forest03	75.6%(295)	88.5%(295)	outdoor01	80.4%(241)	88.1%(241)
forest04	72.9%(231)	87.7%(231)	outdoor02	86.3%(263)	90.6%(263)
indoor01	70.9%(211)	81.2%(211)	outdoor03	69.6%(201)	87.1%(201)
indoor02	68.7%(281)	73.4%(281)	outdoor04	70.9%(181)	88.2%(181)
indoor03	71.0%(197)	77.2%(197)	outdoor05	72.1%(253)	80.1%(253)

Table 4. Comparison of our method with covariance matrix-based method (Cov). FAR indicates false alarm rates.

	FAR/#of frames			FAR/#of frames	
	Cov	Our method		Cov	Our method
neg01	0.0%(304)	0.0%(304)	neg06	10.2%(249)	2.3%(249)
neg02	51.9%(216)	53.1%(216)	neg07	0.0%(233)	0.0%(233)
neg03	0.0%(258)	0.0%(258)	neg08	0.0%(347)	0.0%(347)
neg04	5.7%(266)	3.5%(266)	neg09	3.7%(219)	0.8%(219)
neg05	4.3%(318)	0.0%(318)	neg10	0.0%(206)	2.7%(206)

5   Conclusion

In this paper, we propose a new video-based method for fire detection by combining image saliency detection and convolutional neural networks. We use saliency detection algorithm to extract flame candidate regions, which are refined by the fire-confidence and saliency-confidence measures. Compared with color-model methods, saliency detection can adapt to various fire scenes and achieve better detection results generally. In addition, we apply convolutional neural networks to the extracted candidate regions for automatically learning effective features and classification, and the experimental results show that our method outperforms covariance matrix-based method in fire detection.

Acknowledgments. This work is supported by the Fundamental Research Funds for the Central Universities in China (No. 20720170056), the open funding project of State Key Laboratory of Virtual Reality Technology and Systems, Beihang University (Grant No. BUAAVR-14KF-01), and the Science and Technology Project of Quanzhou City (No. 2015G62).

References

1. Chen, T.H., Wu, P.H., Chio, Y.C.: An early fire-detection method based on image processing, In: Proceedings of the IEEE International Conference on Image Processing, ICIP04, vol. 3, pp. 1707–1710 (2004)
2. Celik, T.T., Demirel, H.: Fire detection in video sequences using a generic color model. Fire Safety J. **44**, 147–158 (2009)
3. Miao, L., Wang, A.: Video flame detection algorithm based on region growing. In: 2013 6th International Congress on Image and Signal Processing (CISP), pp. 1014–1018 (2013)
4. Foggia, P., Saggese, A., Vento, M.: Real-time fire detection for videosurveillance applications using a combination of experts based on color, shape, and motion. IEEE Trans. Circuits Syst. Video Technol. **25**, 1 (2015)
5. Horng, W.B., Peng, J.W., Chen, C.Y.: A new image-based real-time flame detection method using color analysis. In: Networking, Sensing and Control, Proceedings, 2005 IEEE, pp. 100–105 (2005)
6. Celik, T.: Fast and efficient method for fire detection using image processing. ETRI J. **32**, 881–890 (2010)
7. Chen, W.H.P.: A new image-based real-time flame detection method using color analysis. Proc. IEEE Netw. **32**, 100–105 (2005)
8. Rangan, M.K., Rakesh, S.M., Sandeep, G.S.P., Suttur, C.S.: A computer vision based approach for detection of fire and direction control for enhanced operation of fire fighting robot. In: 2013 International Conference on Control, Automation, Robotics and Embedded Systems (CARE), pp. 1–6 (2013)
9. Wirayuda, T.A.B., Sthevanie, F., Widowati, S.: Fire color detection using color look up and histogram analysis. In: 2013 International Conference of Information and Communication Technology (ICoICT), pp. 134–139 (2013)
10. Guang, L.Z., Yang, Y., Hua, J.X.: Flame detection algorithm based on a saliency detection technique and the uniform local binary pattern in the YCbCr color space. Signal Image Video Process. **10**, 1–8 (2015)
11. Habiboğlu, Y.H., Günay, O., Çetin, A.E.: Covariance matrix-based fire and flame detection method in video. Mach. Vis. Appl. **23**, 1103–1113 (2012)
12. Wu, R., Wang, B., Wang, W., Yu, Y.: Harvesting discriminative meta objects with deep CNN features for scene classification. In: Proceedings of the IEEE International Conference on Computer Vision, pp. 1287–1295 (2015)
13. Cao, Z., Principe, J.C., Ouyang, B., Dalgleish, F.: Marine animal classification using combined CNN and hand-designed image features. in Oceans (2015)
14. Huynh, C.K., Le, T.S., Hamamoto, K.: Convolutional neural network for motorbike detection in dense traffic. In: IEEE Sixth International Conference on Communications and Electronics (2016)
15. Girshick, R., Donahue, J., Darrell, T., Malik, J.: Rich feature hierarchies for accurate object detection and semantic segmentation. In: Computer Science, pp. 580–587 (2014)
16. Goferman, S., Zelnikmanor, L., Tal, A.: Context-aware saliency detection. IEEE Trans. Pattern Anal. Mach. Intell. **34**, 1915–1926 (2012)
17. Cheng, M.M., Zhang, G.X., Mitra, N.J., Huang, X., Hu, S.M.: Global contrast based salient region detection. In: Computer Vision and Pattern Recognition, pp. 409–416 (2011)
18. Perazzi, F. Krähenbühl, P., Pritch, Y., Hornung, A.: Contrast based filtering for salient region detection. In: IEEE Conference on Computer Vision and Pattern Recognition, pp. 733–740 (2012)

19. Wang, K., Lin, L., Lu, J., Li, C., Shi, K.: PISA: Pixelwise image saliency by aggregating complementary appearance contrast measures with edge-preserving coherence. IEEE Trans. Image Process. **9**, 2115–2122 (2015)
20. Krizhevsky, A., Sutskever, I., Hinton, G.E.: Imagenet classification with deep convolutional neural networks. Adv. Neural Inf. Process. Syst. **25**, 2012 (2012)

Leveraging Convolutions in Recurrent Neural Networks for Doppler Weather Radar Echo Prediction

Sonam Singh[1]($\boxtimes$), Sudeshna Sarkar[2], and Pabitra Mitra[2]

[1] Advanced Technology Developement Centre, Indian Institute of Technology,
Kharagpur 721302, WB, India
`sonam@iitkgp.ac.in`
[2] Department of Computer Science and Engineering, Indian Institute of Technology,
Kharagpur 721302, WB, India

Abstract. Precipitation forecasting for short duration is an important problem in weather prediction. In this work, we propose a deep learning based approach for precipitation forecasting using Doppler weather radar data. Our approach uses convolutions within recurrence structure in vanilla recurrent neural networks exploiting both spatial and temporal dependencies in the data. We show that this approach can be applied for fine grained precipitation forecast with similar accuracy as that of complex models while reducing the model size by 4 times. Results are presented on the task of echo state prediction and skill scores for rainfall estimates on the data from Seattle, WA, USA as well as from cross testing the model, trained on Seattle data, on unseen data from Albany, NY, USA.

1 Introduction

Estimating precipitation for short and long duration is an important part of weather prediction. Doppler Weather Radar (DWR) is commonly used for precipitation nowcasting as an input for Numerical Weather Prediction(NWP) models for better accuracy. DWR provides high resolution echo maps covering a large area which can be converted into rainfall estimates and for hydrometeor classification etc. These estimates can be used for planning by civil authorities, as well as in critical services like disaster management in emergency situations. Data from DWR is also used in calibration and forecasts from NWP.

Prediction of future state of radar echo maps has been studied recently from machine learning point of view and can be formulated as a sequence forecasting problem [9]. Spatio temporal prediction from deep learning perspective is a relatively new area and recent works addressing this problem use some combination of Convolution Neural Network (CNN) and Recurrent Neural Network (RNN) structures [1,8,9]. Commonly used architectures in RNNs for learning long term dependencies use some variant of Long Short Term Memory (LSTM)[3]. Combination of convolutions with recurrent structures is a hybrid learning approach for

© Springer International Publishing AG 2017
F. Cong et al. (Eds.): ISNN 2017, Part II, LNCS 10262, pp. 310–317, 2017.
DOI: 10.1007/978-3-319-59081-3_37

spatio-temporal sequence learning. In these approaches, convolutions are either used before recurrence units or within recurrence. These hybrid structures can also be stacked to learn hierarchical spatio-temporal dependencies.

In this work, we propose to use convolutions within vanilla recurrent units for a sequence to sequence forecasting problem of DWR echo state prediction and show that even with sharing of convolution kernels in plain recurrent structure of vanilla RNN, we can predict radar echo maps with high accuracy. We compare against strong baselines and test our model on Seattle, USA radar echo dataset. We want to see how transferable the model is which is learnt on Seattle data. To test the Seattle trained model's generalization ability, we test it on Albany, NY data also, which model hasn't seen during training and report results on it too.

2   Related Work

Short duration precipitation prediction, also known as precipitation nowcasting, is commonly performed using some variant of physical model which incorporate the characteristics of regional atmosphere and geography of the area. Earlier work like Li et al. [5] uses motion vectors from Doppler radar while Madapaka et al. [6] uses a Lagrangian extrapolation based method. Current methods use flow based techniques for this problem [2].

From machine learning point of view, recent work from Shi et al. [9] formulate the problem of precipitation nowcasting as a sequence to sequence forecasting. They used convolutions in LSTM to predict radar echo states for nowcasting in Hong Kong region.

Using convolutions and recurrent units together have been explored from different perspectives and for different tasks. Recently, Nicolas et al. [1] used convolutions in GRU (Gated Recurrent Unit) for various tasks like activity recognition, video captioning etc. Shi et al. [9] and Nicolas et al. [1] both use convolutions *within* recurrence where convolution kernels are shared across both the space and time. Ji et al. [4] uses 3D kernels with convolutions before recurrence starts.

3   Motivation

Our motivation for this work is to see if convolutions can work in vanilla recurrent units where we use convolution operations instead of usual multiplication operation in recurrent network. Using convolutions with vanilla recurrent network reduces parameters by 4 times as compared to using with LSTM structure. This leads to faster training and reduced overall model size for efficient deployment to resource critical environments.

We want to explore such an approach for the task of DWR echo state prediction which is used for high resolution precipitation forecasting.

4   Our Approach

4.1   Preliminaries

Recurrent Neural Network. Recurrent neural network has shared weights across time to capture dependency and have an implicit notion of ordering within the input.

Vanilla RNN has recurrent structure for input $\mathcal{X}_t$, at time t, where weights $\mathcal{W}_{ih}, \mathcal{W}_{hh}, \mathcal{W}_{hy}$ are shared across time.

$$h_t = \sigma(\mathcal{W}_{ih}\mathcal{X}_t + \mathcal{W}_{hh}h_{t-1} + b_{ih}) \tag{1}$$

$$y_t = f(\mathcal{W}_{hy}h_t + b_{hy}) \tag{2}$$

where $\mathcal{X}_t, h_t, y_t$ are the input, hidden state and output at time t. Usually $\sigma(x) = \tanh(x)$ and $f(x) = sigmoid(x)$. $W_{ih}, W_{hh},$ and b_{hy}, b_{hh} are the weight matrices and biases respectively.

Convolution Neural Network. A convolution neural network (CNN) has kernels shared across space in the input.

$$X_s = \mathcal{X}_t * \mathcal{W}_s \tag{3}$$

where $\mathcal{W}_s$ is the shared kernel, X_s is the feature map generated using convolution operator $*$ from input $\mathcal{X}_t$.

4.2   Proposed Hybrid Structure

Convolutional Recurrent Neural Network. We combine both of these approaches in a hybrid structure which exploits both spatial and temporal strengths of the above methods.

In Convolutional Recurrent Neural Network, recurrent structure is,

$$h_t = \sigma(\mathcal{W}_{ih} * \mathcal{X}_t + \mathcal{W}_{hh} * h_{t-1} + b_{ih}) \tag{4}$$

$$y_t = f(\mathcal{W}_{hy} * h_t + b_{hy}) \tag{5}$$

where $*$ is the convolution operator used in CNN. Similar to RNN weights, here *kernels* are shared across time.

In convolutional recurrent structure (Eqs. 4, 5), inputs ($\mathcal{X}_t$), hidden states (h_t) and outputs (y_t) all are $3D$ tensors of dimensions (channels, height, width).

Similar to vanilla recurrent neural networks, such convolutional recurrent structure can be stacked hierarchically to form layers. To maintain the consistency between consecutive layers,

$$channels_o(\mathcal{L}_{l-1}) = channels_i(\mathcal{L}_l) \tag{6}$$

where $channels_o = output\ channels$ and $channels_i = input\ channels$, $\mathcal{L}_l = l^{th}$ layer.

The architecture used in this paper is shown in Fig. 1. We refer feature maps and hidden states interchangeably and both mean the same in the description below. Details of the architecture are described below,

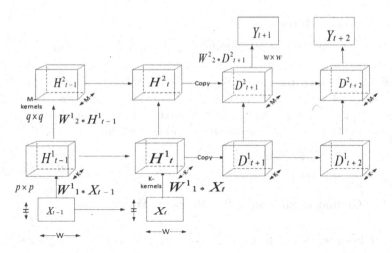

Fig. 1. Multi-layer convolutional RNN encoder-decoder architecture.

1. The architecture shown is for 2 hidden layers only. It may be extended to more layers easily.
2. In our task, both the input and output are sequences of images with 1 channel (Grey scale) which we show in this Fig. 1. Though the same structure can be used for many channels (RGB etc.).
3. Number of time steps are t, t−1 in input and t+1, t+2 for the output. It may be different for encoder and decoder.
4. The last hidden states, H_t^1 and H_t^2 are copied from encoder to decoder.
5. In decoder, D_{t+1}^1 and D_{t+1}^2 are the copied states from encoder and D_{t+1}^2 is conditioned on D_{t+1}^1 too.
6. p, q, w are the kernel sizes in input to hidden, hidden to hidden and hidden to output layers respectively.
7. Number of feature maps for layer 1 is K and layer 2 is M.
8. In decoder, Y_{t+1} and Y_{t+2} are the outputs from the model.
9. W_1^1 denotes input to hidden layer kernels in encoder. Similarly, W_2^2 denotes hidden to output kernels in decoder.

5 Doppler Weather Radar Echo State Prediction

5.1 Problem Details

Radar Echo Prediction. In this problem, we attempt to estimate future states of the radar echo using previous states. The number of future states corresponds to the length of prediction, and time interval between the states sampling in past data which may vary from 5 min to 20 min on a given day. The predicted radar echo states can be converted to rainfall rate using Marshall-Palmer [7] formula: $\mathcal{Z} = aR^b$ where $\mathcal{Z}$ = Reflectivity (dBZ), R = Rainfall rate (mm/h), a, b = constants. We have a = 200, b = 1.6 from American Meteorological Society[1].

---

[1] http://glossary.ametsoc.org/wiki/Marshall-palmer-relation.

5.2   Evaluation Metrics

We discuss some metrics used for evaluating prediction performance.

Binary Cross Entropy Loss. Since, radar echo states are predicted with normalized pixel values between $(0, 1)$, we use binary cross entropy loss which reduces the pixel level deviation between predicted and ground truth. We use binary cross entropy loss for optimization and evaluation.

$$loss = -\frac{1}{\sum_{s,t,p}} \overset{sequences\ time\ pixels}{\sum_s \sum_t \sum_p} T_{s,t,p} \log P_{s,t,p} + (1 - T_{s,t,p}) \log(1 - P_{s,t,p}) \quad (7)$$

where T = Ground Truth Frame, P = Predicted Frame.

Binary Classification Metrics. As our task is to predict high resolution rainfall estimates, we evaluate our predictions using binary classification metrics also.

We use a threshold of 2 mm/h to convert rainfall rates to binary values $\{0, 1\}$ representing absence of rainfall (N) and rainy events (P) respectively. For tp = true positive (predicted = P, truth = P), tn = true negative (predicted = N, truth = N), fp = false positive (predicted = P, truth = N) and fn = false negative (predicted = N, truth = P), we evaluate using precision, recall and F1 score.

6   Experimental Detail

6.1   Datasets

We use Doppler radar data from Seattle, USA and Albany, NY, USA[2]. The Albany dataset is used for cross evaluating our proposed approach to see whether the model trained on dataset can be used for predictions at other locations.

For the current task, we consider only reflectivity at 10 time steps ahead for prediction and input of past 10 time steps. Time duration between consecutive radar echo varies from 5–15 min.

Seattle Dataset: We select top 83 rainy days (similar strategy as in [9]) between year 2008 to 2015 from Seattle, USA (NEXRAD code: KATX). Using overlapping strategy for splitting, we have 10,577 sequences with 20 radar scans in each sequence. We use 7000 for training, 2000 for testing and 1577 for validation.

Albany Dataset: For robust testing, we cross-evaluate our best model which is trained on Seattle data only and test it on Albany dataset and compare against baselines. For this dataset, we select the top 64 rainy days data from Albany, NY (NEXRAD code: KENX) between year 2009 to 2014. We use the same splitting strategy as we do for Seattle dataset. This dataset has 12046 sequences with 20 radar scans in every sequence. The Albany dataset is used for cross evaluation of the models only.

---

[2] https://aws.amazon.com/noaa-big-data/nexrad/.

6.2   Methodology

DWR data has reflectivity echoes at different elevation angles. We use mean projection to lowest elevation which captures the reflectivity variation along the altitude. We resize echo maps from 720×1832 (polar) to 100×100 (cartesian). This projection preserves the overall motion and shape information.

We use 10 Doppler radar images of 100×100 for input and predict for the next 10 radar images of the same size. We optimize using RMSProp [10] with *learning rate* $= 0.001$.

We use Eulerian persistence as a trivial baseline which involves taking last frame as the predicted frame for all time steps. We evaluate Convolutional RNN (Conv-RNN) in single layer and stacked structure using different kernel sizes and number of feature maps and compare against Convolutional LSTM (Conv-LSTM).

7   Results

7.1   Seattle Dataset

We experiment with varying number of feature maps to (32, 64 and 128) and kernel sizes 3×3, 5×5, 9×9 for Conv-RNN. We use the same kernel size across all convolution operations for a given model. The same kernel size gives increasing receptive field when used in hierarchical structure.

Table 1. Comparison of Conv-RNN in stacked structure with different layer sizes and baseline models. $p \times p$ denotes kernel size p (same kernel used across all layers) and $Conv - Model - M - N-$ denotes two layers with number of feature maps = M and N respectively corresponding to that particular '*Model*'.

Model	Aggregate loss
Eulerian persistence	0.060
Conv-RNN- 32-(3×3)	0.0443
Conv-RNN- 128 -(9×9)	0.0457
Conv-RNN-64-64-(5×5)	0.0440
Conv-RNN-32-64-(3×3)	0.0433
Conv-RNN-64-64-(3×3)	**0.0430**
Conv-LSTM-64-64-(3×3)	**0.0429**

Results for different architectures are shown in Table 1 and binary classification metrics shown in Fig. 2. In Fig. 2, horizontal axis is the number of time steps in future ahead. We show some of the predictions for sampled time steps ($t = 0, 3, 6, 8$) from a test sequence in Fig. 3.

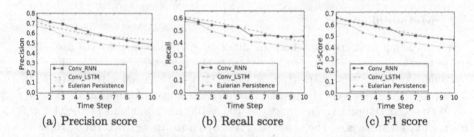

(a) Precision score   (b) Recall score   (c) F1 score

Fig. 2. Evaluation metrics for Seattle dataset

Fig. 3. Model predictions from Seattle dataset. Top row is ground truth and bottom row is the corresponding predictions.

7.2   Cross Testing on Albany Dataset

We take our best models i.e. Conv-LSTM-64-64-(3×3) and Conv-RNN-64-64-(3×3) respectively, trained on Seattle dataset and test it on Albany dataset. Table 2 shows the loss and skill scores for this dataset.

Table 2. Model Cross-evaluation on Albany dataset

Model	Loss (Binary cross entropy)	Precision	Recall	F1 score
Conv-RNN-64-64-(3×3)	**0.0439**	**0.60**	**0.51**	**0.55**
Conv-LSTM-64-64-(3×3)	0.0437	0.61	0.52	0.56

8   Discussion

Results from Table 1 suggest that multi layer Conv-RNN outperforms single layer architecture with high margin. Conv-RNN with 3×3 kernel performs best among all kernel sizes which is intuitive as our task require high spatial precision for prediction. The difference between Conv-RNN-64-64-(3×3) and Conv-LSTM-64-64-(3×3) is 0.0001 in aggregate loss while the number of parameters is reduced by 4 times.

Figure 2 shows binary classification metrics. The result is in line with our intuition that accuracy of the Conv-RNN model overall decreases with increasing number of future states ahead. Conv-RNN and Conv-LSTM models have precision, recall and F1-score with a deviation of 7% with major deviation in last time steps.

Our cross testing experiment on Albany dataset (Table 2) suggest that Conv-RNN is able to learn general features from Doppler weather radar phenomenon. Though, we cannot claim that it will generalize well on other places too due to limited scope of these experiments which needs to be investigated further.

9 Conclusion

In this paper we proposed Convolution-RNN. Convolutional structures in vanilla RNN use lesser parameters compared to other hybrid approaches. One benefit of convolutions in recurrence is that it explicitly encodes the spatio-temporal correlations using kernels shared across both time and space.

The problem of weather radar echo prediction, we attempt here, involves complex interdependencies among various factors like topography, atmospheric conditions etc. This make it really difficult to accurately predict the future states. But high spatial and temporal correlation exist between successive states which can reduce uncertainty in predictions. We attempt to exploit these correlations using conv-RNN architecture proposed in this paper.

References

1. Ballas, N., Yao, L., Pal, C., Courville, A.C.: Delving deeper into convolutional networks for learning video representations. CoRR abs/1511.06432 (2015). http://arxiv.org/abs/1511.06432
2. Bowler, N.E., Pierce, C.E., Seed, A.: Development of a precipitation nowcasting algorithm based upon optical flow techniques. J. Hydrol. **288**(1), 74–91 (2004)
3. Hochreiter, S., Schmidhuber, J.: Long short-term memory. Neural Comput. **9**(8), 1735–1780 (1997)
4. Ji, S., Xu, W., Yang, M., Yu, K.: 3D convolutional neural networks for human action recognition. IEEE Trans. Pattern Anal. Mach. Intell. **35**(1), 221–231 (2013)
5. Li, L., Schmid, W., Joss, J.: Nowcasting of motion and growth of precipitation with radar over a complex orography. J. Appl. Meteorol. **34**(6), 1286–1300 (1995)
6. Mandapaka, P.V., Germann, U., Panziera, L., Hering, A.: Can lagrangian extrapolation of radar fields be used for precipitation nowcasting over complex alpine orography? Weather Forecast. **27**(1), 28–49 (2012). http://dx.doi.org/10.1175/WAF-D-11-00050.1
7. Marshall, J.S., Palmer, W.M.K.: The distribution of raindrops with size. J. Meteorol. **5**(4), 165–166 (1948)
8. Pinheiro, P.H., Collobert, R.: Recurrent convolutional neural networks for scene labeling. In: ICML, pp. 82–90 (2014)
9. Shi, X., Chen, Z., Wang, H., Yeung, D., Wong, W., Woo, W.: Convolutional LSTM network: a machine learning approach for precipitation nowcasting. CoRR (2015). http://arxiv.org/abs/1506.04214
10. Tieleman, T., Hinton, G.: Lecture 6.5-rmsprop: divide the gradient by a running average of its recent magnitude. COURSERA: Neural Netw. Mach. Learn. **4**, 2 (2012)

Neuroadaptive PID-like Fault-Tolerant Control of High Speed Trains with Uncertain Model and Unknown Tracking/Braking Actuation Characteristics

Q. Song[1,2] and T. Sun[1,2(✉)]

[1] State Key Lab of Rail Traffic Control and Safety, Beijing Jiaotong University,
Beijing, China
qsong@bjtu.edu.cn
[2] 706 Institute of the Second Academy of CASIC, Beijing, China
suntaohe4379@163.com

Abstract. The work focus on the position and velocity tracking control problem of high speed train with uncertain system model and external disturbance as well as unknown traction/braking actuation characteristics. Neuroadaptive Proportion-integral-derivative (PID)-like fault-tolerant control algorithms are developed to achieve uniformly ultimately bounded (UUB) stable position/velocity tracking control of high speed train by using a well defined smooth function. Unlike the traditional PID control, the resultant control scheme is of PID structure and able to deal with unknown system parameters and nonlinearities and actuator failures without the need for any "trial and error" process to determine the PID gains. The effectiveness of the proposed control strategy is confirmed by theoretical analysis and numerical simulations.

Keywords: Neural network · PID-like high speed train · Input nonlinearities · Actuator failure

1 Introduction

Safe and reliable operation of high speed train rely on advanced automatic train operation system (ATO). The core problem of ATO is how to real-time control the train system to get the control objectives that consist of making the train track the pre-planned 'ideal' speed-position curve under different environments via certain control algorithms. As a simple and reliable control method with easy adjustment, Proportion-integral-derivative (PID)-based controllers are widely used in engineering systems consisting of vehicle systems [1–4]. Besides, fuzzy control was also proposed for train systems these years [5,6]. Neural network(NN) attracts more and more attention in intelligent control system design, owing to its unique capabilities in approximating any nonlinear function with arbitrary precision if the NN is constructed properly, and there are some successful applications of NN in industrial systems with nonlinearities and uncertainties [7,8].

© Springer International Publishing AG 2017
F. Cong et al. (Eds.): ISNN 2017, Part II, LNCS 10262, pp. 318–325, 2017.
DOI: 10.1007/978-3-319-59081-3_38

However, the system model or system parameters are required as a priori in some of the methods. In [9], a data-based neroadaptive control method is developed, where traction/braking dynamics and the uncertain nature of the resistance coefficients are addressed explicity.

It is noted that, as a nonlinear nonaffine system, it is very difficult to get the precise system model of high speed train, and there are nonlinearities in basic resistance. Also, the nonlinear impact on system dynamics becomes increasingly significant. Besides, precise determination of the key parameters such as train mass and rotary mass coefficient as well as resistance coefficients is very difficult, if not impossible, to obtain in practice because such coefficients depend on many factors such as vehicle type, structure, train speed, weather condition, wind speed, wind direction, friction between vehicle surface and air, wheel and rail, etc. [10]. Furthermore, actuation faults may occur during the system operation, it is important to address the fault-tolerant control issue explicitly. Evidently, the velocity and position tracking control problem of high speed train becomes interesting yet challenging when concurrently considering the uncertainties and nonlinearities.

Motivated by the work of [11], we develop a neuroadaptive PID-like control method for high speed train in the presence of modeling uncertainties and unknown tracking/braking actuation characteristics and unexpected actuator failures.

2   Modeling and Problem Statement

By Newton's law, the dynamic motion equation for each vehicle can be established as $(1 + r_i)m_i\ddot{x}_i = \lambda_i f_{ai} - f_{di} + f_{in_{i-1}} - f_{in_i}$, where $r_i \geq 0$ represents the rotary mass coefficient of the ith vehicle. During the whole operation, r_i is not always considered [13]. m_i is the mass of the ith vehicle. x_i denotes the distance between the center of the ith vehicle and the reference point. $\lambda_i \geq 0$ is a distribution constant determining the power/braking effort of the ith vehicle. f_{ai} is the actual traction/braking force of the ith vehicle and the designed traction/braking force f_i is not identical anymore, instead, they are related to through $f_{ai} = \rho_i \psi_i(f_i) + \upsilon_i$. $\psi_i(f_i)$ is traction/braking force with unknown actuation characteristics, which has two typical models- asymmetric nonsmooth saturation with unknown slope and asymmetric nonsmooth saturation with dead zone [12]. $0 \leq \rho_i \leq 1$ is the "powering/braking health indicator" for the ith vehicle, which has the same physical meanings as in [9]. To address the control design problem, we consider the case that some or all actuators suffer from partial actuation failures, that is, $0 < \rho_i \leq 1$. $|\upsilon_i| \leq \bar{\upsilon}_i < \infty$ is the uncertain part caused by the actuator failure. f_{di} consists of mechanical resistance and aerodynamic resistance which can be described as $f_{di} = a_0 + a_1\dot{x}_i + a_2\dot{x}_i^2 + f_{ri} + f_{ci} + f_{ti}$, where a_0, a_1 and a_2 are the resistance coefficients for the ith vehicle, f_{ri}, f_{ci} and f_{ti} are additional resistances. Significant nonlinearities and uncertainties are involved in f_{di}, particularly for high-speed and long distance operations, which should explicitly be addressed in the control design for high speed trains. However, it is reasonable to assume that f_{ci}, f_{ti}, and f_{di} are bounded. f_{in_i} is the in-train force from the couplers connecting the adjacent vehicles and obviously $f_{in_0} = 0$ and $f_{in_n} = 0$ because no such in-train force exists for the first and the last vehicle. Actually, it is extremely difficult to model or precisely measure f_{in_i} due to the nonlinear and elastic nature of the couplers. [9] has tried to develop control algorithms to compensate the in-train forces based on some assumptions. In the work, the constraints and assumptions are no longer needed.

For simple notation, we rewrite the above model equations as:

$$(I + R)M\ddot{X} = \Lambda F_a - F_d + (T - I)F_{in} \qquad (1)$$

and $F_a = \rho\Psi(F) + \Upsilon$, where $R = diag(r_i)$, $M = diag(m_i)$, $\Lambda = diag(\lambda_i)$ is the power/braking effectiveness distribution matrix, $F_a = [f_{a1}, f_{a2}, \ldots, f_{an}]^T$, $F_d = [f_{d1}, f_{d2}, \ldots, f_{dn}]^T$ and $F_{in} = [f_{in_1}, f_{in_2}, \ldots, f_{in_n}]^T$. $\dot{X}$ is the velocity vector, $\ddot{X}$ is the acceleration vector, and $\begin{bmatrix} 0_{1*(n-1)} & 1 \\ I_{(n-1)*(n-1)} & 0_{(n-1)*1} \end{bmatrix}$, with I being the identity matrix. $\rho = diag\{\rho_i\}$, $\Psi(F) = [\psi_1(f_1), \psi_2(f_2), \ldots, \psi_n(f_n)]^T$, and $\Upsilon = [v_1, v_2, \ldots, v_n]^T$. $\|\Upsilon\| \leq \bar{v} < \infty$ because of $|v_i| \leq \bar{v}_i < \infty$, and $\bar{v}$ is an unknown constant. For all of the above symbolic description, $i = 1, 2, \ldots, n$. For simplicity, let $\bar{M} = (I + R)M$. It is noted that $\bar{M}$ is reversible, then the above model (1) can be reformed as

$$\ddot{X} = \bar{M}^{-1}\Lambda\rho\Psi(F) + \bar{M}^{-1}\Lambda\Upsilon + \bar{M}^{-1}[-F_d + (T - I)F_{in}] \qquad (2)$$

The control objective is to design control force $F = [f_1, f_2, \ldots, f_n]$ so that, for any given desired velocity-displacement $\dot{X}^* - X^*$ pair, it is ensured that $\|E\| \leq e_0 < \infty$ and $\|\dot{E}\| \leq e_1 < \infty$ for some small constants e_0 and e_1 related to control precision (uniformly ultimately confined within a small compact set containing the origin), where $E = X - X^*$ and $\dot{E} = \dot{X} - \dot{X}^*$ denote the position tracking error and velocity tracking error respectively. $X = [x_1, x_2, \ldots, x_n]^T$ and $\dot{X} = [\dot{x}_1, \dot{x}_2, \ldots, \dot{x}_n]$ are the displacement vector and velocity vector respectively, and $\dot{X}^*$ and X^* are the desired velocity and position, assumed to be smooth and bounded, produced from the train operation planning unit.

Remark 1. It should be emphasized that the precise information of r_i is difficult to measure precisely, and considering r_i makes sense only the vehicle acceleration/deceleration is not zero [13]. Besides, precise determination of train mass and resistance coefficients is very difficult. It is important and challenging to develop a control method without using the precise information of these parameters and coefficients.

3   Control Design and Stability Analysis

To facilitate the PID-like controller design, we define a filtered variable first, i.e., $S = \dot{E} + \beta E$, in which $\beta = diag(\beta_i)$ for $i = 1, 2, \ldots, n$, and $\beta_i > 0$ is chosen by the designer. It ban be readily shown that the boundedness of S implies the boundedness of E and $\dot{E}$ [14]. E and $\dot{E}$ are defined as before. In order to carry out the analysis, a generalized error $\bar{S}$ is defined as $\bar{S} = S + \alpha \int_0^t S d\tau$, with $\alpha = diag(\alpha_i)$, $\alpha_i > 0$, $i = 1, 2, \ldots, n$. Before establishing the main results, the following lemma is needed. In light of the definition S and $\bar{S}$, one has from the system model (2) that

$$
\dot{\bar{S}} = \bar{M}^{-1} \Lambda \rho \Psi(F) + \bar{M}^{-1} \Lambda \Upsilon + \bar{M}^{-1}[-F_d + (T - I)F_{in}] \\
+ L(X, X^*, t) + \alpha S \tag{3}
$$

in which $L(X, X^*) = -\ddot{X}^* + \beta(\dot{X} - \dot{X}^*)$.

Remark 2. Clearly, the relationship between the applied control $\psi_i(f_i)$ and the control input f_i has a sharp corner when f_i reaches the saturation value or break point [12]. In order to develop the PID-like technique, the saturation is approximated by a smooth function defined as $g_i(f_i) = \frac{f_{gi} e^{\omega_i f_i} - f_{li} e^{-\omega_i f_i}}{e^{\omega_i f_i} + e^{-\omega_i f_i}}$ in which the parameter $\omega_i > 0$ is chosen by designer, and different values of ω_i may lead to different approximation to $\psi_i(f_i)$, and $f_{gi} > 0$ and $f_{li} > 0$ are the uncertain saturation value of f_i.

Then with the help of $g_i(f_i)$, $\psi_i(f_i)$ can be expressed as $\psi_i(f_i) = g_i(f_i) + \delta_i(f_i)$, where $\delta_i(f_i) = \psi_i(f_i) - g_i(f_i)$ is the approximate error and it is a bounded function in time and its bound can be obtained as $|\delta_i(f_i)| = |\psi_i(f_i) - g_i(f_i)| \leq max\{f_{gi}, f_{li}\} + f_{li} = \delta_{i0}$, in which δ_{i0} is an uncertain positive constant. $i = 1, 2, \ldots, n$. Therefore, it can be established that $\Psi(F) = G(F) + \Delta(F)$, where $G(F) = [g_1(f_1), g_2(f_2), \ldots, g_n(f_n)]^T$, and $\Delta(F) = [\delta_1(f_1), \delta_2(f_2), \ldots, \delta_n(f_n)]^T$. With $|\delta_i(f_i)|$, it can be obtained that $||\Delta(F)|| \leq \delta_0$, where δ_0 is an unprecise positive constant. Furthermore, since $G(F)$ is a smooth function of F, it can be obtained $G(F) = G(0) + G(\xi)F$, with using Lagrange's mean value theorem,

where $G(\xi) = \begin{bmatrix} \frac{\partial g_1}{\partial f_1}|_{\xi_{11}} & \cdots & \frac{\partial g_1}{\partial f_n}|_{\xi_{1n}} \\ \vdots & \ddots & \vdots \\ \frac{\partial g_n}{\partial f_1}|_{\xi_{n1}} & \cdots & \frac{\partial g_n}{\partial f_n}|_{\xi_{nn}} \end{bmatrix}$, if $f_j \geq 0$, $\xi_{ij} \in (0, f_j)$, and, if $f_j < 0$,

$\xi_{ij} \in (f_j, 0)$. In view of $\Psi(F)$ and $G(F)$ with using the generalized error dynamic Eq. (3), we get that

$$
\dot{\bar{S}} = \bar{M}^{-1} \Lambda \rho G(\xi) F + \bar{M}^{-1} \Lambda \rho G(0) + \bar{M}^{-1} \Lambda \rho \Delta(F) + \bar{M}^{-1} \Lambda \Upsilon \\
+ \bar{M}^{-1}[-F_d + (T - I)F_{in}] + L(X, X^*) + \alpha S \tag{4}
$$

Let $H = \bar{M}^{-1} \Lambda \rho \Delta(F) + \bar{M}^{-1} \Lambda \Upsilon + \bar{M}^{-1}[-F_d + (T - I)F_{in}] + L(X, X^*) + \alpha S$, then (4) can be readily expressed as

$$
\dot{\bar{S}} = \bar{M}^{-1} \Lambda \rho G(\xi) F + H(X, X^*, F) \tag{5}
$$

where $G(0) = 0$ is used. Note that although H is a nonlinear function of (X, X^*, F) and totally unavailable, it satisfies $||H|| \leq ||\bar{M}^{-1} \Lambda \rho|| \delta_0 + ||\bar{M}^{-1} \Lambda|| \bar{\upsilon} + ||\bar{M}^{-1}[-F_d + (T - I)F_{in}]|| + ||L(X, X^*)|| + \alpha ||S|| = \varpi(X, X^*)$.

Remark 3. It should be pointed out that it is not advisable to use NNs to approximate H directly, because H is a vector nonlinear function of F. If we use NNs to approximate H directly, it may not only create notorious algebra-loop problem, but also involve heavy computation when the controlled system is of higher dimension. In order to avoid this problem, NNs are employed online to approximate the scalar nonlinear function $\varpi(X, X^*)$ which is interdependent of F.

RBF NNs are popular for their simplicity, fast learning, and universal approximation properties. In order to develop the neuroadaptive PID-like control, RBF NNs are used to approximate the upper bound of H as: $\varpi = W^T \Phi(Z) + \kappa(Z)$. $\Phi(Z) = [\phi_1(Z), \phi_2(Z), \ldots, \phi_N(Z)]^T \in R^N$ is the basic function with $Z = [X, X^*]^T$ (measurement noises are not considered in this paper), and $W \in R^N$ is the optimal constant weight vector. $\phi_k(.) = e^{-\sum_{j=1}^{q}(z_j - c_{kj})^2/2\sigma_{kj}^2}$ ($k = 1, 2, \ldots, N$), where $Z = [z_1, \ldots, z_q]^T$ is the input vector of NN. $C_{kj} = [c_{k1}, \ldots, c_{kq}]^T$ and $Q_{kj} = [\sigma_{k1}, \ldots, \sigma_{k1}]^T$ are the center states and standard deviations of Gaussian associated with each element of the input vector, respectively, and N is the number of hidden-layer neurons. By the universal approximation theory, it is reasonable to assume that the NN reconstruction error $|\kappa| \le \kappa_0 < \infty$. For control design in the forllowing subsection, further treatment on ϖ is needed as $|\varpi| \le \mu(1 + \Phi(Z))$, with $\mu = max\{\|W^T\|, \kappa_0\}$ is a non-negative constant. Let $\Theta = \bar{M}^{-1} \Lambda \rho G(\xi)$, then (5) can be expressed as $\dot{S} = \Theta F + H$.

With the definition of $\bar{M}^{-1}$, Λ, ρ, and $G(\xi)$, it can be shown that Θ is an unknown and time-varying $n \times n$ square matrix, which make the neuroadaptive PID-like fault-tolerant control design possible. Furthermore, define $\aleph = (\Theta + \Theta^T)/2$, and $\bar{\aleph} = (\Theta - \Theta^T)/2$. It can be seen that $\aleph^T = \aleph$, and $\bar{\aleph}^T = -\bar{\aleph}$, which implies $\aleph$ is symmetric and $\bar{\aleph}$ is skew symmetric. As most existing works in addressing MIMO systems, $\aleph$ is either positive definite or negative definite here, which guarantees that the system considered in the work is controllable. Without loss of generality, we consider that $\aleph$ is positive definite, such that for all X is the domain of interest, there exists some unknown positive constant ζ satisfying $0 < \zeta \le min\{eig(\aleph)\}$. The analysis above leads to the following theorem.

Theorem. Consider the train dynamics as described by (1). If the following control algorithm is applied

$$F = k\alpha E(0) - k(\alpha + \beta)E - k\alpha\beta \int_0^t E d\tau - k\dot{E} \tag{6}$$

and $k = c_0 + c_1 \hat{\mu}(1 + \Phi(Z))^2$, with $\dot{\hat{\mu}} = -c_2 \hat{\mu} + c_1(1 + \Phi(Z))^2 \|\bar{S}\|^2$, where α and β are defined as before. $c_0 > 0$, $c_1 > 0$ and $c_2 > 0$ are chosen by the designer. $\hat{\mu}$ is the estimation of $\mu = max\{\|W\|, \kappa_0\}$. Then the velocity and position tracking error E and $\dot{E}$ is ensured to be ultimately uniformly bounded (UUB), and all the internal signals in the system are bounded, and the control signal is continuous and smooth everywhere.

Proof. Consider the Lyapunov function candidate as

$$V = \frac{1}{2}\bar{S}^T\bar{S} + \frac{1}{2\zeta}(\mu - \zeta\hat{\mu})^2 \tag{7}$$

It can be shown with the proposed control algorithms and $|\varpi| \leq \mu(1+\Phi(Z))$, $\dot{\bar{S}}$, S and $\bar{S}$ that

$$\dot{V} \leq -k\bar{S}^T\Theta\bar{S} + \mu(1 + \Phi(Z))\|\bar{S}\| - \dot{\hat{\mu}}(\mu - \zeta\hat{\mu}) \tag{8}$$

Upon using the definition of $\aleph = (\Theta + \Theta^T)/2$ and $\bar{\aleph} = (\Theta - \Theta^T)/2$ and $0 < \zeta \leq min\{eig(\aleph)\}$, we further have

$$\dot{V} \leq -(c_0 + c_1\hat{\mu}(1 + \Phi(Z))^2)\zeta\|\bar{S}\|^2$$
$$+\mu[c_1(1 + \Phi(Z))^2\|\bar{S}\|^2 + \frac{1}{4c_1}] - \dot{\hat{\mu}}(\mu - \zeta\hat{\mu}) \tag{9}$$

where $\frac{1}{2}\bar{S}^T\bar{\aleph}S = 0$, $(1+\Phi(Z))\|\bar{S}\| \leq c_1(1+\Phi(Z))^2\|\bar{S}\|^2 + \frac{1}{4c_1}$ and the aforementioned unpdating scheme for k were employed. Upon substituting in $\hat{\mu}$, regrouping, and completing the square, it is not difficult to further express $\dot{V}$ as

$$\dot{V} \leq -2c_0\zeta\frac{\|\bar{S}\|^2}{2} - c_2\frac{(\mu - \zeta\hat{\mu})^2}{2\zeta} + \frac{c_2\mu^2}{2\zeta} + \frac{\mu}{4c_1}$$
$$\leq -\iota_1 V + \iota_2 \tag{10}$$

where $\iota_1 = min\{2c_0\zeta, c_2\}$, $\iota_2 = \frac{c_2\mu^2}{2\zeta} + \frac{\mu}{4c_1} < \infty$. And $\hat{\mu}(\mu - \zeta\hat{\mu}) \leq \frac{\mu^2 - (\mu - \zeta\hat{\mu})^2}{2\zeta}$ is used. Meanwhile, because $\hat{\mu}(\mu - \zeta\hat{\mu}) \leq \frac{\mu^2}{4\zeta}$ and $\hat{\mu}(\mu - \zeta\hat{\mu}) \leq \frac{\mu^2}{2\zeta}$ with using the fundamental inequality, the following inequalities holds,

$$\dot{V} \leq -c_0\zeta\|\bar{S}\|^2 + \iota_2 \tag{11}$$

and

$$\dot{V} \leq -c_0\zeta\|\bar{S}\|^2 + \iota_3 \tag{12}$$

with $\iota_3 = \frac{c_2\mu^2}{4\zeta} + \frac{\mu}{4c_1}$.

From wich it holds that $\dot{V} < 0$ if $\bar{S}$ is outside of either the compact regions $\Omega_1 = \{\|\bar{S}\| \leq \sqrt{\iota_2/c_0\zeta}$ or $\Omega_2 = \{\|\bar{S}\| \leq \sqrt{\iota_3/c_0\zeta}$. Because $\iota_2 > \iota_3$, Ω_1 encloses Ω_2. Then the system error trajectory may move in or out of Ω_2 (the small region), but once inside the set Ω_1, it cannot go out of it. Thus UUB tracking is ensured with the proposed control scheme, therefore, E and $\dot{E}$ are UUB by the definition of S and $\bar{S}$ by using the Lemma in [14].

Remark 4. It is noted that the proposed neuroadaptive PID-like algorithm contains proportional term, integral term and differential term. This PID-like algorithm gives a better convergence property than using integral term alone or both proportional and integral terms.

4   Simulation Studies

To test the performance of the proposed control strategies, simulation tests are carried out on a train similar to CRH-5 with eight vehicles. The travel distance tested in the simulation covers two acceleration phases, four cruise phases and three braking phases. Curve, slope and tunnel are considered in the operation condition. The goal is to make the actual velocity and position track the desired velocity and desired position with high precision, respectively. With three fading actuators, the neuroadaptive PID-like fault tolerant control algorithms are tested and the results are presented in the figure, from which one can observe that the proposed control scheme performs well even if some of the actuator lose their effectiveness during the system operation (Fig. 1).

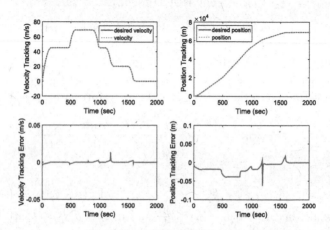

Fig. 1. Velocity and position tracking process and errors.

5   Conclusions

The problem of velocity and position tracking control of high speed train was studied in this paper. Uncertain mass and rotary mass coefficient, uncertain resistance, and unknown tracking/braking actuation characteristics including input nonlinearities and actuator failures were considered explicitly. Neuroadaptive PID-like fault tolerant control was proposed based on Lyapunov stability theory. The attraction feature of the developed control lies in its simplicity in design and implementation. The independence of the PID-like controller from the high speed train system model renders it relatively insensitive to system model uncertainties and perturbations as well as actuation faults.

Acknowledgments. This work is supported by National Natural Science Foundation (NNSF) of China under Grant 61503021, the Talent Fund (No. 2015RC048) and the State Key Laboratory Program (No. RCS2015ZT003).

References

1. Romero, J.G., Ortega, R., Donaire, A.: Energy shaping of mechanical systems via PID control and extension to constant speed tracking. IEEE Trans. Autom. Control **61**, 3551–3556 (2016)
2. Moradi, M., Fekih, A.: Adaptive PID-sliding-mode fault-tolerant control approach for vehicle suspension systems subject to actuator faults. IEEE Trans. Veh. Technol. **63**, 1041–1054 (2014)
3. Chen, X.Q., Ma, Y.J., Hou, T., Cai, R.C.: Study on the speed control of high-speed train based on predictive fuzzy PID control. J. Syst. Simul. **26**, 191–196 (2014)
4. Wang, X.Q., Song, J.C., Wushi, H.Q.: A study of fuzzy PID control algorithm for high speed train's eiectric hydraulic braking system. Mach. Des. Manuf. **9**, 22–24 (2013)
5. Gao, B., Dong, H.R., Zhang, Y.X.: Speed adjustment braking of automatic train operation system based on fuzzy-PID switching control. In: 6th International Conference on Fuzzy Systems and Knowledge Discovery, pp. 577–580. IEEE Press (2009)
6. Yang, H., Zhang, K.P., Liu, H.E.: Online regulation of high speed train trajectory control based on T-S fuzzy bilinear model. IEEE Trans. Intell. Transp. Syst. **17**, 1496–1508 (2016)
7. Wang, T., Gao, H.J., Qiu, J.B.: A combined adaptive neural network and nonlinear model predictive control for multirate networked industrial process control. IEEE Trans. Neural Netw. Learn. Syst. **27**, 416–425 (2016)
8. Wai, R.J., Yao, J.X., Lee, J.D.: Backstepping fuzzy-neural-network control design for hybrid maglev transportation system. IEEE Trans. Neural Netw. Learn. Syst. **26**, 302–317 (2015)
9. Song, Q., Song, Y.D.: Data-based fault-tolerant control of high speed trains with traction/braking notch nonlinearities and actuator failures. IEEE Trans. Neural Netw. **22**, 2250–2261 (2011)
10. Schaefer, H.H.: The comparison of formulae on train resistance and locomotive adhesion coeffcient in various countries. Foreign Diesel Locomot. **2**, 35–43 (1989)
11. Wang, Y.J., Song, Y.D., Krstic, M., Wen, C.Y.: Fault-tolerant finite time consensus for multiple uncertain nonlinear mechanical ssystems under single-way directed communication interactions and actuation failures. Automatica **63**, 374–383 (2016)
12. Liu, X.Y., Song, Y.D., Song, Q.: Fault-tolerant control of dynamic systems with unknown control direction-input nonlinearities-actuator failures. In: Decision and Control and European Control Conference, pp. 4973–4978. IEEE Press, Orlando (2011)
13. Huang, W.Y., Sun, Z.Y.: Noticeable problems of braking calculation for high speed train. Railw. Locomot. Car **26**, 24–27 (2006)
14. Slotine, J.J., Li, W.: Applied Nonlinear Control. Prentice-Hall, Upper Saddle River (1991)

A Programmable Memristor Potentiometer and Its Application in the Filter Circuit

Jinpei Tan, Shukai Duan[(⊠)], Ting Yang, and Hangtao Zhu

School of Electronics and Information Engineering,
Southwest University, Chongqing, China
duansk@swu.edu.cn

Abstract. Based on the electrical properties of memristor, the spintronic memristor is introduced into analog circuit design. Through programming the memristor by pulse signals, a programmable memristor potentiometer is designed in this paper. The potentiometer has the advantages of simple structure, small volume and continuous variable dynamic range. Based on the potentiometer above, this paper presents an amplifier circuit whose magnification is precisely adjustable and an active memristor-filter whose cut-off frequency and phase are adjustable as well. The circuit is characterized with smaller size, higher integration and lower power consumption, and its correctness and feasibility are verified by SPICE simulation. The research in this paper will promote the application of memristor in analog circuit, in addition, the performance of programmable memristor-analog circuit is satisfactory, which provides a new way for the integrated circuit design in the future.

Keywords: Meminductor · Model comparison · Variation control · Manufacture

1 Introduction

In 1971, Chua proposed a clear definition of memristor, predicted that memristor is the fourth basic circuit element except resistor, inductor and capacitor, and analyzed some unique circuitry properities of memristor in detail [1–4].

However, in 2009, Chen Yiran proposed a spintronic memristor model with current threshold, which is similar to the actual memristor [5]. It does not only have the advantages of a common HP model, but it can also effectively overcome its shortcomings: The rapid change of resistance greatly improves the speed of data storage; the antijamming capability is so strong that only if the current density passing the device is larger than the threshold current density, will the resistance change and then effectively avoid the influence of signal noise [6].

In order to achieve adjustable cut-off frequency, the potentiometer is generally used to replace the resistor in traditional analog filter circuits by adjusting the resistance of the potentiometer. Now, analog mechanical potentiometer and digital potentiometer are the two kinds of potentiometers widely used [7]. However, low stability of mechanical potentiometer makes it not easy to realize automatic control, meanwhile, due to the discrete resistance and narrow passband of digital potentiometer, both greatly affect the accuracy and stability of analog filter. Memristance variation is nonlinear and

© Springer International Publishing AG 2017
F. Cong et al. (Eds.): ISNN 2017, Part II, LNCS 10262, pp. 326–335, 2017.
DOI: 10.1007/978-3-319-59081-3_39

continuous, and its precision is higher than the digital potentiometer. So, because of the requirement for higher accuracy of cut-off frequency, the memristor-based filter circuit structure is simpler. The memristive control signal is pulse voltage, whose control and adjustability of memristance are more convenient compared with programming control digital potentiometer [10, 11].

As a result, in this paper the digital potentiometer based on spintronic memristor is designed by introducing it into the analog circuit, and then it is applied to multiple precision adjustable amplifying circuits and the MC active filter with continuously adjustable cut-off frequency and amplitude can be realized [14, 15].

2   Spintronic Memristor-Based Potentiometer

2.1   Spintronic Memristor

The spintronic memristor has a variety of physical structures, in this paper the model build on the basis of promoting the magnetic domain wall technical theory is discussed [7, 8]. As shown in Fig. 1 (Fig. 1(a) is the structure of spintronic memristor. Figure 1(b) is the equivalent circuit).

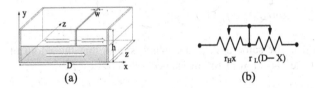

(a)   (b)

Fig. 1. The structure and equivalent circuit of spintronic memristor.

In Fig. 1(a), D, h, z are the length, height and width of the device respectively, which are 1000 nm, 7 nm and 10 nm. w is the width of the domain wall. Under the effect of voltage, the domain wall moves, and the length in the two magnetization directions change, so the memristance of the whole device changes correspondingly. r_L and r_H denote the memristance of per unit length of segment of low- and high-resistance states respectively. By ignoring the width of the domain wall in Fig. 1(b), the total memristance is calculated by:

$$M(x) = [r_H x + r_L(D - x)] \tag{1}$$

x is the distance of the domain wall movement. Its velocity v is proportional to the current density J, the relation is:

$$v = \frac{dx}{dt} = \Gamma v \cdot J = \frac{\Gamma v}{h \cdot z} \cdot \frac{dq}{dt} \tag{2}$$

Γv is the scaling factor, the magnitude is related to the structure of the device and the nature of the material. The expression is: $\Gamma v = \frac{P \mu_B}{e M_S}$. P is magnetic susceptibility of the material, u_B is Bonr magneton, e is elementary charge and M_S is saturation magnetization. e and u_B are constants. The magnitude of P and M_S is only related to the material.

Substituting Eq. (2) into Eq. (1), we can get the relationship of the memristance and charge as follows:

$$M(q) = [r_L \cdot D + (r_H - r_L)\Gamma q(t)] \tag{3}$$

Here $\Gamma = \frac{\Gamma v}{h \cdot z}$, which is the ratio of the scaling factor and the cross-sectional area of the device. By considering the effect of domain wall on the memristance, then Eq. (1) is written as [8]:

$$M(x) = \left[r_H \cdot \left(x - \frac{w}{2} \right) + R_{dw} + r_L \cdot \left(D - x - \frac{w}{2} \right) \right] \tag{4}$$

R_{dw} and w are the memristance and width of domain wall respectively. Supposing that the domainwall is located the middle of whole width, the resistance per unit length changes linearly from r_L to r_H, then:

$$\begin{aligned} M(x) &= \left[r_H \cdot \left(x - \frac{2}{w} \right) + (r_H + r_L) \cdot \frac{2}{w} + r_L \cdot \left(D - x - \frac{w}{2} \right) \right] \\ &= [r_H \cdot x + r_L \cdot (D - x)] \end{aligned} \tag{5}$$

Equation (5) is the same as Eq. (1), which shows that the width of domain wall does not affect the variation of the resistance when $0 < x < D$.

2.2   Spintronic Memristor-Based Design of the Digital Potentiometer

In order to observe the typical characteristic curve of spintronic memristor, a Sine signal, whose amplitude is 35 V and frequency is 5 MHz, is applied to the SPICE model of spintronic memristor. The V-I characteristic curve and the change curve of memristance are obtained and shown in Fig. 2 [9]. From Fig. 2(a), we can find that excited by Sine signal voltage, the V-I characteristic curve of spintronic memristor has hysteresis effect. As it shown in Fig. 2(b), it can be verified that the memristance is kept constant when the voltage is smaller than 7 V, but when the voltage is between 7 V and 35 V, whether the memristance varies with the voltage is determined by the magnitude of memristance at that time.

Applying different pulse signals with different voltage amplitudes and different duty cycles to the SPICE model of spintronic memristor, as shown in Fig. 3, different pulse inputs lead to different variations of memristance. Here, D is the duty cycles and U is pulse signal voltage.

According to the curves, we can get the following conclusions: under the condition of the same pulse duration, the larger the pulse voltage amplitude is, the greater the resistance changes; in the case of consistent pulse voltage amplitude, the longer the pulse duration is, the greater the resistance changes.

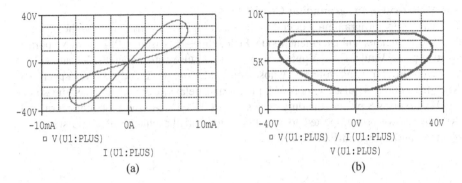

Fig. 2. The characteristic curve of spintronic memristor (a) V-I characteristic curve (b) Change curve of memristance.

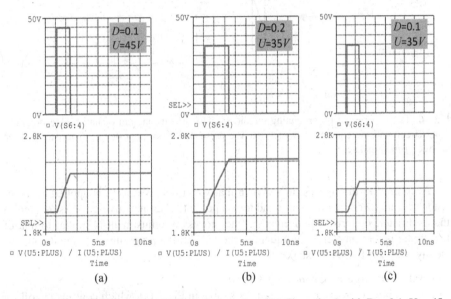

Fig. 3. The pulse characteristics of spintronic memristor. (a) The pulse 1 with D = 0.1, U = 45; (b) The pulse 2 with D = 0.2, U = 35; (c) The pulse 3 with D = 0.1, U = 35.

2.3 The Programming Circuit of Spintronic Memristor

To realize the programmable operation in the spintronic memristive circuit, this paper designs the following two programming circuits: The half-bridge programming circuit, which is used to program the memristance with a port connected with ground. It has simple circuit structure and obvious effects. The whole-bridge programming circuit, which is used to program the memristance with two ports connected with ground. Although the circuit structure is more complex, its application is more widespread.

The Half-Bridge Programming Circuit

The half-bridge programming circuit is shown in Fig. 4(a), which controls the off-on of the two P-channel enhancement mode Field. effect transistors (PFET) by exporting positive pulse Vpulse1 and Vpulse2 from the two pulse signal sources. V1 and V2 are the given programming voltages, whose amplitudes are 35 V and −35 V respectively. After the source Vpulse1 exports 11 pulses and delays a period of time, the source Vpulse2 exports 8 pulses (the duty cycles are all 0.2). Due to the switch effect of FET, the pulse waveform imported to the memristor and the change of memristance are obtained and shown in Fig. 4(b).

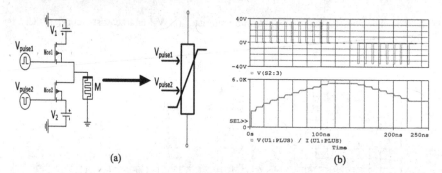

(a)   (b)

Fig. 4. The half-bridge programming circuit. (a) The circuit diagram of half-bridge programming circuit and the equivalent model; (b) The pulse waveform imported to memristor and the change of memristance.

According to the figures, it is shown that when positive pulse is applied to memristor, memristance increases in a ladder shape, but when negative pulse is applied to the memristor, memristance decreases. The process of adjustment is similar to that of the digital potentiometer. Consequently, the half-bridge programming circuit can efficiently program the memristance to obtain the resistance required by the circuit.

The Whole-Bridge Programming Circuit

The whole-bridge programming circuit is shown in Fig. 5(a), which controls the off-on of the four PFETs in the same way as half-bridge programming circuit, V1 and V2 are same too. In this circuit, after which, the source Vpulse1 exports 15 pulses and delays a period of time, the source Vpulse2 exports 11 pulses (the duty cycles are all 0.2). Due to the switch effect of FET, the pulse waveform imported to memristor and the change of memristance are obtained and shown in Fig. 5(b).

According to the figures, the programming effects of the whole- and half- bridge programming circuits are the same, which can both adjust memristance greatly. For the two programming circuits mentioned above, we can obtain different resistance by importing different number of pulses. Then Table 1 is obtained.

According to the Table 1, different memristance can be obtained by importing different number of pulses. By this way, the uncertainty of memristance is easily converted into the control of pulse number, the aim of programming memristance can

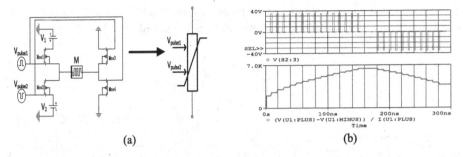

(a)   (b)

Fig. 5. The whole-bridge programming circuit (a) the circuit diagram of whole-bridge programming circuit and the equivalent model (b) the pulse waveform imported to memristor and the change of memristance.

Table 1. The corresponding memristance of different numbers of input pulse

The number of pulse	Memristance/K
0	2.00
5	3.960
10	5.220
15	6.235
20	7.108
25	7.882
30	8.588
35	9.239
40	9.58

be realized through a convenient method. In addition, the change of memristance caused by a single pulse can be obtained by setting the programming voltage and duration of pulse. That is to say, the accuracy of the digital potentiometer ba sed on spintronic memristor is adjustable. To conclude, the random memristance can be obtained in a wide range of memristance.

3   The Application of Spintronic Memristor in Programmable Circuit

3.1   Programmable Operational Amplifier Based on Spintronic Memristor

To collect the signal, a front-end circuit needs to be set to amplify or reduce the sampled signal, i.e. operational amplifier circuit. A novel programmable operational amplifier is designed in this paper, the circuit structure is shown in Fig. 6(a), in which the resister R and R1 are 1 K and 10 K respectively. The same as the common operational amplifier, the input signal is applied to the positive port, and the negative port is connected with the spintronic memristor and feedback resistance. According to virtual open and virtual short, the magnification is calculated as [12, 13]:

$$A_v = 1 + \frac{R_1}{M} \tag{6}$$

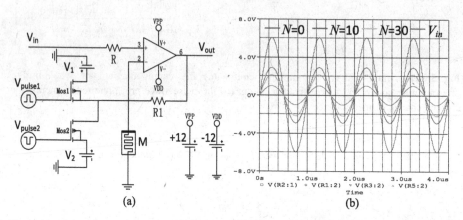

(a)   (b)

Fig. 6. Programmable operational amplifier based on spintronic memristor (a) The circuit diagram (b) The result of different number of pulses. (Color figure online)

Memristance varies by adjusting the bridge programming circuit, thus the output with different magnifications is obtained, which is similar to the property of input waveform. When importing 10 and 30 positive pulses (the duty ratio is 0.2), the input and output waveform is shown in Fig. 6(b), in which the green is input, and the yellow, blue, and red are the output of 30 pulses, 10 pulses, and 0 pulse respectively. The amplifier gain of the amplifier is obtained by the constant memristance variation, as is shown in Table 2. Compared with common programmable gain amplifier circuit, the proposed operation amplifier can realize adjustable constantly gain, which is not limited by the current like the common digital potentiometer anymore. Its application range is wide under the condition of fulfilling the need of adjustable gain.

Table 2. The corresponding magnification of the amplifier with different number of pulses

The number of pulse	Memristance/K	Magnification
0	2	6.00
5	3.96	3.525
10	5.22	2.916
15	6.235	2.604
20	7.108	2.407
25	7.882	2.269
30	8.588	2.164
35	9.239	2.082
40	9.85	2.015

3.2   The Programmable Active Low-Pass Filter Based on Spintronic Memristor

Through adding a spintroinc memristor-based phase proportion amplifier on the output end of MC low-pass filter, a first-order active low-pass filter is obtained and shown in Fig. 7(a), where $C = 10\mu F$, $R = 10K\Omega$. Because of the high input impedance and low output impedance, the load capacity is reinforced.

From the figures, when $\omega = 0$, the Passband voltage gain A0 is the ratio of output voltage Vout and input voltage Vin. In this circuit, the Passband voltage gain equals to the voltage gain AVF in non-inverting proportion amplifier, that is:

$$A_0 = A_{VF} = 1 + \frac{R}{M_2} \tag{7}$$

the amplitude-frequency response equation is obtained as follows:

$$|A(jw)| = \frac{|V_{out}(jw)|}{|V_{in}(jw)|} = \frac{A_0}{\sqrt{1 + \left(\frac{w}{w_0}\right)^2}} \tag{8}$$

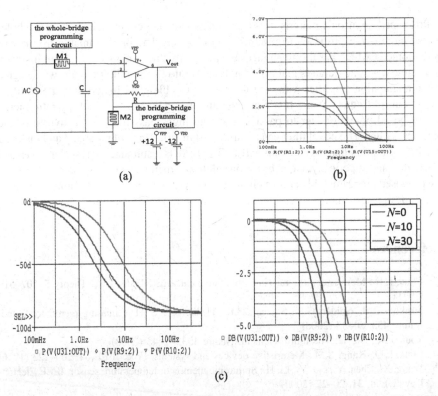

(a)

(b)

(c)

Fig. 7. The programmable active low-pass filter based on spintronic memristor (a) The circuit diagram; (b) The gain curve; (c) The amplitude- and phase-frequency response.

From Eqs. (7) and (8), once the memristance M2 changes, A0 changes correspondingly, so that the filter's passband gain changes. The cut-off frequency varies with memristance M1 which is consistent with MC passive low-pass filter. The SPICE simulation result is given in Fig. 7(b). Keeping memristance M2 constant, we change the memristance M1, different number of pulses are imported to M1 by programming circuit, and the corresponding amplitude and phase frequency curve are shown in Fig. 7(b). It is suggested that the cut-off frequency and phase of first-order active low-pass filter vary with the difference of input pulse's number. However, keeping memrictance M1 constant, when importing different number of pulses to M2, the filter's passband gain curve is shown in Fig. 7(c). It is shown that the gain changes with the change of pulse's number.

To conclude, the filtering property of the first-order active low-pass filter based on programmable spintronic memristor is determained by two memristors. Through which the programming circuit exports different number of pulses to adjust memristance, the filter's cut-off frequency, phase response and gain change, so that the proposed filter can perfectly select the job characteristics under different situations, resulting in wider application and higher efficiency.

4 Conclusions

In this paper, spintronic memristor is introduced to analog circuit. The half- and whole-bridge circuits are proposed to program memristance. The programmable memristive potentiometer which has the advantages of simple structure, small size and continuously adjustable dynamic range is realized. Applying it into the amplifier circuit, we design a novel memristive amplifier with continuously adjustable gain. In addition, a new filter, whose cut-off frequency and phase are continuously adjustable, is built by introducing the proposed potentiometer to passive low- and high- pass filter and active low- and high- pass filter. The novel filter can control the dynamic range of passband and make up the drift of input signal to a certain extent. The SPICE result and analysis have verified its validity and superiority. Here, based on spintronic memristor with good performance, the proposed programmable analog circuit can provide new idea for the future integrated circuit design and improve memristive application in analog circuit.

References

1. Chua, L.O.: Memristor - the missing circuit element. IEEE Trans. Circ. Theory **5**, 507–519 (1971)
2. Strukov, D.B., Snider, G.S., Stewart, D.R., Williams, R.S.: The missing memristor found. Nature **453**, 80–83 (2008)
3. Tour, J.M., He, T.: The fourth element. Nature **453**, 42–43 (2008)
4. Chua, L.O., Kang, S.M.: Memristive devices and systems. Proc. IEEE **64**, 209–223 (1976)
5. Wang, X., Chen, Y., Gu, Y., Li, H.: Spintronic memristor temperature sensor. IEEE Electron Device Lett. **31**, 20–22 (2009)

6. Chen, Y., Li, H., Wang, X.: Spintronic devices: from memory to memristor. In: International Conference on Communications, Circuits and Systems, pp. 811–816 (2010). http://ieeexplore.ieee.org/xpl/mostRecentIssue.jsp?punumber=5570016
7. Wang, X., Chen, Y.: Spintronic memristor devices and application. In: Design, Automation & Test in Europe Conference & Exhibition (DATE), pp. 667–672 (2010). http://ieeexplore.ieee.org/xpl/mostRecentIssue.jsp?punumber=5450668
8. Chen, Y., Wang, X.: Compact modeling and corner analysis of spintronic memristor. In: IEEE/ACM International Symposium on Nanoscale Architectures, pp. 7–12 (2009)
9. Adzmi, A.F., Nasrudin, A., Abdullah, W.F.H., Herman, S.H.: Memristor spice model for designing analog circuit. In: 2012 IEEE Student Conference on Research and Development (SCOReD), pp. 78–83 (2012). http://ieeexplore.ieee.org/xpl/mostRecentIssue.jsp?punumber=6516133
10. Shin, S., Kim, K., Kang, S.M.: Memristor applications for programmable analog ICs. IEEE Trans. Nanotechnol. **10**, 66–274 (2009)
11. Sozen, H., Cam, U.: On the realization of memristor based RC high pass filter. In: 2013 8th International Conference on Electrical and Electronics Engineering (ELECO), pp. 45–48 (2013). http://ieeexplore.ieee.org/xpl/mostRecentIssue.jsp?punumber=6709575
12. Yener, S.C., Mutlu, R., Kuntman, H.: A new memristor-based high-pass FilterAmplifier: its analytical and dynamical models. In: 2014 24th International Conference on Radioelektronika (RADIOELEKTRONIKA), pp. 1–4 (2014). http://ieeexplore.ieee.org/xpl/mostRecentIssue.jsp?punumber=6822756
13. Ascoli, A., Tetzlaff, R., Corinto, F., Mirchev, M., Gilli, M.: Memristor-based filtering applications. In: 2013 14th Latin American Test Workshop (LATW), pp. 1–6 (2013). http://ieeexplore.ieee.org/xpl/mostRecentIssue.jsp?punumber=6554180
14. Bi, X., Zhang, C., Li, H., Chen, Y., Pino, R.E.: Spintronic memristor based temperature sensor design with CMOS current reference. In: Design, Automation & Test in Europe Conference & Exhibition (DATE), pp. 1301–1306 (2012). http://ieeexplore.ieee.org/xpl/mostRecentIssue.jsp?punumber=6171057
15. Mahmoudi, H., Sverdlov, V., Selberherr, S.: Domain-wall spintronic memristor for capacitance and inductance sensing. In: 2011 International Conference on Semiconductor Device Research Symposium (ISDRS), pp. 1–2 (2011). http://ieeexplore.ieee.org/xpl/mostRecentIssue.jsp?punumber=6128902

Rapid Triangle Matching Based on Binary Descriptors

Min Tian and Qiu-Hua Lin[✉]

School of Information and Communication Engineering,
Dalian University of Technology, Dalian 116023, China
qhlin@dlut.edu.cn

Abstract. Geometric constraints have been widely applied to image matching to gain additional advantages over feature points. A rapid triangle matching (RTM) algorithm was such an algorithm for matching triangles formed by three feature points using 38-dimensional floating-point descriptors. The RTM was faster than SIFT, but it was still hard to meet the real-time requirement. As such, we designed two kinds of binary descriptors including FREAK and rBRIEF used in ORB to replace the floating-point descriptors of RTM, and compared the improved RTM algorithms with original RTM algorithm and SIFT based on simulated and actual binocular images. The results demonstrate that our algorithms greatly improve the speed as well as the precision and matching score. Furthermore, our algorithms can match additional points compared to SIFT and the original RTM when applied to structural scenes.

Keywords: Image matching · RTM · Binary descriptors · FREAK · rBRIEF

1 Introduction

Image matching algorithms have been widely used in various applications including object recognition, image retrieval and visual navigation. Local feature points are frequently used by matching algorithms at the feature point extraction and description stage [1, 2]. The local feature points generally include blobs (e.g. SIFT [3] and SURF [4, 5]) and corners (e.g. Harris [6] and Fast [7, 8]). Descriptors can be divided into floating-point descriptors (e.g. SIFT and SURF) and binary descriptors (e.g. BRIEF [9], rBRIEF of ORB [10] and FREAK [11]). While the floating-point descriptors show matching precision advantages, the binary descriptors provide a very efficient method for time-constrained applications with good matching accuracy [12].

Recently, people have been paying more attention to adding geometric constraints and geometric features to image matching in an effort to gain additional advantages over purely using feature points. Triangular constraint was one of them. Zhang et al. utilized the angle and normalized length information of triangles to describe the corners after detecting corners and constructing triangle net [13]. Liu and An decomposed each triangle into six sub-triangles and obtained the relative moment invariant of each part after triangulation. Subsequently, the six descriptors were used for calculating the similarity of two triangles [14]. Yang et al. linked the k nearest neighbors of a given corner to build triangle chain and matched them based on comparing the angles and

© Springer International Publishing AG 2017
F. Cong et al. (Eds.): ISNN 2017, Part II, LNCS 10262, pp. 336–344, 2017.
DOI: 10.1007/978-3-319-59081-3_40

lengths' ratio of triangles [15]. Cao proposed a rapid triangle matching (RTM) algorithm in [16]. The RTM algorithm not only used the length and angle information but also combined the grey information to describe triangles. Compared with SIFT, RTM reduced the dimension of descriptors so that the speed was increased. However, it was still hard to meet the real-time requirement. As such, the aim of this study is to improve the speed of RTM algorithm with no degradation in precision. We designed two kinds of binary descriptors including FREAK (512 bits) or rBRIEF (256 bits) of ORB to describe the triangle. The efficacy was tested using simulated and actual binocular images. The results demonstrated that our algorithms improved the speed largely and the precision in most cases.

The rest of this study is organized as follows. Section 2 introduces the floating-point descriptors of original RTM algorithm. In Sect. 3, we present the improved algorithms in detail. Section 4 describes experimental conditions and evaluation indexes. Section 5 shows the results of experiments. Finally, conclusions and future works are given in Sect. 6.

2 Floating-Point Descriptors in RTM Algorithm

In RTM algorithm, every triangle was described by a 38-dimensional floating-point descriptor as follows:

$$Desc\_F = [ShapeDesc\_F; BlockDesc\_F] = \left[l_1', l_2', l_3', \theta_1', \theta_2', \theta_3'; f_1, f_2, \cdots, f_{32}\right] \quad (1)$$

where the floating-point descriptor $Desc\_F$ consists of a shape descriptor $ShapeDesc\_F$ and a block descriptor $BlockDesc\_F$. Variables l_1', l_2', l_3' represent three normalized lengths of the triangle; $\theta_1', \theta_2', \theta_3'$ are three angles which were balanced with l_1', l_2', l_3' in the weights; $f_1, f_2, \cdots, f_{32}$ denote a 32-dimensional SIFT descriptor. Thus, the 38-dimensional descriptor includes 6-dimensional shape descriptor $(l_1', l_2', l_3', \theta_1', \theta_2', \theta_3')$ and 32-dimensional block descriptor $(f_1, f_2, \cdots, f_{32})$.

The description of RTM is shown in Fig. 1(a), where A, B and C are the three feature points forming the triangle $\triangle ABC$; D is the centroid of the triangle; $\overrightarrow{DE}$ is the dominant orientation that is the counterclockwise direction of the longest length $(\overrightarrow{CA})$ in the triangle. The description radius is a certain proportion of the longest length in the triangle. The three lengths and angles of the triangle are all normalized as the 6-dimensional shape descriptor. Based on D, $\overrightarrow{DE}$ and the description radius, the 32-dimensional SIFT descriptor is computed as the block descriptor.

3 Improved RTM Algorithm with Binary Descriptors

In order to improve the speed of original RTM algorithm with the floating-point descriptors, we utilized the binary descriptors, including 512 bits FREAK and 256 bits rBRIEF in ORB, to describe the triangle respectively. For convenience, we call them

RTM$_{\text{FREAK}}$ and RTM$_{\text{rBRIEF}}$ respectively. Similarly, we call the original RTM as RTM$_{\text{SIFT}}$. More precisely, we constructed the binary descriptor for a triangle as follows:

$$Desc\_B = [ShapeDesc\_B; BlockDesc\_B] = \left[l_1'', l_2'', l_3'', \theta_1'', \theta_2'', \theta_3''; b_1, b_2, \cdots, b_N\right] \quad (2)$$

where the binary descriptor $Desc\_B$ is also formed by a binary shape descriptor $ShapeDesc\_B$ and a binary block descriptor $BlockDesc\_B$; l_1'', l_2'', l_3'' and $\theta_1'', \theta_2'', \theta_3''$ represent the binary values of l_1', l_2', l_3' and $\theta_1', \theta_2', \theta_3'$ in Eq. (1); $b_1, b_2, \cdots, b_N$ are N bits binary string defined by FREAK or rBRIEF descriptor, N can be 512 or 256.

A binary string F was formed as:

$$F = b_N b_{N-1} \cdots b_1 = \sum_{0 \leq a < N} 2^a T(P_a) \quad (3)$$

where a is the ath bit, N is the desired size of the descriptor, and

$$T(P_a) = \begin{cases} 1 & if\,(I(P_a^{r_1}) - I(P_a^{r_2})) > 0, \\ 0 & otherwise, \end{cases} \quad (4)$$

where P is a sampling field; $P_a^{r_1}$ means the first field in the ath pair of sampling fields and $P_a^{r_2}$ is the second; $I(P_a^{r_1})$ refers to the smoothed intensity of $P_a^{r_1}$.

Figure 1(b) shows the sampling pattern of FREAK. There are 43 sampling points including the centroid D of $\triangle ABC$ and 7 concentric circles with 6 sampling points for each circle. Thus, 43 sampling fields are built based on 43 sampling points with certain radius. According to Eqs. (3) and (4), there will be C_{43}^2 pairs of binary tests, which will lead to a large descriptor. When we use 512-bit descriptor, 512 pairs with low correlations (means of 0.5) are selected to form $b_1, b_2, \cdots, b_{512}$.

In addition, we used the rBRIEF descriptor to describe the triangle as shown in Fig. 1(c). In $\triangle ABC$, the description patch is $w_p \times w_p$ and each sampling field (P) is a $w_s \times w_s$ sub-window. We set $w_p = 31$ and $w_s = 5$, there are $C_{(31-5+1)^2}^2$ binary tests according to Eqs. (3) and (4). Let $N = 256$, a greedy search was used to find 256 low correlated tests with means near 0.5. Hence, the 256 bits will be $b_1, b_2, \cdots, b_{256}$.

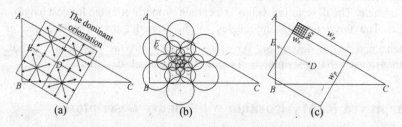

(a)   (b)   (c)

Fig. 1. Illustration of description in RTM$_{\text{SIFT}}$, RTM$_{\text{FREAK}}$ and RTM$_{\text{rBRIEF}}$ (a) SIFT description in RTM$_{\text{SIFT}}$. (b) Sampling pattern of FREAK in RTM$_{\text{FREAK}}$. Every circle represents a sampling field. (c) Approach of rBRIEF description in RTM$_{\text{rBRIEF}}$. The description patch is $w_p \times w_p$ and every sub-window ($w_s \times w_s$) is a sampling field.

The main steps of our improved RTM algorithms are as follows:

(1) Detect and select feature points for two images to be matched. The feature points can be blobs or corners. Steady points can be selected by setting certain thresholds.
(2) Construct triangulation according to the Delaunay [17] algorithm and remove triangles with too large or small sides.
(3) Describe the triangles with binary descriptors, referring to Eqs. (2) and (3).
(4) Match the images by computing the ratio of the closest neighbor Hamming distance between two descriptors for the two images to the second-closest neighbor Hamming distance.
(5) Extract feature points corresponding to all matched triangles.
(6) Eliminate the error matching points using RANSAC [18] algorithm.

4 Experimental Conditions and Evaluation Indexes

4.1 Image Datasets and Experimental Conditions

Experimental images included simulated and actual binocular images. The simulated image datasets were provided by Mikolajczyk et al. [19] of the Oxford University. Since RTM was suitable for matching images with relatively small changes in scale and viewpoint, we chose the first and second images of the six sets (bikes, boat, graf, leuven, ubc, wall) from the datasets. The actual binocular images were from the KITTI Vision Benchmark Suite [20]. We selected synchronous frames from them. Visual Studio 2010 and OpenCV2.4.9 were development tools. All the experiments have been taken on the platform with Intel Core i7-4790 processor, 3.6 GHz frequency and 16 GB memory. It is worth noticing that the parameters were consistent for all experiments. We detected and selected 200 Harris [6] points for all algorithms and ran every experiment 10 times to compute the average time.

4.2 Evaluation Indexes

We used the matching speed, precision [19] and matching score [21] to evaluate our proposed algorithms. The matching speed was specifically evaluated using the average total time and average description time. The total time was the time from detecting feature points to outputting the correct matches, the description time was the time consumed by constructing descriptors for all feature points, and the average total time and average description time were defined by averaging the total time and the description time over all experimental runs, respectively.

The precision is computed as follows:

$$precision = \frac{\#\,correct\,matches}{\#\,matches} \tag{5}$$

where $\#\,matches$ is the number of total matching points; $\#\,correct\,matches$ is the number of correct matches after eliminating the wrong matches.

The matching score is represented as follows:

$$matching\ score = \frac{\#\,correct\,matches}{\min(n1, n2)} \qquad (6)$$

where $n1$ and $n2$ are the numbers of detected feature points in the pair of images to be matched, respectively.

5 Experimental Results

We compared our proposed RTM_FREAK and RTM_rBRIEF with RTM_SIFT and SIFT based on the simulated and actual images in terms of speed (average total time and average description time), precision and matching score. Finally, we showed the additional feature points that our algorithms matched.

5.1 Simulated Images

Figure 2 shows the average total time (Fig. 2(a)) and average description time (Fig. 2 (b)) consumed by SIFT, RTM_SIFT, RTM_FREAK and RTM_rBRIEF based on six sets of images. We can see that in one case (bikes), the average total time and average description time for RTM_SIFT were longer than those for SIFT. Our algorithms solved the problem by using binary descriptors instead of float-point descriptors. More precisely, for the average total time, RTM_FREAK was twice as fast as RTM_SIFT in the best case and 1.3 times faster in the worst case; RTM_rBRIEF was 2.3–4.2 times faster than RTM_SIFT. When comparing the average description time, FREAK description was 1.6 times faster than SIFT description and rBRIEF was 18 times faster than SIFT. On average, referring to the last columns in Fig. 2(a) and (b), RTM_rBRIEF is the fastest, followed by RTM_FREAK, RTM_SIFT, and SIFT.

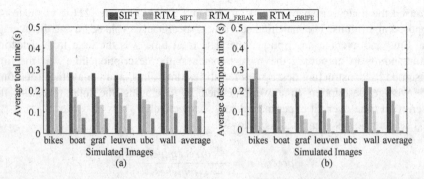

Fig. 2. Comparison of SIFT, RTM_SIFT, RTM_FREAK and RTM_rBRIEF based on average total time (a) and average description time (b). The last column is the average over the six sets.

The results of precision (Fig. 3(a)) and matching score (Fig. 3(b)) for the four algorithms are demonstrated in Fig. 3. We see that RTM_FREAK and RTM_rBRIEF were mostly better than RTM_SIFT in terms of the precision, and achieved similar or slightly better matching scores than RTM_SIFT. The improvements were mainly due to the increase of description dimension, as compared to RTM_SIFT. Note that SIFT obtained the best precision and matching score for the simulated images. Due to no constraints were added to SIFT, it yielded more feature points than RTM-related algorithms which imposed triangle constraints upon the feature points.

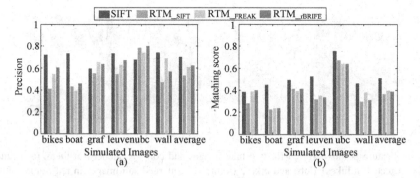

Fig. 3. Comparison of SIFT, RTM_SIFT, RTM_FREAK and RTM_rBRIEF in terms of precision (a) and matching score (b). The last column is the average over the six sets.

5.2 Binocular Images

Table 1 includes comparative results for the four algorithms in terms of the average total time, the average description time, the precision, and the matching score. These results were much similar with those for the simulated images. RTM_SIFT, RTM_FREAK and RTM_rBRIEF were 1.25–3 times faster than SIFT, RTM_FREAK and RTM_rBRIEF were 1.1–2.2 times faster than RTM_SIFT, and RTM_rBRIEF were the fastest. The precision and matching score of RTM_FREAK and RTM_rBRIEF were higher than RTM_SIFT. As for the precision and matching score, RTM_rBRIEF and RTM_FREAK achieved similar performance compared with SIFT.

Table 1. Comparison of SIFT, RTM_SIFT, RTM_FREAK and RTM_rBRIEF in terms of average total time, average desciption time, precision and matching score

	SIFT	RTM_SIFT	RTM_FREAK	RTM_rBRIEF
Average total time (s)	0.25	0.20	0.19	0.09
Average description time (s)	0.18	0.11	0.09	0.01
Precision	0.51	0.39	0.44	0.60
Matching score	0.31	0.26	0.31	0.36

5.3   Additional Feature Points Matched by RTM Algorithms

By taking advantage of triangle constraints, RTM algorithms can match additional feature points which SIFT may skip. We used the bikes images to show the results. Figure 4(a) shows the corner points detected in the images. The sub-images in red frame of Fig. 4(a) are enlarged and shown in Fig. 4(b) for identification. It is worth noticing that we used the homography matrix of the two images to compute the correspondences [19], as shown in Fig. 4(c). The correspondences were ground truth for all algorithms.

(a)   (c)

Fig. 4. Feature points detected for two bike images and correspondences of them. (a) Feature points detected in bikes1 (left) and bikes2 (right). (b) Enlarged sub-images in red frames of (a). (c) Correspondences of feature points computed using homography matrix. (Color figure online)

.Figure 5 shows the matched feature points for SIFT, RTM_$_{SIFT}$ and RTM_$_{rBRIEF}$. Compared with the correspondences displayed in Fig. 4(c), we can see that these results in Fig. 5 were correct. However, the matches were mostly different for the three algorithms. RTM_$_{rBRIEF}$ yielded much more matching corners than RTM_$_{SIFT}$, while RTM_$_{SIFT}$ matched more corners than SIFT, suggesting that our algorithms can match additional points compared to SIFT and the original RTM when applied to structural scenes.

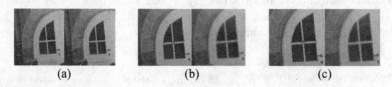

(a)   (b)   (c)

Fig. 5. The matched feature points for different algorithms. (a) SIFT. (b) RTM_$_{SIFT}$. (c) RTM_$_{rBRIEF}$.

6   Conclusions

This study improved the RTM algorithm by designing large dimension binary descriptors for triangles. The experimental results using both simulated and actual images showed that our improved algorithms were better than original RTM in terms of speed, precision, and matching scores. When applied to structural images, the proposed RTM can match additional feature points compared to the original RTM and SIFT. Dealing with images with large scale variance deserves our future work.

References

1. Kim, S., Yoo, H., Sohn, K.: Robust corner detector based on corner candidate region. In: 8th IEEE Conference on Industrial Electronics and Applications, pp. 1620–1626. IEEE Press, Melbourne (2013)
2. Mair, E., Hager, G.D., Burschka, D., Suppa, M., Hirzinger, G.: Adaptive and generic corner detection based on the accelerated segment test. In: Daniilidis, K., Maragos, P., Paragios, N. (eds.) ECCV 2010. LNCS, vol. 6312, pp. 183–196. Springer, Heidelberg (2010). doi:10.1007/978-3-642-15552-9_14
3. Lowe, D.G.: Object recognition from local scale invariant features. In: 7th IEEE International Conference on Computer Vision, pp. 1150–1157. IEEE Press, Kerkyra (1999)
4. Bay, H., Tuytelaars, T., Gool, L.: SURF: speeded up robust features. In: Leonardis, A., Bischof, H., Pinz, A. (eds.) ECCV 2006. LNCS, vol. 3951, pp. 404–417. Springer, Heidelberg (2006). doi:10.1007/11744023_32
5. Bay, H., Ess, A., Tuytelaars, T., Van Gool, L.: Speeded-up robust features (SURF). J. Comput. Vis. Image Underst. **110**(3), 346–359 (2008)
6. Harris, C., Stephens, M.: A combined corner and edge detector. In: 4th Alvey Vision Conference, Manchester, pp. 147–151 (1988)
7. Rosten, E., Drummond, T.: Machine learning for high-speed corner detection. In: Leonardis, A., Bischof, H., Pinz, A. (eds.) ECCV 2006. LNCS, vol. 3951, pp. 430–443. Springer, Heidelberg (2006). doi:10.1007/11744023_34
8. Rosten, E., Porter, R., Drummond, T.: Faster and better: a machine learning approach to corner detection. J. IEEE Trans. Pattern Anal. Mach. Intell. **32**(1), 105–119 (2010)
9. Calonder, M., Lepetit, V., Strecha, C., Fua, P.: BRIEF: binary robust independent elementary features. In: Daniilidis, K., Maragos, P., Paragios, N. (eds.) ECCV 2010. LNCS, vol. 6314, pp. 778–792. Springer, Heidelberg (2010). doi:10.1007/978-3-642-15561-1_56
10. Rublee, E., Rabaud, V., Konolige, K., Bradski, G.: ORB: an efficient alternative to SIFT or SURF. In: IEEE International Conference on Computer Vision, pp. 2564–2571. IEEE Press, Barcelona (2011)
11. Alahi, A., Ortiz, R., Vandergheynst, P.: FREAK: fast retina keypoint. In: IEEE Computer Vision and Pattern Recognition, pp. 510–517. IEEE Press, Providence (2012)
12. Miksik, O., Mikolajczyk, K.: Evaluation of local detectors and descriptors for fast feature matching. In: 21st International Conference on Pattern Recognition, pp. 2681–2684. IEEE Press, Tsukuba (2012)
13. Zhang, H., Wang, L., Jia, R.: Scale-invariant global sparse image matching method based on Delaunay triangle. In: 6th International Symposium on Multispectral Image Processing and Pattern Recognition, p. 7495. SPIE, Yichang (2009)
14. Liu, Z., An, J.: A new algorithm of global feature matching based on triangle regions for image registration. In: 10th International Conference on Signal Processing, pp. 1248–1251. IEEE Press, Beijing (2010)
15. Yang, S., Wei, E., Wang, Y.: Matching triangle chain codes. In: 6th International Conference on Image and Graphics, pp. 290–296. IEEE Press, Hefei (2011)

16. Cao, J.C.: Fast image matching algorithm based on new feature description. Thesis, Dalian University of Technology, Dalian, China (2015)
17. Lawson, C.L.: Transforming triangulations. J. Discret. Math. **3**(4), 3–365 (1972)
18. Fischler, M.A., Bolles, R.C.: Random sample consensus: a paradigm for model fitting with applications to image analysis and automated cartography. J. Commun. ACM **24**(6), 381–395 (1981)
19. Mikolajczyk, K., Schmid, C.: A performance evaluation of local descriptors. J. IEEE Trans. Pattern Anal. Mach. Intell. **27**(10), 1615–1630 (2005)
20. Geiger, A., Lenz, P., Stiller, C., Urtasun, R.: Vision meets robotics: the KITTI dataset. J. Int. J. Robot. Res. **32**(11), 1231–1237 (2013)
21. Mikolajczyk, K., Tuytelaars, T., Schmid, C., Kadir, T., Van Gool, L.: A comparison of affine region detectors. J. Int. J. Comput. Vis. **65**(1–2), 43–72 (2005)

An Improved Symbol Entropy Algorithm Based on EMD for Detecting VT and VF

Yingda Wei[1,2], Qingfang Meng[1,2(✉)], Haihong Liu[1,2], Jin Zhou[1,2], and Dong Wang[1,2]

[1] The School of Information Science and Engineering,
University of Jinan, Jinan 250022, China
ise_mengqf@ujn.edu.cn
[2] Shandong Provincial Key Laboratory of Network Based
Intelligent Computing, Jinan 250022, China

Abstract. In this paper the improved symbol entropy algorithm based on empirical mode decomposition (EMD) was proposed to detect ventricular tachycardia (VT) and ventricular fibrillation (VF). The original symbol entropy arithmetic needed longer time series to distinguish VT and VF by high accuracy while the algorithm we proposed can distinguish VT and VF in shorter time series by high accuracy. Otherwise the execution time of new arithmetic was shorter than original algorithm. The classification accuracy of original arithmetic was 93.5%, and the improved arithmetic was 97.75%. The computer time of feature was 33.32 times less than original.

Keywords: Symbol entropy · Empirical mode decomposition · Ventricular tachycardia · Ventricular fibrillation

1 Introduction

VT and VF were both life-threatening arrhythmia. Treatment protocols of VT and VF were different. The VF patients need to be electric shocked. The VT patient need drug therapy. The VF was misinterpreted ad VT, that's life-threatening. If VT is incorrectly interpreted as VF, the patients' hearts will be shocked and damaged. There is clinical research significance to use an efficient detecting method to distinguish VT from VF [1].

In recent years, the VT and VF detecting methods mainly include time domain analysis [2], frequency-domain analysis [3] and nonlinear analysis [4]. There were overlaps of the heart rate between VT and VF, so according to the heart rate to distinguish the VT and VF while occur high error rate. To overcome this problem, scholars had proposed many quantitative analysis methods [5–9]. But these methods had certain limitations. With the rapid development of neural computing, scholars applied neural networks to distinguish VF and VT. This can distinguish two class by nonlinear [10]. But the execution time of the algorithm was high relatively.

Wang and Chen [11] proposed a method based on Symbol dynamics to research VF and VT. The research showed that the symbol sequence entropy suddenly showed decreased, the patients most likely to enter ventricular tachycardia in the sample, which is an important sample for clinical treatment of patients.

© Springer International Publishing AG 2017
F. Cong et al. (Eds.): ISNN 2017, Part II, LNCS 10262, pp. 345–352, 2017.
DOI: 10.1007/978-3-319-59081-3_41

Empirical Mode Decomposition (EMD) [9] was adaptive processing method for nonlinear and nonstationary signals. This method was especially suitable for the nonlinear analysis of non-stationary signal processing. Defined as the instantaneous frequency, a finite set of band limited signals called the intrinsic mode function (IMF) is decomposed into the original signal.

In this paper, we proposed an improved symbol entropy based on EMD for detecting VT and VF. The method displayed a higher accuracy rate in the classification of VF and VT with shorter time series than original symbol entropy. Furthermore, the execution time of the algorithm was shorter than original symbol entropy.

2 The Method of Symbol Entropy and EMD

2.1 Empirical Mode Decomposition

EMD is an adaptive signal processing method. This method is especially suitable for processing signal which is nonlinear and nonstationary. Research showed that the ECG is non-stationary signals, so we can use EMD to decompose the VF and VF.

Given a signal x(t), we use EMD to decompose x(t). We can get IMF1(t)-IMFn(t) and a residue which is marked as r. The x(t) can be expressed by Eq. (1).

$$x(t) = \sum_{i=1}^{M} IMF_m(t) + r \tag{1}$$

Algorithm steps:

Step 1: Set $r(t) = g(t) = x(t), j = 0, i = 1$.
Step 2: Get the *ith* IMF.
    (1) Set $h_0(t) = r_{i-1}(t), j = 1$.
    (2) The local-maximum and local-minimum values of $h_{j-1}(t)$ are obtained.
    (3) Fitting envelope of upper and lower $e_l(t)$ and $e_u(t)$ by cubic spline interpolation respectively technology.
    (4) Calculate the average value of the envelope $e_l(t)$ and $e_u(t)$, marked as $m(t)$:

$$m(t) = \frac{e_l(t) + e_u(t)}{2} \tag{2}$$

    (5) $h_j(t)$ is defined:

$$h_j(t) = h_{j-1}(t) - m_{j-1}(t) \tag{3}$$

    (6) If $h_j(t)$ satisfies two basic conditions, set $IMF_i(t) = h_j(t)$;

If $h_j(t)$ doesn't satisfies two basic conditions, then $j = j+1$, go to *step2-(2)*.

Step 3: Cauculate $r_i(t) = r_{i-1}(t) - IMF_i(t)$.
Step 4: If the num of local-maximums and local-minimums is greater than 2, set $i = i+1$ and go to *step 2*, else stop.

2.2 Symbol Entropy

Symbol entropy is a kind of time series analysis method. Symbol entropy originates from chaotic time series analysis, symbolic dynamics theory and information theory. Signal's fundamental characteristics can be reacted by symbol entropy. The VT/VT's effective information mainly distributed in the low frequency band.

Given that VF/VT's sample is marked as X, and $X = (x_1, x_2, \ldots, x_i, \ldots, x_N)$, the N is the length of sample.

Algorithm steps:

Step 1:

By the specified rule, we converted original time series into symbol time sequence, which markes as $S = (s_1, s_2, \ldots, s_i, \ldots, s_N)$. The rule we used was as followed. If the average value is greater than x, we mark as '0'. If the average value is less than x, we mark as '1'. The mathematical model was followed.

$$s(i) = \begin{cases} `0' & x(i) \leq mean(X) \\ `1' & x(i) > mean(X) \end{cases} \tag{4}$$

$i = 1, 2, \ldots, N$, and $x(i)$ represents the *ith* point's value.

Step 2:

We put 3 adjacent symbols into a group which we called "3-bit word". Given that the symbolic time series S = "111100101...", therefore S can be grouped by operation of sliding window and get S'{111,111,110,100,....}. Each "3-bit word" contained '0' and '1'. The number of different combinations of "3-bit word" was 2^3. Through the above operation, we get $S' = \{s'_1, s'_2 \ldots, s'_i, \ldots s'_M\}$, and $M = N-3 + 1$.

Step 3:

We coding each 3-bit word of S' by decimal and the coded sequence is marked as D = {d1, d2,..., dM}. Then we statistical histogram of D, and the histogram was marked as H = {h0, h1,..., h7}. And calculated frequency of occurrence of h_i by pi = hi/M. The symbol entropy was defined by Eq. 5.

$$En = -\sum_{i=0}^{7} p(h'_i) \log_2 p(h'_i) \tag{5}$$

2.3 Improved Symbol Entropy

The 0–1 symbol series of symbol entropy retains little detail information. In order to retain more details, we proposed a multi-valued symbol entropy. The symbol series of multi-valued symbol entropy contains variety of symbols rather than two.

Given that VF/VT's sample is marked as X, define $X = (x_1, x_2, \ldots, x_i, \ldots, x_N)$, and sample length was N.

Step 1:

By the specified rule, we converted original time series into symbol time sequence, which markes as $S = (s_1, s_2, \ldots, s_i, \ldots, s_N)$. First, to find the maximum value M_u of X and the minimum value M_d of X, and get $|M| = M_u - M_d$. Second, confirm the symbol space Ω, and the size of Ω is L. Last, according to M_u, M_d and M, get the symbol sequence S.

$$S(i) = \begin{cases} \Omega_{M-1} & if \ \frac{M_dL + |M|(L-1)}{L} \leq x_i \leq \frac{M_dL + |M|L}{L} \\ \Omega_{M-2} & if \ \frac{M_dL + |M|(L-2)}{L} \leq x_i < \frac{M_dL + |M|(L_1)}{L} \\ \quad \vdots \\ \Omega_1 & if \ \frac{M_dL + |M|}{L} \leq x_i < \frac{M_dL + 2|M|}{L} \\ \Omega_0 & if \ M_d \leq x_i < \frac{M_dL + |M|}{L} \end{cases} \tag{6}$$

$i = 1, 2, \ldots, N$, and $x(i)$ represents the *ith* point's value.

Step 2:

We can get the symbol sequence S, and count for each symbol in S. The d_i is the count Ω_i. Therefor get $D = \{d_1, d_2, \ldots, d_i, \ldots, d_L\}$. And the Ω_i probability $p_i = \frac{d_i}{sum(D)}$. The probability vector $H = \{p_1, p_2, \ldots, p_i, \ldots, p_L\}$.

Setp 3:

The Symbol Entropy was defined as:

$$En = -\sum_{i=0}^{L} p_i log_2 p_i \tag{7}$$

Calculate the symbol entropy.

We decompose VT and VF into n sub-signals, which is IMF1-IMFn. And calculate symbol entropy respectively. The Fig. 1 shows the algorithm flow. According to the symbol entropy, distinguish the VT and VF. The classification was operated by threshold.

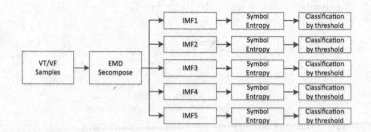

Fig. 1. Algorithm flow

3   Experimental Results and Analysis

3.1   Data Description

We selected the data from MIT-BIH Database and CU Database. The sampling frequency of MIT-BIH Database and CU Database are both 250 Hz. We extracted 100 VF samples and 100 VT samples from CU Database and MIT-BIH Database respectively. All the data's mean and variance were normalized.

The field waveform of VT was plotted in Fig. 2. Ans the field waveform of VT was plotted in Fig. 3.

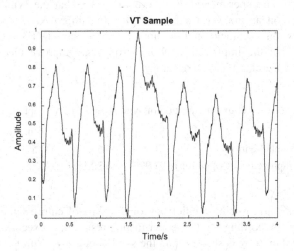

Fig. 2. VT samples

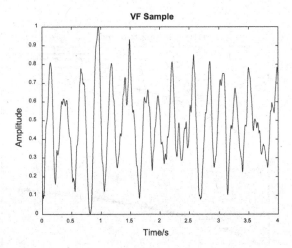

Fig. 3. VF samples

3.2 Experimental Results and Analysis

We have experiment when n = 6, the classification accuracy of IMF3-IMF5 is higher than other IMFS. When n = 4, the symbol entropy of the classification accuracy is 93.5%.

While the classification accuracy of the improved symbol entropy is lower than the former. But when calculate the symbol entropy and the improved symbol entropy of IMF3 + IMF4 + IMF5, the classification accuracy of the improved symbol entropy is 97.75%. The conclusions are drawn as follows: the classification accuracy of the improved symbol entropy is greater than the original symbol entropy.

According to the experimental result, we can know that the symbol entropy has higher classification accuracy on sub-signal, but the improved symbol entropy has higher classification accuracy on the sum of sub-signals (IMF3 + IMF4 + IMF5). Table 1 shows that the improved symbol entropy can improve the sensitivity, the specification and the classification accuracy.

Table 1. The result of classification of symbol entropy and improved symbol

	Sensitivity	Specificity	Accuracy
Symbol entropy	91.87%	95.82%	93.5%
Improved symbol entropy	97.99%	97.51%	97.75%

Now compare the execution efficiency of the symbol entropy and the improved symbol entropy. According to the experimental result, the execution time of the improved symbol entropy is shorted than the symbol entropy. With the increase of the length of the sample, the execution time of symbol entropy and the improved symbol entropy are all increased. Because the symbol entropy needs "3-bit word", it is time-consuming. The time-consuming comparison is shown in Fig. 4.

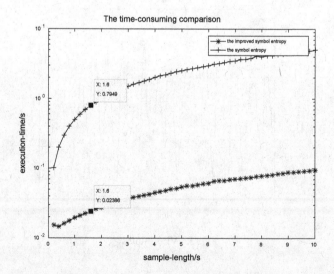

Fig. 4. The time-consuming comparison

With the growth of the sample length, the classification accuracies present different performances. When the time of duration of the sample is less than 0.5 s, the classification accuracies are all less than 90%. When the time of duration of the sample is less than 3 s, the classification accuracy of the improved symbol entropy improves rapidly. But the symbol entropy's classification accuracy growths slowly and less than the improved symbol entropy. Figure 5 shows this experimental result. It follows that the improved symbol entropy can distinguish VT from VF in higher classification accuracy with the small sample whose duration between 1 s–4 s than the symbol entropy.

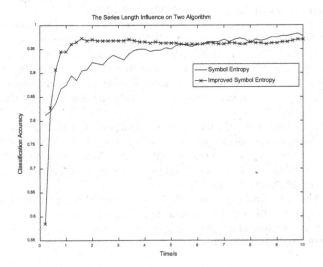

Fig. 5. The series length influence on two algorithm

4 Conclusions

To distinguish VF and VT with high classification accuracy has clinical research significance. We proposes an improved method based on symbol entropy. The method can detect VF and VT using less time-points. From the experimental results, the computation of the improved method is greater than original symbol entropy. The proposed method can be embedded into Automated External Defibrillator (AED) to defibrillate VF in real-time. This can contribute to the timely rescue of the patient's life. Certainly there is a weakness of this paper that the improved symbol entropy method should be tested and verified by other data-sets of VF and VT.

Acknowledgment. This work was supported by the National Natural Science Foundation of China (Grant No. 61671220, 61640218, 61201428), the Shandong Distinguished Middle-aged and Young Scientist Encourage and Reward Foundation, China (Grant No. ZR2016FB14), the Project of Shandong Province Higher Educational Science and Technology Program, China (Grant No. J16LN07), the Shandong Province Key Research and Development Program, China (Grant No. 2016GGX101022).

References

1. Kong, D.R., Xie, H.B.: Use of modified sample entropy measurement to classify ventricular tachycardia and fibrillation. Measurement **44**(4), 653–662 (2011)
2. Amann, A., Tratnig, R., Unterkofler, K.: Detecting ventricular fibrillation by time-delay methods. IEEE Trans. Biomed. Eng. **54**(2), 174–177 (2009)
3. Arafat, M.A., Chowdhury, A.W., Hasan, M.K.: A simple time domain algorithm for detection of ventricular fibrillation in electrocardiogram. Signal Image Video Process. **23**(3), 221–228 (2009)
4. Abbas, R., Aziz, W., Arif, M.: Prediction of ventricular tachyarrhythmia in ECG using neuro-wavelet approach. Emerg. Technol. **34**(5), 82–87 (2012)
5. Thakor, N.V., Zhu, Y.S., Pan, K.Y.: Ventricular tachycardia and fibrillation detection by a sequential hypothesis testing algorithm. IEEE Trans. Biomed. Eng. **37**(9), 837–843 (1990)
6. Zhang, X.S., Zhu, Y.S., Thakor, N.V., Wang, Z.Z.: Detecting ventricular tachycardia and fibrillation by complexity measure. IEEE Trans. Biomed. Eng. **46**(5), 548–555 (1999)
7. Zhang, H.X., Zhu, Y.S., Xu, Y.H.: Complexity information based analysis of pathological ECG rhythm for ventricular tachycardia and ventricular fibrillation. Int. J. Bifurcat. Chaos **12**(10), 2293–2303 (2002)
8. Zhang, H.X., Zhu, Y.S.: Qualitative chaos analysis for ventricular tachycardia and fibrillation based on symbol complexity. Med. Eng. Phys. **23**(8), 523–528 (2001)
9. Owis, M.I., Abou-Zied, A.H., Youssef, A.B., Kadah, Y.M.: Study of features based on nonlinear dynamical modeling in ECG arrhythmia detection and classification. IEEE Trans. Biomed. Eng. **49**(7), 733–736 (2002)
10. Sun, Y., Chan, K.L., Krishnan, S.M.: Life-threatening ventricular arrhythmia recognition by nonlinear descriptor. Biomed. Eng. Online **29**(2), 503–511 (2010)
11. Wang, J., Chen, J.: Symbol dynamics of ventricular tachycardia and ventricular fibrillation. Phys. A **389**(10), 2096–2100 (2010)

A Selective Transfer Learning Method for Concept Drift Adaptation

Ge Xie[1], Yu Sun[1], Minlong Lin[2], and Ke Tang[1(✉)]

[1] School of Computer Science and Technology,
University of Science and Technology of China, Hefei 230027, China
{xiege,sunyu123}@mail.ustc.edu.cn, ketang@ustc.edu.cn
[2] Tencent Company, Shenzhen 518057, China
minlonglin@tencent.com

Abstract. Concept drift is one of the key challenges that incremental learning needs to deal with. So far, a lot of algorithms have been proposed to cope with it, but it is still difficult to response quickly to the change of concept. In this paper, a novel method named Selective Transfer Incremental Learning (STIL) is proposed to deal with this tough issue. STIL uses a selective transfer strategy based on the well-known chunk-based ensemble algorithm. In this way, STIL can adapt to the new concept of data well through transfer learning, and prevent negative transfer and overfitting that may occur in the transfer learning effectively by an appropriate selective policy. The algorithm was evaluated on 15 synthetic datasets and three real-world datasets, the experiment results show that STIL performs better in almost all of the datasets compared with five other state-of-the-art methods.

Keywords: Incremental learning · Transfer learning · Concept drift

1 Introduction

In the age of big data, incremental learning is becoming more and more important. However, the distribution of training data in incremental learning may be nonstationary over time, especially for real-world applications, which is called the concept drift phenomenon [1]. Despite the significance of concept drift for studying incremental learning, the concept change is difficult to be handled effectively and quickly due to the unpredictable dynamic environments.

In recent years, three categories of approaches have been proposed to cope with concept drift, including sliding window [2,3], drift detection [4] and ensemble method [5–7]. Among them, Ensemble method is the only one which preserves the history knowledge and use it to reinforce the learning of new circumstances. Transfer-based ensemble learning is a new type of ensemble method combining transfer learning with the traditional chunk-based ensemble algorithm, which was introduced by TransferIL [8]. TransferIL could make preserved historical models adapt to the new coming data by transfer learning.

© Springer International Publishing AG 2017
F. Cong et al. (Eds.): ISNN 2017, Part II, LNCS 10262, pp. 353–361, 2017.
DOI: 10.1007/978-3-319-59081-3_42

In this paper, a novel method is proposed called Selective Transfer Incremental Learning (STIL). In the new method, all the preserved historical models are not transferred blindly as TransferIL does, because the massive transfer operations are time-consuming and some inappropriate transfer may cause negative transfer and overfitting. Moreover, when the number of models in ensemble reaches the maximum threshold, STIL will replace the least relevant model with the new model learned from recent data chunk, which is more reasonable than the traditional accuracy policy.

The rest of this paper is organized as follows. Section 2 reviews related work and describes specific problems to be solved. Section 3 presents STIL in details, and the experimental studies of STIL are shown in Sect. 4. Finally, Sect. 5 concludes the paper.

2   Related Work

There are three types of methods for handling concept drift in the data stream, including sliding window [2,3], drift detection [4], and ensemble method [5–7]. The sliding window maintains the most recent examples to assist learning from current data, while drift detection will remind to rebuild the current model once concept drift is detected in the data stream, otherwise the model is updated directly. Different from the above two methods, ensemble method preserves some of the historical knowledge learned from the past data and uses it to boost the learning of new data.

In ensemble methods, there is a popular algorithm called chunk-based ensemble. Chunk-based ensemble means every model in the ensemble is trained from one chunk of data in the data stream [6]. Once the model number in ensemble exceeds the limit, one of the existing models should be replaced by the new model. There have been some algorithms handling concept drift based on the idea of chunk-based ensemble. Typical examples are Streaming Ensemble Algorithm (SEA) [5] and Accuracy Updated Ensemble (AUE2) [6]. SEA is one of the early-stage chunk-based algorithm, which uses a simple majority voting policy to combine the outputs of all base models. Accuracy Updated Ensemble (AUE2) uses accuracy-based weighting mechanisms for combining outputs, and the preserved base models are updated incrementally with the new data as well.

For chunk-based ensemble learning, due to the existing of concept drift, historical knowledge may have inconsistent information with the knowledge of current circumstance. Hence, a transfer-based ensemble learning strategy is proposed in TransferIL [8]. Transfer-based ensemble learning transfers the preserved historical models using current data chunk first, instead of using them directly in the ensemble, to adapt the models to the new circumstance. The transfer operation in TransferIL will keep the related knowledge of the old model and update the unrelated part using new data. Then, TransferIL combines the transferred models and the newly trained model to form the final ensemble model.

Despite the good adaptability, there may be some inappropriate transfer in TransferIL, which may cause negative transfer and overfitting problems. On the other hand, the massive transfer operation is time-consuming. Hence, transfer all the preserved historical models blindly is unreasonable.

3   The Proposed Approach

In this paper, a new method named STIL is proposed. STIL will not transfer all the preserved historical models as TransferIL does. Instead, it adopts a selection policy that picks up two categories of historical models which not doing the transfer operation to avoid the negative transfer and overfitting problems. The two categories are the less relevant model with the new coming data and the less transfer effective historical model.

3.1   Method

The Less Relevant Model. When the source domain is not related to the target domain, brute-force transfer may be unsuccessful and may degrade the performance of learning in the target domain [9]. This is called negative transfer in the transfer learning area. In transfer-based ensemble learning, when the historical model is almost not related to the new coming data chunk, there will be very few useful knowledge that could be used in the new circumstance. Hence it makes no sense to do the transfer operation.

To find the less relevant model of the current new data chunk, firstly, STIL uses current data chunk to get a model through training, then uses Q-statistic [10], as shown in (1), to compute the correlation between every preserved historical model and the newly trained model.

$$Q_{f_i, f_j} = \frac{N^{11}N^{00} - N^{01}N^{10}}{N^{11}N^{00} + N^{01}N^{10}}. \tag{1}$$

where f_i and f_j are two classifiers. $N^{y_i y_j}$ is the number of examples for which the classification result is y_i by f_i and y_j by f_j. If classifier recognizes the example correctly then $y = 1$, and otherwise $y = 0$. The result of Q_{f_i, f_j} varies between -1 and 1. If the two classifiers are statistically independent then $Q_{f_i, f_j} = 0$; if the correctness of two classifiers is exactly the same then $Q_{f_i, f_j} = 1$; else if they are totally opposite, then $Q_{f_i, f_j} = -1$ [11].

Hence, in STIL, when the value of Q-statistic is close to 0, the computed history model is a less relevant model to the new chunk of data.

The Less Transfer Effective Model. If the operation of transfer has less effect on the preserved historical model, which means this model fits the new chunk of data well, there will be no need to do the time-consuming operation of transfer. On the other hand, if the operation of transfer had been done in this situation, it would make the model too complicated and may cause overfitting.

STIL finds the less transfer effective model by computing the gain of transfer. As the implementation in TransferIL, decision tree is used as the base model.

STIL makes data chunk fall into the leaf nodes of the model, and counts the number of leaf nodes which satisfy the condition of change if transfer happened. The change could be label change or continue splitting of the leaf node, as introduced in TransferIL. Then STIL sets the proportion of such leaf nodes in all the leaves as the standard of measuring transfer gain, as shown in (2).

$$TG = (N_{labelchange} + N_{splitting})/M. \tag{2}$$

where TG is the predicted gain of transfer, $N_{labelchange}$ and $N_{splitting}$ are the number of leaves which satisfy the condition of label change and continue splitting of leaf, respectively. M is the total number of leaves.

By predicting the gain of transfer, STIL avoids the transfer learning process of the less transfer effective model. What's more, in fact, the operation of putting data chunk to fall into the leaf nodes had been done when computing the classification result in Eq. (1) for finding the less relevant model.

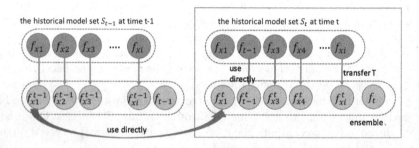

Fig. 1. Illustration of how to construct the alternative model. In this example, at time step t, f_{x1} and f_{t-1} are assumed as the less relevant models to the new data. Obviously, f_{t-1} is the newly added model at last time step $t-1$, which replaced f_{x2}. Moreover, f_{x3} is assumed as the less transfer effective model at time t.

Construction of Alternative Model. For the two kinds of non-transfer models, a strategy is also proposed for STIL to construct alternative models in ensemble. Some existing model is used as the replacement of the original transferred model, as shown in Fig. 1.

For the less relevant model to current data chunk, the model is completely out-of-date to new circumstance, so it will not be used in the ensemble for predicting. However, if it is deleted, the model number in the ensemble will decrease, which may also hurt the performance. STIL uses a trick to solve this problem by using the less relevant historical model's transferred model produced during the last time step $t-1$, such as f_{x1}^t does in Fig. 1. The effect of such operation will be using the transferred model created by adapting the historical model to its most recently relevant data chunk, which can make the historical model adapt to the new circumstance and avoid transferring in unrelated historical model and data. But if it is unfortunate that the less relevant model is the new added one at last time step $t-1$, which will only occur in suddenly concept drift, there is no choice but to maintain this old model, such as f_{t-1}^t does in Fig. 1.

For the less transfer effective model, STIL simply maintains the historical model in the ensemble, such as $f_{x_3}^t$ does in Fig. 1.

3.2 Algorithm

The pseudo code of STIL is presented in Algorithm 1. The lines 4–12 describe the operation for different kind of historical models. Lines 14–18 show the process of how to choose the preserved historical model set. Here a new strategy of removing the least relevant historical model is employed. The relevant computation result for line 4 can be used directly here. Moreover, the relevant strategy is more reasonable than the traditional accuracy policy.

Algorithm 1. STIL

Input: D_1, D_2, ..., D_t: the divided chunk in data stream, S_t: the set of preserved historical models at time t, m: the max number of preserved historical models

Output: F_t: the ensemble model at time t

1: **for** all the data chunks D_t **do**
2: $\quad$ $f_t \leftarrow$ new component classifier trained by D_t
3: $\quad$ **for** every model f_i in S_t **do**
4: $\quad\quad$ **if** f_i is a less relevant model **then**
5: $\quad\quad\quad$ **if** f_i is not the new added model in time t-1 **then**
6: $\quad\quad\quad\quad$ $f_i^t \leftarrow f_i^{t-1}$
7: $\quad\quad\quad$ **else**
8: $\quad\quad\quad\quad$ $f_i^t \leftarrow f_i$
9: $\quad\quad$ **else if** f_i is a less transfer effective model **then**
10: $\quad\quad\quad$ $f_i^t \leftarrow f_i$
11: $\quad\quad$ **else**
12: $\quad\quad\quad$ $f_i^t \leftarrow$ transfer model f_i with D_t
13: $\quad\quad$ $\omega_i^t \leftarrow$ evaluate the model f_i^t
14: $\quad$ **if** $|S_{t-1}| < m$ **then**
15: $\quad\quad$ $S_t \leftarrow S_{t-1} \cup f_t$
16: $\quad$ **else**
17: $\quad\quad$ $f_w \leftarrow$ the least relevant model in S_t
18: $\quad\quad$ replace f_w with f_t in S_t
19: $\quad$ $F_t = (\sum_i w_i^t f_i^t + f_t)/(\sum_i w_i^t + 1)$

---

4 Experiment

4.1 Experimental Setting

The accuracy and speed of STIL were compared with five state-of-the-art algorithms namely TransferIL [8], SEA [5], Learn^{++}.NSE [7], AUE2 [6] and TIX [12].

They are all proposed to handle the concept drift problem in incremental learning, SEA, Learn++.NSE and AUE2 are chunk-based ensemble methods, while TIX treats history knowledge as new features.

The compared algorithms, in general, can accommodate any type of base models. To be fair, the base model of all algorithms was set as a decision tree. In SEA, Learn++.NSE and TIX the traditional decision tree model CART was used. However, the base model needs to be updated incrementally in AUE2, so an online decision tree named Hoeffding tree [14] was used instead.

The limit of preserved models was set as 25 in SEA, AUE2, TransferIL and STIL according to the suggestion in [5]. In STIL, two parameters were defined, k_1 represents the number of the less relevant models and k_2 represents the number of the less transfer effective models. In the experiment we set $k_1 = 8$ and $k_2 = 5$.

The experiment was performed on 15 synthetic datasets and three real-world datasets. Synthetic datasets create the changing concept along time by modifying certain parameters continuously. Based on previous research, five different kinds of widely used synthetic datasets SEA [5], ROT [6,7], CIR [4], SIN [4] and STA [16] were employed in our research and represented five different types of concept drift. Three public real-world datasets Covertype [15], PokerHand [15] and Electricity [4,6] were also used to evaluate the algorithms.

Table 1. Average accuracy (%) of every chunk (± indicates the standard deviation of the accuracy for each chunk) for the tested algorithms. The values in boldface indicate the highest accuracy on the data stream. The last two rows show the result of the Friedman test and Wilcoxon test (with a 0.05 significance level)

Data	STIL	TransferIL	SEA	Learn++.NSE	AUE2	TIX
SEA200A	**96.18±2.89**	94.80±3.03●	86.31±11.43●	89.07±5.13●	94.66±4.94	87.77±3.97●
SEA200G	**95.20±2.54**	94.15±2.55●	88.90±10.02●	90.02±4.98●	94.58±3.80	86.90±4.26●
SEA500G	**97.26±1.99**	96.39±1.66●	89.37±10.17●	91.10±3.45●	95.02±4.05●	88.85±2.54●
ROT200A	**71.89±14.14**	71.59±14.21	37.88±18.17●	62.19±11.49●	52.72±9.99●	65.02±11.45●
ROT200G	**73.86±13.81**	72.36±14.48	54.61±17.45●	63.41±12.84●	55.43±9.76●	64.97±12.16●
ROT500G	**85.27±11.67**	83.92±12.61	69.81±14.29●	74.77±11.57●	74.34±11.34●	76.98±10.44●
CIR200A	**86.02±3.26**	84.90±4.00●	79.90±10.10●	81.33±5.79●	82.21±5.29●	78.98±5.30●
CIR200G	**86.22±3.03**	84.90±3.61●	83.86±8.60●	83.27±4.38●	84.06±4.63●	80.04±4.65●
CIR500G	**87.18±2.16**	86.49±2.20●	84.32±8.43●	83.60±3.00●	84.87±4.36●	80.20±2.70●
SIN200A	**83.40±3.28**	82.58±3.14●	65.78±8.44●	78.02±4.00●	71.91±3.69●	77.05±3.91●
SIN200G	**84.04±3.09**	82.49±3.16●	74.12±8.11●	79.07±3.27●	74.32±5.48●	77.23±3.70●
SIN500G	**85.79±2.07**	85.14±1.90●	73.76±7.80●	80.67±2.51●	78.22±4.71●	78.36±2.84●
STA200A	89.23±5.01	89.50±2.32	70.07±21.57●	82.74±7.88●	86.06±10.93●	**89.85±2.01**
STA200G	89.27±3.86	89.54±2.48	76.01±15.71●	83.56±7.54●	86.58±9.06	**89.77±2.32**
STA500G	89.38±3.54	**90.00±1.34**	76.43±15.35●	84.78±7.25●	87.47±7.73	**90.00±1.34**
Covertype	90.65±10.08	**91.44±8.55**○	71.46±15.14●	84.11±12.45●	87.09±8.74●	88.32±9.15●
PokerHand	52.44±1.69	51.95±1.79●	**56.36±2.54**○	45.86±1.87●	51.31±1.79●	47.23±1.73●
Electricity	**76.74±8.27**	74.90±8.34	72.35±13.99	75.54±8.09	76.47±8.70	73.55±8.72
Friedman-test	1.39	2.11	5.28	3.50	3.67	4.00
Wilcoxon-test	-	10-7-1	16-1-1	17-1-0	13-5-0	14-4-0

4.2 Experimental Result

The results in terms of accuracy are shown in Table 1. They indicate that STIL achieved the highest accuracy value in almost all of the datasets, compared with the other five state-of-the-art algorithms. Although STIL did not achieve the best results in five different datasets: STA200A, STA200G, STA500G, Covertype and PokerHand, it still got competitive results.

The nonparametric Friedman test was conducted on the results in terms of accuracy. The test statistic $F_F = 6.3125$ and the critical value for $\alpha = 0.05$ is 2.32, hence we reject the null-hypothesis that there is no difference among the performance of all the tested algorithms. We also compute the critical difference(CD) chosen by the Nemenyi test, and got $CD = 1.78$, which means that STIL performs significantly better than SEA, Learn[++].NSE, AUE2 and TIX. Moreover, the Wilcoxon rank-sum test gives a detailed information to the comparison of pairs of algorithms, as shown in the last row of Table 1. Comprehensively, STIL performs significantly better than any of the compared algorithms.

Table 2. Runtime of each algorithm (unit:second)

Data	STIL	TransferIL	SEA	Learn[++].NSE	AUE2	TIX
SEA200A	1.40e2	2.08e2	3.17e1	2.09e1	5.09e0	2.06e1
SEA200G	1.36e2	2.16e2	3.20e1	1.97e1	5.73e0	2.04e1
SEA500G	3.20e2	4.80e2	7.27e1	2.64e1	1.21e1	2.52e1
ROT200A	1.65e2	2.58e2	9.00e1	8.41e1	6.51e0	3.14e1
ROT200G	1.63e2	2.84e2	1.03e2	8.40e1	6.70e0	3.13e1
ROT500G	3.87e2	6.07e2	2.63e2	1.03e2	1.53e1	3.69e1
CIR200A	1.41e2	2.57e2	3.42e1	1.99e1	4.91e0	2.73e1
CIR200G	1.38e2	2.24e2	4.14e1	2.40e1	4.83e0	3.12e1
CIR500G	3.20e2	4.80e2	9.68e1	2.94e1	1.25e1	3.38e1
SIN200A	1.55e2	2.46e2	3.83e1	2.26e1	4.93e0	2.43e1
SIN200G	1.49e2	2.41e2	3.90e1	2.19e1	4.90e0	2.28e1
SIN500G	3.26e2	5.27e2	9.59e1	2.91e1	1.07e1	2.80e1
STA200A	1.10e2	1.76e2	3.84e1	2.89e1	5.88e0	2.14e1
STA200G	1.24e2	1.75e2	5.04e1	2.10e1	5.99e0	2.12e1
STA500G	3.27e2	6.14e2	9.32e1	2.57e1	1.68e1	2.36e1
Covertype	1.39e3	1.43e3	1.55e3	1.02e3	5.41e2	9.66e2
PokerHand	7.27e3	9.45e3	8.83e2	7.75e3	2.86e2	6.09e3
Electricity	6.58e1	7.61e1	2.83e1	1.53e1	1.22e1	5.49e0

The runtime results are shown in Table 2. The experiment was run on a laptop with 8 GB RAM and a CPU which is Intel i5 with 2 cores, 2.7 GHz. In comparison to TransferIL, STIL could reduce the runtime almost to all the datasets by only picking up sub-models to transfer. However, due to the additional selecting process and the high costs of transfer operation in unselected models, the runtime is still more than the other algorithms except for TransferIL.

5   Conclusion

In this paper, we propose a selective transfer method for incremental learning named STIL, which only transfers some of the preserved history models. The selective policy of STIL can avoid negative transfer by not transferring less relevant models and avoid overfitting by not transferring less transfer effective models. The experimental results show that STIL could not only reduce runtime compared with transfer all the preserved old models, but also raise up the performance to make it achieve the highest accuracy in almost all the datasets used in the empirical studies.

Acknowledgments. This work was supported in part by the National Natural Science Foundation of China under Grant 61329302 and Grant 61672478, and in part by the Royal Society Newton Advanced Fellowship under Grant NA150123.

References

1. Gama, J., Žliobaitė, I., Bifet, A., Pechenizkiy, M., Bouchachia, A.: A survey on concept drift adaptation. ACM Comput. Surv. **46**(4), 44 (2014)
2. Bifet, A., Gavalda, R.: Learning from time-changing data with adaptive windowing. In: Proceedings of the SIAM International Conference on Data Mining, pp. 443–448 (2007)
3. Hulten, G., Spencer, L., Domingos, P.: Mining time-changing data streams. In: KDD, pp. 97–106 (2001)
4. Gama, J., Medas, P., Castillo, G., Rodrigues, P.: Learning with drift detection. In: Bazzan, A.L.C., Labidi, S. (eds.) SBIA 2004. LNCS, vol. 3171, pp. 286–295. Springer, Heidelberg (2004). doi:10.1007/978-3-540-28645-5_29
5. Street, W.N., Kim, Y.: A streaming ensemble algorithm (SEA) for large-scale classification. In: KDD, pp. 377–382 (2001)
6. Brzezinski, D., Stefanowski, J.: Reacting to different types of concept drift: the accuracy updated ensemble algorithm. IEEE Trans. Neural Netw. **25**(1), 81–94 (2014)
7. Elwell, R., Polikar, R.: Incremental learning of concept drift in nonstationary environments. IEEE Trans. Neural Netw. **22**(10), 1517–1531 (2011)
8. Sun, Y., Tang, K.: Incremental learning with concept drift: a knowledge transfer perspective. In: Gong, M., Pan, L., Song, T., Zhang, G. (eds.) BIC-TA 2016. CCIS, vol. 681, pp. 473–479. Springer, Singapore (2016). doi:10.1007/978-981-10-3611-8_43
9. Pan, S.J., Yang, Q.: A survey on transfer learning. IEEE Trans. Knowl. Data Eng. **22**(10), 1345–1359 (2010)
10. Yule, G.U.: On the association of attributes in statistics: with illustrations from the material of the childhood society, &c. Phil. Trans. **194**, 257–319 (1900)

11. Kuncheva, L.I., Whitaker, C.J., Shipp, C.A., Duin, R.P.: Is independence good for combining classifiers? Pattern Recognit. **2**, 168–171 (2000)
12. Forman, G.: Tackling concept drift by temporal inductive transfer. In: SIGIR, pp. 252–259 (2006)
13. Minku, L., Yao, X.: DDD: a new ensemble approach for dealing with concept drift. IEEE Trans. Knowl. Data Eng. **24**(4), 619–633 (2012)
14. Domingos, P., Hulten, G.: Mining high-speed data streams. In: KDD, pp. 71–80 (2000)
15. Bache, K., Lichman, M.: UCI Machine Learning Repository. http://archive.ics.uci.edu/ml (2013)
16. Schlimmer, J.C., Granger, R.H.: Incremental learning from noisy data. Mach. Learn. **1**(3), 317–354 (1986)

On the Co-absence of Input Terms
in Higher Order Neuron Representation
of Boolean Functions

Oytun Yapar[1,2(✉)] and Erhan Oztop[1]

[1] Ozyegin University, Istanbul, Turkey
oytun.yapar@ozu.edu.tr
[2] Vestel Electronics, Manisa, Turkey

Abstract. Boolean functions (BFs) can be represented by using polynomial functions when −1 and +1 are used represent True and False respectively. The coefficients of the representing polynomial can be obtained by exact interpolation given the truth table of the BF. A more parsimonious representation can be obtained with so called polynomial sign representation, where the exact interpolation is relaxed to allow the sign of the polynomial function to represent the BF value of True or False. This corresponds exactly to the higher order neuron or sigma-pi unit model of biological neurons. It is of interest to know what is the minimal set of monomials or input lines that is sufficient to represent a BF. In this study, we approach the problem by investigating the (small) subsets of monomials that cannot be absent as a whole from the representation of a given BF. With numerical investigations, we study low dimensional BFs and introduce a graph representation to visually describe the behavior of the two-element monomial subsets as to whether they cannot be absent from any sign representation. Finally, we prove that for any n-variable BF, any three-element monomial set cannot be absent as a whole if and only if all the pairs from that set has the same property. The results and direction taken in the study may lead to more efficient algorithms for finding higher order neuron representations with close-to-minimal input terms for Boolean functions.

Keywords: Boolean function · Higher order neuron · Sigma-pi neuron model · Polynomial sign representation · Weight elimination · Minimum fan-in representation

1 Introduction

When −1 and +1 are used to represent True and False respectively, a Boolean function is identified by a real valued vector function $f:\{-1, 1\}^n \to \{-1, 1\}$. In particular, a unique polynomial function can be constructed by using Lagrange interpolation to realize f. This polynomial may have up to 2^n terms, i.e. monomials, and its coefficients correspond to the weights of a higher order neuron or sigma-pi units (Giles and Maxwell 1987, Schmitt 2005) that compute f. In general, in a sigma-pi unit the output of the neuron is obtained after the application of an activation function. When this is chosen as the *sign* function the weights of the input lines are no more unique;

© Springer International Publishing AG 2017
F. Cong et al. (Eds.): ISNN 2017, Part II, LNCS 10262, pp. 362–370, 2017.
DOI: 10.1007/978-3-319-59081-3_43

moreover, not all input lines are necessary to compute the BF. In particular, a small number of input lines (i.e. fan-in of the neuron) are desirable from a computational point of view. This might also have implications for the neural organization of the brain.

In this paper, we focus our attention on the infeasibility of the simultaneous absence of the set of monomials from any sign representation given a BF to represent. We think that the regularities observed then can be used to derive efficient algorithms for representing all or particular class of Boolean functions compactly with higher order neurons. In the literature, there have been several studies for developing algorithms to find low fan-in solutions to given BFs (Ghosh and Shin 1992, Guler 2001, Oztop 2009) and mathematical results indicating that it is always possible to represent an n-variable BF with a higher order neuron that has at most 0.75×2^n input lines. In other words, at least 25% of the weights of higher order neuron can be zeroed; however, this bound is not tight, and thus algorithmic and theoretical improvements are needed (Oztop 2006). For example, recently it has been shown that with 11 monomials any 5-variable Boolean function; and with 26 monomials any 6-variable Boolean function can be sign-represented (Sezener and Oztop 2015).

Higher Order Neurons and Polynomial Sign Representation

Higher order neurons or sigma-pi units are neural models that compute the following function

$$f(x_1, x_2, \cdots, x_n) = sign\left(\sum_{i=1}^{2^n} w_{S_i} \prod_{k \in S_i} x_k\right) \text{ where } S_i \subset \{1, 2, \cdots, n\}$$

The full representation capacity of this unit is easily obtained by adopting a highly symmetrical vector notation for the product term and assignments. This gives raise to so called Sylvester-type Hadamard matrix $\mathbf{D}_n$ (Siu et al. 1995) for n-dimensional Boolean functions, which sets up a one-to-one mapping between positive real vectors and solving weights (i.e. polynomial coefficients). $\mathbf{D}_n$ comes with handy properties, such as recursive definition and orthogonality properties (i.e. $\mathbf{D}_n \mathbf{D}_n = 2^n \mathbf{I}$). We drop the subscript n when it is clear from the context. Noting that Boolean functions can be considered as 2^n dimensional vectors according to the adopted order, it can be shown that for a given n-variable BF f (or in vector form $\mathbf{f}$, which we use from now on) all the solutions are determined with $\mathbf{a}^T = \mathbf{k}^T diag(\mathbf{f})\mathbf{D}$ where $\mathbf{k}^T = [k_1, k_2, \cdots, k_{2^n}]^T > \mathbf{0}$ (Oztop 2006). This means that for having a higher order neuron represent $\mathbf{f}$ with minimum number of input terms (i.e. monomials), we must find $k_i > 0$ values so as to have maximum number of zero elements in $\mathbf{a}$. As each element of $\mathbf{a}$ is the coefficient of a unique monomial we often talk of 'zeroability' of a monomial.

Definition (Zeroability): A set of monomials $(m_{z_1}, m_{z_2} \ldots, m_{z_r})$ are called zeroable for a BF $\mathbf{f}$, if there exists a positive vector k such that $\mathbf{a}^T = \mathbf{k}^T diag(\mathbf{f})\mathbf{D}$ results in zeros at the positions corresponding the monomials, i.e. $a_{z_1} = a_{z_2} \ldots = a_{z_r} = 0$.

2   Zeroability Patterns of Monomials in the Higher Order Neuron Representation of Boolean Functions

2.1   Equivalent Classes of Polynomial Boolean Functions

It is known that the transformations 1–5 below do not change the threshold density of a Boolean function (i.e. the minimum number of monomials that would be sufficient to sign represent it) (Sezener and Oztop 2015).

1. Negation of input variables (e.g., f (x1, x2, x3) → f (x1, x2, −x3))
2. Permutation of input variables (e.g., f (x1, x2, x3) → f (x2, x1, x3))
3. Negation of the output (e.g., f (x1, x2, x3) → −f (x1, x2, x3))
4. XORing an input variable with other variables (e.g., f (x1, x2, x3) → f (x1, x2 ⊕ x3 ⊕ x1, x3))
5. XORing the function with input variables (e.g., f (x1, x2, x3) → x1 ⊕ x2 ⊕ f (x1, x2, x3))

These transformations were used in the spectral classification of BFs introduced by Edwards (1975). The first three of these transformations are more common and usually called NPN (negation-permutation-negation) transformations. Any combination of these transformations creates so called equivalence classes over BFs. Since any pair of functions from an equivalence class can be converted to each other by transformations 1–5, the zeroability properties of the monomials can be studied by looking at one representative function from each equivalence class. Thus, in this study, we identify each equivalent class with one of its member function and study the properties of these functions. A natural labeling system for functions is used as illustrated in Table 1 by adopting a fixed ordering over the function arguments.

Table 1. The function output vector [−1 1−1 1−1 1−1 1] can be mapped (with t(b) = (1−b)/2) to the binary number 10101010 that is 0xaa in hexadecimal notation.

Arguments			Function
x_3	x_2	x_1	value
1	1	1	−1
1	1	−1	1
1	−1	1	−1
1	−1	−1	1
−1	1	1	−1
−1	1	−1	1
−1	−1	1	−1
−1	−1	−1	1

2.2   Numerical Investigations on Monomial Zeroability

Assume that we're given a Boolean function f and are asked to find a higher order neuron representation without the monomials $(m_{z_1}, m_{z_2}..., m_{z_r})$. Then the zeroability of these monomials induces a set of linear equations to be solved with positivity constraints.

As $\mathbf{a}^T = \mathbf{k}^T diag(f)\mathbf{D} = \mathbf{k}^T\mathbf{Q}$ for any $\mathbf{k}^T = [k_1, k_2, \cdots, k_{2^n}]^T > \mathbf{0}$ is a solution, we must ensure $\mathbf{k}^T\mathbf{R} = 0$ where $\mathbf{R}$ is submatrix of $\mathbf{Q}$ composed of columns $q_{z_1}, q_{z_2}\ldots, q_{z_r}$ that correspond to the monomials $m_{z_1}, m_{z_2}\ldots, m_{z_r}$. So we have r equations with 2^n positive unknowns. If no positive $\mathbf{k}$ exists to satisfy $\mathbf{k}^T\mathbf{R} = 0$ then it is concluded that the problem is infeasible without $m_{z_1}, m_{z_2}\ldots, m_{z_r}$. If such $\mathbf{k}$ exists, $\mathbf{a}^T = \mathbf{k}^T\mathbf{Q}$ will be the weights of the higher order neuron that represents $\mathbf{f}$. The (difficult) question of interest is to find the maximum r so as to have $\mathbf{k}^T\mathbf{R} = \mathbf{0}$ satisfiable. Finding the maximum r for a given BF can be performed through an easy yet very costly exhaustive search over all the column subsets of $\mathbf{Q}$ (i.e. subsets of system of linear equations $\mathbf{0}^T = \mathbf{k}^T\mathbf{Q}$). This approach can only be applied to low dimensional problems due to combinatorial explosion of the number of subsets. Even the help of equivalence classes to reduce the number of BFs becomes irrelevent after 6 or more variable functions. So, it would be of great value if we could exploit the smaller (un)feasible subsets of the linear equation system $\mathbf{0}^T = \mathbf{k}^T\mathbf{Q}$ to infer about larger subsets. Towards this end, we have first explored all subsets of linear equations in three and four dimensions for all functions and obtained a full picture of the zeroability patterns of all monomial subsets. For this we used the following exhaustive search algorithm.

Algorithm 1.
1. **Input: f**:function ($2^n x1$ vector)
2. **Output:**
   a. zeroable_counter
   b. zeroable_index_list (number of zeroable subsets, and the list of index sets that achieve this)
3. **Initialization:**
   a. Dimension of the problem: $n = \log_2(length(\mathbf{f}))$
   b. Compute $\mathbf{D}_n$ ($2^n \times 2^n$ Slyvester-type Hadamard matrix)
   c. Potential coeffients of the system of linear equations: $\mathbf{Q} = diag(\mathbf{f})\mathbf{D}_n$
   d. zeroable_counter[i]=0;   for all $1 \leq i \leq 2^n$
   e. zeroable_index_list[i] ={} for all $1 \leq i \leq 2^n$
4. **for** $r = 1$ **to** 2^n
   a. for all r-column subset $\mathbf{R}$ of $\mathbf{Q}$ ($\mathbf{R} = [\mathbf{Q}(z_1), \mathbf{Q}(z_2),\ldots, \mathbf{Q}(z_r)]$ for $\{z_1,\ldots,z_r\} \subset 2^{\{1,2,\cdots,n\}}$
      i.   **if** $\mathbf{k}^T\mathbf{R} = 0$ feasible (check with Linear Programming)
      ii.   zeroable_counter(r) = zeroable_counter(r) + 1
      iii.   add $\{z_1, z_2,\ldots, z_r\}$ to the zeroable_index_list(r) list
      iv.   **endif**
   b. **endfor**
5. **if** zeroable_counter(r) == 0
      break, since there cannot be anymore zeroable solution
   6. **end**
   7. **endfor**
8. **return** zeroable_counter, zeroable_index_list

3-Variable Boolean Functions: According to adopted equivalence class, there are essentially three types of 3-variable Boolean functions, i.e. there are only three equivalence classes. For each of these classes, the number of zeroable subsets as obtained by Algorithm 1 is shown in Table 2. This zeroability pattern necessarily covers all the 3-variable BFs since all functions are equivalent to one of these functions in terms of their zeroability pattern. Note that the first function class (0xaa) covers the functions that can be represented with single input lines, i.e. those that have minimum threshold density equal to 1. So these functions are essentially the monomials themselves (e.g. $f(x_1, x_2, x_3) = x_2 x_3$ or $f(x_1, x_2, x_3) = x_1$)

Table 2. Number of zeroable subsets for each function class is given for each subset cardinality

Function label	1-sized subsets	2-sized subsets	3-sized subsets	4-sized subsets	5-sized subsets	6-sized subsets	7-sized subsets	8-sized subsets
0xaa	7	21	35	35	21	7	1	0
0xab	8	21	35	28	0	0	0	0
0xac	8	22	28	17	4	0	0	0
Total	8	28	56	70	56	28	8	1

4-Variable Boolean Functions: According to adopted equivalence class above, there are essentially only 8 different 4-variable Boolean functions. For each of these classes, the number of zeroable subsets is shown in Table 3. Again, this result necessarily reflects the zeroability patterns for all 4-variable BFs.

Table 3. Number of zeroable subsets for each function class is given for each subset cardinality

Function label	1-sized subsets	2-sized subsets	3-sized subsets	4-sized subsets	5-sized subsets	6-sized subsets	7-sized subsets	8-sized subsets
0xaa55	15	105	455	1365	3003	5005	6435	6435
0xab55	16	105	455	1365	3003	5005	6435	6420
0xbb55	16	113	483	1414	2996	4690	5426	4573
0xaba5	16	117	521	1551	3156	4356	4236	3084
0xaaff	16	114	484	1375	2772	4092	4488	3663
0xaba4	16	119	546	1675	3388	4113	3490	2124
0xab12	16	120	560	1740	3492	4077	2910	1425
0xac90	16	120	560	1760	3648	4096	1600	0
Total	16	120	560	1820	4368	8008	11440	12870

Function label	9-sized subsets	10-sized subsets	11-sized subsets	12-sized subsets	13-sized subsets	14-sized subsets	15-sized subsets	16-sized subsets
0xaa55	5005	3003	1365	455	105	15	1	0
0xab55	4900	2688	840	0	0	0	0	0
0xbb55	2724	1085	259	28	0	0	0	0
0xaba5	1684	672	144	0	0	0	0	0
0xaaff	2200	946	276	49	4	0	0	0
0xaba4	928	256	32	0	0	0	0	0
0xab12	400	61	6	0	0	0	0	0
0xac90	0	0	0	0	0	0	0	0
Total	11440	8008	4368	1820	560	120	16	1

2.3   Pairwise Zeroability of Monomials from Sign Representations

As introduced before, for a given BF **f**, a sign representing polynomial or the weights of a higher order neuron to represent **f** can be find by an arbitrary choice of **k > 0** with

$\mathbf{a}^T = \mathbf{k}^T diag(\mathbf{f})\mathbf{D}$ where $\mathbf{a}$ is weights of the higher order neuron (or the coefficients of the sign representing polynomial). Fourier-Motzkin elimination is a procedure for eliminating the variables of a given inequality system and thus can be used to assess the infeasibility of the system (Chandru 1993). In particular, FM elimination can be applied on a selected set of columns $\mathbf{C} = \{c_{i_1}, c_{i_2}, \ldots c_{i_r}\}$ of $\mathbf{Q} = diag(\mathbf{f})\mathbf{D}$ to yield a $\mathbf{Q_{FM}}$ which can be easily converted to a sign-representation where $m_{i_1}, m_{i_2}, \ldots, m_{i_r}$ are zero by taking the row sum of $\mathbf{Q_{FM}}$ ($\mathbf{a}^T = \mathbf{1}^T\mathbf{Q_{FM}}$) (Oztop 2006). To find a minimal sign representation then one needs to find a maximum cardinality index set $\mathbf{C}$ such the FM elimination zeroes the selected columns. However searching over index sets is of $O(2^{2^n})$ running time complexity for n-variable BFs. Therefore, it would be very beneficial in terms of computational load if smaller index sets can be used to construct larger zeroable index sets. To this end we have obtained this result:

Theorem: For any n-variable BF $\mathbf{f}$, let $\mathbf{Q} = diag(\mathbf{f})\mathbf{D}$ and take any three columns from $\mathbf{Q}$, if all columns cannot be zeroed with FM then there must be a pair of columns from those three that cannot be zeroed by FM.

Proof: We prove this by exhaustive search on all possible unique $r \times 3$ matrices with $1 \leq r \leq 8$. First we make this simple observation.

Observation: Note that although the space of 8×3 matrices with elements ± 1 appears to be large (2^{24}), when duplicate rows are eliminated and the row order is disregarded there can be only 255 possible unique matrices; to be concrete $\binom{8}{r}$ many $r \times 3$ sized unique matrices is possible with $1 \leq r \leq 8$ (i.e. the non-empty subset of all possible 3-bit patterns).

Assume now, we try to eliminate any 3 columns from $\mathbf{Q} = diag(\mathbf{f})\mathbf{D}$ using FM elimination. The eliminability of the columns is equivalent to the infeasibility of $\mathbf{Q_0a} > 0$ where $\mathbf{Q_0}$ is a $2^n \times 3$ sub-matrix of $\mathbf{Q}$. As $\mathbf{Q_0}$ will at most have 8 unique rows, the infeasibility of $\mathbf{Q_0a} > 0$ is equivalent to the infeasibility of $\mathbf{Ra} > 0$ where $\mathbf{R}$ is a sub-matrix of $\mathbf{Q_0}$ composed of unique rows of $\mathbf{Q_0}$. Therefore regardless of problem dimension n, we need to check only a small number of possible $\mathbf{R}$ matrices. Due to the observation above, there can be only 255 many such $\mathbf{R}$ matrices. So, checking the correctness of the claim of the theorem for all possible $\mathbf{R}$ matrices will complete the proof. We do this by doing the search on a computer with the following algorithm:

Algorithm 2
1. claim = *true*;
2. for all possible unique $\mathbf{R}$ made up of ± 1 (of size $r \times 3$ where $1 \leq r \leq 8$) do
   a. Apply FM elimination to the columns of $\mathbf{R}$
   b. if FM cannot eliminate all 3-columns
         i.   For each pair of columns from $\mathbf{R}$, apply FM elimination
         ii.   If any pair can be eliminated, a counterexample is found so set claim= *false*
   c. endif
3. endfor
4. return *claim*

The execution of the code indeed does not find any counterexample (i.e. Claim is returned as true) thus the theorem is proven.

Corollary 1: For any n-variable BF, take any three-element monomial set M. If all 2-subsets of M can be zeroed then so M must be zeroable.

Proof: Assume the contrary, i.e. M cannot be zeroed; but then due to Theorem, not all 2-subsets of M can be zeroed, a contradiction.

Corollary 2: For a BF **f**, if a sign representation cannot be constructed without the monomials $M = \{m_1, m_2, m_3\}$ then if follows that **f** cannot be sign-represented without either $\{m_1, m_2\}$ or $\{m_2, m_3\}$ or $\{m_1, m_3\}$.

Proof. Given the premise assume the contrary case that sign-representations are possible without any of the 2-subsets of M. This means that all 2-pair subsets of M are zeroable, which implies the zeroability of M due to Corollary 1. Thus a sign representation without $\{m_1, m_2, m_3\}$ can be constructed. A contradiction.

It is tempting to ask whether this kind of statement be made for more number of monomials. Unfortunately, this appears to not work as we found counterexamples for 4 and 5 variable Boolean functions. For example consider, the 4-variable Boolean function **f** defined by $f(x_1, x_2, x_3, x_4) = (x_1 \, AND \, x_2) \, XOR \, (x_3 \, AND \, x_4)$ which has label 0x111e in hex and (0001000100011110 in binary), and happens to be a Bent function. If one is not allowed to use the monomials $M = \{x_1, x_2, x_1 x_3, x_2 x_3\}$ then it is not possible to construct a sign representation for **f**; however it is possible to construct sign representations when one of the elements in M is permitted in the sign representation[1].

Pair-Wise Zeroability and Incompatibility Graph

We have noted that for finding a higher order neuron representation with minimal number of input lines (or a minimal sign representation) for a n-variable BF one needs to search over all possible monomial subsets. It may be possible to develop methods to speed this process by looking at pairwise zeroability for some Boolean function classes. We define a formal graph based on the notion of non-zeroability.

Definition (Incompatibility Graph). For an n-variable BF f, define the graph $G_f = (V, E)$ with vertices V and edges E. Each vertex is identified by a monomial of the form $m_i = \prod_{k \in S_i} x_k$ where $S_i \subset \{1, 2, \cdots, n\}$, and E contains edge e_{ij} if only if m_i and m_j cannot be zeroed together (i.e. there does not exist $\mathbf{k} > 0$ such that $\mathbf{k}[m_i \quad m_i] = 0$ where m_i and m_j are the vector representations of m_1 and m_2).

With this definition we can give this simple Lemma.

Lemma. Given an n-variable Boolean function f, 3-vertex independent sets of G_f are always zeroable.

---

[1] Example sign representation with $\{x_1, x_2, x_1 x_3\}$ absent: $f(x1, x2, x3, x4) = 689 - 689 x_1 x_2 + 689 x_3 + 1056 x_2 x_3 - 689 x_1 x_2 x_3 + 689 x_4 + 977 x_1 x_4 + 977 x_2 x_4 - 689 x_1 x_2 x_4 - 689 x_3 x_4 - 977 x_1 x_3 x_4 - 977 x_2 x_3 x_4 + 689 x_1 x_2 x_3 x_4$.

Proof. In an independent set any two vertex is not connected; therefore, any to monomials in this set can be zeroed due to the definition of G_f. Thus the result follows from Corollary 1.

The above Lemma suggest that when searching for zeroable monomial subsets we should start from independent sets, and avoid dense subsets such as cliques. The problems of finding maximum independent set and maximum clique problems are computationally equivalent. If S is a maximum independent set in a graph G then it will be a maximum clique in the complementary graph of G. In general, the problem of finding maximum independent sets (and so is finding maximum cliques) is np-hard but there exists several efficient heuristic algorithms for solving these problems (e.g. Busygin et al. 2002). Consider the graph G_f for 0xaaff shown in Fig. 1 (left); it has a 4-clique (and the maximum independent set size is 12), meaning that the monomials $\{1, x_1, x_4, x_1 x_4\}$ cannot be eliminated together. It turns out that the minimum monomial sign representation for this function can be constructed from this set, namely by taking 3 out of these 4 monomials (shown by computer search). Looking at the function incompatibility graph of 0xaba5 (Fig. 1, right), we see that monomial $x_1 x_3$ cannot be zeroed with three others, although they do not form a clique. In this case, it turns out that all minimum monomial sign representations must include $x_1 x_3$ and a two-subset from $\{x_1, x_1 x_4, x_1 x_3 x_4\}$ plus two other monomials (shown by computer search). Overall these observations suggest that incompatibility graph contains important hints on the structure of the minimal sign representation. A final note on the incompatibility graph is that all the Boolean functions from a single equivalence class as defined via the transformations 1–5 of Sect. 2.1 have isomorphic incompatibility graphs. However, the reverse of this remark is not true.

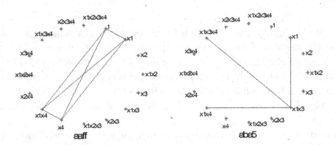

Fig. 1. Incompatibility graphs of two example 4-variable Boolean functions is shown

3   Conclusion

In this paper our efforts for finding algorithms to construct low fan-in higher order neuron representations of Boolean function are presented. In contrast to the earlier works, we try to find out the indispensability of a group of input lines, i.e. monomials together under a given BF. It seems that existence of certain monomials say, set A allow the elimination of others set B, but absence of A + B renders the representation impossible. This is curious because for example for a given BF their projection on the monomial vectors (i.e. the spectral coefficients) does not give a clue on which monomials are 'connected' in this sense. Especially this is more so for Bent functions whose spectral coefficients have the same magnitude, and which seem to emerge in complex forms as dimension increases (e.g. see Mesnager 2016). To this end, we first investigated low dimensional 'simultaneous zeroability' patterns. It is trivially true that if A is set of monomials that cannot be zeroed, then for any super set B of A cannot be zeroed. What is desired, is something in the other direction; i.e. to be able to infer the zeroability of larger sets from smaller ones. In fact, with numerical investigations in lower dimension (3 and 4) we have found that the zeroability of three monomial pairs that mutually intersect can be used to infer the zeroability of the union of them. We then questioned whether this can be a general pattern for all n-variable BFs. With an elementary proof, indeed we could prove this general result. Finally, to visualize the monomial pair non-zeroability we introduced the notion of incompatibility graph for a BF which may be used to find zeroable monomial candidates for low fan-in representations of BFs. Intriguingly, all the functions from a single equivalence class (as defined in the text) must necessarily have isomorphic compatibility graphs. However, the reverse is not true: isomorphic graphs does not necessarily map to a single equivalence class. This suggests that compatibility graph can be used to classify the equivalence classes into different types. Future work should be directed towards how to utilize the incompatibility graph and enrich it for obtaining compact higher order neuron representations (i.e. sign-representations with small number of monomials) for Boolean functions. Furthermore, the identification of the Boolean functions that are amenable to fast minimal sign-representation construction may be also a fruitful research direction.

References

Busygin, S., Butenko, S., Pardalos, P.M.: A heuristic for the maximum independent set problem based on optimization of a quadratic over a sphere. J. Comb. Optim. **6**(3), 287–297 (2002)

Chandru, V.: Variable elimination in linear constraints. Comput. J. **36**(5), 463–470 (1993)

Ghosh, J., Shin, Y.: Efficient higher order neural networks for classification and function approximation. Int. J. Neural Syst. **3**(4), 323–350 (1992)

Giles, C.L., Maxwell, T.: Learning, invariance, and generalization in high-order neural networks. Appl. Opt. **26**(23), 4972–4978 (1987)

Guler, M.: A model with an intrinsic property of learning higher order correlations. Neural Netw. **14**(4–5), 495–504 (2001)

Mesnager, S.: On constructions of bent functions from involutions. In: 2016 IEEE International Symposium on Information Theory (2016)

Oztop, E.: An upper bound on the minimum number of monomials required to separate dichotomies of $\{-1, 1\}^n$. Neural Comput. **18**(12), 3119–3138 (2006)

Oztop, E.: Sign-representation of Boolean functions using a small number of monomials. Neural Netw. **22**(7), 938–948 (2009)

Schmitt, M.: On the capabilities of higher-order neurons: a radial basis function approach. Neural Comput. **17**(3), 715–729 (2005)

Sezener, C.E., Oztop, E.: Minimal sign representation of Boolean functions: algorithms and exact results for low dimensions. Neural Comput. **27**(8), 1796–1823 (2015)

Siu, K.Y., Roychowdhury, V., Kailath, T.: Discrete Neural Computation. Englewood Cliffs, Prentice Hall (1995)

A Genetic Approach to Fusion of Algorithms for Compressive Sensing

Hanxu You[⊠] and Jie Zhu

Department of Electronic Engineering, Shanghai Jiao Tong University (SJTU),
Shanghai, People's Republic of China
gongzihan@sjtu.edu.cn
http://www.sjtu.edu.cn/

Abstract. Inspired by the data fusion principle, we proposed a genetic approach to fusion of algorithms for CS to improve reconstruction performance. Firstly, several compressive sensing reconstruction algorithms (CSRAs) are executed in parallel to provide their estimates of the underlying sparse signal. Next, genetic algorithm is used to fuse these estimates for achieving a new estimate that is better than the best of these estimates. The proposed approach provides flexible design of fitness function and mutation strategy of genetic algorithm, and various participating CSRAs can be used to recover the sparse signal. Experiments were conducted on both synthetic and real world signals. Results indicate that the proposed approach has three advantages: (1) it performs well even when the dimension of measurements is very low, (2) reconstruction performance is better than any participating CSRAs, and (3) it is comparable or even superior to other fusion algorithm like FACS.

Keywords: Compressive sensing · Reconstruction algorithm · Data fusion principle · Genetic algorithm

1 Introduction

For compressive sensing (CS), lots of sparse reconstruction algorithms have been proposed in recent years [4]. Because the performance of those algorithms is associated with several factors like sparsity level, dimension of measurements and statistical characteristics of the underlying sparse signal, so that estimates of compressive sensing reconstruction algorithms (CSRAs) differ from each other due to their different strategies [5,7,11].

According to data fusion principle, the relevant information from two or more data sources can be combined in intelligent ways into a single one that provides a more accurate description than any of the individual data sources [10]. In CS

H. You—work was supported by National Natural Science Foundation of China (grant Nos. 61371147, and 11433002), and Shanghai Academy of Spaceflight Technology (grant No. SAST2015039).

© Springer International Publishing AG 2017
F. Cong et al. (Eds.): ISNN 2017, Part II, LNCS 10262, pp. 371–379, 2017.
DOI: 10.1007/978-3-319-59081-3_44

framework, there are two different ways to get multiple estimates to be fused. The first way is using a same CSRA to obtain multiple estimates via different sparse basis [9,15]. The other way is to obtain multiple estimates by using different CSRAs [1–3]. In [1], Amba et al. developed a fusion based scheme termed fusion of algorithms for compressed sensing (FACS) and it has been shown that fusion of the estimates of a set of CSRAs result in an estimate better than the best one in this set.

Since the work by Ambat et al. [1,2] used only a simple least-squares based approach to identify true elements of union of support sets estimated by participating CSRAs, here we try to propose a new fusion strategy for further performance improvement. In this paper, we introduce a genetic approach to the fused based scheme which is referred to as genetic fusion of reconstruction algorithms (GFRA) for seeking out more true elements of union set than prior works did, especially at small compression ratio. Genetic algorithm (GA) is commonly used to generate high-quality solutions to optimization and search problems by relying on bio-inspired operators such as mutation, crossover and selection [14]. In GFRA, the participating CSRAs provide estimates of the sparse signal independently in the first stage and then these estimates are fused by genetic algorithm in the second stage. In GFRA, individuals, fitness function and mutation strategy are customized for the procedure of genetic algorithm.

The rest of this article is organized as follows. In Sect. 2, we briefly introduce CS theory. And then we prove that the union of support sets is likely to contain more number of true elements than any individual support set in Sect. 3. In Sect. 4, GFRA is proposed to improve reconstruction performance. Experiments were conducted on both synthetic and real world signals, and results are analyzed in Sect. 5. Finally, we draw our conclusions in Sect. 6.

2   Compressive Sensing Theory

Compressed sensing is a signal processing theory for efficiently acquiring and reconstructing a sparse signal, by finding solutions to underdetermined linear systems [6,8].

Consider a linear system of equations

$$y = \Phi x + n \tag{1}$$

where x is an $N \times 1$ sparse signal and y is the $M \times 1$ ($M \ll N$) measurement. Φ is an underdetermined $M \times N$ measurement matrix and n is the additive measurement noise. The problem is to estimate the signal x, subject to the constraint that x is a K-sparse signal ($K \leq M$). K-sparse implies that only K elements of x are non-zero and the rest are zero, i.e. $||x||_0 = K$. Sparsity level K plays a significant role in CS reconstruction, for this reason, many CSRAs such as OMP [16], CoSaMP [13], etc. usually require a prior value of K. The estimate of a CSRA consists two important components: (1) the indices of non-zero elements (known as support set denoted by $\mathcal{S}$, and $|\mathcal{S}| = K$ which denotes its cardinality), and (2) the magnitudes of the non-zero elements. Estimates of different CSRAs have different support sets, and a better estimate of the support set leads to a better CS reconstruction performance.

3   Union of Support Sets

As mentioned before, the support sets estimated by different CSRAs may contain different subsets of the true support set. Hence the union of these support sets is likely to contain more number of true elements than any individual support set [1].

Consider a Gaussian sparse signal (GSS) with length of $N = 500$ and sparsity level of $K = 20$. The GSS is measured by a Gaussian random measurement matrix at different ratio $Ratio = M/N = 0.12, 0.14, 0.16$ and 0.18. And then we employ OMP algorithm and SP algorithm [7] to recover this sparse signal independently. The true support set of sparse signal is denoted by $\mathcal{S}^t$ and $|\mathcal{S}^t| = K$. Support set provided by OMP is denoted by $\hat{\mathcal{S}}_o$, and the set of true elements in $\hat{\mathcal{S}}_o$ can be obtained by calculating an intersection $\hat{\mathcal{S}}_o^t = \hat{\mathcal{S}}_o \cap \mathcal{S}^t$. Similarly, we use $\hat{\mathcal{S}}_s$ and $\hat{\mathcal{S}}_s^t = \hat{\mathcal{S}}_s \cap \mathcal{S}^t$ to represent support set and the set of true elements in $\hat{\mathcal{S}}_s$, respectively. Let $\hat{\mathcal{S}}_u^t$ denote the true elements in the union of the support sets, we have $\hat{\mathcal{S}}_u^t = \hat{\mathcal{S}}_o^t \cup \hat{\mathcal{S}}_s^t$. According to the assumption that the union set is likely to contain more number of true elements than any individual support set, we have $|\hat{\mathcal{S}}_u^t| \geq \max\{|\hat{\mathcal{S}}_o^t|, |\hat{\mathcal{S}}_s^t|\}$ and $|\hat{\mathcal{S}}_u^t| \leq K$. Tests were carried out $N_t = 1000$ times to reduce the effects of the Gaussian random measurement matrix. The average values of $|\hat{\mathcal{S}}_o^t|$, $|\hat{\mathcal{S}}_s^t|$ and $|\hat{\mathcal{S}}_u^t|$ at all range of $Ratio$ are illustrated in Fig. 1.

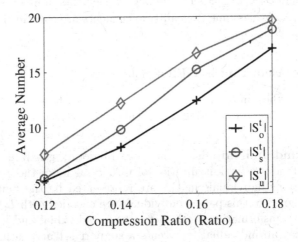

Fig. 1. Average number of true elements estimated by OMP, SP and the union set

From Fig. 1, we can find that the number of true elements in union set is closer to the actual value K than that of any individual support set at all range of *Ratio*. For exemple, at $Ratio = 0.16$, average $|\hat{\mathcal{S}}_o^t|$ is 12.4 and average $|\hat{\mathcal{S}}_s^t|$ is 15.2, while the average number of true elements in the union set is 16.7. The results shown in Fig. 1 prove that union set always contains at least as many true elements as the best individual support set.

4   Genetic Fusion of Reconstruction Algorithms

As mentioned before, we can achieve a better CS reconstruction performance if we can identify more true elements of the union set. In this section, we introduce a genetic approach to fusion of algorithms which is referred to as genetic fusion of reconstruction algorithms (GFRA) for identifying as many true elements as possible. The proposed GFRA has two main algorithmic stages. In the first stage, several participating CSRAs are executed in parallel, independently. Any CSRA can be used in GFRA scheme. Then, in the second stage, the support sets estimated by the participating CSRAs are combined together to form a new union set and genetic algorithm is adopted to identify true elements. Once the true indices of non-zero elements are obtained by genetic algorithm, the magnitudes of the non-zero elements can be efficiently calculated accordingly.

4.1   Obtaining Union Set

For i-th CSRA used by GFRA scheme, the recovered sparse signal is denoted by $\hat{x}_i$ and support set of $\hat{x}_i$ is denoted by $\hat{\mathcal{S}}_i = supp(\hat{x}_i)$. As we employ $P \geq 2$ different CSRAs in the scheme, a union of P support sets, denoted by $\hat{\mathcal{S}}_u$ can be obtained through $\hat{\mathcal{S}}_u = \cup_{i=1}^P \hat{\mathcal{S}}_i$. The number of elements in union set is denoted by $|\hat{\mathcal{S}}_u| = L > K$. We assume that all true elements are contained in union set, so that we have $\mathcal{S}^t \subset \hat{\mathcal{S}}_u$.

4.2   Genetic Algorithm in GFRA Scheme

After obtaining the union of support sets, we try to find out all of the true elements in the union set by using genetic algorithm (GA).

Population and Individuals. Firstly, we randomly generate a population $N_{ind} = 80$ individuals for initialization of GA. d_j denotes the j-th individual in the population. Individuals in GA are represented by the binary alphabet $\{0, 1\}$. As designed in this paper, individuals are encoded with L bits. Figure 2 illustrates the transformational rule between an individual and a support set. We can see that an individual represents a support set and each one support set is a subset of the union set. We use $\hat{\mathcal{S}}_j^d$ to denote the support set that is corresponding to d_j, and $|\hat{\mathcal{S}}_j^d| = K$.

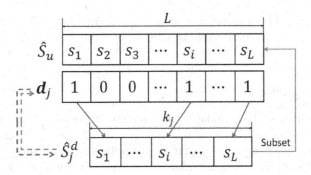

Fig. 2. The transformational rule between an individual and a support set

Objective and Fitness Function. An objective function is used to provide a measure of how individuals have performed and a fitness function is normally used to transform the objective function value into a measure of ralative fitness. To calculate the fitness value of each individual, we should transform the individuals into support sets according to the transformational rule (see Fig. 2).

An objective value of each support set can be obtained by

$$f(d_j) = \|y - \Phi_{\hat{S}_j^d} \Phi_{\hat{S}_j^d}^\dagger y\|_2^2 \tag{2}$$

where $\Phi_{\hat{S}_j^d}^\dagger$ is the pseudo-inverse of $\Phi_{\hat{S}_j^d}$. Φ_S denotes the column sub-matrix of where the indices of the columns are the elements of the set S. In the case of (2), the most fit individuals will have the lowest numerical value of objective function. The objective value is non-negative, so that we can give the most fit individuals the biggest fit value according to the fitness function (3).

$$F(d_j) = \frac{1}{f(d_j) \sum_{j=1}^{N_{ind}} \frac{1}{f(d_j)}} \tag{3}$$

Once individual has been assigned a fitness value, it can be selected from the population with the probability $p_{si} = F(d_j)$ to conduct crossover. In our paper, the typical roulette wheel selection method and multi-point crossover method were used for selection and crossover, respectively.

Mutation Strategy. In our problem, a common used mutation strategy is unsuitable. After the process of crossover, $k_j = |\hat{S}_j^d|$ would be smaller than sparsity level K, equal to K or bigger than K. To keep the value of k_j equal to K, we have to adopt a new mutation strategy of GA. The new matution strategy used here obeys three rules: (1) when $k_j < K$, $K - k_j$ random 0 s would be replaced by 1 s, (2) when $k_j = K$, a random 0 and a random 1 would swap their positions with a mutation probability p_{ma}, and (3) when $k_j > K$, $k_j - K$ random 1 s would be replaced by 0 s. The mutation probability p_{ma} is set to

0.2 in our paper. This mutation strategy can ensure that each one individual includes K 1 s, which means that the number of elements in each support set is equal to K.

4.3   Reconstruction

As the fitness values of a population may remain static for a number of generations before a superior individual is found, so that after a pre-specified number of generations which we refer as number of iteration N_{iter}, genetic procedure will be terminated. The best individual of the newest generation was chosen and transformed to a candidate support set $\hat{S}_{gene}$. In [1], FACS obtained the final support set $\hat{S}_{facs} = supp(\hat{x}_{\hat{S}_u}^K)$, where $\hat{x}_{\hat{S}_u}^K$ is the best K-sparse approximation of $\hat{x}_{\hat{S}_u}$. x_S denotes the sub-vector formed by those elements of whose indices are listed in the set S.

As we know that a support set can be transformed to an individual in genetic algorithm, so that we compare the fitness value of $\hat{S}_{gene}$ with that of $\hat{S}_{facs}$ to find the best true support set of $\hat{x}$ by (4).

$$\hat{S}^t = \begin{cases} \hat{S}_{gene} \text{ if } F(\hat{S}_{gene}) \geq F(\hat{S}_{facs}) \\ \hat{S}_{facs} \text{ Otherwise} \end{cases} \tag{4}$$

We know that the recovered sparse signal $\hat{x}$ is consist of non-zeros values and zero values, and we can have non-zero values $\hat{x}_{\hat{S}^t}$ through (5).

$$y = \Phi_{\hat{S}^t} x_{\hat{S}^t} + n \tag{5}$$

At last we have the true support set $\hat{S}^t$ and reconstruction of GFRA $\hat{x}$. Note that, any CSRA without any modification can be used as a participating algorithm in GFRA as same as FACS. The number of CSRAs is not limited and the performance of GFRA mostly depends on the information that is contained in $\hat{S}_u$ and $\hat{S}^t$. Those true elements that are not in the union set can not be found due to the fusion strategy of GFRA. Experimental results indicated that when the *Ratio* is small, GFRA performed better than FACS due to that GFRA can identify more true elements than FACS did.

5   Experiments

We evaluated the performance of GFRA using GSS (synthetic signal) and ECG signals (real-world signals). CSRAs like SP, OMP, and BP [17] are adopted as participating algorithms. Since that BP would not estimate the support set directly, so that we chose the indices of the K largest magnitudes of signal as the estimated support set. An average error ratio (AER) which is defined as $AER = \sum_{i=1}^{N_t} \|x_i\|_2^2 / \sum_{i=1}^{N_t} \|x_i - \hat{x}_i\|_2^2$ was used to evaluate the reconstruction performance. x_i and $\hat{x}_i$ denote the original and recovered signal in i-th trial, and $N_t = 10000$ was the total number of trials in the experiments. N_{iter} in genetic algorithm was set to 20 for all experiments. When SP and OMP were used as participating algorithms, the reconstruction results of GFRA and FACS were denoted as GFRA (SP, OMP) and FACS (SP, OMP), respectively. When BP was also added to the group of participating algorithms, GFRA (SP, OMP, BP) and FACS (SP, OMP, BP) were used to denote the corresponding results.

5.1 Experiment on GSS

As the experiment in section (3), we conducted experiment using GSS to test the signal reconstruction for small values of *Ratio* in GFRA. We followed the simulation setup used in Sect. 5 of [1]. The dimension of GSS was $N = 500$ and sparsity level was $K = 20$. All of K non-zero values were chosen from $N(0,1)$ independently and randomly located. The signal to measurement noise ratio (SMNR) is set to 20 dB. The dimension of measurements was $M = N \times Ratio(0.14 < Ratio < 0.24)$. Figure 3 shows the AER results for GSS. From Fig. 3, BP performed better than SP and OMP at small *Ratio*, while SP and OMP outperformed BP at big *Ratio*. At all range of *Ratio*, GFRA gave a better result than the best CSRA. So that when the prior knowledge of sparse signal is unknown, GFRA can also have a good performance. The best improvements were obtained at small *Ratio* because the difference of the numbers of true elements in the union set and individual support sets was biggest at that time. When *Ratio* was big enough, the values of GFRA (SP, OMP) and GFRA (SP, OMP, BP) would be the same as that of the best participating CSRA.

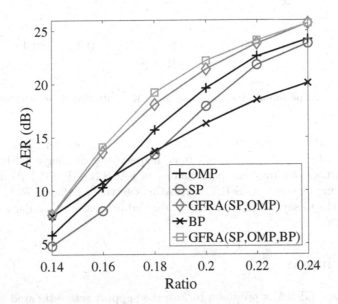

Fig. 3. The AER results for GSS

5.2 Experiment on ECG Signals

We compare the performance of GFRA with FACS to show the advantages of our proposed method. We conducted experiments on ECG signals selected from MIT-BIH Arrhythmia Database [12]. ECG signals are compressible and good for sparse decompositions. We processed $N = 1024$ samples of ECG and assumed the sparsity level $K = 128$. Compression ratio is from 0.3 to 0.4 with an increments of 0.02. The performance comparison of GFRA with FACS in terms of AER is shown in Fig. 4.

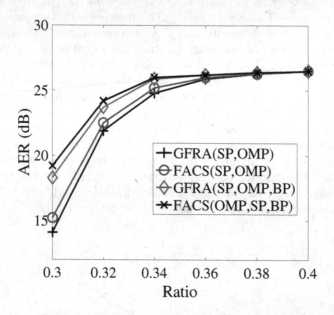

Fig. 4. The performance comparison of GFRA with FACS on ECG signals

GFRA (SP, OMP) gave 0.9 dB and 0.1 dB AER improvement over FACS (SP, OMP) at *Ratio* = 0.3 and *Ratio* = 0.36, respectively. Similarly, GFRA (OMP, SP, BP) further improved the performance of FACS (OMP, SP, BP). It may be noted that impovement of GFRA over FACS is getting smaller as *Ratio* becomes bigger. That's to say, GFRA and FACS resulted in an almost similar AER when *Ratio* ≥ 0.36.

6 Conclusion

In this paper, GFRA is proposed to fuse the support sets estimated by several CSRAs and leads to a better compressive sensing reconstruction performance. Participating CSRAs in GFRA need not require any modification and any CS reconstruction algorithms can be used. GFRA also results in seamless scalability and robustness, because genetic fusion strategy can be replaced with other well designed fusion strategies to obtain advanced performance. From the experimental results, GFRA not only shows the advantages over participating CSRAs, but also outperforms other fusion algorithm like FACS.

References

1. Ambat, S.K., Chatterjee, S., Hari, K.V.S.: Fusion of algorithms for compressed sensing. IEEE Trans. Signal Process. **61**(61), 3699–3704 (2013)
2. Ambat, S., Chatterjee, S., Hari, K.S.: Fusion of algorithms for compressed sensing. In: IEEE International Conference on Acoustics, Speech and Signal Processing, pp. 5860–5864 (2013)
3. Ambat, S.K., Chatterjee, S., Hari, K.V.S.: Progressive fusion of reconstruction algorithms for low latency applications in compressed sensing. Signal Process. **97**(7), 146–151 (2014)
4. Baraniuk, R.: Compressive sensing. In: Conference on CISS 2008 Information Sciences and Systems, pp. 118–120 (2008)
5. Baraniuk, R.G., Cevher, V., Duarte, M.F., Hegde, C.: Model-based compressive sensing. IEEE Trans. Inf. Theory **56**(4), 1982–2001 (2010)
6. Candès, E.J., Wakin, M.B.: An introduction to compressive sampling. IEEE Signal Process. Mag. **25**(2), 21–30 (2008)
7. Dai, W., Milenkovic, O.: Subspace pursuit for compressive sensing signal reconstruction. IEEE Trans. Inf. Theory **55**(5), 2230–2249 (2008)
8. Donoho, D.L.: Compressed sensing. IEEE Trans. Inf. Theory **52**(4), 1289–1306 (2006)
9. Elad, M., Yavneh, I.: A plurality of sparse representations is better than the sparsest one alone. IEEE Trans. Inf. Theory **55**(10), 4701–4714 (2009)
10. Haghighat, M., Abdel-Mottaleb, M., Alhalabi, W.: Discriminant correlation analysis: Real-time feature level fusion for multimodal biometric recognition. IEEE Trans. Inf. Forensics Secur. **11**(9), 1984–1996 (2016)
11. Ji, S., Xue, Y., Carin, L.: Bayesian compressive sensing. IEEE Trans. Signal Process. **56**(6), 2346–2356 (2008)
12. Moody, G.B., Mark, R.G.: The impact of the MIT-BIH arrhythmia database. IEEE Eng. Med. Biol. Mag. **20**(3), 45–50 (2001)
13. Needell, D., Tropp, J.A.: CoSaMP: iterative signal recovery from incomplete and inaccurate samples. Appl. Comput. Harmonic Anal. **26**(3), 301–321 (2008)
14. Schmitt, L.M.: Asymptotic convergence of scaled genetic algorithms to global optima. Genet. Algorithms Evol. Comput. **11**, 157–200 (2006)
15. Starck, J.L., Candes, E.J.: Very high quality image restoration by combining wavelets and curvelets. In: Proceedings of SPIE - The International Society for Optical Engineering, pp. 9–19 (2001)
16. Tropp, J.A., Gilbert, A.C.: Signal recovery from random measurements via orthogonal matching pursuit. IEEE Trans. Inf. Theory **53**(12), 4655–4666 (2007)
17. Wei, L., Vaswani, N.: Regularized modified bpdn for noisy sparse reconstruction with partial erroneous support and signal value knowledge. IEEE Trans. Signal Process. **60**(1), 182–196 (2010)

Industrial Oil Pipeline Leakage Detection Based on Extreme Learning Machine Method

Honglue Zhang[1], Qi Li[1(✉)], Xiaoping Zhang[2], and Wei Ba[3]

[1] School of Control Science and Engineering, Dalian University of Technology,
Dalian 116024, China
201281121@mail.dlut.edu.cn, qili@dlut.edu.cn
[2] Beijing Special Engineering Design and Research Institute,
Beijing 100028, China
zhangxp@163.com
[3] Dalian Scientific Test and Control Technology Institute, Dalian 116013, China
bawei_dut@126.com

Abstract. Pipeline transportation plays a significant role in modern industry, and it is an important way to transport many kinds of oils and natural gases. Industrial oil pipeline leakage will cause many unexpected circumstances, such as soil pollution, air pollution, casualties and economic losses. An extreme learning machine (ELM) method is proposed to detect the pipeline leakage online. The algorithm of ELM has been optimized based on the traditional neural network, so the training speed of ELM is much faster than traditional ones, also the generalization ability has become stronger. The industrial oil pipeline leakage simulation experiments are studied. The simulation results showed that the performance of ELM is better than BP and RBF neural networks on the pipeline leakage classification accuracy and speed.

Keywords: Pipeline leak detection · ELM · Neural networks · Signals classification

1 Introduction

In modern industry society, pipeline transportation is one of the main methods of transporting products like gas, oil products and so on, which has played an important role in the protection of energy demand of industrial and social and economic development. Since the day that pipeline transportation technology was invented, pipeline leakage has always been a big problem which consistently bothered everyone who involved in this field. After the pipeline leakage, the petroleum products will not only bring to the enterprise loss, but also will pollute the environment. So the pipeline leak detection technology has an important role of protecting the natural environment and people's life safety. At the same time, the development of pipeline leak detection technology could reduce the economic loss, enhance the competitive ability of enterprises.

As one of the most popular methods of pipeline leak detection, negative pressure wave method (NPW) is very convenient and effective [1]. Thus this paper also uses the

© Springer International Publishing AG 2017
F. Cong et al. (Eds.): ISNN 2017, Part II, LNCS 10262, pp. 380–387, 2017.
DOI: 10.1007/978-3-319-59081-3_45

negative pressure wave method for pipeline leak detection. An extreme learning machine method (ELM) is proposed to detect the pipeline leakage online. ELM is a kind of single-hidden layer feedforward neural networks algorithm. ELM can randomly select the number of hidden layer neurons and types of them in the network, to construct different learning algorithms. After randomly selecting the weights of input layer and hidden layer neurons deviation, to obtain the weights of hidden layer output by analysing. ELM algorithm has a lot advantages such as great generalization ability, fast learning speed [2, 3]. The experiments are based on industry pipeline transportation data, the sampling frequency of the pipe pressure locale acquisition signals is 20 Hz, from 15 h of continuous signals. The acquisition signals contains small leakage signals, normal signals and valve adjusting signals.

Back propagation (BP) is a kind of multilayer feedforward networks that has been trained by error back propagation algorithm, it is one of the most widely used neural network models nowadays. The topological structure of BP neural network model contains input, hide layer, and output layer. The learning process composed by forward propagating of signals and backward propagating of error [4, 5].

Radial basis function (RBF) methods are fundamental tools for interpolating scattered data especially in multidimensional spaces. Its excellent function approximation capability makes it widely used in pattern recognition, economic forecasts, and other fields [6, 7]. So in this paper, we compared the classification accuracy and speed of ELM with the BP, RBF methods in the simulation experiments.

This paper is organized as follows. In Sect. 2, we introduce the ELM algorithm. In Sect. 3, the performance of the ELM is compared with BP, RBF on the mixed pressure signals classification of oil pipeline through negative pressure wave theory. Finally, the conclusion is given in Sect. 4.

2   Extreme Learning Machine Algorithm

The model of the single-hidden layer feedforward neural networks is shown in the Fig. 1. Output of single-hidden layer feedforward neural networks (SLFNs) which has $\hat{N}$ hidden layer neurons is:

$$f_{\hat{N}}(x) = \sum_{i=1}^{\hat{N}} \beta_i G(w_i, b_i, x), x \in R^n, w_i \in R^n, \beta_i \in R^m \tag{1}$$

In formula (1), $G(w_i, b_i, x)$ is the ith hidden layer neutron corresponding output of input x. $\beta = \left[\beta_{i1}^T, \beta_{i2}^T, \cdots, \beta_{im}^T\right]^T$ is the linking weight vector between ith hidden layer neutron and output neutron.

While activation function $g(x)$ is the additive neutron, the corresponding output of ith hidden layer neutron is:

$$G(w_i, b_i, x) = g(w_i \cdot x + b_i), b_i \in R \tag{2}$$

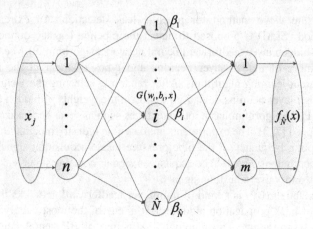

Fig. 1. Single-hidden layer feedforward neural networks model

In formula (2), $w_i = [w_{i1}, w_{i2}, \cdots, w_{in}]^T$ is the weight vector between ith hidden layer neutron and input neutron, b_i is the error of ith hidden neutron.

While activation function is the RBF neutron, the corresponding output is:

$$G(w_i, b_i, x) = g(b_i \|x - w_i\|), b_i \in R^+ \tag{3}$$

In formula (3), w_i is the kernel of radial basis function and b_i is the influencing factor (width), R^+ is positive real numbers set.

To N any input samples (x_j, t_j), $x_j = [x_{j1}, x_{j2}, \cdots, x_{jn}] \in R^n$, $t_j = [t_{j1}, t_{j2}, \cdots, t_{jm}] \in R^m$, given $\hat{N}$ hidden neutron and activation function $G(w_i, b_i, x)$, there must be the presence of β_i, w_i, b_i, to make SLFNS approach N sample points without error.

$$\sum_{i=1}^{\hat{N}} \beta_i G(w_i, b_i, x_j) = t_j, j = 1, 2, \cdots, N \tag{4}$$

Formula (4) could also be written by matrix form:

$$H\beta = T \tag{5}$$

In formula (5),

$$H(w_1, \cdots, w_{\hat{N}}, b_1, \cdots, b_{\hat{N}}, x_1, \cdots, x_N) = \begin{pmatrix} G(w_1, b_1, x_1) & \cdots & G(w_{\hat{N}}, b_{\hat{N}}, x_1) \\ \vdots & \ddots & \vdots \\ G(w_1, b_1, x_N) & \cdots & G(w_{\hat{N}}, b_{\hat{N}}, x_N) \end{pmatrix}_{N \times \hat{N}}$$

$$\beta = [\beta_1^T, \cdots, \beta_{\hat{N}}^T]_{\hat{N} \times m}^T \quad T = [t_1^T, \cdots, t_N^T]_{N \times m}^T$$

H is the hidden layer output matrix of this neural network. The ith column of H is the corresponding output of ith hidden layer neutron of input $x_1, x_2, \cdots, x_N$.

For a single hidden layer forward neural network, ELM is available for any infinitely differentiable activation function, thereby expand the choice space of forward neural network activation function. ELM is different from the traditional function approximation theory, the input layer weight w_i and the hidden layer error b_i could be chosen randomly. To a feedforward neural networks, during network training, it's not necessary to adjust the input layer weight value or hidden layer error, since those parameters had been determined and network training started, hidden layer output matrix H keeps invariant.

We may summarize the ELM algorithm as:

Given a training sample set $\{(x_j, t_j) | x_j \in R^n, t_j \in R^m, j = 1, \cdots, N\}$, activation function $g(x)$, the number of hidden neutron is $\hat{N}$.

Step 1: randomly set the input layer weight w_i and error b_i, $i = 1, \cdots, \hat{N}$.

Step 2: compute the hidden layer output matrix H.

Step 3: compute the output layer weight β, $\hat{\beta} = H^+ T, T = [t_1, \cdots, t_N]^T$. H^+ is the Moore-Penrose generalized inverse of H.

3 Simulation Experiment

3.1 Industrial Oil Pipeline Leakage Detection System

The diagram of the industrial oil pipeline leakage detection system is shown in Fig. 2 and the sample of the negative pressure signals is shown in Fig. 3. In the Fig. 2,

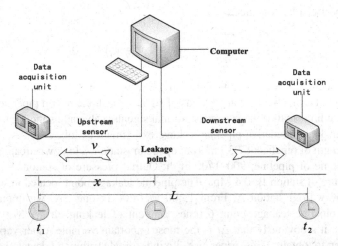

Fig. 2. Diagram of the industrial oil pipeline leakage detection system

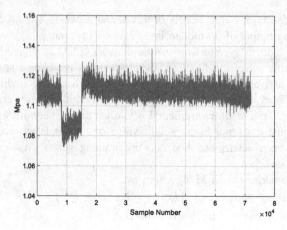

Fig. 3. Negative pressure signals

x: Distance between upstream detection station and leakage point;
L: Distance between upstream detection station and downstream detection station;
a: Speed of negative pressure wave;
v_0:   Speed of liquid;
Δt: Time difference between upstream and downstream receiving the leakage signals.

It can be concluded:

$$\frac{x}{a - v_0} - \frac{L - x}{a + v_0} = \Delta t \tag{6}$$

Thereby it can be concluded:

$$x = \frac{1}{2a} \left[L(a - v_0) + (a^2 - v_0^2) \Delta t \right] \tag{7}$$

In this section, we use the negative pressure wave theory to judge the leakage exiting or not. Normally, there are 3 kinds of signals constitute the oil pipeline pressure signals, including small leakage signals, normal signals, valve adjusting signals. In this paper, we make a simulative pipeline leakage experiment as follows:

Total length of pipeline: 67.3 km from upstream station A to downstream station B; Freight volume of pipeline: 900–1400 m³/h; Output pressure of station A: 1.0 Mpa; Input pressure of station B: 0.4 Mpa; The pipeline leakage point located in where was 42.2 km away from station A. From 13:00 to 18:00 in one day, we drained oil in simulative pipeline leakage point 6 times, amount of leaking oil are 37.9 tons. By formula (7), it's obviously that Δt is the most important variable, other variables are much easier to obtain. This paper use 3 kinds of algorithm to detect the leaking moment, to obtain a preciser Δt, thereby the leakage point can be located precisely.

3.2 Simulation Experiment Results

We make experiments based on MATLAB 2015b, the sampling frequency of the pipe pressure locale acquisition signals is 20 Hz, from 15 h of continuous signals, there were extracted by the extraction width of 15000 to 200 groups of signals, and the feature values were extracted by the formulas of the signal characteristic, there were totally extracted 8 kinds of feature values, which has formed a 200 × 8 characteristic matrix.

And the feature values were extracted by the formulas of the signal characteristic, there were totally extracted 8 kinds of feature values, including mean value, variance, standard deviation, peak value, root mean square value, mean square amplitude, margin factor, kurtosis factor, which has formed a 200 × 8 characteristic matrix to be the experimental database. We use the former 120 vectors to form training set, the latter 80 vectors to form testing set. Both sets contain 3 kinds of signals and there is no two same vector exits in both sets. In this paper, we did 10 times repeated experiments, obtained the average results after calculating.

Experimental results based on the BP, RBF and ELM methods are shown in the Table 1 and Figs. 4 and 5. Obviously, as it is showed in Fig. 5 and Table 1, ELM is better in prediction of testing accuracy than other 2 kinds of algorithms, it only has 4 error which mistakenly regards a leakage situation as non-leakage situation. The BP and RBF are not as much good as ELM, they have 8 errors separately. In those 8 errors, BP and RBF both mistakenly regards 4 leakage situations as non-leakage situations,

Table 1. Data comparison of 4 kinds of algorithms

Algorithm type	Testing accuracy	Test amount	Error amount	Training time/s	Testing time/s	Total time/s
BP	90.00%	80	8	1.4040	0.1092	1.5132
RBF	90.00%	80	8	0.9360	0.5148	1.4508
ELM	95.00%	80	4	0.1248	0.0312	0.1560

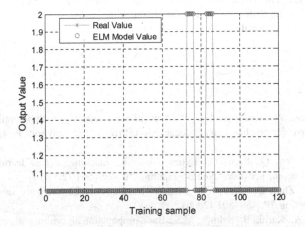

Fig. 4. ELM prediction of training set

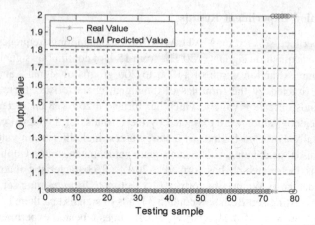

Fig. 5. ELM prediction of testing set

and regards 4 non-leakage situations as leakage situations. ELM is also fast, its training time and testing time are both much shorter than the others, the total time of ELM is only 0.1560 s.

4   Conclusion

In this paper, we use the BP, RBF, ELM methods to analyze the pipeline pressure mixed signals for leakage detection through the negative pressure theory. The acquisition signals contain small leakage signals, normal signals and valve adjusting signals. The experimental results show that, compared with BP, RBF, the method of ELM has a higher testing set classification accuracy and a much shorter total classification time, ELM is much more suitable for industrial pipeline leak detecting.

Acknowledgment. This work is supported by the National Natural Science Foundation of China (61403058), the PetroChina Innovation Foundation (2014D-5006-0601), and the Fundamental Research Funds for the Central Universities of China.

References

1. Lu, W., Liang, W., Zhang, L., Liu, W.: A novel noise reduction method applied in negative pressure wave for pipeline leakage localization. Process Saf. Environ. Prot. **104**, 142–149 (2016)
2. Huang, G., Zhu, Q., Siew, C.: Extreme learning machine: a new learning scheme of feedforward neural networks. Neurocomputing **70**, 489–501 (2006)
3. Huang, G., Zhu, Q., Siew, C.: Extreme learning machine: theory and applications. Neurocomputing **70**, 489–501 (2006)
4. Hameed, A.A., Karlik, B., Salman, M.S.: Back-propagation algorithm with variable adaptive momentum. Knowl.-Based Syst. **114**, 79–87 (2016)

5. Leema, N., Nehemiah, H.K., Kannan, A.: Neural network classifier optimization using Differential Evolution with Global Information and Back Propagation algorithm for clinical datasets. Appl. Soft Comput. **49**, 834–844 (2016)
6. Kindelan, M., Bayona, V.: Application of the RBF meshless method to laminar flame propagation. Eng. Anal. Bound. Elem. **37**, 1617–1624 (2013)
7. Uddin, M.: RBF-PS scheme for solving the equal width equation. Appl. Math. Comput. **222**, 619–631 (2013)
8. Laurentys, C.A., Bomfim, C.H.M., Menezes, B.R., Caminhas, W.M.: Design of a pipeline leakage detection using expert system: a novel approach. Appl. Soft Comput. **11**, 1057–1066 (2011)
9. Mandal, S.K., Chan, T.S., Tiwari, M.K.: Leak detection of pipeline: an integrated approach of rough set theory and artificial bee colony trained SVM. Expert Syst. Appl. **39**, 3071–3080 (2012)
10. Rad, J.A., Kazem, S., Parand, K.: Optimal control of a parabolic distributed parameter system via radial basis functions. Commun. Nonlinear Sci. Numer. Simul. **19**, 2559–2567 (2014)
11. Liang, W., Kang, J., Zhang, L.: Leak detection for long transportation pipeline using a state coupling analysis of pump units. J. Loss Prev. Process Ind. **26**, 586–593 (2013)
12. Sun, J., Xiao, Q., Wen, J., Wang, F.: Natural gas pipeline small leakage feature extraction and recognition based on LMD envelope spectrum entropy and SVM. Measurement **55**, 434–443 (2014)

Audio Source Separation from a Monaural Mixture Using Convolutional Neural Network in the Time Domain

Peng Zhang[1], Xiaohong Ma[1(✉)], and Shuxue Ding[2]

[1] School of Information and Communication Engineering,
Dalian University of Technology, Dalian, China
`zhangpeng2014@mail.dlut.edu.cn`, `maxh@dlut.edu.cn`
[2] School of Computer Science and Engineering, University of Aizu, Fukushima, Japan
`sding@u-aizu.ac.jp`

Abstract. Audio source separation from a monaural mixture, which is termed as monaural source separation, is an important and challenging problem for applications. In this paper, a monaural source separation method using convolutional neural network in the time domain is proposed. The proposed neural network, input and output of which are both time-domain signals, consists of three convolutional layers, each of which is followed by a max-pooling layer, and two fully-connected layers. There are two key ideas behind the time-domain convolutional network: one is learning features automatically by the convolutional layers instead of extracting features such as spectra; the other is that the phase can be recovered automatically since both the input and output are in the time domain. The proposed approach is evaluated using the TSP speech corpus for monaural source separation, and achieves around 4.31–7.77 SIR gain with respect to the deep neural network, the recurrent neural network and nonnegative matrix factorization, while maintaining better SDR and SAR.

Keywords: Monaural source separation · Convolutional neural network · Deep learning

1 Introduction

Monaural source separation is a fundamental and important problem in the field of speech processing. It could be a previous procedure to some advanced speech applications. For instance, separating a speech from background noise can improve the performance of speech recognition [10], and separating a music from singing voice can enhance the accuracy of chord recognition [1]. However, monaural source separation is difficult since it is an underdetermined inverse problem.

Nonegative matrix factorization (NMF) based method [5] decomposes the amplitude of time-frequency spectrogram to nonegative bases and actives.

© Springer International Publishing AG 2017
F. Cong et al. (Eds.): ISNN 2017, Part II, LNCS 10262, pp. 388–395, 2017.
DOI: 10.1007/978-3-319-59081-3_46

For monaural source separation problem, a NMF-based method learns the dictionary of each sources using train data firstly, and then separates the real mixture by fixing the base to the dictionary [8].

Recently, a number of end-to-end methods based on the deep neural network (DNN) were proposed to solve the monaural source separation problem. Wang et al. introduced DNN to perform binary classification for speech separation [12] and suggested that the ideal ratio mask (IRM) should be preferred over the ideal binary mask (IBM) in terms of speech quality [11]. Huang et al. [2,3] used DNN and the recurrent neural network (RNN) to minimize the reconstruction loss of the spectra of two premixed speaking's by embedding the IRM into the loss function. Wang and Wang proposed joint training with an inverse fast Fourier transform (IFFT) layer to reconstruct the time-domain signal directly [13]. Williamson et al. [14] proposed a framework for performing monaural speech separation in the complex domain and showed that enhancing of the phase spectrum of noisy speech leads to perceptual quality improvements.

All the mentioned DNN-based methods used the time-frequency representation as the input of the networks. In other words, features, e.g. short time Fourier transform coefficients and Mel Frequency Cepstrum Coefficients, were extracted firstly and then were passed to the neural network [2,13].

Differing from above approaches, in this paper, we explore the convolutional neural network (CNN) with time-domain audio signals as the inputs, which is referred to as time domain-CNN for short, for monaural speech separation. There are two key ideas behind the time domain-CNN: (1) it can learn features automatically by convolutional layers of CNN instead of extracting particular features manually, and (2) it does not need to deal with the phase recovery since both the input and output are time-domain signals.

This paper is organized as follows. Section 2 discusses the relation to previous works. Section 2 introduces the proposed method. Section 3 presents the experimental settings and results. Finally we make a conclusion in Sect. 4.

2 Proposed Method

2.1 Why CNN

Let us consider the discrete Fourier transform used in the feature extracting in almost all the audio processing methods:

$$X(k) = \sum_{n=0}^{N-1} x(n) e^{-j2\pi \frac{k}{N} n} \tag{1}$$

where k is the frequency-domain scale. For a given k, Eq. (1) means the cumulative sum of product of the signal and the trigonometric functions. In fact, Eq. (1) computes the correlation between the signal and the base function. On the other hand, in neural network, the output of one neuron in a 1-Dimensional (1-D) convolutional layer is as follows:

$$x = a \left(\sum_{i=0}^{M-1} z\left(i\right) * m\left(i\right) + b \right) \qquad (2)$$

where m is the filter kernel, b is the bias, $a(\cdot)$ denotes the activation function, and M is the length of the filter kernel. Ignoring the activation function and bias, the Eqs. (1) and (2) are very similar. This means that the filter kernel can play similar role with the base function. So using time domain-CNN in source separation can combime the feature extraction and separation as a whole system. Moreover, the filters are learnable and each convolutional layer is followed by a pooling layer, which make feature learning of CNN more targeted and powerful.

2.2   Framework

The proposed framework is a two-stage method which supervised training firstly and then separating with the trained network. The illustration of the framework is shown in Fig. 1(a).

In the training stage, as an input, two pieces of training speech are mixed instantaneously with 0 dB SNR. The original sources are the target of the network. We adopt the Euclidean distance as the cost function of the network, hence the source separation is treated as a regression problem.

In the recovery, i.e., testing stage, the mixture is inputed to the network and then the estimation of the two sources can be obtained from the network's outputs. Noteworthy, the outputs are time-domain signals directly, therefore there is no need to do inverse STFT and the phase recovery problem does not exist.

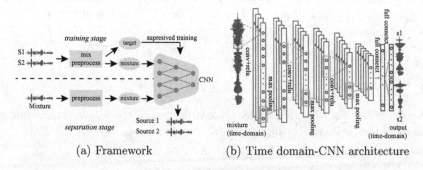

(a) Framework   (b) Time domain-CNN architecture

Fig. 1. Illustration of the proposed framework and network architecture.

2.3   Architecture

1-D convolutional layer is adopted in the CNN architecture, hence the kernel filters are 1-D dimensional. As shown in Fig. 1(b), the time domain-CNN consists of three convolutional layers, each of which is followed by a max-pooling layer, and two fully-connected layers.

An input sample of the network is one frame of time-domain audio signal of which length is set to 1024. The sample is transported to the first convolutional layer of which the filter kernel length should be longer than the following two convolutional layers. We use the rectified linear unit (ReLU) [6] as the nonlinear activation function of the convolutional layer. After ReLU, a max pooling layer, kernel and stride length of which are both 2, is adopted to downsample. For the sake of concision, we refer the collection of the convolutional, ReLU and the max pooling layer to conv-layer-group. The network contains three conv-layer-groups in total and the filter kernel length reduces layer by layer. The length and number of 1-D convolutional kernel in the network's convolutional layers are given by (75, 96)-(29, 96)-(15, 128) in our experiments. The following two layers are fully-connected layers which contain 2048 neurons each layer. The first fully-connected layer is activated by tanh function while the last one is a linear output layer of which the front 1024 values of neuron output denote the source 1, and the others denote source 2.

2.4 Pre- and Post-processing

Training Pre Processing. In the training phase, there are three pre-precessing procedures, dividing to frames with window function, data regularization, and data gain.

The audio signal is 1-D sequence satisfying short-time stationary, hence it can be divided to a series of frames. Meanwhile, a window function may be used for each frame in order to reduce spectrum leakage.

The data regularization is a common and useful method to accelerate optimization convergence in the field of machine learning. For the time domain-CNN, unlike computer vision tasks which normalize the data to (0, 1), we normalize the audio data to $(-1, 1)$ as follow:

$$\mathbf{y}_i = \frac{\mathbf{s}_i}{\max(\mathbf{s}_i)}, i = 1, 2 \tag{3}$$

where $\mathbf{s}_i$ denotes i-th source and $\mathbf{y}_i$ denotes the i-th normalized source.

The data gain is a technique to generate data set with more samples from relatively small data set. It plays an important role when the training data are not very large. Here we perform data gain by *loopshift mixing* as follow:

$$\mathbf{x} = \overset{W-1}{\underset{k=0}{\varPhi}} \left(\varPsi\left(\mathbf{y}_1, k \cdot \tau\right) + \mathbf{y}_2 \right) \tag{4}$$

where $\varPhi$ denotes stacking W vectors into one and τ is the circular shift length. W amounts to L/τ, and L means the length of audio source vector $\mathbf{s}$. $\varPsi$ denotes the loopshift operator:

$$\varPsi(\mathbf{y}, \beta)_i = \begin{cases} \mathbf{y}\left(i + \beta\right) & 1 \le i \le L - \beta \\ \mathbf{y}\left(i - L + \beta\right) & L - \beta < i \le L \end{cases} \tag{5}$$

Testing Pre- and Post-processing. In the testing phase of separation, data should do the same pre-processing, like applying window function to the frames and data normalization, as what have been done in the training phase. Particularly, in the separation phase, we need more overlap between the neighbor frames than that in the training phase.

In the post processing procedure, overlap corresponds to averaging separation predictions of more frames at each time-bin and a more accurate separation result can be expected. The recovery at point t can be obtained by:

$$\tilde{s}_i(t) = \frac{1}{T} \sum_{j=1}^{T} \hat{s}_i(t, j) \tag{6}$$

where $\hat{s}_i(t, j)$ denotes the value of j-th frame containing the point t for the source i. T denotes the number of frames containing the point t

3   Experiments

3.1   Experimental Setup

We evaluate the performance of the proposed approach for the monaural speech separation task using the TSP corpus [4]. There are 1444 utterances, with average length 2.372 s, spoken by 24 speakers (half male and half female). We choose four speakers, FA (female), FB (female), MC (male), and MD (male), from the TSP speech database. After concatenating together 60 sentences for each speaker, we use 80% of the signals for training, 10% for development, and 10% for testing. The signals are downsampled to 16 kHz. The time domain-CNN networks are trained on three different mixing cases: FA versus MC, FA versus FB, and MC versus MD. These setups are the same as Huang's setting in [3]. Since FA and FB are female speakers while MC and MD are male, the latter two cases are expected to be more difficult due to the similar frequency ranges from the same gender. The separation performance is assessed using the Signal-to-Distortion Ratio (SDR), the Signal-to-Interference Ratio (SIR) and the Signal-to-Artifacts Ratio (SAR) in decibels (dB), according to the BSS-EVAL toolbox [9] and its Python reimplementation [7]. SDR indicates the overall quality of each estimated source compared to the target, while SIR reveals the amount of residual crosstalk from the other sources and SAR is related to the amount of musical noise.

3.2   Experiments Results

We compared the proposed approach with NMF method, and the DRNN+ discrim architectures, one with spectra features and another with log-mel features which obtained the best results in Huang's experiments [3]. The experiment parameters configuration for the proposed time domain-CNN is depicted as follows. In the training phase, the frames are taken with half-overlapping hamming window of 1024 sample. The circular shift length for loopshift mixing is set to

$\tau = 10000$. We train the time domain-CNN network using stochastic gradient descent (SGD) method with mini batchsize 256 and learning rate 0.01. In separation phase, the frame hop is set to 128 while the frame length and window are the same as training settings.

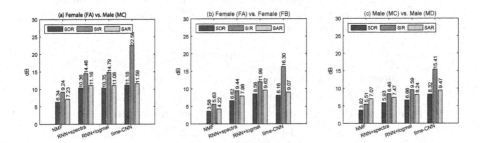

Fig. 2. TSP speech separation results. Three settings are considered. (a) Female vs. Male, (b) Female vs. Female, and (c) Male vs. Male. From the left to the right, the results are obtained from NMF, the DRNN+discrim architectures with spectra features and log-mel features, and the proposed method.

The speech separation results of cases, FA versus MC, FA versus FB, MC versus MD, are shown in Fig. 2(a), (b), and (c), respectively. For FA versus MC case, the proposed time domain-CNN approach improves SIR by 7.77 dB to Huang's DNN-based methods [3] with the best settings, while achieves a little bit of better SDR and SAR performance. For the FA versus FB case, our approach achieves 4.31 dB SIR gain with respect to the DRNN-discrim-logmel, although SDR and SAR descend around 0.4 dB. For the MC versus MD case, our approach outperforms the comparison method in SDR, SIR, and SAR by 1.66 dB, 5.82 dB, and 1.23 dB respectively.

On the whole, our approach has outstanding performance in SIR so that achieves a better separation results with less interference and crosstalk in the recovered sources. Meanwhile, the proposed approach maintaines better SDR and SAR. The sound examples are available online.[1]

To examine the performance of the using of CNN architecture, the results of full-connect DNN and time domain-CNN networks with different convolutional layer parameters are summarized in the Table 1. The time domain-DNN network has two or three full-connect layers of 2048 neurons with time-domain input.

As shown in Table 1, the time domain-DNN cannot achieve the separation goal almost since its low SDR, SIR and SAR performances. Meanwhile the time domain-CNN works well. Furthermore, a 3-conv-layers network achieves better performance with respect to the 2-conv-layers CNN network. The length of convolutional filter kernel is an important parameter for time domain-CNN. Experimental results demonstrate that time domain-CNN with a longer kernel length achieves better separation performance while consuming more training

---

[1] http://pengzhxyz.github.io/bss-time-cnn.

time. These experiment results prove that the convolutional layer is crucial for separation in time-domain and learning feature using the time domain-CNN performs well.

Table 1. FA versus MC separation results of time domain-DNN and time domain-CNN networks. The first column denotes the kernel length of each convolutional layers. E.g. (75-29-15) means the length of three convolutional layers are 75, 29, and 15 respectively.

Method	SDR	SIR	SAR	Training time
Time domain-DNN (2 hide layers)	1.67	2.54	11.14	0.12 h
Time domain-DNN (3 hide layers)	1.64	2.54	10.94	0.13 h
Time domain-CNN (75-29)	9.32	17.69	10.12	1.10 h
Time domain-CNN (75-29-15)	10.36	20.57	10.92	1.23 h
Time domain-CNN (75-39-19)	10.85	20.93	11.36	1.83 h
Time domain-CNN (75-55-27)	11.18	22.56	11.59	2.50 h
Time domain-CNN (55-29-15)	10.46	20.82	10.97	1.35 h
Time domain-CNN (95-29-15)	10.76	21.42	11.22	1.31 h
Time domain-CNN (95-55-27)	10.94	22.21	11.37	2.46 h

4   Conclusion

In this paper, a monaural source separation method using convolutional neural network in the time domain is proposed. The input and output of the network are both in the time domain, and there are three convolutional layers and two full-connected layers in total. It has two major advantages, it can learn features by network automatically and can recover phase automatically. The experimental results demonstrate that the proposed approach outperforms the DNN-based methods with manual feature extraction (spectra or log-mel features). Our approach achieves 4.31–7.77 dB SIR gain with respect to Huang's DRNN network [3], while maintaining better SDR and SIR performance. The further works would be using of big data and exploring what does the time domain-CNN network learn.

Acknowledgments. This work is supported by the National Natural Science Foundation of China under Grant 61071208.

References

1. Huang, P.S., Chen, S.D., Smaragdis, P., Hasegawa-Johnson, M.: Singing-voice separation from monaural recordings using robust principal component analysis. In: 2012 IEEE International Conference on Acoustics, Speech and Signal Processing (ICASSP), pp. 57–60. IEEE (2012)

2. Huang, P.S., Kim, M., Hasegawa-Johnson, M., Smaragdis, P.: Deep learning for monaural speech separation. In: 2014 IEEE International Conference on Acoustics, Speech and Signal Processing (ICASSP), pp. 1562–1566. IEEE (2014)
3. Huang, P.S., Kim, M., Hasegawa-Johnson, M., Smaragdis, P.: Joint optimization of masks and deep recurrent neural networks for monaural source separation. IEEE/ACM Transa. Audio Speech Lang. Process. **23**(12), 2136–2147 (2015)
4. Kabal, P.: TSP Speech Database. McGill University, Database Version 1(0), 09–02 (2002)
5. Lee, D.D., Seung, H.S.: Learning the parts of objects by non-negative matrix factorization. Nature **401**(6755), 788–791 (1999)
6. Nair, V., Hinton, G.E.: Rectified linear units improve restricted Boltzmann machines. In: Proceedings of the 27th International Conference on Machine Learning (ICML-10), pp. 807–814 (2010)
7. Raffel, C., McFee, B., Humphrey, E.J., Salamon, J., Nieto, O., Liang, D., Ellis, D.P., Raffel, C.C.: mir_eval: a transparent implementation of common MIR metrics. In: Proceedings of the 15th International Society for Music Information Retrieval Conference, ISMIR. Citeseer (2014)
8. Smaragdis, P., Raj, B., Shashanka, M.: A probabilistic latent variable model for acoustic modeling. In: Advances in Models for Acoustic Processing, NIPS 148, 8–1 (2006)
9. Vincent, E., Gribonval, R., Févotte, C.: Performance measurement in blind audio source separation. IEEE Trans. Audio Speech Lang. Process. **14**(4), 1462–1469 (2006)
10. Vinyals, O., Ravuri, S.V., Povey, D.: Revisiting recurrent neural networks for robust ASR. In: 2012 IEEE International Conference on Acoustics, Speech and Signal Processing (ICASSP), pp. 4085–4088. IEEE (2012)
11. Wang, Y., Narayanan, A., Wang, D.: On training targets for supervised speech separation. IEEE/ACM Trans. Audio Speech Lang. Process. **22**(12), 1849–1858 (2014)
12. Wang, Y., Wang, D.: Towards scaling up classification-based speech separation. IEEE Trans. Audio Speech Lang. Process. **21**(7), 1381–1390 (2013)
13. Wang, Y., Wang, D.: A deep neural network for time-domain signal reconstruction. In: 2015 IEEE International Conference on Acoustics, Speech and Signal Processing (ICASSP), pp. 4390–4394. IEEE (2015)
14. Williamson, D.S., Wang, Y., Wang, D.: Complex ratio masking for joint enhancement of magnitude and phase. In: 2016 IEEE International Conference on Acoustics, Speech and Signal Processing (ICASSP), pp. 5220–5224. IEEE (2016)

Input Dimension Determination of Linear Feedback Neural Network Applied for System Identification of Linear Systems

Wenle Zhang[(✉)]

Department of Engineering Technology, University of Arkansas at Little Rock,
Little Rock, AR 72204, USA
wxzhang@ualr.edu

Abstract. In the application of a linear neural network (LNN) to linear system identification and parameter estimation, it is important to determine the input dimension of the LNN so that the identification can be performed efficiently. In the LNN for linear system identification, both the input and output data are taken as input of the LNN. The output data are delayed and are fed-back to input of the LNN. The input dimension determination is to determine the right number of past inputs should be applied to its input and the right number of past outputs should be fed-back to its input also. The sampled input and output data are used to train the LNN. The performance errors are collected during training and are used in the evaluation by Akaike's Information Criterion to determine the input dimension. The advantage of LNN method is its simplicity and effectiveness. Satisfactory results from simulation are provided to show the effectiveness of the proposed algorithm.

Keywords: System identification · Linear neural network · Input dimension

1 Introduction

Linear system identification is about how to build a model that best match the structure and system parameters of the actual system from the measured input and output data samples. Structure for a single input and single output linear discrete system is simply the order. Model order estimation has been widely researched, e.g., ratio of determinant of covariance matrix method [16], Akaike's Information Criterion [1], Instrumental variable Product Moment method [14]. After the order is identified, the parameter estimation can be done by a variety of available traditional estimation methods, for example, the maximum likelihood, the LS and instrumental techniques [7, 13].

Lately, neural network methods [5, 8] are studied and applied for system identification. Popular neural networks for this purpose are within the type of recurrent or feedback neural networks, such as, feedback Backpropagation [11] and Hopfield neural network [6]. However, most neural network for system identification methods are for nonlinear systems [3, 9, 10, 12]. Other neural networks have also been used, such as: Hopfield [2, 4], Support Vector and Self-Organizing Map. A Generalized ADALINE neural network based method for linear systems was proposed by the author [17].

© Springer International Publishing AG 2017
F. Cong et al. (Eds.): ISNN 2017, Part II, LNCS 10262, pp. 396–404, 2017.
DOI: 10.1007/978-3-319-59081-3_47

However, input dimension determination of the neural network for this purpose has been less studied.

Presented in this work is a method for determining the input dimension of a linear neural network (LNN) for system identification of linear discrete systems. For efficient system identification, especially for parametric system identification, the input dimension of such network has to be determined so that a mapping to the system order can be established, that is, the input dimension is equal to the sum of number of poles (Order of the system) and number of zeros (Less than the order) of the system. In the LNN for linear system identification, not only the input data are used as its input but also the output data. Both the input and output data are delayed and output data are fed-back to input. The input dimension determination is to determine the right number of past inputs be applied to its input and the right number of past outputs be fed-back to its input also. The measured input and output data set are used to train the LNN. The performance errors are collected during training and are used in the evaluation by Akaike's Information Criterion to determine the input dimension. The advantage of the proposed algorithm is that it is computational simple. Simulation is performed to show the effectiveness the proposed algorithm. The rest of paper is organized as: Sect. 2 describes the order and parameter estimation problem of linear discrete systems, Sect. 3 presents the LNN based method for the input dimension determination, Sect. 4 provides the results of some simulation cases and the last section is for the conclusion.

2 Linear System Identification

The linear discrete time single input and single output system is considered as stable and observable. The output and input data are measured for N samples over an interval of time and formulated into an input-output pair data set: $\{u(kT), y(kT)\}$, where $k = 1$, $2, ..., N$, and T is a constant – the period of sampling, so the data set can be simplified as $\{u(k), y(k)\}$. Then the below difference equation can be used to model the system,

$$y(k) = -a_1 y(k-1) - a_2 y(k-2) - \ldots - a_{na} y(k-n_a) + b_1 u(k-1) + b_2 u(k-2)$$
$$+ \ldots + b_{nb} u(k-n_b) + e(k) \tag{1}$$

or in the form of a polynomial difference equation,

$$y(k) = -P(z^{-1})y(k) + Q(z^{-1})u(k) + e(k)$$
$$P(z^{-1}) = a_1 z^{-1} + a_2 z^{-2} + \ldots + a_{na} z^{-na}, \text{ and} \tag{2}$$
$$Q(z^{-1}) = b_1 z^{-1} + b_2 z^{-2} + \ldots + b_{nb} z^{-nb}$$

where z^{-1} is the operator for unit delay, e is the noise, a_i and b_j are system coefficients, $i = 1, 2, ..., n_a$ and $j = 1, 2, ..., n_b$. n_a and n_b are order parameters, $n_b \leq n_a$, n_a is the order of the system.

3   Input Dimension Determination of the LNN

The main method is build the LNN for the system model in (1), whose number of input is $m = n_b + n_a$, and n_a is number of delayed outputs and n_b the number of delayed inputs, and whose input weights are corresponding to the coefficient b_j and a_i. By adjusting the order n_b and n_a and training the LNN, the training performance error are collected and evaluated to determine proper values for n_b and n_a, thus determining m.

The configuration of our LNN for system identification is indicated in Fig. 1, note the bias term is not needed, each system input is fed to a series of tapped delay line to form multiple inputs and the system output is also delayed through a series of tapped delay line to be fed-back to the input. Then the inputs to the LNN are $u_1 \sim u_m$, and the output a can be calculated as,

$$a = \sum_1^m w_i u_i = \mathbf{w}^T \mathbf{u} \tag{3}$$

where $\mathbf{u}$ is the input vector and $\mathbf{w}$ is the weight vector,

$$\mathbf{u} = [u_1 u_2 \ldots u_m]^T = [u(k-1) \ldots u(k-n_b) - y(k-1) \ldots - y(k-n_a)]^T \tag{4}$$

$$\mathbf{w} = [w_1 w_2 \ldots w_m]^T = [b_1 b_2 \ldots b_{nb} a_1 a_2 \ldots a_{na}]^T \tag{5}$$

By arranging both $\mathbf{u}$ and $\mathbf{w}$ in this manner, obviously, the input dimension is given by,

$$m = n_a + n_b$$

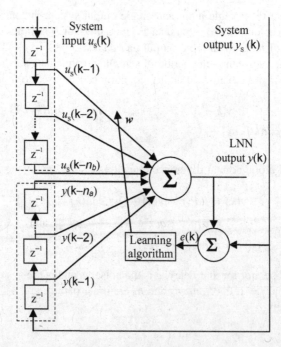

Fig. 1. LNN configuration

3.1 LNN Training

From Fig. 1 and Eq. (4), the training data set can be formed as,

$$\{u(1), t(1)\}, \ldots, \{u(k), t(k)\}, \ldots, \{u(N), t(N)\} \tag{6}$$

where input vector and the target at time step k are given as,

$$u(k) = [u(k-1)\ldots u(k-n_b)-y(k-1)\ldots-y(k-n_a)]^{\mathrm{T}}$$
$$t(k) = y(k)$$

The LNN learning algorithm is developed from the original LMS algorithm. Training is performed by applying each data pair over the entire data set as shown in (6), calculating the error of the output from the target and adjusting the weights by minimization of the following error function (the Mean Square Error, denoted as MSE),

$$J(w) = \frac{1}{N}\sum_1^N e^2(k) \tag{7}$$

where k is the time step and $e(k)$ is given by,

$$e(k) = t(k)-a(k) \tag{8}$$

where $t(k)$ is the desired output at time k and $a(k) = w^{\mathrm{T}}(k)u(k)$ as in (3).

In LMS learning, the MSE in (7) is approximated by the instantaneous squared error at time k,

$$J(w) \approx e^2(k) = (t(k)-a(k))^2 = (t(k)-w^T(k)u(k))^2 \tag{9}$$

This replacement works because (7) is minimized if squared error at each step $e^2(k)$ is minimized. Now, the gradient of J can be found as,

$$\nabla J(w) = \nabla\left(e^2(k)\right) = 2e(k)\frac{\partial e(k)}{\partial w} = -2e(k)u(k) \tag{10}$$

Then, the weight adjustment in LMS learning algorithm is given in the steepest descent (the negative gradient) direction by,

$$w(k+1) = w(k) + \Delta w(k) = w(k) + 2\eta e(k)u(k) \tag{11}$$

where η is the learning rate parameter, a small positive value.

3.2 Input Dimension Determination

Here determination of input dimension is to calculate the two order values k_a and b_b (versus true order values n_a and n_b) respectively. Ideally, if there is some a priori information for the system under identification, then experiential guess can be used to determine a proper range of values for n_a and n_b, and the estimation for k_a and k_b should be limited within the range $n_a \pm C_1$ and $n_b \pm C_1$, here C_1 is the range parameter for each order value and it should be a small integer, e.g. 3 or 4. When there is no a priori information is available, the estimation should start from 1 and increase to C_2 (a number similar to C_1 but just a little bigger such as 6). In order to estimate the input dimension, in a way similar to the Akaike's Information Criterion (AIC) in traditional order estimation method, the LNN will be trained by LMS learning on each order pair (k_a and k_b) in the range as discussed above. During training for all the order pairs, the average squared training errors (MSE) are calculated as by Eqs. (6) and (7). Then the AIC value for each pair k_a and k_b can be established as,

$$\text{AIC}(k_a, k_b) = N log \left(\frac{\sum_1^N e^2(k)}{N} \right) + 2(k_a + k_b) \tag{12}$$

where log is the natural log and N the sample size of the measured data. According to Akaike, once $k_a = n_a$ and $k_b = n_b$, the AIC value will be minimized. Then the model order pair can be chosen as $(n_a, n_b) = \min(\text{AIC}(k_a, k_b))$.

The Input Dimension Determination method for the LNN can be organized into an algorithm, termed IDD algorithm:

  (i)   Select a small value for η, e.g. 0.1. Set weight w to uniformly distributed small randomly generated numbers in the range, say $(-0.1, +0.1)$.
 (ii)   Set $k_a = 1$, $k_b = 1$
(iii)   Perform training on LNN for $k = 1$ to N
 (iv)   Evaluate $\text{AIC}(k_a, k_b) = N\log(1/N \sum_1^N e^2(k)) + 2(k_a + k_b)$
  (v)   Increment k_b until k_a, this makes sure $n_b \leq n_a$ in (2)
 (vi)   Increment k_a until C_2
(vii)   Let $(n_a, n_b) = \min(\text{AIC}(k_a, k_b))$
(viii) Let $[b_1 \, b_2 \, ... \, b_{nb}] = [w_1 \, w_2 ... w_{nb}]$
 (ix)   Let $[a_1 \, a_2 \, ... a_{na}] = [w_{nb+1} \, w_{nb+2} ... w_{na+nb}]$

3.3 Implementation of the IDD Algorithm

The IDD algorithm can be readily programmed with any high level language. We implemented it in MATLAB. Like many methods using neural network, the trial and error process is used to select suitable values for the η for solving different problems. A rule of thumb is to start η with a small value such as 0.01 and increase gradually, say by 0.01. It is important that η should be kept smaller than the upper bound $1/\lambda_{max}$, where λ_{max} is the maximal eigenvalue of the mean input data correlation matrix E $[uu^T]$, given by [15]. It should be also noted that the system under estimation should be stable, meaning all the roots of Eq. (1) are within unit circle.

4   Simulation Results

Simulation of a couple of example systems that often appear in the literature has done to show the effectiveness of the proposed algorithm. The AIC criteria are calculated to determine the input dimension from the order parameters, and parameter estimates are obtained from the corresponding the weight vector.

Example 1. The difference equation of a 2nd order linear system is,

$$y(k) = 1.5y(k-1) - 0.7y(k-2) + 1.0\,u(k-1) + 0.5u(k-2)$$

The input signal is a PRBS and the output is calculated for N = 1000 samples. The first 50 samples of input and output are drawn in Fig. 2. The AIC matrix are obtained as in the following for learning rate $\eta = 0.01$, where first 300 samples are skipped to avoid transient period of LMS training, as seen in parameter trajectories indicated in Fig. 3 for the determined order.

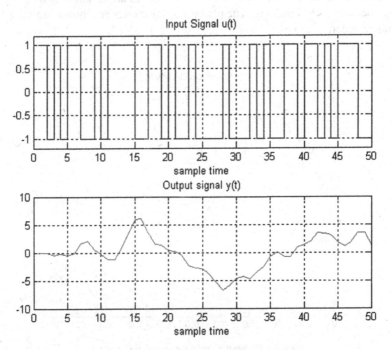

Fig. 2. Input and output signals for Example 1

In the AIC matrix k_a is the row number and k_b is the column number Minimum AIC is found as, AIC(2,2) = −4532.2, then the order and parameters should be selected as,

$$n_a = 2, n_b = 2; \text{ so } m = 4.$$

$$[b_1 \; b_2 \; a_1 \; a_2] = [1.0000 \; 0.5000 \; -1.5000 \; 0.7000]$$

$$AIC_{4,4} = \begin{pmatrix} 800.65 & \infty & \infty & \infty \\ -928.17 & -4532.2 & \infty & \infty \\ -677.23 & -2331.8 & -4424.6 & \infty \\ 483.85 & -2117.2 & -3851.9 & -4241.8 \end{pmatrix}$$

Here, the input dimension is found to be $m = n_a + n_b = 4$, and the parameter estimates are very close to the true values. Notice that AIC(3, 3) is small and the order is 3 by 3, but no significant reduction. The training trajectories for this case are shown in Fig. 3.

Example 2. A third order system given in [7],

$$y(k) = 1.0\,y(k-1) + 0.15\,y(k-2) - 0.35\,y(k-3) + 1.0\,u(k-1) + 0.5\,u(k-2)$$

The simulation is done with N = 1000 samples on a PRBS input signal. The AIC matrix is obtained for $\eta = 0.01$, where first 300 samples are similarly skipped to avoid transient period of LMS training. The parameter trajectories are shown in Fig. 4 for the determined order.

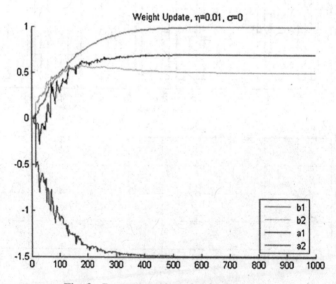

Fig. 3. Parameter trajectories for Example 1

The Minimum AIC is found as, AIC(3, 2) = −5533.4, then the order and parameters are,

$$n_a = 3, n_b = 2 \rightarrow m = 5$$

$$[b_1\, b_2\, a_1\, a_2\, a_3] = [1.0000\,0.5082\,-0.9917\,-0.1626\,0.3554]$$

Notice that the estimated parameters are slightly off. The reason could be multi-fold: the magnitude of the input vector is large even the error is smaller after training convergence, the level of persistent excitation of the input applied is not 100% and the existence of roots of the system (1) is near the unit circle.

$$\text{AIC}_{5,5} = \begin{pmatrix} -132.38 & \infty & \infty & \infty & \infty \\ -3681.6 & -2224.0 & \infty & \infty & \infty \\ -2392.9 & -5533.4 & -4761.9 & \infty & \infty \\ -2412.1 & -3829.6 & -5187.0 & -5119.7 & \infty \\ -2148.7 & -3885.1 & -4140.7 & -4319.2 & -4843.0 \end{pmatrix}$$

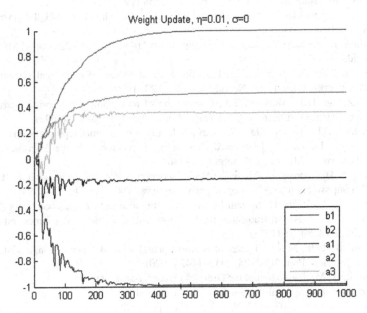

Fig. 4. Parameter trajectories for Example 2

5   Conclusions

Presented in this work is a method for determining the input dimension of the LNN used for linear system identification. The training of the neural network is performed on different order combinations, i.e., k_b past inputs and k_a past outputs are applied as inputs to the network, so input dimension is the sum: $k_a + k_b$. The mean square errors are calculated for each order pair during training of the network. The order of the two system polynomials are then determined as the pair (k_a, k_b) corresponding to the minimum AIC and polynomial coefficients are estimated as the weights at the same time as order obtained. The input dimension is then the sum of (k_a, k_b). The performance of the proposed method is shown by the simulation results.

References

1. Akaike, H.: A new look at the statistical model identification. IEEE Trans. Autom. Control **19**, 716–723 (1974)
2. Atencia, M., Sandoval, G.: Gray box identification with hopfield neural networks. Rev. Inv. Oper. **25**, 54–60 (2004)
3. Bhama, S., Singh, H.: Single layer neural network for linear system identification using gradient descent technique. IEEE Trans. Neural Netw. **4**, 884–888 (1993)
4. Chu, S.R., Shoureshi, R., Tenorio, M.: Neural networks for system identification. IEEE Control Syst. Mag. **10**, 31–34 (1990)
5. Haykin, S.: Neural Networks – A Comprehensive Foundation, 2nd edn. Prentice Hall, Upper Saddle River (1999)
6. Hopfield, J.: Neural networks and physical systems with emergent collective computational abilities. Proc. Natl. Acad. Sci. USA **79**, 2554–2558 (1982)
7. Ljung, L.: System Identification – Theory for the User, 2nd edn. Prentice-Hall, Upper Saddle River (1999)
8. Mehrotra, K., Mohan, C., Ranka, S.: Elements of Artificial Neural Networks. MIT Press, Cambridge (1997)
9. Narendra, K.S., Parthasarathy, K.: Identification and control of dynamical systems using neural networks. IEEE Trans. Neural Netw. **1**, 1–27 (1990)
10. Qin, S.Z., Su, H.T., McAvoy, T.J.: Comparison of four neural net learning methods for dynamic system identification. IEEE Trans. Neural Netw. **2**, 52–262 (1992)
11. Rumelhart, D.E., Hinton, G.E., Williams, R.J.: Learning internal representations by error propagation. In: Parallel Distributed Processing: Explorations in the Microstructure of Cognition, vol. I, MIT Press, Cambridge (1986)
12. Sjöberg, J., Hjalmerson, H., Ljung, L.: Neural networks in system identification. In: The 10th IFAC Symposium on SYSID, Copenhagen, Denmark, vol. 2, pp. 49–71 (1994)
13. Söderström, T., Stoica, P.: System Identification. Prentice Hall, Englewood Cliffs (1989)
14. Wellstead, P.E.: An instrumental product moment test for model order estimation. Automatica **14**, 89–91 (1978)
15. Widrow, B., Lehr, M.A.: 30 years of adaptive neural networks: perceptron, madaline, and backpropagation. Proc. IEEE **78**, 1415–1442 (1990)
16. Woodside, C.M.: Estimation of the Order of Linear Systems. Automatica **7**, 727–733 (1971)
17. Zhang, W.: System identification based on a generalized ADALINE neural network. In: Proceedings of the 2007 ACC, New York City, pp. 4792–4797 (2007)

Joining External Context Characters to Improve Chinese Word Embedding

Xianchao Zhang[1], Shike Liu[1], Yuangang Li[2,3], and Wenxin Liang[1(✉)]

[1] School of Software Technology, Dalian University of Technology,
Dalian 116024, China
{xczhang,wxliang}@dlut.edu.cn, shikeliu@mail.dlut.edu.cn
[2] Shanghai University of Finance and Economics, Shanghai 200433, China
[3] Goldpac Limited, Zhuhai 519070, China
gary.li@goldpac.com

Abstract. In Chinese, a word is usually composed of several characters, the semantic meaning of a word is related to its composing characters and contexts. Previous studies have shown that modeling the characters can benefit learning word embeddings, however, they ignore the external context characters. In this paper, we propose a novel Chinese word embeddings model which considers both internal characters and external context characters. In this way, isolated characters have more relevance and character embeddings contain more semantic information. Therefore, the effectiveness of Chinese word embeddings is improved. Experimental results show that our model outperforms other word embeddings methods on word relatedness computation, analogical reasoning and text classification tasks, and our model is empirically robust to the proportion of character modeling and corpora size.

Keywords: Word embeddings · Neural network · NLP

1 Introduction

In natural language processing, word representation is always a heated topic. Recently, distributed word representation, also known as word embeddings, is proved to be effective in capturing both semantic and regularities in language [2,10,12], and experimentally superior to discrete representations [1]. Word embeddings have benefited natural language processing in many tasks including entity recognition and disambiguation [7,15], syntactic parsing [13], word sense disambiguation [5], semantic composition [16] and knowledge extraction [9].

Existing word embeddings models extract vectors representing word meaning by relying on the distributional hypothesis, namely, these models typically learn word embeddings according to the contexts of words in large-scale corpora. However, in Chinese, the semantic meaning of a word is also related

This work was supported by NSFC (No. 61632019) and 863 project of China (No. 2015AA015403).

© Springer International Publishing AG 2017
F. Cong et al. (Eds.): ISNN 2017, Part II, LNCS 10262, pp. 405–415, 2017.
DOI: 10.1007/978-3-319-59081-3_48

to its composing characters and external context characters. Take a Chinese sentence "科技 改变 生活" (technology changes life) for example, the external context characters of target word "改变" (change) are "科" (department), "技" (technique), "生" (grow) and "活" (live). Recently, a number of researchers have demonstrated the usefulness of exploiting the internal structure of words and modeling the basic units, such as morphemes in English or characters in Chinese [3,6]. Botha and Blunsom [3] introduced a log-bilinear model which used addition as composition function to derive word vectors from morpheme vectors. Chen et al. [6] extended their idea, assuming that the word semantics is composed of the meanings of its characters and its particular meaning, and constructed a character-enhanced word embeddings model (CWE). In Chinese, CWE outperformed the original word-based models in varieties tasks.

However, integrating only character semantics into word semantics causes some problems. First, CWE tends to produce similar word embeddings for words with common characters, e.g. the relatedness of word pair "歌手" (singer) and "水手" (sailor) are overestimated due to having the common character ("手"). Second, CWE updates character embeddings directly using the updating expression of word embeddings, this limits the quality of character embeddings.

To solve the above problems and optimize Chinese word embeddings that score robustly across multiple test sets, in this paper we propose a novel external context characters-enhanced Chinese word embedding model (ECWE). Our main contribution are summarized as follows:

- We take advantages of both internal characters and external context characters to make more relevance among isolated characters and weakens the influence of internal characters to word semantic, thus solving the problem of similar word embeddings for words with common characters in CWE.
- In ECWE, we generate a new updating expression for character embeddings, this updating expression ensures that character embeddings contain more semantic information.
- Extensive experimental results show the superiority of our proposed model in terms of effectiveness and robustness.

2 Related Work

In recent years, many researches are proposed to find a better method on learning word embeddings. Some researchers explore sub-word units and how they can be used to compose word embeddings. Collobert et al. [7] used extra features such as capitalization to enhance their word vectors, however, it can not generate high-quality word embeddings for rare words. Sun et al. [14] proposed SEING model, which consider that these words which contain common morphemes have similar semantic, it requires word embeddings in line with the context distribution and internal morphemes distribution, that destroyed the distribution hypothesis.

Some other works try to join external knowledge, Botha et al. [3] proposed a scalable method for integrating compositional morphological representations into a log-bilinear language model. Chen et al. [4] presented an approach which first initializes the word sense embeddings through learning sentence level embeddings from WordNet glosses, then the initialized word embeddings are used to generate word embeddings. Cotterell et al. [8] exploited existing morphological resources that can enumerate a word's component morphemes, then combine the latent variables represent embeddings of morphemes to create words embeddings. However, these models are mostly sophisticated and task-specific, they are non-trivial to be applied to other scenarios.

ECWE presents a simple and general way to integrate the internal characters knowledge and external contexts knowledge to learn Chinese word embeddings which are capable in various tasks.

3   Preliminaries

3.1   CBOW

Continuous bag-of-words model (CBOW) [10] aims at predicting the target word, given context words in a sliding window. In this model, each word $w \in W$ is associated with the vector $v_w \in \mathbb{R}^d$, where W is the word vocabulary and d is the vector dimension. Formally, the objective of CBOW is to maximize the average log probability, shown as Eq. (1),

$$L(\theta) = \frac{1}{M} \sum_{w \in W} \log p(w|Context(w)),\tag{1}$$

where M is the size of word vocabulary and $Context(w)$ is the context words of w. CBOW formulates the probability $p(w|Context(w))$ using a softmax function as Eq. (2),

$$p(w|Context(w)) = \frac{\exp(\mathbf{x}_w^\top \cdot v_w)}{\sum_{w' \in W} \exp(\mathbf{x}_w^\top \cdot v_{w'})},\tag{2}$$

where $\mathbf{x}_w$ is the average of all context word vectors of the target word w, and $\mathbf{x}_w$ is defined by Eq. (3).

$$\mathbf{x}_w = \frac{1}{2K} \sum_j v_j,\tag{3}$$

where $j = w - K, \ldots, w - 1, w + 1, \ldots w + K$

In Eq. (3), K is the context window size of a target word. In order to make the learning model more effective, the techniques of hierarchical softmax and negative sampling [11] are used.

3.2   CWE

CBOW treats each word as a basic unit and fails to capture the internal structure of words. CWE [6] considers internal character embeddings to improve the effectiveness of Chinese word embeddings. The key idea of this model is to represent the word with both word embeddings and character embeddings:

$$v_w = v_w + \frac{1}{N_w} \sum_{k=1}^{N_w} c_k^w, \tag{4}$$

where v_w is the word embedding of w, N_w is the number of characters in w, c_k^w is the embedding of the k-th character in w.

While CWE outperformed the original word-based models in many tasks, there are some problems in CWE. As show in Eq. (4), CWE represents a word with word embeddings and character embeddings together. When CWE updates word embeddings and character embeddings, it uses the same updating expression as Eq. (5), which limit the quality of character embedding.

$$v := v + \eta \sum_{u \in \{w\} \cup NEG(w)} \frac{\partial L(w, u)}{\partial \mathbf{x}_w} \tag{5}$$

In Eq. (5), v indicates the word embeddings or character embeddings which need to be updated, η is the learning rate.

4   Our Proposed Framework

4.1   ECWE

As we discussed above, the semantic meaning of a Chinese word is related to its composing characters and contexts, and CWE can not obtain excellent character embeddings. So, in order to make the characters contain richer semantic information, we consider the external context characters. We represent a character using the distribution of its context words, with this method, put characters into the word semantic space, and modeling characters more effectively. In order to achieve the joint training of characters and words, we optimize the conditional probability $L(\theta)$ of context words to target word w and context characters to target word w at the same time, the objective of ECWE as is shown as Eq. (6).

$$L(\theta) = \frac{1}{M} \sum_{w \in W} [(1 - \beta)\log p(w|Context(w)) + \beta \log p(w|\widehat{Context}(w))] \tag{6}$$

$\widehat{Context}(w)$ is the context characters of w, and β is an adjustable weight parameter, which indicates the proportion of character modeling. In order to optimize the calculation, we formulate the probability $p(w|Context(w))$ and $p(w|\widehat{Context}(w))$ using negative sampling, as shown in Eqs. (7) and (8).

$$p(w|Context(w)) = \prod_{u \in \{w\} \cup NEG(w)} [\sigma(\mathbf{x}_w^\top \theta^u)]^{L^w(u)} \cdot [1 - \sigma(\mathbf{x}_w^\top \theta^u)]^{1-L^w(u)} \tag{7}$$

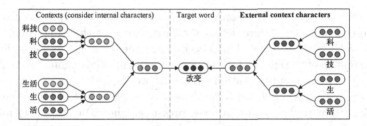

Fig. 1. The architectures of ECWE. Here "科技 (technology) 改变 (change) 生活 (life)" is a word sequence. The word "科技" (technology) is composed of characters "科" (department) and "技" (technique), and the word "生活" (life) is composed of characters "生" (grow) and "活" (live). (Color figure online)

In Eq. (7), $NEG(w)$ is the negative sampling set. $L^w(u)$ is the label of a sample u, $L^w(u) = 1$ when $u = w$, and $L^w(u) = 0$ when $u \neq w$. θ^u is the parameter vector representation.

$$p(w|\widehat{Context}(w)) = \prod_{u \in \{w\} \cup NEG(w)} [\sigma(\mathbf{c}_w^\top \theta^u)]^{L^w(u)} \cdot [1 - \sigma(\mathbf{c}_w^\top \theta^u)]^{1-L^w(u)} \quad (8)$$

In Eq. (7), $\mathbf{x}_w$ is the average of all context word vectors of the target word w with internal character embeddings. Chinese characters does not have semantics, but when they combined to words, they will have different semantics in different words. To solve the problem of Chinese characters ambiguous, we use the solution proposed in [6]. We keep three embeddings(c^B, c^M, c^E) for each character c, corresponding to its three types of positions('Begin','Middle','End') in a word. Formally, $\mathbf{x}_w$ is represented as Eq. (9).

$$\mathbf{x}_w = \frac{1}{2K} \sum_j (v_j + \frac{1}{N_j}(c_1^B + \sum_{k=2}^{N_j-1} c_k^M + c_{N_j}^E)) \quad (9)$$
$$\text{where} \quad j = w - K, \ldots, w-1, w+1, \ldots w+K$$

In Eq. (8), $\mathbf{c}_w$ is the average of all context characters vectors of the target word w. Similarly, we need solve character ambiguous.

$$\mathbf{c}_w = \frac{1}{2K} \sum_j \frac{1}{N_j} \sum_{k=1}^{N_j} (c_1^B + \sum_{k=2}^{N_j-1} c_k^M + c_{N_j}^E) \quad (10)$$
$$\text{where} \quad j = w - K, \ldots, w-1, w+1, \ldots w+K$$

The framework of ECWE is shown in Fig. 1. Word embeddings (yellow boxes in figure) and character embeddings (green boxes) are composed together to get new embeddings (orange boxes), the new embeddings are combined together to get the embeddings (gray box) for the prediction of the target word. In addition, the character embeddings are composed together to get new embeddings, which are combined together for predicting the target word at the same time.

In ECWE, both each word and each character have its corresponding vector. Different vector representations have the same dimension. From Eqs. (9) and (10), the vector representation of context words and characters have different expressions, thus, in the update phase, the word embeddings and character embeddings have different updating expressions, shown as Eqs. (11) and (12) respectively.

$$v(\widetilde{w}) := v(\widetilde{w}) + \eta \sum_{u \in \{w\} \cup NEG(w)} \frac{\partial L(w, u)}{\partial \mathbf{x}_w}, \quad \widetilde{w} \in Context(w) \qquad (11)$$

$$v(\widetilde{c}) := v(\widetilde{c}) + \eta \sum_{u \in \{w\} \cup NEG(w)} \frac{\partial L(w, u)}{\partial \mathbf{c}_w}, \quad \widetilde{c} \in \widehat{Context}(w) \qquad (12)$$

There are many words in Chinese which do not exhibit semantic compositions from their characters. Such as single-morpheme multi-character words, transliterated words, and entity names. CWE proposes to neglect characters when learning these words, and artificially building a word list to store these words. This step is important to the effectiveness of CWE, but it's far more complicated. Our proposed model introduce external context characters, weaken the effect of internal characters, so we need not to collect non-compositional words. Therefore, the complex process of building the list can be removed.

4.2   Initialization and Optimization

We randomly initialize both word and character embeddings. And we use stochastic gradient descent (SGD) to optimize ECWE. Gradients are calculated using the back-propagation algorithm.

4.3   Complexity Analysis

Table 1 compares the complexity of CBOW, CWE and ECWE. These models are all based on Neural Netword approaches.

In the table, the dimension of vector is d, the word vocabulary size is $|W|$, the character vocabulary size is $|C|$, and the number of character positions in a word is $P = 3$. The window size is $2K$, the corpus size is M, the average number of characters of each word is N, and the computational complexity of negative sampling for each target word is F.

From the complexity analysis, we can observe that, compared with CWE, the ECWE model has same parameters. ECWE have additional computation complexity $O(2KMN)$, because in parameters update phase, we need more time to compute the updating expression of character embeddings, this ensures getting better character embeddings.

Table 1. Model complexity.

Model	Model parameters	Computational complexity
CBOW	$|W|d$	$2KMF$
CWE	$(|W| + P|C|)d$	$2KM(F + N)$
ECWE	$(|W| + P|C|)d$	$2KM(F + 2N)$

5   Experiments

In this section, we first describe our experimental settings, including datasets and model parameters. Then we compare our models with baseline methods on three tasks: (1) Word relatedness computation, (2) Analogical Reasoning, and (3) Document classification. Last, We deeply discuss the robustness of ECWE.

5.1   Datasets and Settings

We use Wikipedia Chinese corpus[1] for learning word embeddings. This corpus has 182 million words. The word vocabulary size is 457 thousand and the character vocabulary size is 9 thousand. We use ICTCLAS[2] toolkit for word segmentation.

We introduce CBOW [11], GloVe [12] and CWE [6] as baseline methods. We set vector dimension as 200 and context window size as 5, and default set $\beta = 0.5$. For optimization, we use 5-word negative sampling and the initial learning rate is 0.05.

5.2   Word Relatedness Computation

In this task, each model is required to compute semantic relatedness of given word pairs. The performance of each model is measured as the correlations between results of models and human judgements. In this paper, we select two datasets which was provided in [6], in the following abbreviated as ws296 and ws240. In ws240, there are 240 pairs of Chinese words and human-labeled relatedness scores. Of the 240 word pairs, the words in 227 word pairs have appeared in the learning corpus. In ws296, the words in 279 word pairs have appeared in the learning corpus.

We compute the Spearman correlation ρ between relatedness scores from a model and the human judgements for comparison. The relatedness score of two words are computed via cosine similarity of word vectors.

The evaluation results of ECWE and baseline methods on ws240 and ws296 are shown in Table 2. From the evaluation results, we observe that: ECWE significantly outperforms all baseline methods on these both datasets.

---

[1] https://dumps.wikimedia.org/zhwiki/latest/.
[2] http://ictclas.nlpir.org.

Table 2. Evaluation results on ws240 and ws296 ($\rho \times 100$).

Model	ws240		ws296	
	227 pairs	240 pairs	279 pairs	296 pairs
CBOW	56.05	51.09	59.09	54.83
Glove	55.70	48.91	47.96	43.01
CWE	55.77	51.52	60.22	55.86
ECWE	**57.44**	**52.90**	**61.36**	**57.68**

5.3 Analogical Reasoning

This task quantitatively evaluate the linguistic regularities between pairs of word representations. The task consists of questions like "男人 (man) is to 女人 (woman) as 父亲 (father) to ?". To answer such question, we need to find a word w such that its vector x is close to vec (女人) − vec (男人) + vec (父亲) according to the cosine similarity. The question is judged as correctly answered only if x is exactly the answer in the evaluation set. The evaluation metric for this task is the percentage of questions answered correctly. We use Chinese analogy dataset from [6]. The dataset contains 1,124 analogies and 3 analogy types: capitals of countries (687 groups); states/provinces of cities (175 groups); and family words (240 groups). Table 3 shows the evaluation results on analogical reasoning.

Table 3. Evaluation accuracies (%) on analogical reasoning

Model	Total	Capital	State	Family
CBOW	71.51	71.79	72.06	70.33
GloVe	74.04	74.00	76.89	**72.12**
CWE	71.62	68.40	**81.86**	69.67
ECWE	**74.56**	**75.21**	75.86	71.15

From Table 3, we observe that: ECWE performs better than all baseline methods on average. GloVe also performs good, but it's based on matrix decomposition, compared with ECWE, its computational complexity is too high, especially when the corpus size is large.

5.4 Document Classification

In this task, we regard the average of the word vector representations in that document as the document representation, which can be evaluated with document classification. We use Logistic regression models for the classification task.

Table 4. Results on document classification ($\rho \times 100$).

Model	Accuracy
CBOW	80.53
GloVe	83.12
CWE	81.34
ECWE	**84.73**

We run experiments on the dataset Fudan corpus[3] of text categorization. This dataset contains 20 categories and about 9804 documents and is split into training set and test set with 1:1. Each document belongs to only one category. From Table 4, we can observe that ECWE method performs significantly better than all baseline methods.

5.5   The Robustness of ECWE

In order to further show the effectiveness and robustness of ECWE, we adjust corpus size and the parameter β in Eq. (5), analyze the effect of ECWE in different settings.

In ECWE, β indicates the proportion of character modeling. For comparison, we adjust the proportion of character modeling in CWE model. $\beta = 0$ indicates these models only modeling on words, such that, all these models are equivalent to CBOW. The results on ws296 show in Fig. 2(a). In Fig. 2(b), We list the results of CBOW and CWE and ECWE on ws296 with various corpus size from 10 MB to 500 MB.

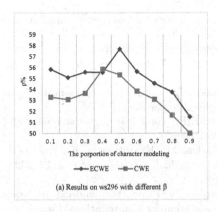

(a) Results on ws296 with different β

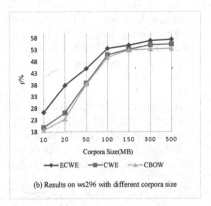

(b) Results on ws296 with different corpora size

Fig. 2. Results on ws296 task with different setting.

---

[3] http://www.datatang.com/data/44139.

From Fig. 2(a), we observe that: Overall, gradually increase the proportion of character modeling, the evaluation accuracy on ws296 of ECWE model rises in the first stage, and then decrease. CWE model has a similar tendency, but the performance is always worse than ECWE. From Fig. 2(b), we observe that: ECWE can quickly achieve much better performance than CWE and CBOW when the learning corpus is still relatively small. The reason is that, ECWE takes advantage of external context characters to contact the original isolated characters and expand the contexts, therefore, promotes the word embeddings contain more semantic information.

6   Conclusion

In this paper we propose a compositional neural language models which incorporates internal characters and external context characters for obtaining high-quality Chinese word embeddings. We investigate the effectiveness of our model in three tasks. ECWE consistently and significantly performs better on word relatedness computation, analogical reasoning and document classification tasks. We demonstrate the robustness of ECWE across different settings. These indicate the necessity of considering external characters information for Chinese word representations.

References

1. Baroni, M., Dinu, G., Kruszewski, G.: Don't count, predict! A systematic comparison of context-counting vs. context-predicting semantic vectors. In: ACL (1), pp. 238–247 (2014)
2. Bengio, Y., Ducharme, R., Vincent, P., Jauvin, C.: A neural probabilistic language model. J. Mach. Learn. Res. **3**, 1137–1155 (2003)
3. Botha, J.A., Blunsom, P.: Compositional morphology for word representations and language modelling. In: ICML, pp. 1899–1907 (2014)
4. Chen, T., Xu, R., He, Y., Wang, X.: Improving distributed representation of word sense via wordnet gloss composition and context clustering. Association for Computational Linguistics (2015)
5. Chen, X., Liu, Z., Sun, M.: A unified model for word sense representation and disambiguation. In: EMNLP, pp. 1025–1035. Citeseer (2014)
6. Chen, X., Xu, L., Liu, Z., Sun, M., Luan, H.B.: Joint learning of character and word embeddings. In: IJCAI, pp. 1236–1242 (2015)
7. Collobert, R., Weston, J., Bottou, L., Karlen, M., Kavukcuoglu, K., Kuksa, P.: Natural language processing (almost) from scratch. J. Mach. Learn. Res. **12**, 2493–2537 (2011)
8. Cotterell, R., Schütze, H., Eisner, J.: Morphological smoothing and extrapolation of word embeddings. In: Meeting of the Association for Computational Linguistics, pp. 1651–1660 (2016)
9. Lin, Y., Liu, Z., Sun, M., Liu, Y., Zhu, X.: Learning entity and relation embeddings for knowledge graph completion. In: AAAI, pp. 2181–2187 (2015)
10. Mikolov, T., Chen, K., Corrado, G., Dean, J.: Efficient estimation of word representations in vector space. Computer Science (2013)

11. Mikolov, T., Sutskever, I., Chen, K., Corrado, G.S., Dean, J.: Distributed representations of words and phrases and their compositionality. In: Advances in Neural Information Processing Systems, pp. 3111–3119 (2013)
12. Pennington, J., Socher, R., Manning, C.D.: Glove: global vectors for word representation. In: EMNLP, vol. 14, pp. 1532–1543 (2014)
13. Socher, R., Bauer, J., Manning, C.D., Ng, A.Y.: Parsing with compositional vector grammars. In: ACL (1), pp. 455–465 (2013)
14. Sun, F., Guo, J., Lan, Y., Xu, J., Cheng, X.: Inside out: two jointly predictive models for word representations and phrase representations. In: AAAI, pp. 2821–2827 (2016)
15. Turian, J., Ratinov, L., Bengio, Y.: Word representations: a simple and general method for semi-supervised learning. In: Proceedings of the 48th Annual Meeting of the Association for Computational Linguistics, pp. 384–394. Association for Computational Linguistics (2010)
16. Zhao, Y., Liu, Z., Sun, M.: Phrase type sensitive tensor indexing model for semantic composition. In: AAAI, pp. 2195–2202 (2015)

Enhancing Auscultation Capability in Spacecraft

Jin Zhou and Chiman Kwan[✉]

Signal Processing, Inc., Rockville, MD 20850, USA
ferryzhou@gmail.com, chiman.kwan@signalpro.net

Abstract. In this research, we developed an adaptive filtering system for enhancing auscultation performance in noisy environments such as spacecraft and International Space Station (ISS). The system uses a stethoscope, a microphone, and an adaptive filter. Four filtering algorithms (least mean square (LMS), normalized LMS (NLMS), recursive least square (RLS), and our patented algorithm known as Frequency-domain Minimum Square Error with length N and Signal Detection (FMSENSD)) were implemented and compared. Extensive experiments using actual data collected by several commercial stethoscopes clearly demonstrated the performance of the system under noisy conditions up to 79 dBA.

Keywords: Auscultation enhancement · Noisy spacecraft · Adaptive filter

1 Introduction

NASA is planning to have a medical suite in spacecraft for future long duration manned missions to Mars. One of the capabilities in the suite is accurate auscultation in noisy environments. Typical noise level in spacecraft can reach 73 dBA and beyond. Due to high noise environment caused by fans and pumps in the spacecraft, conventional stethoscopes do not work well in noisy environment beyond 66 dBA [1]. To reliably assess the health condition of astronauts, novel and high performance technology for auscultation is needed.

There are several potential approaches to enhancing the auscultation performance. First, beamforming using various array configurations [2–4, 9–14] can be applied. Although this approach may have great potential, it will involve a complete redesign of stethoscopes. This is undesirable based on NASA's evaluation. Second, speech enhancement techniques using a single sensor [17, 18] have advanced over the past two decades and can be used. However, based on our past experience in speech enhancement in high noise environments such as battlefield, this approach only works well in stationary noise environment. Moreover, the color noise artifacts are very undesirable for medical diagnostics. The third approach is to use one additional microphone or another sensor [6, 15, 16] to pick up the background noise and we found this approach to be promising because (1) it can deal with non-stationary noise; (2) conventional stethoscopes can be used without any modification.

© Springer International Publishing AG 2017
F. Cong et al. (Eds.): ISNN 2017, Part II, LNCS 10262, pp. 416–426, 2017.
DOI: 10.1007/978-3-319-59081-3_49

We first applied LMS [5], NLMS [5], RLS [5], and FMSENSD [6] algorithms to the data collected by Wyle Inc., which is a contractor of NASA Johnson Space Center. FMSENSD was developed by our team recently and a patent was granted [6]. There were 3 stethoscopes in Wyle's data and the noise environment was 66 dBA. We found that RLS performed better than LMS and NLMS, and FMSENSD performed better than RLS. In general, the signal-to-noise (SNR) improvement was 5 to 10 dB. Second, in order to investigate how well our algorithm will perform in noise conditions beyond 66 dBA, we set up our own testbed, which can emulate noisy conditions at 45, 66, 73, and 79 dBA. Data were collected under the above 4 noisy conditions using 6 internal body sounds, 2 stethoscopes (Thinklabs and 3M), and 4 signal levels. Third, we thoroughly evaluated the performance of RLS and FMSENSD using our own data. It was found that FMSENSD performed the best. SNR improvement was 10 to 20 dB under various noisy conditions.

This paper is organized as follows. Section 2 briefly review two key algorithms: RLS and FMSENSD even though we have implemented two other algorithms (LMS and NLMS). Section 3 summarizes all the experimental results. We first summarize the application of RLS and FMSENSD to data supplied by NASA. We then describe how we set up our own testbed and the comparative results. Two commercial stethoscopes were used in our experiments. Three noise levels (66 dBA, 73 dBA, and 79 dBA) were investigated. Extensive experiments demonstrated that our own algorithm outperformed RLS in almost all experimental conditions by 10 to 20 dB. Finally, some concluding remarks and future research directions will be mentioned in Sect. 4.

2 Technical Approach

Several time-domain adaptive filter algorithms were applied to process the various data sets. These include LMS (Least Mean Square), NLMS (Normalized Least Mean Square), and RLS (Recursive Least Squares) [5]. Since RLS performed better than LMS and NLMS, we only briefly summarize RLS and FMSENSD algorithms below in order to save some space.

2.1 RLS Filter

RLS is a well-known adaptive filter and has been widely used in many applications. It is a time-domain filtering approach, meaning that no Fourier transform is needed. The adaptation is done with a matrix of coefficients and hence the convergence speed is faster than other time-domain filters such as LMS and NLMS. The principle of RLS is shown in Fig. 1. Of course, this principle is applicable to all other filters. Essentially, the adaptive filter adjusts the phase and amplitude of the noise and cancels the corresponding noise component in the stethoscope.

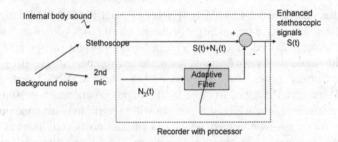

Fig. 1. Principle of RLS.

2.2   FMSENSD Algorithm

In contrast, FMSENSD stands for frequency mean square error n-coefficient with signal detection and works in the frequency domain. FMSENSD was developed by us in this project and a patent has been approved [6]. There are several key steps. First, it computes spectrogram of both the stethoscope signals and the noise channel signals. Second, it estimates noise only regions. Third, it estimates filter coefficients using noise only regions. Finally, it generates filtered signals by performing subtraction. The principle of FMSENSD is shown in Fig. 2. Although there are 4 steps, the computational complexity is quite low.

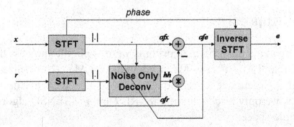

Fig. 2. Principle of FMSENSD. x is the stethoscope signal and r is the reference mic signal.

FMSENSD is very efficient and converges fast. FMSENSD has better preservation of signal details and introduces less distortions. It also works well in reverberant environments. We estimate the channel response hh with noise only section, i.e.

$$afr_i(o) * hh = afx_i(o)$$

where o is the indices of noise only frames. The noise indices are estimated based on the strength of estimated signal. The detailed algorithm is as follows:

Algorithm FMSENSD (See Fig. 2 for notations)
Input: x the mixed sound, r the reference sound (noise only)
Output: e the filtered sound
Parameters: *nwin (window length), noverlap (overlap size), nfft, N (filter length)*
Algorithm:

1. $fx = spectrogram(x, nwin, noverlap, nfft); afx = abs(fx)$ //compute spectral magnitudes of mixed signal
2. $fr = spectrogram(r, nwin, noverlap, nfft); afr = abs(fr)$ //compute spectral magnitudes of reference signal
3. for each band i // estimate convolution vector using all signals
4. $AFR_i = Toeplitz(afr_i, N)$
5. $hh = (AFR_i^T AFR_i)^{-1} AFR_i^T afx_i$, $afy_i = afr_i * hh$
6. end
7. $afe = afx - afy$ //compute residual
8. $nafe = smooth(afe.^2)$
9. $T = median(nafe)$
10. $o = nafe < T$ // extract noise only regions
11. for each band i // estimate convolution vector again using noise only signals
12. $AFR_i = Toeplitz(afr_i(o), N)$
13. $hh = (AFR_i^T AFR_i)^{-1} AFR_i^T afx_i(o)$, $afy_i = afr_i * hh$
14. end
15. $afe = \max(0, afx - afy)$ // compute residual
16. $e = overlap\_add(afe, angle(fx), nwin, noverlap, nfft)$ //reconstruct time domain signals

---

3   Experimental Results

3.1   Algorithm Evaluation Using Wyle's Data

The data were recorded by Wyle Inc. and provided to us by NASA. The data were recorded in both quiet and noisy environments. Since our goal is to evaluate the performance of adaptive filters in noisy environments, we only focused on the noisy data, which has an actual noise level of 66 dBA. Four types of stethoscopes were used in recording: (a) Andromed iStethos; (b) Welch-Allyn Meditron; (c) Cardionics e-Scope II; and (d) 3M 4000. Two types of body sounds were recorded: heart and lung. Note that only the first three types of data contain both stethoscope outputs and noise data (recorded by a separate microphone). In our experiments, we applied adaptive filtering techniques to the first three types of data.

We applied two different adaptive filters to Wyle data that contain 3 stethoscopes: (a) Andromed iStethos; (b) Welch-Allyn Meditron; (c) Cardionics e-Scope II. This is because the recordings have two channels: one saves the stethoscope data and one saves the data from a microphone which recorded the background noise. We first compared the performance of the LMS, NLMS, and RLS and found that RLS performed the best among these time-domain filters [9]. We then focused on the comparison of FMSENSD (frequency-domain adaptive filter) and RLS (time-domain adaptive filter). In FMSENSD, we set window length 1,024, overlap 960 and filter length 4. In RLS, the filter length was 128. The 3M 4000 stethoscope does not have a second channel to record the background noise.

Figure 3 shows sample FMSENSD results for 3 types of commercial stethoscopes mentioned earlier. One heart sound and one lung sound were recorded. The RLS results are not shown here due to page limitation. From the figures, we can observe that the filtered results have much better signal-to-noise (SNR) than original signals, especially for Fig. 3(b) and (c). Figures 4, 5 and 6 show the comparison charts of SNR of different types of data. In Fig. 4, results from 4 heart and 9 lung recordings using

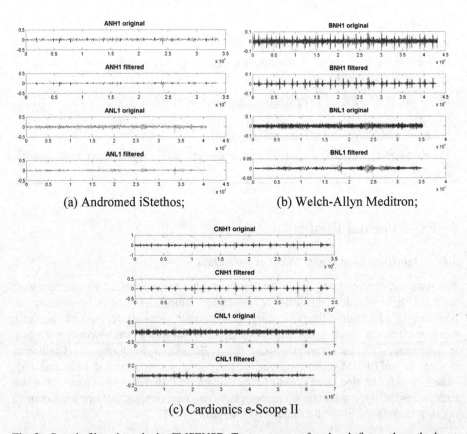

(a) Andromed iStethos;   (b) Welch-Allyn Meditron;

(c) Cardionics e-Scope II

Fig. 3. Sample filtered results by FMSENSD. Top two rows of each sub-figure show the heart sounds (original and filtered) and bottom two rows of each sub-figure show the lung sound (original and filtered).

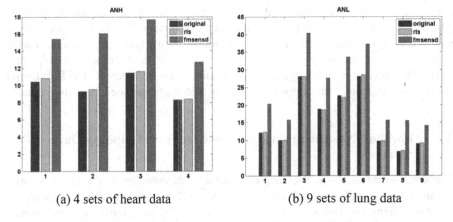

(a) 4 sets of heart data   (b) 9 sets of lung data

Fig. 4. SNR charts for type-A stethoscope (Andromed iStethos) data.

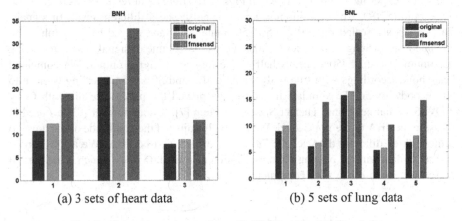

(a) 3 sets of heart data   (b) 5 sets of lung data

Fig. 5. Comparison charts of type-B (Welch-Allyn Meditron) data.

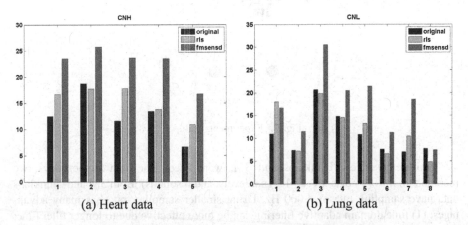

(a) Heart data   (b) Lung data

Fig. 6. Comparison charts of type-C (Cardionics e-Scope II) data.

type-A stethoscope were displayed. RLS improves the SNR, but our FMSENSD performed even better. In Fig. 5, results from 3 heart and 5 lung recordings using type-B stethoscope were displayed. In Fig. 6, results from 5 heart and 8 lung recordings using type-C stethoscope were displayed. In all of the above figures, we observe that FMSENSD always provides significant SNR gain, while RLS cannot. The FMSENSD average gain is about 8 dB. Subjective listening also confirmed the above observations.

3.2   Algorithm Evaluation Using Data Collected from Testbed in Our Laboratory

The comparative study in Sect. 3.1 used actual data collected under 66 dBA environment. It is natural to ask whether FMSENSD still outperforms RLS in even higher noise environments. To answer this question, we will need to build our own testbed.

Experimental Setup

The whole experiment setup is shown in Fig. 7. To simulate different levels of signal to noise ratio (SNR), we set 4 different levels of noise and 4 different levels of signals. The signals are scaled by 100%, 75%, 50% and 25%, and played to a headphone to emulate heart or lung sound. Here, 100% means the volume of signal is adjusted to the maximum of output. 50% means half of the maximum signal output. We compared headphone recordings with true body sound and found 50% signal scaling is close to actual body sound. To simulate the noise, we played the pink noise file with Coby 300W 5.1-Channel Home Theater Speaker System [7]. A sound meter [8] was used to measure the dBA of the noise level. We manually adjusted the amplitude of the noise to achieve three different dBA levels: 79 dBA, 73 dBA and 66 dBA. We also recorded one set of data without playing the noise, which is labeled 45 dBA, which is the normal noise level in an office with computers.

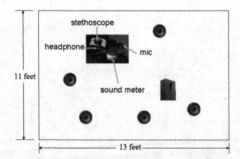

Fig. 7. Experimental setup.

Two stethoscopes, Thinklabs and 3M, were used. For both stethoscopes, we recorded 96 files (4 noise levels, 4 signal levels, and 6 sounds). Both 3M and Thinklabs data have sampling rate of 4,000 Hz. Using smaller sampling rate has many advantages: (1) time domain adaptive filtering can be more effective due to longer filter time; (2) data size is smaller and (3) processing is faster. Since the stethoscope signal are low pass filtered (usually the bandwidth is less than 1500 Hz), 4000 Hz sampling rate is enough to retain all the information.

From the recordings, we observed that 3M has better SNR than Thinklabs due to its built-in noise canceling technology. At 79 dBA noise level, all the signals are buried in noise.

Data Using Thinklabs Stethoscope

Since we have the ground truth heart and lung sounds, we can easily generate the SNR. The SNR comparison charts are shown in Figs. 8, 9 and 10. We observe that both FMSENSD and RLS significantly improve SNR; FMSENSD is significantly better than RLS. In general, FMSENSD can achieve at least 20 dB gain.

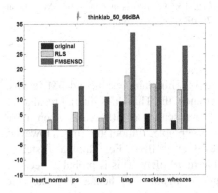

Fig. 8. Comparison charts of Thinklabs data at 66 dBA (signal volume: 50%).

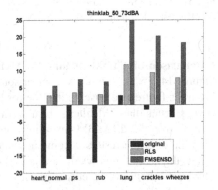

Fig. 9. Comparison charts of Thinklabs data at 73 dBA (signal volume: 50%).

Data Using 3M Stethoscope

The SNR comparison charts are shown in Figs. 11, 12 and 13. We can observe that FMSENSD significantly improves SNR and RLS can only improve a little bit; FMSENSD is significantly better than RLS. In most cases, FMSENSD can achieve at least 10 dB gain.

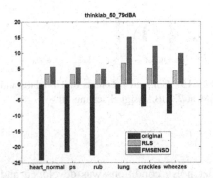

Fig. 10. Comparison charts of Thinklabs data at 79 dBA (signal volume: 50%).

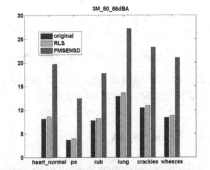

Fig. 11. Comparison charts of 3M data at 66 dBA (signal volume 50%).

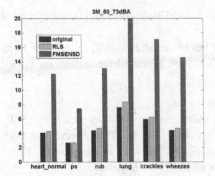

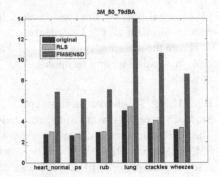

Fig. 12. Comparison charts of 3M data at 73 dBA (signal volume 50%).

Fig. 13. Comparison charts of 3M data at 79 dBA (signal volume 50%).

Comparison of Thinklabs and 3M Stethoscopes

Figures 14, 15 and 16 show comparison charts of Thinklabs results and 3M results. From the figures we observe that the for lung sounds, the Thinklabs filtered results are always better than 3M filtered results. In the signal scaling of 50% case, when noise level is over 66 dBA, 3M recordings are not acceptable without filtering. However, after filtering, the results become acceptable even at 79 dBA. For Thinklabs' stethoscope, the heart sound results are not acceptable in 79 dBA. This is probably because Thinklabs' stethoscope cuts more low frequency signals and hence attenuates the magnitude of heart sounds. In contrast, 3M preserves more low frequency signals.

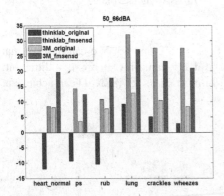

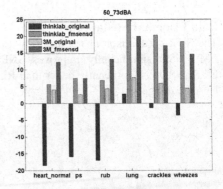

Fig. 14. Comparison charts of Thinklabs and 3M at 66 dBA (signal scaling 50%).

Fig. 15. Comparison charts of Thinklabs and 3M at 73 dBA (signal scaling 50%).

4   Conclusion

A high performance auscultation system under noisy conditions was developed and evaluated. Extensive evaluations using actual data in extremely noisy environments were performed to demonstrate the performance of the proposed system. Two adaptive

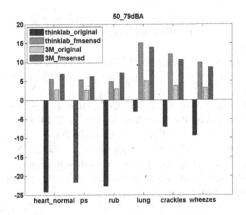

Fig. 16. Comparison charts of Thinklabs and 3M at 79 dBA (signal scaling 50%).

filters were thoroughly evaluated. In extremely noisy conditions (73 dBA and 79 dBA), the raw data were useless without filtering because the SNRs are negative. After filtering, the SNRs have been improved quite significantly. Both subjective and objective evaluations showed that our own algorithm performed a lot better than the conventional RLS filter.

One future direction of our research is to develop a compact (cell phone size), light weight, and low power prototype that can perform real-time auscultation in noisy conditions. The prototype should be standalone in that it is detached from the stethoscope. So if astronauts prefer to use the original stethoscope, they can bypass our device.

Acknowledgement. This research was supported NASA under contract # NNX11CD44P.

References

1. Bacal, K., Rasbury, J., McCulley, P., Ownby, M., Paul, B.: Will a Conventional Stethoscope Function Effectively in the Noisy International Space Station Environment? Research Report. Wyle Inc. (2011)
2. Li, Y., Vicente, L., Ho, K.C., Kwan, C., Lun, D.P.K., Leung, Y.H.: A study of partially adaptive concentric ring array. J. Circ. Syst. Sig. Process. **27**(5), 733–748 (2008)
3. Li, Y., Ho, K.C., Kwan, C.: Design of broad-band circular ring microphone array for speech acquisition in 3-D. In: International Conference on Acoustics, Speech, and Signal Processing (2003)
4. Li, Y., Ho, K.C., Kwan, C.: A novel partial adaptive algorithm for broadband beamforming using concentric circular array. In: IEEE International Conference on Acoustics, Speech and Signal Processing, Montreal, pp. 177–180, 17–21 May 2004
5. Haykin, S.: Adaptive Filter Theory. Prentice Hall, Upper Saddle River (2002)
6. Kwan, C., Zhou, J.: Compact Plug-In Noise Cancellation Device. Patent # 9,117,457 (2015)
7. Speaker System. http://www.radioshack.com/product/index.jsp?productId=4098578
8. Sound Meter. http://www.radioshack.com/product/index.jsp?productId=2103667

9. Vicente, L.M., Ho, K.C., Kwan, C.: An improved partial adaptive narrow-band beamformer using concentric ring array. In: IEEE International Conference on Acoustics, Speech, and Signal Processing (2006)

10. Li, Y., Ho, K.C., Kwan, C., Leung, Y.H.: Generalized partially adaptive concentric ring array. In: IEEE International Symposium on Circuits and Systems ISCAS05, pp. 3745–3748. Kobe, Japan (2005)

11. Xu, R., Mei, G., Ren, Z., Kwan, C., Stanford, V., Aube, J., Rochet, C.: A real time speaker verification demonstration on the smart flow system. In: IEEE International Symposium on Intelligent Multimedia, Video and Speech Processing (2004)

12. Kwan, C., Yin, J., Ayhan, B., Chu, S., Liu, X., Puckett, K., Zhao, Y., Ho, K. C., Kruger, M., Sityar, I.: An integrated approach to robust speaker identification and speech recognition. In: IEEE International Joint Conference on Neural Networks (IEEE World Congress on Computational Intelligence) (2008)

13. Kwan, C., Ho, K.C., Mei, G., Li, Y., Ren, Z., Xu, R., Zhang, Y., Lao, D., Stevenson, M., Stanford, V., Rochet, C.: An automated acoustic system to monitor and classify birds. Eurosip J. Appl. Sig. Process. **1**, 54–64 (2006)

14. Li, Y., Ho, K.C., Kwan, C.: 3-D array pattern synthesis with frequency invariant property for concentric ring array. IEEE Trans. Sig. Process. **54**(2), 780–784 (2006)

15. Xu, R., Ren, Z., Dai, W., Lao, D., Kwan, C.: Multimodal Speech Enhancement in Noisy Environment. In: IEEE International Symposium on Intelligent Multimedia, Video and Speech Processing (2004)

16. Kwan, C., Li, X., Lao, D., Deng, Y., Raj, B., Singh, R., Stern, R.: Voice driven applications in non-stationary and chaotic environment. In: IEEE International Conference on Robotics and Biomimetics (ROBIO) (2005)

17. Kwan, C., Chu, S., Yin, J., Liu, X., Kruger, M., Sityar, I.: Enhanced speech in noisy multiple speaker environment. In: IEEE International Joint Conference on Neural Networks (IEEE World Congress on Computational Intelligence) (2008)

18. Deng, Y., Li, X., Kwan, C., Xu, R., Raj, B., Stern, R., Williamson, D.: An integrated approach to improve speech recognition rate for non-native speakers. INTERSPEECH 2006 - ICSLP, Ninth International Conference on Spoken Language Processing, Pittsburgh, PA, USA (2006)

Underwater Moving Target Detection Based on Image Enhancement

Yan Zhou[1,2(✉)], Qingwu Li[1,2], and Guanying Huo[1,2]

[1] College of Internet of Things Engineering,
Hohai University, Changzhou 213022, China
strangeryan@163.com, huoguanying@163.com,
liqw@hhuc.edu.cn
[2] Key Laboratory of Sensor Networks and Environmental Sensing,
Changzhou 213022, China

Abstract. Motion detection in underwater video scenes is very important for many underwater computer vision tasks, such as target location, recognition and tracking. However, due to the strong optical attenuation and light scattering in water, underwater images are essentially characterized by their poor visibility, especially the low contrast and distorted information. To solve these situations, underwater moving target detection algorithm based on image enhancement is presented. The algorithm improves the contrast and clarity of the target by an adaptive underwater color image enhancement, and then extracts the moving targets by using ViBe background model. Experimental results show that the proposed algorithm can effectively extract the complete moving target by overcoming the impact of underwater environment.

Keywords: Underwater moving target detection · Adaptive image enhancement · ViBe model · Background subtraction

1 Introduction

Underwater moving target detection is an important means for underwater vehicles to acquire underwater target information. Effective underwater moving target detection contributes to many scientific researches and engineering applications, such as marine biology, seabed topography, marine environment monitoring and marine exploration [1]. At present, the common detection methods can be divided into three categories: inter-frame difference, optical flow and background subtraction [2]. Inter-frame difference method [3], which extracts the moving targets by the difference of several adjacent frames, is real-time and simple. However, generally, moving targets extracted are not complete and have the void phenomenon. Optical flow method, which detects moving targets by using its optical flow characteristics over time, has high computational complexity and is sensitive to illumination variation [4].

Background subtraction methods construct a model for the background and compare the background model with the current frame so as to detect the regions where a significant difference occurs. The Gaussian Mixture Model is one of the most popular parametric background subtraction methods [5]. It can handle the multi-modal

© Springer International Publishing AG 2017
F. Cong et al. (Eds.): ISNN 2017, Part II, LNCS 10262, pp. 427–436, 2017.
DOI: 10.1007/978-3-319-59081-3_50

appearance of the background under dynamic environments. However, the parameter estimation of the model may become difficult for noisy images. ViBe (for 'Visual Background Extractor') is a samples-based background subtraction method. Due to the use of memoryless update strategy, spatial information propagation method, and instantaneous initialization technique, it shows an outstanding detection rate and robustness to noise [6, 7].

However, due to the serious underwater interference and dynamic change of underwater scenes, it becomes very difficult to accurately extract underwater moving targets. The strong optical attenuation and light scattering caused by the water medium and suspended particles will obviously reduce the contrast between the target and the background. These low quality video image data seriously hamper the underwater computer vision tasks. In order to solve the problems in underwater moving target detection, this paper proposes an underwater moving target detection algorithm based on image enhancement [8]. An adaptive underwater color video image enhancement algorithm is presented to improve the contrast and clarity of the target and inhibit inhomogeneous illumination. Then, the moving targets are extracted by using ViBe background model. The proposed method has high dynamical adaptability in the underwater target extraction task and strong robustness to the underwater environment. The experiment results prove its efficiency in target detection under the complex underwater optical environments.

Section 2 describes our new underwater moving target detection algorithm. Experimental results are detailed in Sect. 3. Section 4 concludes the paper.

2 Underwater Moving Target Detection Based on Image Enhancement

Underwater environment is complex and dynamic, in which there are plenty of disturbances, e.g. wave, illumination changes, light absorption and scattering. Underwater video scenes are notorious for poor visibility, low contrast, edge-blurring and being full of noise. Therefore, motion detection in underwater video scenes is more difficult than that in air. It is necessary to improve the visual quality of underwater images for subsequent accurate motion detection. So we propose an underwater moving target detection algorithm based on image enhancement. Firstly, an adaptive underwater video image enhancement algorithm inspired by the human visual system (HVS) is used to suppress noise and improve the edge sharpness. Then, underwater moving targets are detected from background model by using ViBe background subtraction algorithm. Experimental results show that the proposed algorithm can extract underwater moving targets accurately and completely.

2.1 Algorithm Flow

The algorithm flow of underwater moving target detection based on image enhancement is shown in Fig. 1.

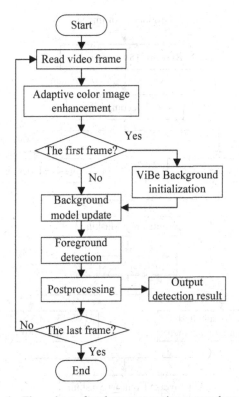

Fig. 1. Flow chart of underwater moving target detection

2.2 Adaptive Underwater Video Image Enhancement

In this paper, an adaptive underwater video image enhancement algorithm is proposed to improve the contrast and clarity of the target. Firstly, the color video image is converted from RGB to HSV color space. Secondly, the multiscale retinex (MSR) approach is used in nonsubsampled contourlet transform (NSCT) domain of V channel in order to eliminate non-uniform illumination, and threshold denoising method is adopted to suppress noise. Thirdly, the luminance masking (LM) and contrast masking (CM) characteristics of the HVS are integrated into NSCT to yield the HVS-based NSCT contrast. Subsequently, a nonlinear mapping function and a nonlinear gain function are designed to manipulate the HVS-based NSCT contrast coefficients and the NSCT lowpass subband coefficients respectively and automatically. Lastly, the enhanced V channel image can be reconstructed from NSCT coefficients, and the enhanced color image is obtained by the conversion from HSV to RGB color space. This adaptive color image enhancement algorithm, which is free of parameters adjusting, can effectively emphasize weak edges, suppress noise, remove uneven illumination and increase the identifiable characteristic information of the target. It will help to improve the accuracy of the subsequent target detection.

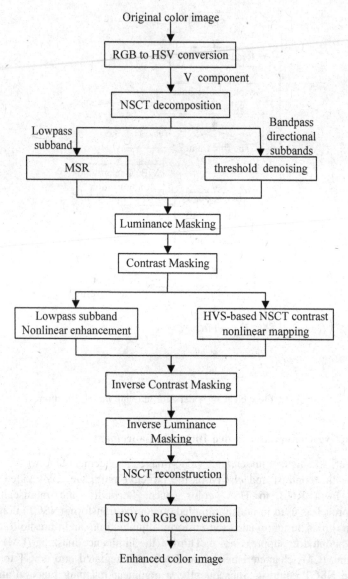

Fig. 2. Block diagram of adaptive underwater video image enhancement algorithm

The block diagram of the adaptive underwater video image enhancement algorithm is shown in Fig. 2. The NSCT is one kind of multiscale and multidirectional geometric transform, so it is able to effectively capture geometry and directional information of images. Furthermore, since each pixel of the NSCT subbands corresponds to that of the original image in spatial domain, we can collect the geometrical information pixel by pixel from the NSCT coefficients. We enhance the color video image in HSV color space by RGB to HSV conversion. The H and S component are kept unchanged, and only V component is handled. After NSCT decomposition on V channel, the lowpass subband, which is nearly noiseless, includes overall contrast information. While bandpass directional subbands contain not only edges but also noise.

Taking into account of the presence of non-uniform illumination in the underwater image, the NSCT lowpass subband is manipulated with MSR algorithm. In this stage, dynamic range is properly compressed to eliminate shadows and uneven illumination.

Because edges correspond to the large NSCT coefficients and noise corresponds to the small NSCT coefficients in bandpass directional subbands, noise can be effectively suppressed by thresholding. Thresholds for each bandpass directional subband can be chosen according to:

$$T_{s,d} = k\sigma\sqrt{\tilde{\sigma}_{s,d}} \tag{1}$$

The noise standard deviation σ of the original image is estimated by using the robust median operator, i.e.,

$$\sigma = median(abs(C))/0.6745 \tag{2}$$

where C refers to the NSCT coefficients in the finest scale. $\tilde{\sigma}_{s,d}^2$ refers to the approximate value of the individual variances at the directional subband indexed by scale s and direction d, which is calculated by using Monte-Carlo simulations.

Subsequently, HVS-based masking model in NSCT domain is constructed to yield the HVS-based NSCT contrast. To obtain HVS-based NSCT contrast, two steps are to be conducted. Firstly, the LM contrast in NSCT domain is measured by

$$C_{LM(s,d)} = \frac{y_{(s,d)}}{|y_{(s,0)}| + c} \tag{3}$$

where $y_{(s,d)}$ is the original NSCT bandpass directional subband indexed by scale s and direction d. $y_{(s,0)}$ is the original NSCT lowpass subband at the sth scale. In an N level NSCT decomposition of an image, $s = N$ denotes the scale after performing NSCT decomposition procedure one time (i.e., the finest scale), and $s = 1$ denotes the scale after performing NSCT decomposition procedure N times (i.e., the coarsest scale). c is a small constant to avoid dividing by 0. $C_{LM(s,d)}$ is the output of LM contrast indexed by scale s and direction d. Secondly, the LM contrast is masked with Contrast Masking to yield the LCM contrast, which is the HVS-based NSCT contrast. The multiscale LCM contrast, which is a function of the LM contrast, is defined as

$$C_{LCM(s,d)} = \frac{C_{LM(s,d)}}{|C_{LM(s-1,d)}|^{0.62} + c} \tag{4}$$

where c is a small constant to avoid dividing by 0.

We also propose a nonlinear mapping function to modify the HVS-based NSCT contrast coefficients at each scale and direction independently and automatically so as to achieve multiscale contrast enhancement. The proposed nonlinear mapping function is given by:

$$\hat{C}_{LCM(s,d)} = s \cdot max(|C_{LCM(s,d)}|) \cdot sign(C_{LCM(s,d)}) \cdot \left[\sin\left(\frac{\pi}{2} \cdot \frac{|C_{LCM(s,d)}|}{max(|C_{LCM(s,d)}|)}\right) \right]^{\sqrt{p}} \quad (5)$$

where

$$p = \frac{\log\left(\frac{mean(|C_{LCM(s,d)}|)}{max(|C_{LCM(s,d)}|)}\right)}{\log\left[\sin\left(\frac{\pi}{2} \frac{mean(|C_{LCM(s,d)}|)}{max(|C_{LCM(s,d)}|)}\right)\right]} \quad (6)$$

$C_{LCM(s,d)}$ is the original HVS-based NSCT contrast coefficient at the subband indexed by scale s and direction d. $\hat{C}_{LCM(s,d)}$ is the modified HVS-based NSCT contrast coefficient. $max(|C_{LCM(s,d)}|)$ denotes the maximum absolute contrast coefficient amplitude at the subband indexed by scale s and direction d. $mean(|C_{LCM(s,d)}|)$ denotes the mean value of absolute contrast coefficient amplitude at the subband indexed by scale s and direction d. Given an N scale NSCT decomposition of an image, $C_{LCM(s,d)}$ includes N-1 scales (i.e., $1 < s \leq N$), according to Eq. (4).

This nonlinear mapping function can well enhance the low-contrast areas, and also avoid over-enhancement of the high-contrast areas simultaneously [8].

Furthermore, the global dynamic range of the image can be adjusted by using one nonlinear gain function in the lowpass subband at the coarsest scale of NSCT decomposition. The nonlinear gain function is defined as follows:

$$\hat{y}_{(1,0)} = max(|y_{(1,0)}|) \cdot sign(y_{(1,0)}) \cdot \left[\sin\left(\frac{\pi}{2} \cdot \frac{|y_{(1,0)}|}{max(|y_{(1,0)}|)}\right) \right]^{q} \quad (7)$$

where

$$q = \frac{\log\left(\frac{mean(|y_{(1,0)}|)}{max(|y_{(1,0)}|)}\right)}{\log\left[\sin\left(\frac{\pi}{2} \frac{mean(|y_{(1,0)}|)}{max(|y_{(1,0)}|)}\right)\right]} \quad (8)$$

$y_{(1,0)}$ is the NSCT lowpass subband coefficient at the first scale (i.e., the coarsest scale). $\hat{y}_{(1,0)}$ is the modified NSCT lowpass subband coefficient at the first scale.

The modified LM contrast can be calculated from the modified LCM contrast by Inverse Contrast Masking:

$$\hat{C}_{LM(s,d)} = \hat{C}_{LCM(s,d)} * (|\hat{C}_{LM(s-1,d)}|^{0.62} + c) \quad (9)$$

The modified bandpass directional subband coefficients of NSCT are calculated from the modified LM contrast by Inverse Luminance Masking:

$$\hat{y}_{(s,d)} = \hat{C}_{LM(s,d)} * \left(\left| \hat{y}_{(s,0)} \right| + c \right) \tag{10}$$

Finally, the enhanced V channel image can be reconstructed from NSCT coefficients, and the enhanced color image is obtained by the conversion from HSV to RGB color space.

2.3 ViBe Model

ViBe is one kind of powerful samples-based background subtraction method. The background model is initialized from only single frame by using a random selection policy. Background model is updated dynamically by using a memoryless update strategy and a neighborhood propagation mechanism. It mainly includes three parts: background modeling, foreground target detection and model updating.

Background Modeling and Initialization. ViBe builds a model for each background pixel with a set of samples instead of with an explicit pixel model. Denote $f(x)$ by the pixel value located at x in the image in a given Euclidean color space, and f_i by a background sample value with an index i. The background pixel located at x is modeled by a collection of N background sample values as follows:

$$M(x) = \{f_1, f_2, \ldots, f_N\} \tag{11}$$

Considering that adjacent pixels share a similar temporal distribution, ViBe initializes the background model from a single frame. Each pixel model contains N values randomly extracted from the spatial neighborhood of each pixel in the first frame. Assume that $t = 0$ denotes the first frame and $N_G(x)$ is a spatial neighborhood of a pixel located at x, thus the model $M^0(x)$ is as follows:

$$M^0(x) = \{f^0(y | y \in N_G(x))\} \tag{12}$$

Foreground Target Detection. If we consider the background subtraction as a classification problem, we want to classify a new pixel as a background or foreground pixel with respect to its neighborhood. To classify a pixel value $f(x)$ according to its corresponding model $M(x)$, we compare it to the closest values within the set of samples by defining a sphere $S_R(f(x))$ of radius R centered on $f(x)$. The pixel value $f(x)$ is then classified as background if the cardinality, denoted $\#T(x)$, of the set intersection between the sphere and the model sample set is greater than or equal to a given threshold $\#min$.

$$\#T(x) = \#\{S_R(f(x)) \cap (f_1, f_2, \ldots, f_N)\} \tag{13}$$

Assume that $t = k$ denotes the k frame and that $f^k(x)$ is the value of the pixel located at x in the k frame, with its corresponding model $M^{k-1}(x)$. To classify a pixel value as foreground or not as follows:

$$f^k(x) = \begin{cases} background, & \#T(x) \geq \#\min \text{ and} \\ & \{|f^k(x) - f_i^{k-1}(x)| \leq R, i = 1, \ldots, N\}, \\ foreground, & \text{else} \end{cases} \tag{14}$$

Update Background Model. ViBe can ensure a smooth decaying lifespan for the samples in the background pixel models by using a memoryless update strategy. The random time subsampling method is used to expand the time windows covered by the background pixel models. A time subsampling weighting factor φ is designed to control the probability of background pixel model updating. A background pixel value has one chance in φ to be selected to update its pixel model and also has the same chance to be selected to update its neighboring pixel model. By using spatial consistency in background samples propagation, a spatial diffusion of information about the background evolution is achieved. Therefore, this background model can adapt to structural evolutions and varying illumination.

3   Experimental Results and Analysis

All the tests are implemented at MATLAB R2011b platform on a PC with 2.4-GHz Intel(R) Xeon(R) CPU and 8-GB RAM. ViBe parameters are set as follows: $N = 20$, $R = 20$, $\#\min = 2$, $\varphi = 16$.

Simulation platform detects moving targets from video scenes by the underwater still camera. Figure 3 is the experimental results of video 'Diver'. The 19th frame in the original video is shown in Fig. 3(a), and its corresponding enhanced video image is shown in Fig. 3(b). It can be seen that the enhanced image is clearer and the contrast

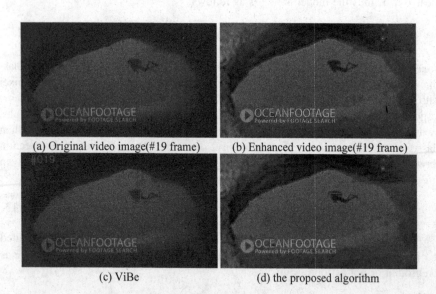

(a) Original video image(#19 frame)	(b) Enhanced video image(#19 frame)
(c) ViBe	(d) the proposed algorithm

Fig. 3. Moving target detection results of video 'Diver'

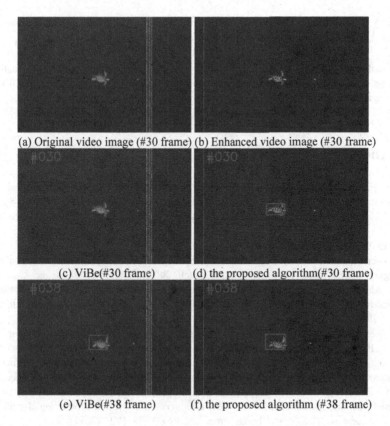

(a) Original video image (#30 frame) (b) Enhanced video image (#30 frame)

(c) ViBe(#30 frame)   (d) the proposed algorithm(#30 frame)

(e) ViBe(#38 frame)   (f) the proposed algorithm (#38 frame)

Fig. 4. Moving target detection results of video 'Turtle'

between the target and the background is higher. Figure 3(c) and (d) are detection results of ViBe and the proposed algorithm, respectively. The visual results show that the proposed algorithm can extract the complete moving diver accurately.

Figure 4 is the experimental results of video 'Turtle'. The 30th frame in the original video is shown in Fig. 4(a), and its corresponding enhanced video image is shown in Fig. 4(b). The turtle in the enhanced image is clearer than that in the original image. Figure 4(c) and (e) are the detection results by using ViBe, and Fig. 4(d) and (f) are the results by using the proposed algorithm. The proposed algorithm can detect the target accurately in 30th frame, but ViBe cannot. ViBe algorithm finds the target until 38th frame, and the detection window is slightly offset.

4   Conclusions

In this paper, we present an underwater moving target detection algorithm based on image enhancement. The proposed algorithm achieves the adaptive underwater color video image enhancement so as to improve the contrast and clarity of the target and inhibit inhomogeneous illumination, and then efficiently extracts underwater moving targets by ViBe model. The experiments show that the proposed algorithm can accurately detect the moving targets under the complex underwater environment.

Acknowledgment. This work was supported by the National Natural Science Foundation of China (No. 41306089), the Natural Science Foundation of Jiangsu Province (No. BK20130240) and Changzhou Science and Technology Program (No. CJ20160055).

References

1. Shen, J., Fan, T., Tang, M., Zhang, Q., Sun, Z., Huang, F.: A biological hierarchical model based underwater moving object detection. Comput. Math. Methods Med. **2014**, 1–8 (2014)
2. Lei, F., Huang, W., Zhang, Z.: Underwater moving object detection based on codebook model. In: Fifth International Conference on Computational and Information Sciences, pp. 1319–1322 (2013)
3. Moscheni, F., Bhattacharjee, S., Kunt, M.: Spatialtemporal segmentation based on region merging. IEEE Trans. Pattern Anal. Mach. Intell. **20**(9), 897–915 (1998)
4. Negahdaripour, S.: Revised definition of optical flow: integration of radiometric and geometric cues for dynamic scene analysis. IEEE Trans. Pattern Anal. Mach. Intell. **20**(9), 961–979 (1998)
5. Stauffer, C., Grimson, W.E.L.: Learning patterns of activity using real-time tracking. IEEE Trans. Pattern Anal. Mach. Intell. **22**(8), 747–757 (2000)
6. Barnich, O., Droogenbroeck, M.V.: ViBE: a powerful random technique to estimate the background in video sequences. In: IEEE International Conference on Acoustics, Speech and Signal Processing -Proceedings, pp. 945–948. Taipei, Taiwan (2009)
7. Barnich, O., Droogenbroeck, M.V.: ViBe: a universal background subtraction algorithm for video sequences. IEEE Trans. Image Process. **20**(6), 1709–1724 (2011)
8. Zhou, Y., Li, Q., Huo, G.: Human visual system based automatic underwater image enhancement in NSCT domain. KSII Trans. Internet Inf. Syst. **10**(2), 837–856 (2016)

Bio-signal and Medical Image Analysis

A New Epileptic Seizure Detection Method Based on Degree Centrality and Linear Features

Haihong Liu[1,2], Qingfang Meng[1,2(✉)], Yingda Wei[1,2], Qiang Zhang[3], Mingmin Liu[1,2], and Jin Zhou[1,2]

[1] School of Information Science and Engineering,
University of Jinan, Jinan 250022, China
ise_mengqf@ujn.edu.cn
[2] Shandong Provincial Key Laboratory of Network
Based Intelligent Computing, Jinan 250022, China
[3] Institute of Jinan Semoconductor Elements Experimentation,
Jinan 250014, China

Abstract. With the increasing incidence of epilepsy, we need to detect the epilepsy with high efficiency to avoid the disease attack. In this paper, we proposed two novel feature extraction methods for automatic epileptic seizure detection with high performance based on the statistic properties of complex network. One is the degree centrality combined with the linear features as the features to classify the epileptic EEG signal. Firstly, we transformed the time series into complex network by using horizontal visibility graph (HVG). Then we extracted the degree centrality of the complex network combined with the fluctuation index and variation coefficient as the three-dimensional features and the classification accuracy is up to 95.98%. To enhance the difference of the degree centrality feature, we put the other new feature. That is the improved degree centrality and chose the improved degree centrality as the single feature to classify the signal. Experimental results showed that the classification accuracy of this single feature is 96.50%.

Keywords: Feature extraction method · Degree centrality · Horizontal visibility graph · Fluctuation index · Variation coefficient · Epileptic seizure detection

1 Introduction

Epilepsy is an ancient disease which influenced the life of patients for a long time. Detection for epilepsy is main work for the scholars from past to present and has a profound significance on the patients. In the beginning, professional doctor detected the epilepsy by judging electroencephalogram with their eyes. But this way consumed a lot of manpower. Later, by studying the signal further researchers utilized nonlinear time series analysis method, which depicts the nonlinear information of original time series to complete the detection of epileptic.

© Springer International Publishing AG 2017
F. Cong et al. (Eds.): ISNN 2017, Part II, LNCS 10262, pp. 439–446, 2017.
DOI: 10.1007/978-3-319-59081-3_51

Some scholars utilized traditional nonlinear characteristic to detect the epileptic EEG. Jing and Takigawa analyzed epileptic EEG and normal EEG by utilizing the correlation dimensions [1] and showed that the correlation dimension of the epileptic EEG is larger than the normal EEG's. Osowski analyzed the epileptic seizure based on the largest Lyapunov exponent [2]. Nurujjaman researched the Hurst exponent [3] of the epileptic EEG and discovered that the normal EEG is uncorrelated whereas the epileptic EEG is long range anti-correlated. Kannathal [4] researched the epileptic by utilizing the approximate entropy and confirmed that different states had different entropies. Y. Song put up with optimized sample entropy [5, 6], which was applied to seizure detection, which removes the calculation redundancy.

Later, the theory of complex network provided a new perspective to study the nonlinear time series. Jie zhang [7] used the theory of complex network to describe pseudo-periodic time series. Zhang and Small [8] put forward the pioneering algorithm. They analyzed the complex network and extracted the degree distribution of complex network as the feature and the experimental results showed that different types of time series have different degree distributions. With the increasing incidence of epilepsy, the classification performance of these methods need to be improved. So we need to explore more efficient methods. In 2008, Lacasa come up with the visibility graph algorithm [9] firstly and this algorithm can transform any time series to complex networks. Some scholars utilized the visibility graph to study exchange rate series [10] and fractional Brownian motions [11, 12]. Later, researchers improved the rule of visibility graph by studying this algorithm and put forward the horizontal visibility graph [13]. The algorithm has also been used widely.

In this paper, we built on complex network perspective and utilized the topological statistical properties of complex network to analyze the time series. The methods we proposed were introduced in detail in prat two.

2 Feature Extraction Methods Based on Degree Centrality and Linear Features

In this section, we presented the conversion algorithm and introduced the feature extraction methods. During this conversion process, we used horizontal visibility graph to complete the construction of complex network. Compared to proximity network, this conversion algorithm omits the selection of the threshold. Then we took the degree centrality feature combined with linear features and the single improved degree centrality feature as the extracted features respectively.

2.1 Horizontal Visibility Graph

Horizontal visibility graph provides a better way to complete the conversion. The specific algorithm of horizontal visibility graph is described below. Firstly, we use $\{x_i\}$ $i \in [1\ M]$ to denote an EEG signal, where x_i is the i_{th} sampling point. The length of time series is M. To construct complex network, we need to construct node set and edge set. In this conversion algorithm, each sampling point of time series presents a node, so the whole time series constitute the node set. Whether there is an edge between two nodes depends on the local convex constraints. As for two sampling point x_i and x_j, if $x_k < \min(x_i, x_j)$ for all k with $t_i < t_k < t_j$, there is an edge between the node x_i and x_j; if not, there is no edge between the node x_i and x_j. We can get the edge set according to this principle. Compared to horizontal visibility graph, the criteria of visibility graph is $x_k < x_j + (x_i - x_j)\frac{t_j - t_k}{t_j - t_i}$. When the values of the two time samples satisfy the rule, there is an edge between the two nodes; if not, there is no edge. Through the conversion algorithm, we can get the adjacency matrix of complex network.

Proximity network is also a kind of conversion algorithm, which is used widely. The computational process of the algorithm is different from the horizontal visibility graph. At first, the time series is segmented and each segment is treated as a node of complex network. Secondly, we achieve the distance matrix by calculating the distance between two nodes. To obtain the adjacent matrix, we need to judge the edge between nodes. The judgement rule is defined below.

$$a_{ij} = \begin{cases} 1, & d_{ij} < th \\ 0, & d_{ij} \geq th \end{cases} \tag{1}$$

The th presents the threshold. If the distance between nodes is greater than the threshold, there is no edge between nodes; if the distance is smaller than the threshold, there is an edge between nodes.

From the above, it's clear that horizontal visibility graph is not involved in selecting any threshold. But the proximity network needs to select the threshold. Thus it can be seen that horizontal visibility graph decreases the subjectivity greatly.

2.2 The Degree Centrality and Improved Degree Centrality

In the theory of complex network, degree centrality describes the center of degree. The degree centrality of complex network is defined as:

$$c_D(x) = \frac{k(x)}{n - 1} \tag{2}$$

As for one node, $k(x)$ is the number of the edges, which connect to this node; n is the total number of nodes in complex network. This property reflects some information of original time series. To further improve the classification accuracy, we modify the degree centrality. The improved degree centrality enhances the degree centrality. The formula is listed below:

$$c_{DD}(x) = \frac{k(x)^2}{n - 1} \tag{3}$$

$k(x)$ describes the degree of the node. This property enhances some mathematical characteristics of feature, so the performance of the classification is better.

2.3   The Fluctuation Index and Variation Coefficient

As for EEG signal, the signal fluctuation during a seizure is larger than the normal. Fluctuation index describes the fluctuation of signal intensity. It is defined as:

$$F = \frac{1}{n} \sum_{i=1}^{n-1} |x(i+1) - x(i)| \tag{4}$$

x(i) stands for the i_{th} point of original signal and n is the length of the signal. The fluctuation index reflects the fluctuation of time series.

Variation coefficient presents the change of signal amplitude. It is denoted as:

$$V_c = \frac{\delta^2}{\mu^2} \tag{5}$$

δ is standard deviation of the EEG signal and μ is the mean of the EEG signal.

3   Experiment Result and Analysis

All the simulations were based on a 2.60 GHz quad-core Inter Pentium processor with 4 GB memory. The code was executed in environment of MATLAB 7.0. Experimental data is from Bonn University, which is in Germany. We select dataset D and dataset E from it, which are digitized at 173.6 samples per second at 12-bit resolution and have 23.6 s time duration. The data, which is from dataset D, describe the interictal periods of epilepsy patients, while the data from dataset E describe the ictal periods epilepsy patients. Dataset D and dataset E contain 100 sets of datum respectively. Each datum has the 4097 sampling points. Experimental data is described in Fig. 1. In this experiment, the length of each sample datum we adopt is 1024.

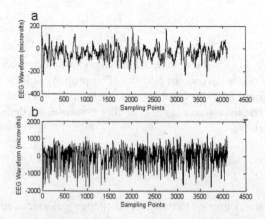

Fig. 1. (a) An interictal EEG sample in dataset D, (b) An ictal EEG sample in dataset E

Degree centrality describes the internal information of the EEG signal and reflects some information of original time series. We calculate it according to the formula 2. So we can obtain the degree centrality of complex network corresponding to the time series. We extract it as the classification feature to complete the classification process. In the Figs. 3 and 5, + presents ictal EEG signal, * presents interictal EEG signal. From the Figs. 2 and 3, we can see that different states have different value range of degree centrality. To some extent, degree centrality feature can distinguish the experimental datum well. As we all know, fluctuation index describes the fluctuation of signal intensity. From the Fig. 1, we can find that the fluctuation of EEG signal during a seizure is larger than the normal. It seems that fluctuation index can contain some information of EEG signal. Variation coefficient is the linear feature which describes the change of signal amplitude. So in order to increase the accuracy rate, we put forward the following feature extraction method. At first, we extract the degree centrality feature combined with fluctuation index and variation coefficient as three-dimensional features. Then we normalize the three-dimensional features and put it to the support vector machine [14–17] to classify the EGG signal. Compared to other existing methods, the method we proposed achieves the higher accuracy rate which is up to 95.98%. The comparison between various methods is described in Table 1. In Table 1, recurrence quantification analysis (RQA) is a kind of nonlinear time series

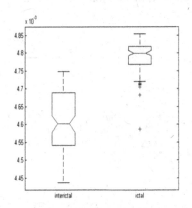

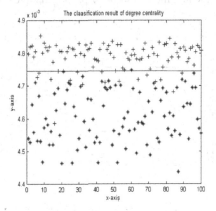

Fig. 2. The boxplot of degree centrality   **Fig. 3.** The distribution graph of degree centrality

Table 1. The classification results of the proposed feature and other features.

Feature	ACC(%)
Approximate entropy +SVM	89.00
Sample entropy +SVM	91.00
Degree centrality +SVM	93.92
RQA + SVM	94.00
Degree centrality +linear +SVM	95.98
Improved degree centrality	96.50

analysis which is based on recurrence plot. From the Table 1, we can see that the feature extraction method has the better performance, which depicts that the three-dimensional features can clearly classify the different state of epilepsy.

To enhance the difference of the degree centrality feature, we analyzed the EEG signal deeply and improved the degree centrality feature. We calculate the square of degree centrality feature, so as to enhance the difference of the degree centrality feature. We can reference to formula 3. After calculating the improved degree centrality, we treat it as the single feature to classify the data. Figure 4 clearly shows that the latter eigenvalues is generally higher than the former. It means that the improved degree centrality feature values have a bigger difference in different state. We can distinguish the EEG signal whether it is during a seizure or the normal according to the improved degree centrality feature. From the Fig. 5, we can find that most points can be classified correctly except for a few points. The straight line in Fig. 5 is the best classification threshold and the classification accuracy is 96.5%. In general, classier can improve the classification efficiency. So, compared with other features the improved degree cen-trality feature as the single feature can obtain better performance in classifying EEG data. Results show the classification accuracy of the improved degree centrality com-bined with SVM would not be higher than the single improved degree centrality. It depicts this feature does well in classifying the EEG signal and the SVM can't enhance the classification result. In the Table 1, different methods are summarized and we can conclude that the methods we proposed have the better performance than other methods.

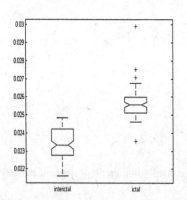

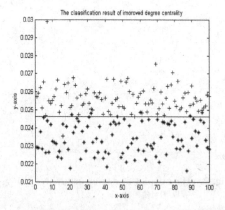

Fig. 4. The boxplot of improved degree centrality

Fig. 5. The classification result of the improved degree centrality

4   Conclusions

In this paper, we put forward two novel feature extraction methods for epileptic seizure detections which have better performance. We build on complex network perspective and utilize the topological statistical properties of complex network to analyze the signal. Firstly, we transform the original time series into complex network by using the horizontal visibility graph algorithm. From the above calculation, we can find horizontal visibility graph omits the choice of threshold and decreases the subjectivity greatly, compared with proximity network. Fluctuation index and variation coefficient are the linear features, which both describe the change of signal. Then we extract the degree centrality combined with fluctuation index and variation coefficient as three-dimensional features. After the three-dimensional features normalized, we use the support vector machine to deal with it. The classification accuracy of the three-dimensional features is 95.98%. To enhance the difference of the degree centrality, we improve the degree centrality feature and extract it as the single feature. The classification result shows the improved degree centrality feature has the better classification performance which is up to 96.5%. In Table 1, we can see the feature extraction methods we proposed achieve higher classification accuracy compared to other existing methods. From experimental results, we can reach the conclusion that the statistic properties of complex network can describe the EEG signal better and we propose new methods to distinguish ictal EEG from interictal EEG. With the increasing incidence of epilepsy, the detection methods are helpful to the diagnosis and treatment of epilepsy patients. In this paper, the research ideas and classification algorithms are of great significance to the study of medical signals and contribute to improve the medical level.

Acknowledgments. This work was supported by the National Natural Science Foundation of China (Grant Nos. 61671220, 61640218, 61201428), the Shandong Distinguished Middle-aged and Young Scientist Encourage and Reward Foundation, China (Grant No. ZR2016FB14), the Project of Shandong Province Higher Educational Science and Technology Program, China (Grant No. J16LN07), the Shandong Province Key Research and Development Program, China (Grant No. 2016GGX101022).

References

1. Jing, H., Takigawa, M.: Topographic analysis of dimension estimates of EEG and filtered rhythms in epileptic patients with complex partial seizures. Biol. Cybern. **83**(5), 391–397 (2000)
2. Osowski, S., Swiderski, B., Cichocki, A., Rysz, A.: Epileptic seizure characterization by Lyapunov exponent of EEG signal. COMPEL: Int J. Comput. Math. Electr. Electron. Eng. **26**(5), 1276–1287 (2007)
3. Nurujjaman, M., Ramesh, N., Iyengar, A.N.S.: Comparative study of nonlinear properties of EEG signals of normal persons and epileptic patients. Nonlin. Biomed. Phys. **3**(1), 6 (2009)
4. Kannathal, N., Lim, C.M., Acharya, U.R., Sadasivan, P.K.: Entropies for detection of epilepsy in EEG. Comput. Methods Programs Biomed. **80**(3), 187–194 (2005)
5. Song, Y., Crowcroft, J., Zhang, J.: Automatic epileptic seizure detection in EEGs based on optimized sample entropy and extreme learning machine. J. Neurosci. Methods **210**, 132–146 (2012)
6. Lake, D.E., Richman, J.S., Griffin, M.P., Moorman, J.R.: Sample entropy analysis of neonatal heart rate variability. Am. J. Physiol.-Regul. Integr. Comp. Physiol. **283**, 789–797 (2002)

7. Zhang, J., Sun, J., Luo, X., Zhang, K., Nakamurad, T., Michael, S.: Characterizing pseudoperiodic time series through the complex network approach. Phys. D **237**, 2856–2865 (2008)
8. Zhang, J., Small, M.: Complex network from Pseudoperiodic time series: topology versus dynamics. Phys. Rev. Lett. **96**, 238701 (2006)
9. Lacasa, L., Luque, B., Ballesteros, F., Luque, J., Nuno, J.C.: From time series to complex networks: the visibility graph. Proc. Natl. Acad. Sci. U.S.A. **105**(13), 4972–4975 (2008)
10. Yang, Y., Wang, J., Yang, H., Mang, J.: Visibility graph approach to exchange rate series. Phys. A: Stat. Mech. Appl. **388**, 4431–4437 (2009)
11. Ni, X.H., Jiang, Z.Q., Zhou, W.X.: Degree distributions of the visibility graphs mapped from fractional Brownian motions and multifractal random walks. Phys. Lett. A **373**, 3822–3826 (2009)
12. Lacasa, L., Luque, B., Luque, J., Nuño, J.C.: The visibility graph: a new method for estimating the Hurst exponent of fractional Brownian motion. EPL **86**, 30001 (2009)
13. Luque, B., Lacasa, L., Ballesteros, F., Liuque, J.: Horizontal visibility graphs: exact results for random time series. Phys. Rev. E **80**, 046103 (2009)
14. Moguerza, J., Muñoz, A.: Support vector machines with applications. Stat. Sci. **21**, 322–336 (2006)
15. Cristianini, N., Shawe-Taylor, J.: An Introduction to Support Vector Machines and Other Kernel-Based Learning Methods. Cambridge University Press, London (2000)
16. Cai, D.-M., Zhou, W.-D., Li, S.-F., Wang, J.-W., Jia, G.-J., Liu, X.-W.: Classification of epileptic EEG based on detrended fluctuation analysis and support vector machine. Acta Biophys. Sinica **27**, 175–182 (2011)
17. Li, S., Zhou, W., Yuan, Q., Geng, S.: Dongmei Cai.: feature extraction and recognition of ictal EEG using EMD and SVM. Comput. Biol. Med. **43**, 807–816 (2013)

A Comparison Between Two Motion-Onset Visual BCI Patterns: Diffusion vs Contraction

Minqiang Huang, Hanhan Zhang, Jing Jin$^{(\boxtimes)}$, Yu Zhang, and Xingyu Wang$^{(\boxtimes)}$

Key Laboratory of Advanced Control and Optimization for Chemical Processes, Ministry of Education, East China University of Science and Technology, Shanghai, People's Republic of China
jinjingat@gmail.com, xywang@ecust.edu.cn

Abstract. Two motion-onset visual patterns are proposed in this paper. The difference between two patterns is the motion pattern of stimuli. One motion pattern is dot diffusion and another motion pattern is dot contraction. The performances of two patterns in terms of classification accuracy, comfort level and interference level (feedback from the participants) are compared with each other. Although no significant difference can be found at the online accuracy ($p = 0.0861$) and comfort level ($p = 0.7804$), the single-trial accuracy of the contraction pattern is significantly higher than the diffusion pattern ($p = 0.0084$) and the interference level is significantly lower than the diffusion pattern ($p = 0.0243$). The results demonstrate that the contraction pattern can decrease the interference from other stimuli and improve the classification accuracy.

Keywords: ERPs · Visual · Motion · P300 · BCI · Classification accuracy · Diffusion · Contraction

1 Introduction

Brain computer interfaces (BCIs) can perform users' commands and intentions of users without relying on the peripheral nerve and muscles [1]. People who suffer amyotrophic lateral sclerosis (ALS) or stroke lose ability to control their muscles and this situation will severely reduce their life quality [1]. BCIs are available for this group to help them interact with external environment. The first visual BCI based on event related potentials (ERPs) which was called P300 speller was proposed by Farwell and Donchin [2]. P300 component is usually evoked in the oddball paradigm, which is also used in the Farwell and Donchin's work. P300 component appears at ~ 300 ms after a rare attended stimuli or event happen [3]. It is an endogenous component of ERPs, which is related to the selective attention and information processing of brain [4]. In [2], some problems such as the double-flash effect and the interference from the adjacent characters (adjacency-distraction errors), blocked the BCI system from better performance. Since then, to improve the property of BCI, many researchers proposed meaningful and practical methods. It has been proved that the stimuli of paradigm had impacts on the performance of BCI [5].

© Springer International Publishing AG 2017
F. Cong et al. (Eds.): ISNN 2017, Part II, LNCS 10262, pp. 447–456, 2017.
DOI: 10.1007/978-3-319-59081-3_52

The double-flash effect would happen if the target flashed consecutively and this might result errors because P300 component could not be evoked or could be overlapped [6]. The adjacency-distraction errors would appear when the non-targets near the target were flashing and the attention of users was distracted by the flashing of non-targets [7].

To reduce the errors, researchers split an 8 × 9 symbol matrix into two 6 × 6 matrixes that the adjacent symbols were not in the same matrix and optimized the flashing interval between targets [8]. This paradigm was called the checkerboard paradigm (CBP) and it was optimized further through avoiding any of eight adjacent items of target flashing simultaneously with target [7].

Another work proposed a new stimulus presentation way which based on the binomial coefficients to improve the performance of BCI [9]. The design in this work also avoided the double-flash errors and improved the practical bit rate comparing with the conventional RC paradigm.

In this study, we explore a novel method based on the work [10] to investigate whether a contraction motion-onset pattern will perform better in items of decreasing the interference from the adjacent stimuli.

The visual motion-onset paradigm was first proposed at 2008 [11]. The brain response elicited by the visual motion-onset paradigm is the motion-onset visual evoked potential (mVEP), which typically contains P1, N2 and P2 components [12]. The main purpose using visual motion-onset stimuli in this study is to explore whether motion-onset patterns will guide the attention of users and affect the performance of BCI systems.

2 Method

2.1 Stimuli

A 6 × 6 character matrix was presented on the screen as shown in the Fig. 1. In the diffusion pattern, instead of the flashing rows or columns, there was a small white dot appearing on the left of the characters which were to be illuminated. The white dot would divide into six small dots with arranging in a hexagonal shape and then enlarging the shape. In the contraction pattern, the motion way was opposite with the diffusion pattern. Six white dots were arranged in a big hexagonal shape on the left of the characters and then the six dots would contract to one dot. The procedures of the motions in two patterns are demonstrated in the Fig. 2. The size of each picture was 80 pixels. The stimuli lasted 120 ms, and the inter-stimuli-interval (ISI) was 200 ms.

2.2 Participants and Experimental Setup

Ten students (four female, age range from 22–25, mean age 23.50 ± 1.27 years old) participated in this study. All subjects have normal or corrected-to-normal vision. All the subjects are right-handed. Before experiments, the subjects signed a written consent about the study.

The EEG data was recorded through 14 electrodes of a 64-channel 'g.EEGcap' and a 16-channel g.USBamp amplifier (Guger Technologies, Graz, Austria). The electrodes were F3, Fz, F4, C3, Cz, C4, CP3, CP4, P3, Pz, P4, P7, P8 and Oz. The EEG data was sampled at 256 Hz and the band pass filter was between 0.1–30 Hz. The notch filter was at 50 Hz. The right mastoid was chosen as the reference and the FPz was chosen as the ground. The impedances of all electrodes were below 10 KΩ.

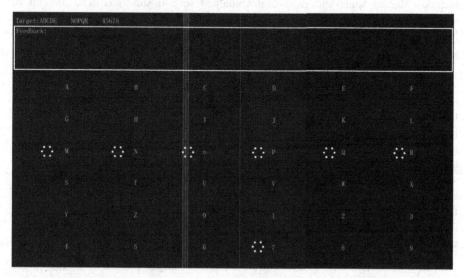

Fig. 1. Screenshot of the visual stimuli. The top characters are the targets in the offline experiments. The 'Feedback' region is used to demonstrate the results in the online experiments. This picture shows one moment during the stimuli presenting.

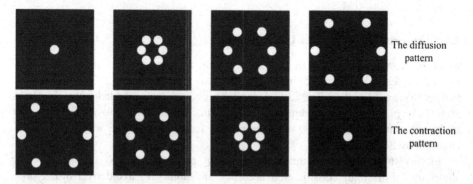

Fig. 2. The details of the motion. The motions in the both patterns are presented through refreshing the four pictures sequentially. Each picture lasts 30 ms and the total time of stimuli on is 120 ms. After four pictures are displayed, a flash (a sub-trial) is completed.

2.3   Experiment Procedure

Before the experiments, subjects were given an oral introduction about the experiments and they were asked to avoid eye blinking and body movements.

The whole experiments included two parts: the offline experiments and the online experiments. In the offline experiments, there were three sessions. Each session contained five runs. The training process of one target was called one run. Each run included sixteen trials and each trial contained several sub-trials. The amount of sub-trials in a trial was determined by an approach proposed by [9]. As shown in the Fig. 1, fifteen target-selections were executed in the offline experiments. After one session was completed, subjects had 3–5 min to rest.

The online experiments were copy spelling tasks. Subjects were asked to spell 36 given characters. The online target-selection results would be printed on the 'Feedback' region (see Fig. 1). Instead of a fixed number of trials in the offline experiments, the number of trials in one run was determined by an adapted stopping method designed in [13]. The number of trials ranged from 2 to 16, which meant that at least two trials were needed to output one target-selection result and the maximum number of trials was sixteen. The criterion to get an output was that if the classification results of two successive trials were same then BCI system decided that this result was the target or if sixteen trials were completed but no successive results appeared and then the result of the last trial (the sixteenth trial) would be regarded as the output.

After the online experiments, subjects needed to answer two questions as two of the indexes of evaluating the property of the patterns. The questions were as following:

(1) How comfortable do you feel in these patterns? Please rate the two patterns in a scale from 0 to 10. The higher score indicates stronger agreement.
(2) Does this pattern make you easy to be interfered by non-target stimuli? Please rate the two patterns in a scale from 0 to 10. The higher score indicates stronger agreement.

2.4   Feature Extraction and Classification

The EEG data obtained from the offline experiments were used to training the classification model. A third order Butterworth band pass filter between 0.5–30 Hz was used to filter the raw EEG data. The EEG data after filter was downsampled from 256 Hz to 36.6 Hz by picking every 7th sample from the data. The time window of EEG data was from each stimulus on to 1000 ms later and the feature vector size was 14 × 36 (14 channels by 36 time points).

In this study, we chose linear discriminant analysis (BLDA) as classification algorithm for its high classification accuracy and ability to avoid over-fitting phenomenon which was caused by the high dimension data [14]. The details of BLDA could be found in [15].

3   Results

3.1   ERPs

The grand average amplitudes of ERPs for two patterns are shown in the Fig. 3. It can be seen that the target amplitudes between two patterns have differences at all electrodes. The obvious differences mainly appear at the central and posterior regions such as electrodes Cz, CP3, CP4 and Pz.

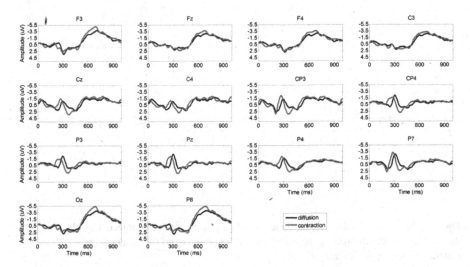

Fig. 3. The grand averaged amplitudes of 14 electrodes across 10 subjects for the diffusion pattern and the contraction pattern. The black line indicates the target amplitude of the diffusion pattern and the red line indicates the target amplitude of the contraction pattern. (Color figure online)

3.2   Offline Classification Accuracy

The offline classification accuracies of two patterns of each subject are demonstrated in the Fig. 4. The accuracies can achieve 100% after several iterations for each subject in both patterns. The group mean accuracy of the contraction pattern is higher than that of the diffusion pattern in the early trials. To analyze the offline accuracies further, the single-trial classification accuracies of two patterns of each subject are computed and displayed in the Fig. 5. A paired t-test is used to valid the difference of the single-trial accuracies between the two patterns. It is found that the single-trial accuracies of the contraction pattern were significantly higher than the diffusion pattern ($p = 0.0084$).

3.3   Online Accuracy

The averaged online accuracies of each subject in two patterns are presented in the Fig. 6. Although no significant difference is shown between two patterns ($p = 0.0861$), the group mean online accuracy of the contraction pattern is higher.

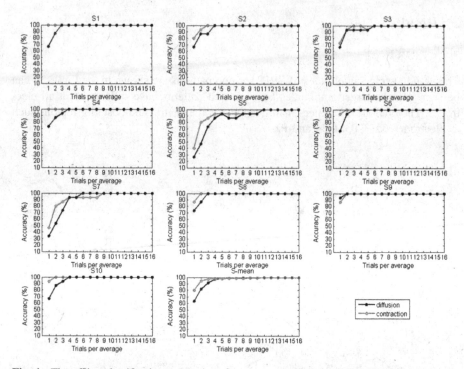

Fig. 4. The offline classification accuracies of two patterns of each subject. The black line was the accuracy of the diffusion pattern and the red line referred to the accuracy of the contraction pattern. (Color figure online)

3.4 Feedback of Questions

The feedback of the questions mentioned before was collected from each subject. The feedback about the comfort level and interference level is displayed in the Fig. 7. No significant difference is found at the comfort level between two patterns ($p = 0.7804$), but the interference level shows a significant difference ($p = 0.0243$). It manifests that the contraction pattern has less interference from the non-target stimuli than the diffusion pattern.

4   Discussion

In this study, two motion-onset patterns are compared in items of ERP waveforms, classification accuracy, comfort level and interference level to investigate whether different motion-onset ways will influence the performance of BCI. The results show that the contractive motion-onset way (the contraction pattern) has better performance than the diffused motion-onset way (the diffusion pattern).

Clear differences can be observed on the N200 component and P300 component. The shorter peak latency is the major difference for the contraction pattern from the diffusion pattern. The morphology of ERPs can be affected by the different experiment

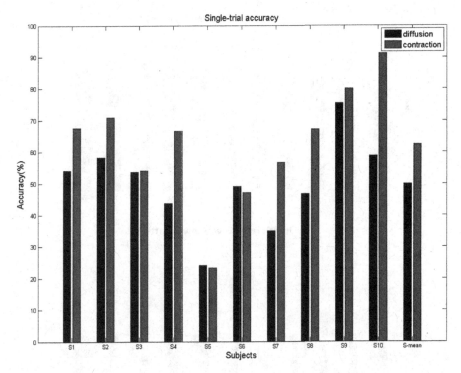

Fig. 5. The single-trial accuracies of each subject. The black bar indicates the single-trial accuracy of the diffusion pattern. The red bar indicates the single-trial accuracy of the contraction pattern. (Color figure online)

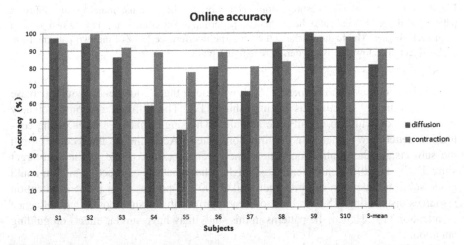

Fig. 6. The online accuracies of each subject. The blue bar indicates the online accuracy of the diffusion pattern. The red bar indicates the online accuracy of the contraction pattern. (Color figure online)

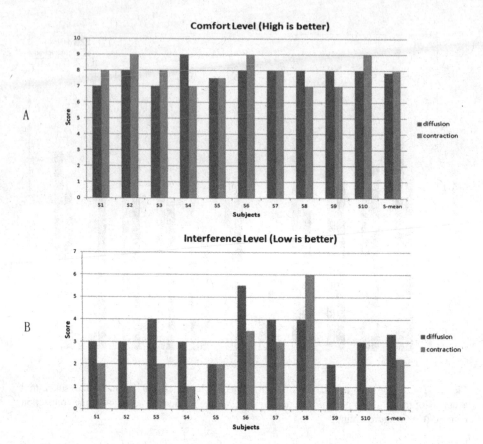

Fig. 7. The scores of the questions from each subject. The blue bar indicates the diffusion pattern and the red bar indicates the contraction pattern. (A) The comfort level of the two patterns from each subject. The higher is the better. (B) The interference level of the two patterns from each subject. The lower is the better. (Color figure online)

conditions and the stimuli property [16]. It was found that P300 latency would increase with the difficulty of stimulus distinguishing [17]. The diffused motion-onset may increase the difficulty to classify the stimulus. The single-trial accuracies of the contraction pattern are higher than the diffusion pattern. According to the feedback from the subjects, in the contraction pattern they were not susceptible to the non-target stimuli as in the diffusion pattern. This might reflect that the contractive motion could guide subjects' attention on the target well. P300 component is an index of attention resources distributing [18]. And it had been approved that visual motion had influence on attention [19]. Hence, the patterns in this study may have similar effects on guiding attention.

In the visual BCI system, the errors coming from the adjacent stimuli are a problem that affects the performance. One of the reasons why subjects could concentrate on the target easily in the contraction pattern than the diffusion pattern might be that the contractive motion led the sight focusing on a smaller region rather than a larger range.

5   Conclusion

In this paper, two motion-onset patterns are compared to explore the effect of the motion way on the BCI performance. The results demonstrate that different motion ways have different influences on the BCI performance and the contractive motion way can help users concentrate on the target easily. The future work is to compare the contraction pattern with other proposed patterns to valid its property.

Acknowledgements. This work was supported in part by the Grant National Natural Science Foundation of China, under Grant Nos. 61573142, 61203127, 91420302, and 61305028. This work was also supported by the Fundamental Research Funds for the Central Universities (WG1414005, WH1314023, and WH1516018) and Shanghai Chenguang Program under Grant 14CG31.

References

1. Birbaumer, N., Cohen, L.G.: Brain–computer interfaces: communication and restoration of movement in paralysis. J. Physiol. **579**, 621–636 (2007)
2. Farwell, L.A., Donchin, E.: Talking off the top of your head: toward a mental prosthesis utilizing event-related brain potentials. Electroencephalogr. Clin. Neurophysiol. **70**, 510–523 (1988)
3. Rakotomamonjy, A., Guigue, V.: BCI competition III: dataset II-ensemble of SVMs for BCI P300 speller. IEEE. Trans. Bio-Med. Eng. **55**, 1147–1154 (2008)
4. Beverina, F., Palmas, G., Silvoni, S., Piccione, F., Giove, S.: User adaptive BCIs: SSVEP and P300 based interfaces. PsychNol. J. **1**, 331–354 (2003)
5. Sellers, E.W., Krusienski, D.J., McFarland, D.J., Vaughan, T.M., Wolpaw, J.R.: A P300 event-related potential brain–computer interface (BCI): the effects of matrix size and inter stimulus interval on performance. Biol. Psychol. **73**, 242–252 (2006)
6. Martens, S., Hill, N., Farquhar, J.: Overlap and refractory effects in a brain? computer interface speller based on the visual P300 event-related potential. J. Neural Eng. **6**, 026003 (2009)
7. Frye, G., Hauser, C., Townsend, G., Sellers, E.: Suppressing flashes of items surrounding targets during calibration of a P300-based brain–computer interface improves performance. J. Neural Eng. **8**, 025024 (2011)
8. Townsend, G., LaPallo, B., Boulay, C., Krusienski, D., Frye, G., Hauser, C., Schwartz, N., Vaughan, T., Wolpaw, J., Sellers, E.: A novel P300-based brain–computer interface stimulus presentation paradigm: moving beyond rows and columns. Clin. Neurophysiol. **121**, 1109–1120 (2010)
9. Jin, J., Allison, B.Z., Sellers, E.W., Brunner, C., Horki, P., Wang, X., Neuper, C.: Optimized stimulus presentation patterns for an event-related potential EEG-based brain–computer interface. Med. Biol. Eng. Comput. **49**, 181–191 (2011)
10. Jin, J., Allison, B.Z., Kaufmann, T., Kübler, A., Zhang, Y., Wang, X., Cichocki, A.: The changing face of P300 BCIs: a comparison of stimulus changes in a P300 BCI involving faces, emotion, and movement. PLoS ONE **7**, e49688 (2012)
11. Guo, F., Hong, B., Gao, X., Gao, S.: A brain–computer interface using motion-onset visual evoked potential. J. Neural Eng. **5**, 477 (2008)

12. Kuba, M., Kubová, Z.: Visual evoked potentials specific for motion onset. Doc. Ophthalmol. **80**, 83–89 (1992)
13. Jin, J., Allison, B.Z., Sellers, E.W., Brunner, C., Horki, P., Wang, X., Neuper, C.: An adaptive P300-based control system. J. Neural Eng. **8**, 292–301 (2011)
14. Jin, J., Sellers, E.W., Zhou, S., Zhang, Y., Wang, X., Cichocki, A.: A P300 brain–computer interface based on a modification of the mismatch negativity paradigm. Int. J. Nerual. Syst. **25**, 1550011 (2015)
15. Hoffmann, U., Vesin, J.M., Ebrahimi, T., Diserens, K.: An efficient P300-based brain–computer interface for disabled subjects. J. Neurosci. Methods **167**, 115–125 (2008)
16. Strüber, D., Polich, J.: P300 and slow wave from oddball and single-stimulus visual tasks: inter-stimulus interval effects. Int. J. Psychophysiol. **45**, 187–196 (2002)
17. Leuthold, H., Sommer, W.: Postperceptual effects and P300 latency. Psychophysiology **35**, 34–46 (1998)
18. Gray, H.M., Ambady, N., Lowenthal, W.T., Deldin, P.: P300 as an index of attention to self-relevant stimuli. J. Exp. Soc. Psychol. **40**, 216–224 (2004)
19. Mattingley, J.B., Bradshaw, J.L., Bradshaw, J.A.: Horizontal visual motion modulates focal attention in left unilateral spatial neglect. J. Neurol. Neurosurg. Psychiatry **57**, 1228–1235 (1994)

Detecting Community Structure Based on Optimized Modularity by Genetic Algorithm in Resting-State fMRI

Xing Hao Huang[1], Yu Qing Song[1], Ding An Liao[1,2], and Hu Lu[3(✉)]

[1] School of Computer Science and Communication Engineering,
Jiangsu University, Zhenjiang, China
893540551@qq.com
[2] Department of Electromechanical Engineering,
ChangZhou Textile Garment Institute, Changzhou, China
czfyandy@163.com
[3] School of Computer Science, Fudan University, Shanghai, China
myluhu@126.com

Abstract. Research has attempted to detect the community structure of the brain network using rs-fMRI data to determine differences in brain networks. Traditional clustering methods used to detect the community structure of the brain network, require a priori specification of cluster numbers. However, the cluster number of the brain network remains unknown. In this paper, we propose a new method, GAcut, to detect the community structure of real-world networks and brain functional networks. Here, genetic algorithm is applied to change the connection between nodes, based on optimized modularity Q, and to automatically detect community structure, realizing true, unsupervised analysis. GAcut was then applied to rs-fMRI data to compare differences between autism spectrum disorders (ASDs) and normal controls. Utilizing modularity Q and NMI as measurement indices for differentiation, some characteristic and meaningful network communities that feature in ASDs.

Keywords: Resting-state network · Community structure · Autism

1 Introduction

The human brain is an extremely complicated system. In recent decades, functional magnetic resonance imaging has provided a technical method for studying and analyzing brain networks. In 1995, Biswal, et al. studied the bilateral motor cortices and found functional connectivity present in the resting state [1]. In recent years, functional integration has become more and more increasingly popular. Scientists have applied investigated functional integration by a variety of methods, such as functional connectivity [1] and ICA [2].

Community detecting is a kind of clustering technology based on the graph model. Almost all real systems possess community structure, such as biological systems, social systems and economic systems [3]. As a significant property in of the complex systems, community structure has for several years been a global research hotspot. Brain The brain

© Springer International Publishing AG 2017
F. Cong et al. (Eds.): ISNN 2017, Part II, LNCS 10262, pp. 457–464, 2017.
DOI: 10.1007/978-3-319-59081-3_53

is one of the most complicated systems in nature, which means it probably possesses community structure. In fact, community structure is an important direction of rs-fMRI research, in which various methods have been used to explore community structure. Van, et al. used Ncut to partition the brain functional network [4]. Wang, et al. applied a hierarchical clustering method to detect community structure [5]. These methods show relatively reasonable results in brain networks, while certain parameters must be set or adjusted certain parameters, such as the number of communities, before segmentation and during the segmentation. In fact, the number of communities in the human brain network remains thus far undefined. Therefore, setting the community numbers and any other parameters in advance could lead to a negative effect on community division.

In this paper, we propose a new community detection method, based on optimized modularity Q [6], and GAcut, which combines the random walk model and a genetic algorithm to detect community structure in networks without any parameters. We applied GAcut to different brain networks in both ASDs and typical controls. Finally, we analyzed the difference in community structures of brain networks that evidence pathologies.

2   Dataset and Preprocessing

2.1   Data Acquisition

The resting-state fMRI datasets used were acquired from the Autism Brain Imaging Data Exchange (http://fcon_1000.projects.nitrc.org/indi/abide/index.html). In this paper, two resting-state fMRI datasets were utilized. Dataset 1 was collected by California Institute of Technology, and includes 19 individuals with ASD and 19 typical controls. Dataset 2 was collected by University of Michigan, and includes 13 individuals with ASD and 22 typical controls. More specific information can be found on the ABIDE website.

2.2   Data Preprocessing

Data preprocessing was performed using DPARSF (Yan and Zang, 2010, http://www.restfmri.net), which is based on SPM8 (http://www.fil.ion.ucl.ac.uk/spm) and REST (http://www.restfmri.net). We ran the software on MATLAB R2012a.

First, we removed the first 10 s in the datasets to make ensure a steady signal. Based on a 6-parameter, rigid body image registration, head motion correction was applied. The functional data were transformed into MNI space, then resampled to $3 \times 3 \times 3$ mm voxel size. To avoid mixing signals between different regions, spatial smoothing was not performed. Temporal filtering (0.01–0.08 Hz) was then applied to the time series of each voxel to reduce the effect of low-frequency drifts and high-frequency noise.

2.3   Correlation Analysis and Network Construction

Applying DPARSF software, the time series for each AAL brain region was extracted [7]. To obtain the correlation matrix, each brain region was treated as a node and

Spearman correlation was used to calculate the correlation coefficients between any two brain regions. In this way, we calculated the correlation matrix for every subject, including datasets 1 and 2. To analyze all the subjects within each group, we computed a group mean correlation matrix including both ASDs and typical controls for each dataset. Each correlation matrix must be converted to an unweighted adjacency matrix by choosing a specific threshold, R. If the absolute value of the correlation matrix is greater than R, it must be assigned to 1, otherwise, to 0. The two datasets were divided into four groups: ASD 1, Controls 1, ASD 2 and Controls 2. To compare the subject networks on different scales, we totally computed 31 complete adjacent matrixes for each group, with the edge numbers from 100 to 400 in steps of 10.

3 GAcut

To obtain a more accurate similarity matrix, every unweighted matrix was converted into a corresponding similarity matrix using the random walk model. During the chromosome coding step, the nearest neighbor method was applied to encode various community structures as chromosomes. From there, the genetic method was applied to optimize modularity Q and to obtain a corresponding community structure. It is unnecessary to set certain parameters in advance, since GAcut can automatically detect community structure and obtain the optimal modularity Q.

3.1 Community Structure Detection

Widely utilized in social network analysis, community structure detection algorithm is a graph partitioning method. The graph model represents a complex network in the abstract by, $G = (V, E)$, where V represents vertices, and E represents the edges between any two vertices. The adjacent matrix, $A = [n, n]$, can be utilized to represent the relationship between different vertices in networks, where n is the number of vertices. The modularity function, Q, has been proposed to assess the quality of a community structure, and to transform community discovery into an optimization problem.

$$Q = \frac{1}{2m} \sum_{ij} (A_{ij} - \frac{k_i k_j}{2m}) \delta(C_i, C_j) \tag{1}$$

where k_i and k_j are the respective degrees of node i and node j. C_i is the community that node i belongs to, while m is the number of edges in the network. δ_{ij} is the Kronecker delta function, which equals 1 when node i and node j are in the same community, and 0 when said nodes are elsewhere. modularity Q lies between 0 and 1. However, for an unweighted network, Q is always between 0.3 and 0.7. A larger Q value indicates stronger community structure, While a value for Q lower than 0.3 can be considered as having no obvious community structure in the networks. In this paper, modularity Q is utilized as the objective function to detect community structure in both the real-world networks and in brain networks.

3.2   Coding and Decoding

To avoid setting parameters in advance, a new encoding and decoding method is proposed. First, the adjacent matrix A is converted into the similarity matrix, S, via the random walk model [8], by which the nearest neighbor node is found next, based on the similarity matrix. The nearest neighbor node is unique, since the similarity matrix is a weighted matrix. We also use the backtracking-forbidden method, to prevent excessive division of the community. For example, if node M is the nearest neighbor to node N, and N is the nearest neighbor to M as well, only one of them is allowed to regard the other node as the nearest neighbor. As well, the remaining node is required to discover the secondary second-nearest neighbor node in the network. As shown in Fig. 1(b), node 1 and node 3 are the nearest neighbor node to each other, since both S(1,3) and S(3,1) are the minimum value in their own rows separately, which means they are quite similar to each other. To avoid excessive partition, node 5 is treated as the nearest neighbor node to node 3, instead of node 1. The backtracking-forbidden method was used in GAcut to code chromosomes.

In the process of decoding, node 1 and node 3 are linked, because node 3 is the nearest neighbor to node 1, while the other nodes are treated in the same way. Each chromosome in the population can represent a certain kind of community, structure. As shown in the Fig. 1(d), two communities are discovered directly by decoding process, and without any parameters in the network.

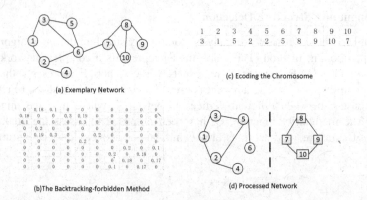

(a) Exemplary Network

(c) Ecoding the Chromosome

(b)The Backtracking-forbidden Method

(d) Processed Network

Fig. 1. Coding and decoding: (a) The exemplary network comprises ten nodes. (b) The backtracking-forbidden method (c) The exemplary network is coded into a chromosome. The first row represents ten nodes. The second row represents the chromosome. (d) The exemplary network is partitioned into two communities by decoding process.

3.3   Genetic Algorithm

Population Initialization. To develop the encoding method used here, numerous chromosomes are produced as the initial population. Different community structures can then be revealed through the decoding method. Every kind of community structure corresponds to a value of modularity Q, while chromosome corresponding to the largest Q value is selected as the parent generation to carry out the following operations.

Crossover and Mutation. Two parents produce two offspring by crossover operation, and new offspring correspond to new community structures and their modularity Q values. In the crossover operation, the chromosomes of the two parents randomly recombine to form offspring. Mutation operation can effectively prevent premature phenomenon and maintain population diversity. In GAcut, we randomly select a gene position, i, then we search its minimum similarity node, j. Because i and j are most likely to be divided into the same community, exchange next occurs between the value of i and the value of j.

4   Results

4.1   Community Structure in rs-fMRI

Modularity Q Analysis. GAcut was applied to four groups of constructed networks which have been constructed: ASD 1, Controls 1, ASD 2 and Controls 2. To demonstrate that brain networks feature modularity, rather than random networks, some comparable random networks with the same number of nodes and edges were also built.

Figure 2 depicts the results for dataset 1, where modularity Q of Controls 1 is slightly greater than that of ASD 1 in most cases; that is the community structure of Controls 1 is stronger than that of ASD 1 to a less extent. However, in Fig. 3, we cannot reach the foregoing conclusion, since the modularity Q values of ASD 2 and Controls 2 are very close. In this sense, there is no reason to distinguish between ASD and Controls solely according to the modularity Q.

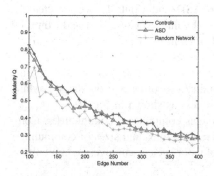

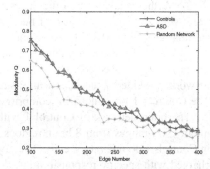

Fig. 2. Modularity Q in dataset 1.   **Fig. 3.** Modularity Q in dataset 2

Recent literature indicates only limited evidence in autism for abnormal resting-state connectivity at the regional level and no evidence for altered connectivity at the whole-brain level [9]. The same conclusion could be drawn from calculating the modularity Q in this paper. Furthermore, the modularity Q values of ASDs and typical controls are larger than the random networks, which indicates that the brain networks have community structure.

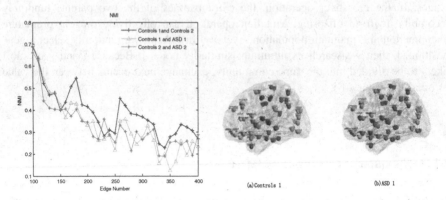

Fig. 4. NMI results and community structure

NMI Analysis. To compare the similarity of community structures in different brain networks, NMI (Normalized Mutual Information) was used as the evaluation standard. The value for NMI lies between 0 and 1. The, while the larger NMI value is, the more similar the two given community structures are. Three kinds of contrast experiments were studied in this paper: (1) Controls 1 and Controls 2 (2) Controls 1 and ASD 1 (3) Controls 2 and ASD 2, for which. The results are shown in Fig. 4. Based on Fig. 4, most situations show NMI of Controls 1 and Control 2 as greater than the NMI of Controls and ASD. This indicates that the differences of community structure exist between typical controls and ASDs.

To further analyze the differences between ASD and Controls, we selected the result with 260 edges in dataset 1 because the NMI values of Controls 1 and Control 2 are relatively high (0.4531) and while the NMI of Controls 1 and ASD 1 is relatively low (0.3517). Furthermore, dataset 1 shows the correspondingly greater difference in Modularity Q between ASD 1 (0.3664) and Controls 1 (0.4203). The ASD 1 brain network is divided 7 modules with 2 small prominent modules and one large module. The Controls 1 brain network comprised 8 modules (as shown in Fig. 4).

The brain network of Control 1 with 260 edges comprises 8 communities. Community size ranges from 8 brain regions to 16 brain regions, and sizes of communities in Controls 1 do not vary greatly. The division of labor is clear-cut; every community is charged with specific responsibilities.

Community sizes in ASD brain networks range from 24 brain regions to 2 brain regions. In comparison, there are two small communities in the ASD brain network, which indicates that these communities are not closely related to the other communities. One of the two modules comprises bilateral caudate nuclei, which are involved primarily in cognitive learning and have a close relation with classification processes. Bilateral caudate nuclei form the regulatory pathway connecting thalamus, amygdala and other nerve nuclei to the frontal and temporal lobe cortices. Studies have indicated that there is an abnormal increase of caudate nuclei in adult ASD patients, which is associated with severity of repetitive stereotyped behavior. In the Controls 1 brain network, however, caudate nuclei are included in a big module with other brain regions, such as lenticular putamen and lenticular pallidum. These brain regions belong to the regions of basal nuclei. The main functions of the basal nuclei are to control independent movement, as well as integration and adjustment of conscious activity and movement reaction. Basal nuclei also participate in memory, emotion, reward learning and other advanced cognitive functions. Many kinds of locomotive and cognitive disorders, including Parkinson's disease and Huntington's disease, are caused by lesions in basal nuclei. The basal nucleus module for ASD lacks a caudate nucleus, which incomplete module may be related to ASD.

A large number of studies have indicated that CSTC (cortico-striatal-thalamic-cortical) circuits play an important role in restricted and repetitive behaviors. Caudate nuclei are clearly separated from the other brain regions in the circuits. Imbalance between direct and indirect pathways in circuits has been confirmed as closely related to restricted and repetitive behaviors in ASD patients. The other small community comprises the superior frontal gyrus and medial orbital, which are related to human emotion cognitive and mental activities.

In addition, there is a large community in the brain network for ASD 1, which comprises most regions of the prefrontal lobe and occipital gyrus, occipital gyrus, inferior occipital gyrus, fusiform gyrus etc. The brain network of Controls 1 does not contain such a large community, but is divided into a number of small communities, suggesting that ASD behaviors may be related to prefrontal regions. Some studies have indicated that the prefrontal lobe region exhibits significant functional change in autism. The community structure supports some conclusions in those researches.

There is another significant feature of community structures in patients with autism: inadequate connection between cerebral hemispheres. For example, the left anterior central gyrus and right anterior central gyrus divide into two different communities, while similar cases also occur in some brain regions of frontal lobe and cuneus. The corpus callosum links both sides of the cerebral hemispheres corresponding to the cortex, and whose main function is to integrate the activities of both hemispheres. The corpus callosum also relates to memory, retrieval, attention, awakening, language, auditory and visual information transmission. A lack of connectivity between the cerebral hemispheres may suggest an abnormality in the corpus callosum. Studies have indicated that autism is associated with a lack of connection in corpus callosum.

5 Conclusion

In this paper, a new genetic algorithm, GAcut, is proposed, which automatically determines, without parameters. The brain networks of both ASDs and typical controls possess obvious community structure, while at the whole-brain level, no clear differences between ASDs and typical controls appear when using modularity Q as indicator. By comparing NMI values, however, we found that NMI values between typical controls are significantly higher than between ASDs and typical controls overall. Finally, we clarify major characteristics of the ASD community structure by comparing results.

Acknowledgements. This study was supported by the National Natural Science Foundation of China (Project No. 61375122 and Project No. 61572239), China Postdoctoral Science Foundation (Project No. 2014M551324). Scientific Research Foundation for Advanced Talents of Jiangsu University (Project No. 14JDG040).

References

1. Biswal, B., Yetkin, F.Z., Haughton, V.M., Hyde, J.S.: Functional connectivity in the motor cortex of resting human brain using echo-planar MRI. Magn. Reson. Med. **34**(4), 537–541 (1995)
2. Kiviniemi, V., Kantola, J.H., Jauhiainen, J., Hyvärinen, A., Tervonen, O.: Independent component analysis of nondeterministic fMRI signal sources. Neuroimage **19**(2), 253–260 (2003)
3. Girvan, M., Newman, M.E.: Community structure in social and biological networks. Proc. Natl. Acad. Sci. **99**(12), 7821–7826 (2002)
4. Van Den Heuvel, M., Mandl, R., Hulshoff Pol, H.: Normalized cut group clustering of resting-state fMRI data. PLoS ONE **3**(4), e2001 (2008)
5. Wang, Y., Li, T.Q.: Analysis of whole-brain resting-state fMRI data using hierarchical clustering approach. PLoS ONE **8**(10), e76315 (2013)
6. Newman, M.E.J., Girvan, M.: Finding and evaluating community structure in networks. Phys. Rev. E: Stat. Nonlin. Soft Matter Phys. **69**(2 Pt 2), 026113 (2004)
7. Tzourio-Mazoyer, N., Landeau, B., Papathanassiou, D., Crivello, F., Etard, O., Delcroix, N.: Automated anatomical labeling of activations in SPM using a macroscopic anatomical parcellation of the MNI MRI single-subject brain. Neuroimage **15**(1), 273–289 (2002)
8. Lu, H., Wei, H.: Detection of community structure in networks based on community coefficients. Phys. A Stat. Mech. Appl. **391**(23), 6156–6164 (2012)
9. Tyszka, J.M., Kennedy, D.P., Paul, L.K., Adolphs, R.: Largely typical patterns of resting-state functional connectivity in high-functioning adults with autism. Cereb. Cortex **24**(7), 1894–1905 (2014)

Pin Defect Inspection with X-ray Images

Hsien-Pei Kao[1], Tzu-Chia Tung[1], Hong-Yi Chen[1],
Cheng-Shih Wong[2], and Chiou-Shann Fuh[1,2(✉)]

[1] Department of Computer Science and Information Engineering,
National Taiwan University, Taipei, Taiwan
jack805201@gmail.com, fuh@csie.ntu.edu.tw
[2] Graduate Institute of Biomedical Electronics and Bioinformatics,
National Taiwan University, Taipei, Taiwan

Abstract. A method of Printed Circuit Board (PCB) pin defect inspection is proposed in this paper. First, we input the pin location image. Then, we align images with Circle Hough Transform (CHT) [4]. Finally, we train cascade classifier with adaptive boosting [3] and Local Binary Pattern (LBP) [5], and detect pin defect to reduce false alarm rate and miss detection rate and thus enhance pin production yield rate.

Keywords: X-ray inspection · Pin · Circle Hough Transform · Adaptive boosting · Cascade training

1 Introduction

In industry, factory automation is getting more important; including manufacturing, transporting, packaging, and defect inspection. Defective products lead to bad brand reputation and product recall and repair cost, so defect inspection plays an important role in gaining profits. Our goal is to decrease false alarm and miss detection rates. However, miss detection may cause defective final product and higher repair cost, and thus has higher weight than false alarm rate. Nowadays, Haar-like feature and Local Binary Pattern (LBP) are two main features used at cascade classifier training; we will compare their differences in next section.

We aim to implement the pin defect inspection with X-ray image. Generally, many factors affect this issue, including X-ray intensity, the distance between X-ray tube and board, the angle between X-ray tube and camera, the intensity of light source, and so on. These factors affect the quality of X-ray images.

In this paper, we will focus on defect inspection by using adaptive boosting with Local Binary Pattern (LBP) feature. Our approach consists of two parts, analysis and detection.

Since pins are physically on Printed Circuit Board (PCB), in different images, pins may not locate at the same location and have different values in x, y axes. Different light source intensities also generate different pin image intensities. This may cause displacement in X-ray image and also affect the image quality. Thus alignment is the first step to carry out, helping to align pin positions.

© Springer International Publishing AG 2017
F. Cong et al. (Eds.): ISNN 2017, Part II, LNCS 10262, pp. 465–473, 2017.
DOI: 10.1007/978-3-319-59081-3_54

Pin X-ray images have many types, including different sizes and directions, and also background pixels. Thus, training has to consider different pin locations, hole sizes, and image intensities to improve performance. Therefore, we adopt Circle Hough Transform (CHT) to clip background pixels and resize to retain only hole and pin region to align our input images.

In defect inspection, we use adaptive boosting to collect a group of inferior classifiers. Each classifier's hit rate should be above 50%, so it is better than random guessing. The higher the hit rate, the more likely it will be chosen.

At last, we aim to improve final defect inspection performance. There are many methods to achieve lower false alarm rate, but we aim to lower miss detection rate due to final defective product and high repair cost. In this problem, we use intersection of two cascades (each cascade with 20 stages) to decrease miss detection rate, because it takes too much time to group two times of inferior classifiers into one cascade (with 40 stages which takes roughly 30 times more computation than 20 stages).

Some important techniques used in our approach will be introduced in Sect. 2. The detailed procedure of our approach will be described in Sect. 3. Our experimental results will be demonstrated in Sect. 4. Section 5 concludes this paper, and the final section is the list of references.

2 Background

2.1 Circle Hough Transform

The Circle Hough Transform (CHT) [4] is a prime algorithm used in Digital Image Processing to find circular objects in a digital image, and also a feature extraction technique to detect circles. CHT is a special form of Hough Transform. CHT aims to find circles in imperfect image inputs. The circle candidates are generated by "voting" in the Hough parameter space and then select the local maxima in an accumulator matrix.

In a two-dimensional space, a circle can be described by:

$$(x - a)^2 + (y - b)^2 = r^2$$

where a and b is the center point of the circle, and radius is represented by r. The parameter space would be three-dimensional that consists of a, b, and r. Moreover, all the parameters that satisfy (x, y) would lie on the surface of an inverted right-angled cone whose apex is at $(x, y, 0)$. In the 3D space, the circle parameters can be identified by the intersection of many tapered appearances which are defined by points on the 2D circle. This process can separate into 2 main stages. One stage is to find the optimal center of circles in a 2D parameter space by fixing radius. Another stage is to find the optimal radius in a one dimensional parameter space (Fig. 1).

An accumulator matrix is introduced to detect the cross point in the parameter space in practice. First, we need to partition the parameter space into "buckets" using a grid and generate an accumulator matrix according to the grid. The element in the accumulator matrix denotes the number of "circles" in the parameter space passing through the corresponding grid cell in the parameter space. The number is also called "voting

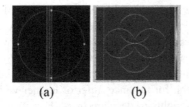

(a)   (b)

Fig. 1. Four points on a circle in (a) the original image [4]. The Circle Hough Transform's result is shown in the right side (b). The intersection of all the circles shows the center of the circle in (a).

number". At first, all elements in the matrix default to zero. For each "edge" point in the original space, we can formulate a circle in the parameter space and increase the voting number of the grid cell where the circle passes through. This process is called "voting". After voting, we can find local maxima in the accumulator matrix. The positions of the local maxima correspond to the circle centers in the original space.

2.2 Adaptive Boosting [3]

Adaptive Boosting is a machine learning meta-algorithm. Many other types of learning algorithms can use it to improve and enhance their capability. According to these learning algorithms, we can combine them into a weighted sum, which stands for the final result of boosted classifier.

Adaptive Boost is sensitive to outliers and noisy data. In some problems, it can be less susceptible to the overfitting problem than other learning algorithms. The single learners can be inferior, but as long as correct rate of each learner is briefly better than random guessing (e.g., correct rate of random guessing is 0.5, and correct rate of single learner is above 0.5 for binary classification), we can prove a superior learner be converged by previous inferior learners.

Adaptive Boost implies a specific method of boosted classifier training. A boost classifier is a classifier in the form

$$F_r(x) = \sum_{t=1}^{T} f_t(x)$$

where each f_t is an inferior learner that takes an object x as input and returns a binary value showing the category of the input object. Similarly, the Tth classifier will be correct if the sample has faith in being correct class and wrong otherwise.

For each case in the training set, each inferior learner generates an output, hypothesis $h(x_i)$. An inferior learner is chosen and given a coefficient α_t such that the sum training error E_t of the resulting t-stage boost classifier is minimized at each iteration t.

$$E_t = \sum_i E[F_{t-1}(x_i) + \alpha_t h(x_i)]$$

where F_{t-1} is the boosted classifier that has been composed of the previous classifier of training, $E(F)$ is a function that count error rate and $f_t(x) = \alpha_t h(x)$ is the inferior learner that is being believed for addition to final output. A weight ω_t is given to each sample in the training set according to the current error $E(F_{t-1}(x_i))$ on that sample at each iteration of the training process. We can use these weights to know the training property of inferior learners, for instance, inferior learner with high weights will split more branches than low weights in decision trees (Fig. 2).

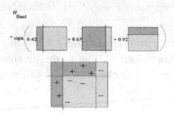

Fig. 2. The schematic diagram of adaptive boosting with three inferior learner [2].

2.3   Visual Descriptor [6]

Visual descriptors or image descriptors, in computer vision, are descriptions of the visual features of the contents in videos, applications, images or algorithms that generate these descriptions. They depict elementary characteristics such as texture, shape, motion or color, and so on.

There are two main groups in Visual Descriptor: General information descriptors: they contain low-level descriptors which give a description about the texture, the shape, the motion or the color, and so on. Specific domain information descriptors: they give information about objects and events in the scene.

2.3.1   Local Binary Pattern (LBP) [5]

Local Binary Patterns (LBP) is a type of visual descriptors used for classification in computer vision. LBP is the particular case of the Texture Spectrum model proposed in 1990.

In its purest way, LBP feature vector is produced by the following pattern: Partition the examined window into cells (e.g. 9×9 pixels for each cell). For each pixel in a cell, compare the center pixel with each of its 8 neighbors. Along the direction of the edge of the circle (i.e. clockwise or counter-clockwise). If the center pixel's value is less than the neighbor's value, assign "1". Else, assign "0" (Fig. 3).

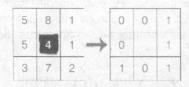

Fig. 3. Follow the pixels along a circle, and give value after comparing each pixel with center pixel [1].

This produces an 8-bit binary number (it is convenient to convert to decimal value). Calculate the histogram of the occurrence of each "number" occurring. This histogram can be seen as a 256-dimensional feature vector, which shows some feature points in the examined window (Fig. 4).

Fig. 4. The result of Lena after running our own Local Binary Pattern [5].

3   Methodology

The flow of our proposed algorithm is as follows (Fig. 5):

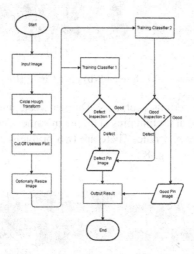

Fig. 5. The flow of our proposed algorithm.

In our approach, our input images have several types. The target we detect may be anywhere and any direction in input image (Fig. 6).

We have to clip some useless region in order to enhance our performance. How do we know where is important and where is ignorable in our original image. Since the hole of the pin is circular, we use Circle Hough Transform and set a particular range to find this special circular hole.

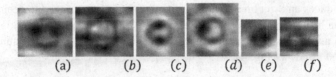

Fig. 6. Different types of input images. (a) 82 × 70 pixels good image. (b) 82 × 69 pixels defect image. (c) 66 × 66 pixels good image. (d) 67 × 67 pixels defect image. (e) 52 × 52 pixels good image. (f) 52 × 52 pixels defect image.

Our goal is to rotate image to make direction uniform. Now we use Microsoft Office Picture Manager to manually rotate 66 × 66 pixels good and defect images counterclockwise 90° to make both pins horizontal (Fig. 7).

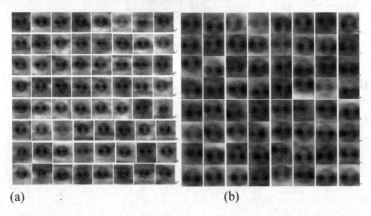

Fig. 7. The result of pin images (a) before and (b) after Circle Hough Transform and clipping.

In this step, we face serious problems. Due to weak intensity of X-ray images, Circle Hough Transform cannot find pins and hole in some images. We have about 200 images failing to find pins and hole. There are two problems: no pins and hole found and incorrect hole position (Fig. 8).

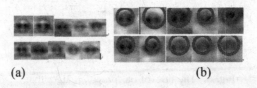

Fig. 8. Circle Hough Transform result. Blue circle shows the Circle Hough Transform output of detected hole and good pin center and radius. Circle Hough Transform cannot detect hole and good pin for defect image thus no center and radius output. (a) Part of 130 images no pins and hole found: 10 images. (b) Part of 60 images with incorrect hole position: 10 images. (Color figure online)

After Circle Hough Transform, there may be several different sizes of image. We can optionally resize images.

Then, it is time to train classifier. At each stage, we choose some adjacent rectangles from pool, and select the highest correct rate from good and defect images.

$$F_r(x) = \sum_{t=1}^{T} f_t(x)$$

Because next stage's learner is related to previous learner, and we expect the increasing of correct rate when our stage gets larger. Under this condition, an inequality is applied as follows

$$F_r(x) \geq F_{r-1}(x)$$

to make sure our correct rate iterates higher (Fig. 9).

Experimental Environment	
1.	OS: Windows 10 Enterprise
2.	CPU: 3.1GHz Intel Core i5
3.	Memory: 8 GB
4.	Platform: Visual Studio 2012 with OpenCV 3.0.0

Step Name	Time Cost
Circle Hough Transform	0.1 second / 1200 images
Cascade Classifier 1 Training Time	52 minutes
Cascade Classifier 2 Training Time	37 minutes
Intersection of 2 Cascade Classifiers	0.001 second
Defect Inspection	0.2 second / 240 images

Fig. 9. Time table of each step.

In Sect. 1, we train 2 cascade classifiers (each with 20 stages) and take intersection to decrease miss detection rate (Fig. 10).

	False Alarm Rate	Miss Detection Rate
Classifier 1	0%	2.5%
Classifier 2	0%	4.1%

Fig. 10. Two cascade classifier miss detection rates and false alarm rates.

Now, we take intersection of two cascade classifiers to enhance our performance (Fig. 11).

At last, we show images mis-classified as good, but it is physically defect. Thus manufacturer knows what kind of images will be identified as a good image even they are defective.

False Alarm Rate	Miss Detection Rate
0%	0.8%

Fig. 11. Miss detection rate and false alarm rate of intersection of two cascade classifiers (originally each with 20 stages).

4   Experimental Result

We have 1430 good images and 1430 defect images. Their base solutions are between 44×44 pixels and 60×60pixels at 8-bit/pixel images.

We divide into two parts, training data and test data (training data have 1190 images and test data have 120 images for both good images and defect images, respectively) (Fig. 12).

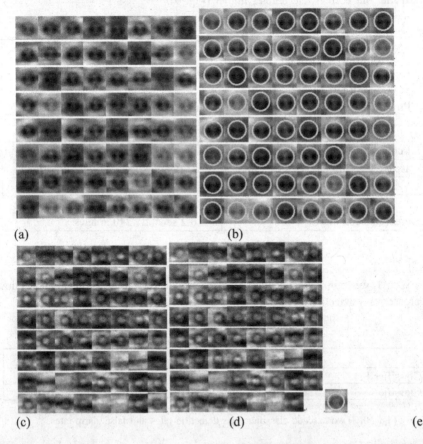

(a)   (b)

(c)   (d)   (e)

Fig. 12. Our final intersection of two cascade classifiers output. Green circle shows the classifier output of detected hole and good pin center and radius. Cascade classifier cannot detect hole and good pin for defect image thus no center and radius output. (a) Part of 120 good input images: 64 images. (b) Part of 120 correctly classified as good images: 64 images. 0 good image mis-classified as defect. (c) Part of 120 defect input images: 64 images. (d) Part of 119 correctly classified as defect images: 63 images. (e) 1 defect images mis-classified as good images (0.8% = 1/120). (Color figure online)

5 Conclusion

Defective products lead to bad brand reputation and product recall and repair cost. Miss detection rate is a critical element for the successful automation of manufacture. The pin defect inspection is a challenging problem due to the complex background. In our work, we propose a method for this purpose based on Circle Hough Transform to clip and resize to enhance images, and cascade classifier training and cascade classifier intersection to detect pin defect. Our experiment result shows that the proposed method can reduce miss detection rate to 0.8% for pin defect inspection.

Acknowledgement. This research was supported by the Ministry of Science and Technology of Taiwan, R.O.C., under grants MOST 103-2221-e-002-188 and 104-2221-e-002-133-my2, and by Test Research (TRI), Liteon, Egistec, Delta Electronics, and Lumens Digital Optics.

References

1. Rosebrock, A.: Local Binary Patterns with Python & OpenCV (2016). http://www.pyimagesearch.com/2015/12/07/local-binary-patterns-with-python-opencv/
2. Read01.com: Machine Learning (2016). https://read01.com/Dngkd.html
3. Wikipedia: AdaBoost (2016). https://en.wikipedia.org/wiki/AdaBoost
4. Wikipedia: Circle Hough Transform (2016). https://en.wikipedia.org/wiki/Circle_Hough_Transform
5. Wikipedia: Local Binary Pattern (2016). https://en.wikipedia.org/wiki/Local_binary_patterns
6. Wikipedia: Visual Descriptor (2016). https://en.wikipedia.org/wiki/Visual_descriptor

A Study on the Effects of Lesions on CA3b in Hippocampus

Babak Keshavarz-Hedayati$^{(\boxtimes)}$, Nikitas Dimopoulos, and Arif Babul

University of Victoria, Victoria V8P 5C2, Canada
{babak,nikitas}@ece.uvic.ca, babul@uvic.ca

Abstract. Hippocampus is a region in the mammalian brain largely responsible for short term memory and spatial navigation. In this paper we present a model to simulate a part of hippocampus called CA3b. We show that our model is capable of producing waveforms in Theta rhythm range. Then, we introduce two types of lesions in the structure and analyze their effect on the collective behavior of neurons in CA3b.

Keywords: Spiking neural networks · Theta · Oscillations · Lesion · CA3 · Hippocampus

1 Introduction

The human brain has always been a most fascinating structure in which, billions of neurons, in conjunction with each other, perform highly complex tasks. Each region in the brain, while communicating with the other regions, is in charge of specific tasks. Hippocampus is one of these areas that is responsible for memory and spatial navigation and its structure has been subject to extensive analyses [1,5,21]. Several of the brain's most prominent rhythms such as gamma and theta can be induced and/or detected in this region [9,15]. For many years, the medial septum was widely believed to be the source of the Theta rhythm generation [4] in the brain. However, recently it was shown that an intact Hippocampus without any external connections is capable of generating a Theta rhythm in vitro [9]. This capability of Hippocampus in generating Theta rhythm independent of the medial septum is studied and shown in other literature as well [5,14]. Our simulation experiments have shown that a "Hippocampal" structure is capable of exhibiting Theta rhythm behavior. Whether the Hippocampus is solely responsible of generating this rhythm or the Theta rhythm is a result of interactions between several regions in the brain [16], is out of the scope of this study. The focus of the work reported here is to study the dynamics of the Theta rhythm as an essential part of the inner workings of the Hippocampus [18].

Hippocampus is divided into 5 main sections: CA1, CA3, Dentate Gyrus, Subiculum and Entorhinal Cortex [1]. Each of these regions posses unique structures which enables them to carry out different tasks compared to the other regions. In the study of rhythms of hippocampus, CA3 with the biggest network of recurrent connections, is of high interest. CA3 itself, comprises 3 sections:

© Springer International Publishing AG 2017
F. Cong et al. (Eds.): ISNN 2017, Part II, LNCS 10262, pp. 474–482, 2017.
DOI: 10.1007/978-3-319-59081-3_55

CA3a, CA3b and CA3c amongst which, CA3b has the highest percentage of recurrent connections [1].

The structure of the brain and its properties have always been the center of attention in many studies. However, because of the limitations of in vivo studies, simulation of the neuronal structures has been a popular alternative. In one of the earliest simulations of the nervous system [6], the stability of the cerebellum, as an essential part in keeping balance in mammals, has been analyzed. Since then, many studies have examined various parts of the nervous system to understand their structural properties [5, 11, 19].

One of the main challenges in simulation of the nervous system is the computational power available. There are billions of neuron cells in the human brain and simulation of such a large structure is a very difficult task. That being said, in recent years with the advances in the architecture of processors, simulation of large-scale neuronal networks has become possible.

From the earliest simulations of Hippocampus [7, 21] to the stability analysis of cerebelum [6] and from the inner dynamics of the working memory [19] to the effects of plasticity in recurrent neuronal networks [23], the main aim is to have a better understanding of the behavior of the nervous system.

Physiological studies have shown that damage to the neuronal structures in human brain, visibly alters the behavior of these structures [2]. Thus, our motivation in simulating the CA3 structure in Hippocampus is to study the effects of lesions on the collective behavior of neurons in CA3 [17]. More specifically, our main interest is to determine the functional resilience of CA3 to lesions. We perform this inspection by analyzing the frequency of the theta waveform as a measure of the functionality of CA3. Theta waveform is a prominent rhythm in Hippocampus; it has been the subject of many studies since its discovery in 1938 [13] and it is believed to be responsible for learning and spatial navigation in mammalian brains [1]. Therefore, in our studies, we focus on the frequency of the Theta rhythm and its variations in response to changes in the (hippocampal) structure.

The rhythmic behavior of CA3 structure using Izhikevich's model was documented in our previous study [12]. However, our simulations did not succeed in demonstrating the presence of theta rhythm (a strong frequency component in the range of 4–10 Hz). For the experiments reported in this study, we replaced the Izhikevich's neuron model with the more physiologically accurate model of Hodgkin Huxley's. In addition, we added spontaneous firing of neurons to all of the neuron cells throughout the structure [3]. In this paper we will continue our study of the behavior of large groups of neuronal networks in Hippocampus examining the impact of the more physiologically relevant Hodgkin-Huxley model in the generation of rhythmic behavior. The overall structure of the model of CA3b region is the one developed in our previous work [12]. In addition, we will investigate the appearance and the dynamics of the rhythmicity as it depends on the strengths (weights) of the interneurons and the presence and distribution of lesions in the structure.

As we shall show the structure is capable of generating rhythmic behavior in the Theta wave region (4–10 Hz) and that this behavior is rather stable for a large variability of interconnection strengths. Furthermore, we shall show that the introduction of lesions starts affecting this behavior when the said lesions become prominent.

In the following section, we will summarize the model of CA3b which was presented in [12]. Then, in Sect. 3, we will present the sequence of simulation experiments we performed using the developed model. Finally, we will summarize our results in Sect. 4

2   The Model of CA3b

In our simulations of CA3b, we utilize 3 distinct classes of neurons: Pyramidal, Basket and Axo-axonic cells. Pyramidal cells are the only excitatory class in the structure and are the main driving force in CA3b. Pyramidal cells outnumber the Basket and Axo-axonic cells by a large margin. Basket and axo-axonic cells are the two inhibitory classes of neurons in the structure and their main responsibility seems to be to control and modulate the firings of Pyramidal cells [12]. Figure 1 shows the structure of the connectivity in CA3b. The recurrent connections between the pyramidal cells, which is a signature of CA3b, and the relationships between the three classes of neurons are shown in Fig. 1.

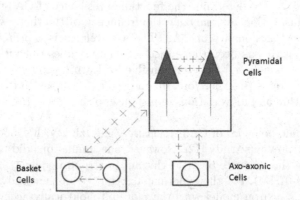

Fig. 1. The connectivity of the various classes of neurons in CA3b. Lines marked with "+" ("−") represent excitation (inhibition) [12].

An in-depth description of the model is presented in [12]. NEST [8] v2.2 was our neuronal simulator of choice to carry out the experiments. In our simulations, we used the Hodgkin Huxley based neuron model modified with Traub channel dynamics [21] which is a most biologically plausible model for the behavior of a neuron. This model describes the behavior of potential of the axon membrane using the three ionic flows in and out of the membrane and an external excitation:

$$C\frac{dv}{dt} = -g_{n0}m^3h(v - V_{Na}) - g_kn^4(v - V_K) - g_r(v - V_r) + I_{ap} \qquad (1)$$

$$\frac{dm}{dt} = \alpha_m(v)(1 - m) - \beta_m(v)m \qquad (2)$$

$$\frac{dn}{dt} = \alpha_n(v)(1 - n) - \beta_n(v)n \qquad (3)$$

$$\frac{dh}{dt} = \alpha_h(v)(1 - h) - \beta_h(v)h \qquad (4)$$

where n is the probability of activation of the potassium channel, m and h are the probability of activation and deactivation of sodium channels, V_K, V_{Na} and V_r are the current equilibrium potentials, C is the capacitance of the membrane, g_{n0}, g_k and g_r are the conductances for each of the three ionic currents, α and β are transition rates between open and closed states of the ionic channels for each flow based on the membrane voltage.

In our simulations, we used the default values of the neuron model as described in [21] and implemented in NEST [8] with a few alterations: we used the stimulus current to tune the firing behavior of our neurons [20]. For the pyramidal cells, the stimulus current (I_e) was tuned to -31.81 pA so the firings of an isolated neuron were limited to 0.2 Hz. This was to ensure that without an external incoming signal, the pyramidal cells fire at the lowest possible frequency. For interneurons, we set this stimulus current to 9.6 pA to create a resting potential of -57 mv [10] when the neuron is not receiving any other input. In addition, for the interneurons, we set the conductances of sodium and potassium channels to 10% of their respective values for the pyramidal cells as suggested in [22].

The initial voltages of all of the neurons in the network were initialized to a random value from a uniform interval of -60 to -35 mv. To simulate the effects of external input and spontaneous firing of neurons themselves, all the excitatory and inhibitory cells randomly fire following a Poisson distribution with a mean of 2.5 times per second [3]. The parameters used in the model are presented in Table 1.

3 Experiments

In our experiments, we were most interested in demonstrating that CA3b is capable of creating a low frequency rhythm in the Theta frequency band. In addition to that, we were interested in investigating how this rhythm changes with the growth of lesions. We define a lesion as an area where the neurons contained within are "dead". However, in our experiments, we consider that axons of other neurons traversing a lesion are unaffected and continue operating normally. The behavior of the structure is observed for 10000 ms to ensure stable behavior of the system. The time a spike is generated by any neuron is recorded during this simulation interval. To exclude the warm-up of the simulations and the early transient effects, the first 2000 ms of each simulation was excluded from the analysis.

Table 1. The parameters of the CA3b model

	Pyramidal	Basket	Axo-axonic
Number of layers	5	1	1
Dimension of the layer (neurons)	333 × 27	107 × 7	85 × 3
Dimension of the layer (μm)	4995 × 405	1605 × 105	1275 × 45
Number of cells in the model	44955	749	255
Number of source cells a pyramidal receives input from	3000	25	7
Number of source cells a basket receives input from	1500	60	0
Number of source cells an AACreceives input from	1500	0	0
Maximum length of the axon (mm)	3	0.825	0.825
Minimum weight of the connection	0.005	−25	−25
Maximum weight of the connection	0.4	−2000	−2000
Connection probability function	$0.54e^{-\frac{d}{20}}$	$0.54e^{-\frac{d}{20}}$	$0.54e^{-\frac{d}{20}}$
Synaptic delay (ms)	1	0.1	0.1
Speed of signal down the axon (m/s)	0.5 [21]	0.2 [22]	0.2

3.1   The Base Model

The firing pattern of the neurons for the model explained above, and a closer look at the progression of these firings in space and time, are shown in Fig. 2(a) and (b) respectively. The horizontal and vertical axes are time in ms and indices of pyramidal neurons in the network respectively. The neurons in adjacent locations in different layers appear next to each other. The frequency spectrum of the firings (DC component is filtered out) is shown in Fig. 2(c). The fundamental frequency of the model in this experiment is 4.062 Hz.

It should be noted that the behavior of the structure is consistent over wide ranges of variability of weights and spread of connections. Even for 50% increase or decrease in the interconnection weights or in the decay rate of the exponential probability functions (used to determine the connections between the neurons), there are no significant changes in the fundamental frequency of the structure. As it can be seen from Fig. 3, the fundamental frequency of the system does not show any significant sensitivity to the changes of the interconnection weights. The subsequent simulation experiments were designed to analyze the effects of lesions on the rhythmic behavior of the structure.

3.2   Lesions: Increase in the Number and/or the Size of the Lesions

We define a lesion as a region in which all of the neurons are dead and do not respond to input or produce any output; However, the traversing axons (that

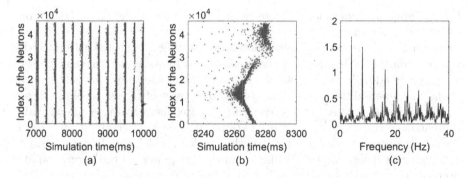

Fig. 2. Firing patterns of pyramidal cells (a), a close up of progression of the firings in the structure (b) and their frequency spectrum (c)

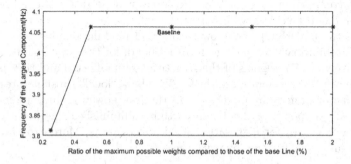

Fig. 3. Sensitivity of the fundamental frequency to the maximum possible weights of the connections

do not emanate from the dead neurons) are not affected. In order to analyze the effects of these lesions on the collective behavior of neurons, we consider two scenarios. In the first scenario, the radius of all of the lesions is constant and similar to each other; however, the number of these lesions varies. In the second scenario, although the radius of all of the lesions change, the number of lesions is constant. In the first scenario, we used a lesion radius of 5 neurons (each lesion includes 69, 9 and 3 pyramidal, basket and AAC neurons respectively) and for the second scenario we set the number of lesions to 10. In both scenarios, we distributed the lesions randomly throughout our structure.

As previously explained, we collected the time series of the occurring spikes and obtained their frequency spectrum. Additionally, we calculated the total power and the ratios of the power of the fundamental frequency with respect to the total power. The results are shown in Fig. 4. The horizontal axes show the ratio of the "dead" neurons to the total population, while the vertical axes represent the fundamental frequency in each experiment. The size of each marker represents the normalized ratio of the power of the fundamental frequency to the total power of the signal including the DC component of the signal. Figure 4a shows experiments where the number of constant-radius lesions increases while

Fig. 4b, presents experiments where, while the number of lesions remains constant, their sizes increase.

From these results, our main observation is that as the lesions grow (the neurons "die out"), the value of the fundamental frequency of the signal slowly increases. In addition to that, we also observe that as the lesions grow, the power of the fundamental frequency compared to the power of the signal diminishes.

For the small lesions (the first scenario), the frequency increases more rapidly compared to the lesions (the second scenario). However, at about 50% dead, the power decreases and no theta rhythm is detectable. For the large lesions, although the Theta rhythm is maintained, the power of the signal rapidly decreases.

Increase in the number of the small lesions seems to have more effect on the behavior of the structure compared to the increase in the radius of the large lesions (in comparisons, we consider the equal or close ratios of dead neurons). The reason might be that a wide spread of small size lesions are more powerful in destroying the underlying synchronizations and modulations between the excitatory and inhibitory neurons. On the other hand, for the large lesions, we think that relatively sizable regions of the structure are unaffected and they maintain the overall working mechanism of the CA3b. These "local" substructures may be the main factor in maintaining the Theta rhythm; however, since fewer neurons are involved, the power of the Theta rhythm is diminished significantly as it is observed in Fig. 4(b) for high ratios of "dead" neurons. More experiments are needed to fully qualify these observations.

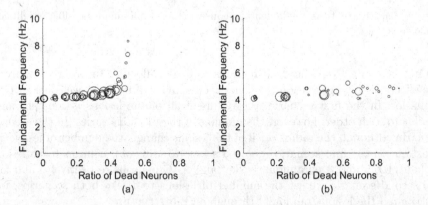

Fig. 4. The fundamental frequency versus the ratio of the dead neurons in the population: increase in the number of lesions with a constant radius of 5 units (left) and increase in the radius of 10 lesions are shown (right)

4   Conclusion and Future Work

In this paper we presented a model for Hippocampus and we showed that it was capable of producing rhythms in the Theta frequency range. We showed that this model is able to produce waveforms in the theta rhythm frequency range. After that, we analyzed the effects of two types of lesions on the collective behavior of neurons in CA3b. We demonstrated that, regardless of the type of the lesion, an increase in the fundamental frequency of the rhythms in the structure can be observed as the size of the lesions grows. In one case, the Theta rhythm is not exhibited when a large number of small lesions exist. When a small number of large lesions exist, the Theta rhythm is manifested albeit a markedly diminished power as the lesions grow, until at a limit, it also stops.

In our subsequent studies, we plan to conduct new experiments to study the effect of the size and the number of lesions on the Theta rhythm. In addition, we plan to incorporate additional neural classes to create a more biologically plausible model of CA3b.

Acknowledgment. Computational support was provided by Compute Canada through WestGrid and the HPC Facility at the University of Victoria. This work was supported in part by The Natural Sciences and Engineering Research Council of Canada (NSERC) and by the Lansdowne Chair in Computer Engineering at the University of Victoria. We would like to thank the nest-initiative for their support in usage of NEST, and Dr. Naznin Virji-Babul for her many insights.

References

1. Andersen, P.: The Hippocampus Book. Oxford Neuroscience Series. Oxford University Press, Oxford (2007)
2. Borich, M., Babul, A.N., Hsiang, Y.P., Boyd, L., Virji-Babul, N.: Alterations in resting-state brain networks in concussed adolescent athletes. J. Neurotrauma **32**(4), 265–271 (2015)
3. Cohen, I., Miles, R.: Contributions of intrinsic and synaptic activities to the generation of neuronal discharges in in vitro hippocampus. J. Physiol. **524**(2), 485–502 (2000)
4. Colgin, L.L.: Mechanisms and functions of theta rhythms. Annu. Rev. Neurosci. **36**(1), 295–312 (2013)
5. Cutsuridis, V., Graham, B., Cobb, S., Vida, I.: Hippocampal Microcircuits: A Computational Modeler's Resource Book. Springer Series in Computational Neuroscience. Springer, Heidelberg (2010)
6. Dimopoulos, N.: A study of the asymptotic behavior of neural networks. IEEE Trans. Circ. Syst. **36**(5), 687–694 (1989)
7. Dimopoulos, N.J., Newcomb, R., Smith, M.: Some observations on oscillations in neural type networks. In: Proceedings of 33rd Annual Conference on Engineering in Medicine and Biology, p. 64 (1980)
8. Gewaltig, M.O., Diesmann, M.: Nest (neural simulation tool). Scholarpedia **2**(4), 1430 (2007)
9. Goutagny, R., Jackson, J., Williams, S.: Self-generated theta oscillations in the hippocampus. Nat. Neurosci. **12**(12), 1491–1493 (2009)
10. Hardie, J.B., Pearce, R.A.: Active and passive membrane properties and intrinsic kinetics shape synaptic inhibition in hippocampal CA1 pyramidal neurons. J. Neurosci. **26**(33), 8559–8569 (2006)

11. Kadmon, J., Sompolinsky, H.: Transition to chaos in random neuronal networks. Phys. Rev. X **5**, 041030 (2015)
12. Keshavarz-Hedayati, B., Dimopoulos, N.J., Babul, A.: An analysis of dynamics of CA3b in hippocampus. In: 2014 International Conference on Embedded Computer Systems: Architectures, Modeling, and Simulation (SAMOS XIV), pp. 375–383 July 2014
13. Konopacki, J.: Theta-like activity in the limbic cortex in vitro. Neurosci. Biobehav. Rev. **22**(2), 311–323 (1998)
14. Leao, R.N., Mikulovic, S., Leao, K.E., Munguba, H., Gezelius, H., Enjin, A., Patra, K., Eriksson, A., Loew, L.M., Tort, A.B.L., Kullander, K.: OLM interneurons differentially modulate CA3 and entorhinal inputs to hippocampal CA1 neurons. Nat. Neurosci. **15**(11), 1524–1530 (2012)
15. Leung, L.S.: Generation of theta and gamma rhythms in the hippocampus. Neurosci. Biobehav. Rev. **22**(2), 275–290 (1998)
16. Pignatelli, M., Beyeler, A., Leinekugel, X.: Neural circuits underlying the generation of theta oscillations. J. Physiol.-Paris **106**(34), 81–92 (2012)
17. Rennó-Costa, C., Lisman, J.E., Verschure, P.F.: A signature of attractor dynamics in the CA3 region of the hippocampus. PLoS Comput. Biol. 10, e1003641 (2014)
18. Roux, F., Uhlhaas, P.J.: Working memory and neural oscillations: alpha-gamma versus theta-gamma codes for distinct WM information? Trends Cogn. Sci. **18**(1), 16–25 (2014)
19. Szatmáry, B., Izhikevich, E.M.: Spike-timing theory of working memory. PLoS Comput. Biol. **6**(8), 1–11 (2010)
20. Traub, R.D., Wong, R.K., Miles, R., Michelson, H.: A model of a CA3 hippocampal pyramidal neuron incorporating voltage-clamp data on intrinsic conductances. J. Neurophysiol. **66**(2), 635–650 (1991)
21. Traub, R., Miles, R.: Neuronal Networks of the Hippocampus, vol. 777. Cambridge University Press, Cambridge (1991)
22. Traub, R.D., Bibbig, A., Fisahn, A., LeBeau, F.E.N., Whittington, M.A., Buhl, E.H.: A model of gamma-frequency network oscillations induced in the rat CA3 region by carbachol in vitro. Eur. J. Neurosci. **12**(11), 4093–4106 (2000)
23. Zenke, F., Agnes, E.J., Gerstner, W.: Diverse synaptic plasticity mechanisms orchestrated to form and retrieve memories in spiking neural networks. Nat. Commun. **6** (2015). Article no. 6922

Enhancement of Neuronal Activity by GABAb Receptor-Mediated Gliotransmission

Taira Kobayashi[1]([✉]), Asahi Ishiyama[1], and Osamu Hoshino[1,2]

[1] Department of Intelligent Systems Engineering, Ibaraki University,
4-12-1 Nakanarusawa, Hitachi, Ibaraki 316-8511, Japan
tairakobayashi.bip@gmail.com,
{15nm902s,osamu.hoshino.507}@vc.ibaraki.ac.jp
[2] Southern Tohoku Research Institute for Neuroscience,
Southern Tohoku General Hospital, 7-115, Yatsuyamada, Koriyama,
Fukushima 963-8563, Japan

Abstract. Transporters, embedded in glial plasma membranes, can regulate ambient GABA levels. We proposed here a neural network model with a GABAergic gliotransmission mechanism, and simulated the model to investigate how the type of GABA receptors on glia affects neuronal information processing. Although the synaptic interneuron-glia signaling via GABAb receptors slows the modulation of ambient GABA levels compared to that via GABAa receptors, it could increase principal cell activity and accelerate their reaction speed to an applied feature stimulus. Our preliminary simulation result suggests that GABAb receptors, generally expressed by glia (astrocytes), may improve the perceptual performance of the sensory cortex.

Keywords: Ambient GABA · GABAergic gliotransmission · Tonic inhibition · Sensory information processing · GABAb receptor

1 Introduction

Gamma-aminobutyric acid (GABA) mediates phasic inhibition by activating intrasynaptic GABA receptors; i.e., GABA receptors in the synaptic cleft. Tonic inhibition occurs when extracellular GABA activates receptors in membranes outside synapses [1]. GABA molecules in extracellular space are referred as "ambient GABA" and GABA receptors in extrasynaptic membranes as "extrasynaptic GABA receptor". Extrasynaptic GABAa receptors were evidenced in the cerebellum as well in the cortex [2]. A presynaptic action potential triggers a release of GABA (one millimolar level) into the synaptic cleft, while ambient GABA is maintained within a range of submicromolar to several micromolar levels. This is sufficient to activate extrasynaptic but not intrasynaptic GABAa receptors, leading to inhibiting neuronal activity in a tonic manner.

Our previous study [3] demonstrated that GABA transporters, embedded in glial plasma membranes, regulated ambient GABA levels. Synaptic interneuron-glia signaling via GABAa receptors contributed to importing (removing) GABA

© Springer International Publishing AG 2017
F. Cong et al. (Eds.): ISNN 2017, Part II, LNCS 10262, pp. 483–490, 2017.
DOI: 10.1007/978-3-319-59081-3_56

from the extracellular space, thereby enhancing neuronal activity. However, a recent study suggested that GABAb but not GABAa receptors are expressed by astrocytes [4]. Employing GABAb receptors on glia will slow the modulation of ambient GABA levels and affect neural activity. The purpose of this study is to examine whether the type of GABA receptors in glial cell membranes has a significant influence on neuronal information processing.

2 Neural Network Model

The neural network model is shown in Fig. 1 and the transporter model is schematically illustrated in the inset. Dynamic evolution of membrane potential of the ith E cell that belongs to cell assembly n is defined by

$$c_m^E \frac{dv_{n,i}^E(t)}{dt} = -g_m^E(v_{n,i}^E(t) - v_{rest}^E) + I_{n,i}^{EE}(t) + I_{n,i}^{EB}(t)$$
$$+ I_{n,i}^{ext}(t) + I_n^{inp}(t), \tag{1}$$

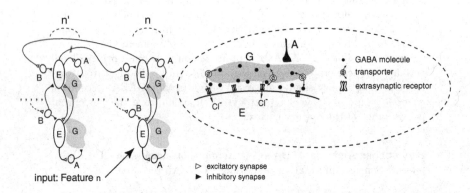

Fig. 1. The neural network model. Each cell assembly ($1 \leq n \leq 8$) comprises excitatory principal cells (E), inhibitory interneurons (A, B), and glial cells (G). The open and filled triangles denote excitatory and inhibitory synapses, respectively. A constant excitatory current is provided to E cells when presented with a feature stimulus as an input; see "Feature n". Inset: A schematic illustration of GABAergic gliotransmission. Transporters on a G cell import or export GABA molecules depending on glial membrane potential. Ambient GABA molecules are accepted by extrasynaptic GABAa receptors and tonically inhibit an E cell.

where $I_{n,i}^{EE}(t)$ is an excitatory synaptic current from other E cells, $I_{n,i}^{EB}(t)$ an inhibitory synaptic current from B cells, $I_{n,i}^{ext}(t)$ an inhibitory nonsynaptic current mediated by ambient GABA via extrasynaptic receptors, and $I_n^{inp}(t)$ an excitatory input current. These currents are defined by

$$I_{n,i}^{EE}(t) = -\hat{g}_{AMPA}(v_{n,i}^E(t) - v_{rev}^{AMPA}) \sum_{j=1(j \neq i)}^{N} w_{n,ij}^{EE}(t) r_{n,j}^E(t), \tag{2}$$

$$I_{n,i}^{EB}(t) = -\hat{g}_{GABA_a}(v_{n,i}^{E}(t) - v_{rev}^{GABA_a}) \sum_{j=1}^{N} w_{n,ij}^{EB} r_{n,j}^{B}(t), \tag{3}$$

$$I_{n,i}^{ext}(t) = -\hat{g}_{GABA_a}(v_{n,i}^{E}(t) - v_{rev}^{GABA_a}) \delta_E r_{n,i}^{ext}(t), \tag{4}$$

$$I_{n}^{inp}(t) = I_{inp}, \tag{5}$$

where δ_E denotes the amount of extrasynaptic GABAa receptors embedded in E cell membrane. Excitatory current I_{inp} is supplied to E cells during a stimulus presentation period. Dynamic evolution of membrane potential of the ith A or B cell that belongs to cell assembly n is defined by

$$c_m^{\alpha} \frac{dv_{n,i}^{\alpha}(t)}{dt} = -g_m^{\alpha}(v_{n,i}^{\alpha}(t) - v_{rest}^{\alpha}) + I_{n,i}^{\alpha E}(t), \qquad (\alpha = A, B) \tag{6}$$

where $I_{n,i}^{\alpha E}(t)$ is an excitatory synaptic current from E cell(s). These currents are defined by

$$I_{n,i}^{AE}(t) = -\hat{g}_{AMPA}(v_{n,i}^{A}(t) - v_{rev}^{AMPA}) w_{n,i}^{AE} r_{n,i}^{E}(t), \tag{7}$$

$$I_{n,i}^{BE}(t) = -\hat{g}_{AMPA}(v_{n,i}^{B}(t) - v_{rev}^{AMPA}) \sum_{n'=1(n' \neq n)}^{M} w_{nn',i}^{BE} r_{n',i}^{E}(t). \tag{8}$$

Dynamic evolution of membrane potential of the ith G cell that belongs to cell assembly n is defined by

$$c_m^{G} \frac{dv_{n,i}^{G}(t)}{dt} = -g_m^{G}(v_{n,i}^{G}(t) - v_{rest}^{G}) + I_{n,i}^{GA}(t), \tag{9}$$

where $I_{n,i}^{GA}(t)$ is an inhibitory synaptic current from an A cell. This current is defined by

$$I_{n,i}^{GA}(t) = -\hat{g}_{GABA_b}(v_{n,i}^{G}(t) - v_{rev}^{GABA_b}) w_{n,i}^{GA} \frac{s_{n,i}^{l}(t)}{s_{n,i}^{l}(t) + K_{ds}}, \tag{10}$$

$$\frac{ds_{n,i}^{l}(t)}{dt} = K_{ac} r_{n,i}^{A}(t) - K_{da} s_{n,i}^{l}(t). \tag{11}$$

In these equations, $r_{n,j}^{E}(t)$ is the fraction of AMPA receptors in the open state triggered by presynaptic action potentials of the jth E cell. $r_{n,j}^{B}(t)$ and $r_{n,j}^{A}(t)$ are the fractions of intrasynaptic GABAa and GABAb receptors in the open state triggered by presynaptic action potentials of the jth B and A cells, respectively. $r_{n,i}^{ext}(t)$ is the fraction of extrasynaptic GABAa receptors, located on the ith E cell, in the open state provoked by ambient GABA. $s_{n,i}^{l}(t)$ is the concentration of activated G-protein with l binding sites and K_{ds} is the dissociation constant of the binding of G-protein on K^+ channels. K_{ac} and K_{da} are G-protein activation and deactivation rates, respectively [5].

AMPA and GABA receptors are described by

$$\frac{dr_{n,j}^E(t)}{dt} = \alpha_{AMPA}[Glu]_{n,j}(t)(1 - r_{n,j}^E(t)) - \beta_{AMPA}r_{n,j}^E(t), \qquad (12)$$

$$\frac{dr_{n,j}^A(t)}{dt} = \alpha_{GABA_b}[GABA]_{n,j}^A(t)(1 - r_{n,j}^A(t)) - \beta_{GABA_b}r_{n,j}^A(t), \qquad (13)$$

$$\frac{dr_{n,j}^B(t)}{dt} = \alpha_{GABA_a}[GABA]_{n,j}^B(t)(1 - r_{n,j}^B(t)) - \beta_{GABA_a}r_{n,j}^B(t), \qquad (14)$$

$$\frac{dr_{n,j}^{ext}(t)}{dt} = \alpha_{GABA_a}[GABA]_{n,j}^{ext}(t)(1 - r_{n,j}^{ext}(t)) - \beta_{GABA_a}r_{n,j}^{ext}(t), \qquad (15)$$

where $[Glu]_{n,j}(t)$, $[GABA]_{n,j}^A(t)$ and $[GABA]_{n,j}^B(t)$ are concentrations of glutamate and GABA in synaptic clefts, respectively. $[Glu]_{n,j}(t) = 1$ mM, $[GABA]_{n,j}^A(t) = 1$ mM, and $[GABA]_{n,j}^B(t) = 1$ mM for 1 ms when the presynaptic jth E, A, and B cells fire, respectively. Otherwise, $[Glu]_{n,j}(t) = 0$ and $[GABA]_{n,j}^A(t) = [GABA]_{n,j}^B(t) = 0$.

Neuronal firing occurs in a probabilistic manner [6], defined by

$$Prob[Y_{n,j}(t); firing] = \frac{1}{1 + e^{-\eta_Y(u_{n,j}^Y(t) - \zeta_Y)}} .(Y = E, A, B) \qquad (16)$$

When a cell fires, its membrane potential is depolarized to -10 mV, which is kept for 1 msec and then reset to the resting potential.

The concentration of ambient GABA around the ith E cell that belongs to cell assembly n is defined by

$$\frac{d[GABA]_{n,i}^{ext}(t)}{dt} = -\gamma_{trn}([GABA]_{n,i}^{ext}(t) - [GABA]_0)$$
$$+ T_G\{[GABA]_{max} - [GABA]_{n,i}^{ext}(t)\}\{[GABA]_{n,i}^{ext}(t) - [GABA]_{min}\}$$
$$\times (v_{n,i}^G(t) - v_{rev}^G), \qquad (17)$$

where γ_{trn} and $[GABA]_0$ are a decay constant and the basal ambient GABA concentration, respectively. T_G determines the modulation rate of ambient GABA concentration. $GABA_{max}$ and $GABA_{min}$ restrict ambient GABA concentration to a maximum and a minimum, respectively. v_{rev}^G is the reversal potential of the GABA transporter. For model parameters and their values, see Table 1 and our previous studies [6–8].

Table 1. List of parameters

Description	Parameter	Value
Membrane capacitance of type K ($K = E, A, B, G$)	c_m^K	$c_m^E = 500, c_m^A = 200,$ $c_m^B = 600, c_m^G = 45 [pF]$
Membrane conductance	g_m^K	$g_m^E = 25, g_m^A = 20,$ $g_m^B = 15, g_m^G = 9 [nS]$
Resting potential	u_{rest}^K	$u_{rest}^E = -65, u_{rest}^A = u_{rest}^B = u_{rest}^G = -70 [mV]$
Maximal conductance for type Z receptors ($Z = AMPA, GABA_a, GABA_b$)	$\hat{g}_Z$	$\hat{g}_{AMPA} = \hat{g}_{GABA_a} = \hat{g}_{GABA_b} = 1.0 [nS]$
Reversal potential	u_{rev}^Z	$u_{rev}^{AMPA} = 0, u_{rev}^{GABA_a} = -80, u_{rev}^{GABA_b} = -95 [mV]$
Number of cell-units within cell assemblies	N	20
Number of cell assemblies	M	8
Synaptic weight from j to i th E cell that belongs to cellasembly n	$\omega_{n,ij}^{E,E}$	2.0
Synaptic weight from j th B to i th E cell	$\omega_{n,ij}^{E,B}$	10.0
Synaptic weight from i th E cell to A cell	$\omega_{n,i}^{A,E}$	40.0
Synaptic weight from i th E cell to B cell between different $(n' \neq n)$ cell assemblies	$\omega_{nn',i}^{B,E}$	35.0
Synaptic weight from i th A to G cell	$\omega_{n,i}^{G,A}$	6.5
Amount of extrasynaptic GABA receptors on E cell	δ_P	8×10^2
Channel opening rate for type Z receptor ($Z = AMPA, GABA_a, GABA_b$)	α_Z	$\alpha_{AMPA} = 1.1 \times 10^6, \alpha_{GABA_a} = 5 \times 10^6, \alpha_{GABA_b} = 90 \times 10^3 [M^{-1}sec^{-1}]$
Channel closing rate	β_Z	$\beta_{AMPA} = 190, \beta_{GABA_a} = 180, \beta_{GABA_b} = 1.2 [sec^{-1}]$
Steepness of sigmoid function for type Y cell ($Y = E, A, B$)	η_Y	$\eta_E = 250, \eta_A = 310, \eta_B = 310$
Threshold of sigmoid function	θ_Y	$\zeta_E = -36, zeta_A = -37, zeta_B = -34 [mV]$
Decay constant for ambient GABA concentration	γ_{trn}	2.5
Basal ambient GABA concentration	$[GABA]_0$	$1 [\mu M]$
Maximal ambient GABA concentration	$[GABA]_{max}$	$3.5 [\mu M]$
Minimal ambient GABA concentration	$[GABA]_{min}$	$0 [\mu M]$
GABA transfer coefficient	T_G	0.7×10^9
Reversal potential of transporter	u_{rev}^G	$-70 [mV]$
Activating rate for G-protain	K_{ac}	$180 [\mu M sec^{-1}]$
Deactivating rate for G-protain	K_{da}	$34 [sec^{-1}]$
Dissociation constant of the binding of G-protain on K^+ channels	K_{ds}	$100 [\mu M^4]$
The number of binding sites	l	4

3   Results

Figure 2A (top) shows how the network responds to a sensory stimulus (Feature 2), whose period is indicated by a horizontal bar. The second and bottom panels present membrane potentials of G cells and ambient GABA concentrations around E cells for respective cell assemblies: $1 \le n \le 4$ (among $1 \le n \le 8$). The E cell responds to the stimulus, evoking a train of spikes (see the top panel: n = 2), which in turn activates the A cell (not shown), hyperpolarizes the G cell (see the second panel: n = 2), reduces the ambient GABA level (see the bottom panel: n = 2), and enhances the stimulus-evoked E cell activity (see the top panel: n = 2).

Figure 2B present those when GABAa instead of GABAb receptors were expressed by G cells. The amount of hyperpolarization in G cell (see the second panel: n = 2) is less than that when the GABAb receptor worked (see the second panel of Fig. 2A: n = 2), resulting in less reduction in ambient GABA levels and thus in weak stimulus-evoked E cell activity. Interestingly, the GABAb receptor accelerates the reaction speed of the network: compare the onsets of stimulus-evoked spike trains in the top panels.

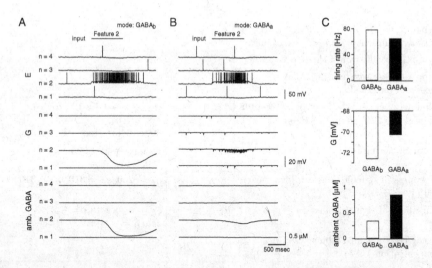

Fig. 2. Response of the network to sensory stimulation. (A) Membrane potentials of E (top), G (middle) cells, and ambient GABA concentrations around E cells (bottom). The GABAb receptor-mediated synaptic A-to-G signaling worked. (B) Those when the GABAa receptor worked. (C) Top: Stimulus-evoked E cell activity. Middle: G cell membrane potential. Bottom: Ambient GABA level. The open and filled bars denote those when the GABAb and GABAa receptor-mediated synaptic A-to-G signaling worked, respectively.

Figure 2C quantitatively shows stimulus-evoked E cell activity (top), G cell membrane potential (middle), and the level of ambient GABA (bottom) for each condition; i.e., when the GABAb (see the open bars) or GABAa (see the filled bars) receptor worked. These results indicate that the increase in stimulus-evoked E cell activity (see the top panel) arises from a decrease in local ambient GABA levels around stimulus-relevant E cells (see the bottom panel), which is triggered by the hyperpolarization of G cells (see the middle panel).

4   Discussion

We proposed a neural network model and simulated the model to investigate how the type of GABA receptors on glia affects neuronal information processing. Synaptic interneuron-glia signaling via GABAb (but not via GABAa) receptors increased stimulus-evoked principal cell activity and accelerated their reaction speed to an applied feature stimulus. Our preliminary simulation result suggests that GABAb receptors, generally expressed by glia (astrocytes), may improve the perceptual performance of the sensory cortex. We will further this study in order to elucidate why the brain employ GABAb (but not GABAa) receptors on glia.

To regulate ambient GABA levels, we employed here the glial transporter model [7]. Various types of gliotransmission have been proposed [9], including release from storage organelles via exocytosis and release from the cytosol via plasma membrane ion channels. For both types, experimental studies suggest that Ca^{2+} is the key regulator. However, less is known about their mechanisms based on which we could construct a GABAergic gliotransmission model. In contrast, the mechanism of GABAergic gliotransmission via transporters can be explained theoretically, which allowed us to construct the glial transporter model.

References

1. Semyanov, A., Walker, M.C., Kullmann, D.M., Silver, R.A.: Tonically active GABA A receptors: modulating gain and maintaining the tone. Trends Neurosci. **27**, 262–269 (2004)
2. Drasbek, K.R., Jensen, K.: THIP, a hypnotic and antinociceptive drug, enhances an extrasynaptic GABAA receptor-mediated conductance in mouse neocortex. Cereb. Cortex **16**, 1134–1141 (2006)
3. Zheng, M., Matsuo, T., Miyamoto, A., Hoshino, O.: Tonically balancing intracortical excitation and inhibition by GABAergic gliotransmission. Neural Comput. **6**, 1690–1716 (2014)
4. Losi, G., Mariotti, L., Carmignoto, G.: GABAergic interneuron to astrocyte signalling: a neglected form of cell communication in the brain. Philos. Trans. R. Soc. Lond. B **369**, 20130609 (2014)
5. Destexhe, A., Mainen, Z.F., Sejnowski, T.J.: Kinetic models of synaptic transmission. In: Koch, C., Segev, I. (eds.) Methods in Neuronal Modeling, pp. 1–25. The MIT Press, Cambridge (1998)
6. Hoshino, O.: GABA Transporter preserving ongoing spontaneous neuronal activity at firing subthreshold. Neural Comput. **21**, 1683–1713 (2009)

7. Hoshino, O.: Regulation of ambient GABA levels by neuron-glia signaling for reliable perception of multisensory events. Neural Comput. **24**, 2964–2993 (2012)
8. Hoshino, O.: Balanced crossmodal excitation and inhibition essential for maximizing multisensory gain. Neural Comput. **26**, 1362–1385 (2014)
9. Sahlender, D.A., Savtchouk, I., Volterra, A.: What do we know about gliotransmitter release from astrocytes? Philos. Trans. R. Soc. Lond. B Biol. Sci. **369**, 20130592 (2014)

The Feature Extraction Method of EEG Signals Based on Transition Network

Mingmin Liu[1,2], Qingfang Meng[1,2(✉)], Qiang Zhang[3],
Dong Wang[1,2], and Hanyong Zhang[1,2]

[1] School of Information Science and Engineering,
University of Jinan, Jinan 250022, China
ise_mengqf@ujn.edu.cn
[2] Shandong Provincial Key Laboratory of Network
Based Intelligent Computing, Jinan 250022, China
[3] Institute of Jinan Semiconductor Elements Experimentation,
Jinan 250014, China

Abstract. High accuracy of epilepsy EEG automatic detection has important clinical research significance. The combination of nonlinear time series analysis and complex network theory made it possible to analyze time series by the statistical characteristics of complex network. In this paper, based on the transition network the feature extraction method of EEG signals was proposed. Based on the complex network, the epileptic EEG data were transformed into the transition network, and the variance of degree sequence was extracted as the feature to classify the epileptic EEG signals. Experimental results show that the single feature classification based on the extracted feature obtains classification accuracy up to 98.5%, which indicates that the classification accuracy of the single feature based on the transition network was very high.

Keywords: Transition network · Variance of degree sequence · Epilepsy EEG automatic detection

1 Introduction

Epilepsy is a chronic recurrent transient brain dysfunction syndrome. At present, complex epilepsy EEG data is enormous in clinical medicine, and the efficiency of artificial classification is low and the accuracy is not high. Therefore, the automatic detection method of epilepsy EEG has important significance for clinical research.

The Hurst exponent of the epileptic EEG was discussed in [1] and the results shown that the normal EEG was uncorrelated whereas the epileptic EEG was long range anti-correlated. Spectral entropy and embedding entropy, which could be used to measure the system complexities, were introduced to epilepsy detection in [2, 3]. Combined with these classification features, the classifiers, such as artificial neural network (ANN) and support vector machine (SVM), had also been widely applied into the epilepsy detection algorithm [4–9]. From these literature, we can conclude that an excellent classification feature not only obtains better classification accuracy but also spends less computational complexity because of it does not need combined with classifier. These advantages are significant for the clinical application.

F. Cong et al. (Eds.): ISNN 2017, Part II, LNCS 10262, pp. 491–497, 2017.
DOI: 10.1007/978-3-319-59081-3_57

Recently, complex networks theory provided a new perspective for nonlinear time series analysis. Zhang and Small [10] proposed an algorithm that transformed the pseudo-periodic time series into complex networks. A bridge between nonlinear time series analysis and complex networks theory had been built. Lacasa et al. [11] first proposed the visibility graph algorithm, which could convert arbitrary time series into a graph. Time series conversion to complex network algorithm made it possible to the application of complex network theory researching time series. Sun et al. [12] took the Rossler chaotic system as an example to give a concrete algorithm for the conversion of nonlinear time series into transition network. Based on the statistical properties of complex networks, Sun et al. gave a detailed analysis of the different periods of the Rossler system, thus converting the nonlinear time series into transition network to maturity. In the paper [13], an improved method for converting nonlinear time series into transition networks was proposed, and the possibility of transforming any time series into transition networks was proved. In paper [14], the transformation of epileptic EEG to proximity network was proposed. According to the statistical characteristics of complex network, the classification of epileptic EEG could be realized by combining classification method.

In this paper, we improve the method that transform nonlinear time series into transition network mentioned by Sun et al. [12], so the operation rate and classification accuracy are raised. According to the statistical characteristics of complex networks, we extract the variance of degree sequence to classify the epileptic EEG and the classification accuracy up to 98.5%. So we improve the accuracy of the automatic detection of epilepsy EEG.

2   The Feature Extraction Method

2.1   A Method of Constructing Complex Network by Time Series

Sun et al. [12] gave a concrete algorithm for the conversion of nonlinear time series into transition network. The method as follows:

- given a time series $\{x_t\}$, take a window of length L, sliding along this series and denote the windowed segment at time t as $\{X_t\}_t$. The ordinal pattern $\pi_\tau = (\tau_1, \tau_1, \tau_1, \ldots\ldots\tau_L)$ of X_t is defined as the permutation of $(1, 2, 3, \ldots\ldots L)$, satisfying $x_{(t-1)+\tau_1} < x_{(t-1)+\tau_2} < x_{(t-1)+\tau_2} < \cdots x_{(t-1)+\tau_2}$. To better capture more details of the system behavior, add amplitude information. Specifically, a pair of symbols were used to describe the segment in a window. One symbol described the amplitude level and the other was the ordinal pattern. The former symbol was obtained by splitting the range of the time series [a, b] into Q equal regions. Each region was labeled by an index. Then each node was symbolized as $\{S_i = \{\alpha, \pi_\tau\}\}_{i=1}$.
- To investigate the transitions among the different states identified by modification, a weighted and directed network was constructed with fixed Q as follows: S_i represents a node in a complex network, thus we got the node set of complex network; Naturally, built the corresponding connections, the link starting from the node

corresponding to S_i ends at the node corresponding to S_{i+1}. The weight W_{ij} of the link directed from node i to j is given by

$$W_{ij} = \#(S_i \rightarrow S_j) \tag{1}$$

$\#(S_i \rightarrow S_j)$ finally, the adjacency matrix $W_{ij} = (w_{ij})_{M*M}$ was used to represent the generated networks.

However, the efficiency of this method was low and classification accuracy rate was not high. The value of sliding window length L had a great influence on classification accuracy. So, in this paper, this method had been improved. The time series was segmented by the maximum value, and then the nodes were constructed. This method eliminated the influence of L and improved the efficiency of operation and the accuracy of classification.

The improved method as follows: Given a time series (x_{ij}). Then, found all the maximum value in the time series. The time series between every two adjacent maximum value was defined as one sub-segment contained a maximum value, labeled (X_m) and L was the length of each sub-segment. The ordinal pattern of was defined as $\pi_\tau = (\tau_1, \tau_1, \tau_1, \ldots \ldots \tau_L)$ of X_m was defined the permutation of $(1, 2, 3 \ldots \ldots L)$. To distinguish same ordinal pattern. but amplitude differently (while maintaining the useful features of the ordinal representation), we proposed the simple modification of adding amplitude information. We first found the maximum X_{max} of each sub-segment and the minimum X_{min} of this time series, then we used rounded operation formula computing the value of M

$$M = \left\lceil \frac{X_{max} - X_{min}}{Q} \right\rceil \tag{2}$$

Finally each sub-segment was symbolized as $S_i = \{M, \pi_\tau\}_{i=1}^k$. Thus we got the node set of complex network. The building the corresponding connections method employed the above method given by Sun and Small. So we got the improved transition network.

2.2 Feature Extraction

The complex network with adjacency matrix W_{ij} was obtained by the transition networks construction algorithm. The geometric topological structure stored the dynamic characteristic information of the original time series, and the characteristics of the epileptic EEG were extracted by studying the statistical properties of the complex network.

The node degree was defined as the number of connected edges between a certain node and the remaining nodes. According to the adjacency matrix W_{ij}, the node degree of the node i was

$$k_i = \sum_{j=1}^{M} a_{ij} = k_{in}^i + k_{out}^i \tag{3}$$

k_{in} was the number of edge connections for other nodes to connect to the first node, k_{out} was the number of edge connections for the first node to join the other nodes.

The structure of complex networks was different in time series with different dynamic structures, In other words, the degree of complex network was different. The variance of degree sequence was considered as an indicator of the network heterogeneity, and it can be calculated as follows

$$\sigma = M^{-1} \sum_K k^2 M(K) - \langle k \rangle^2 \tag{4}$$

where $\langle k \rangle$ was the average degree, and $M(K)$ was the number of nodes, which had degree k. The smaller the value of σ, the consistency of the degree was better, The epileptic seizure EEG owned lower complexity than that of intermittent EEG, and the chaos was weakened. And consistency of complex network's degree based on epileptic seizure EEG was better. So it could be used as epileptic EEG signal classification.

3   Experiment Results and Analysis

In this study, we use a clinical epileptic EEG data set from the University of Bonn, Germany. The epileptic EEG data file contains 100 ictal EEG data and 100 interictal EEG data. Every EEG datum was sampled at a rate of 128 Hz, and has 4096 points and EEG data such as manual or eye movement disturbances were removed. In the experiment, we set the sample length of 512 and 1024, respectively, to construct a complex network. Further more, we evaluate the performance of EEG feature extraction method and epileptic EEG automatic detection algorithm.

Table 1. Result of the feature automatic detection of transition network

Method	Data length	Data length	ACC
Transition network	512	100	95.5
	512	120	96.5
	1024	100	97.5
	1024	120	98.5

Experiments set the Q values are 100 and 120, the data length of 512 and 1024, respectively, the detailed classification results in Table 1. When the Q value is 120 and the data length is 1024, the classification accuracy is the highest, reaching 98.5%. When the Q value is 120, the classification results with data lengths of 512 and 1024 are shown in Figs. 1 and 2, respectively. The classification threshold shows by the solid line in the figure separates the two types of epileptic EEG. The variance of degree sequence in the interictal period is significantly higher than that of the ictal period. This conclusion is consistent with one fact that the complexity of ictal EEG data is lower than that of the interictal EEG data.

As can be seen from the Table 1, Q has little effect on the classification accuracy rate. The proposed method in this paper is good for small data analysis.

Table 2 shows the accuracy of epileptic EEG single feature classification based on transition network and other methods. It can be seen from the table that the accuracy of classification by the extracted single feature based on improved transition network is

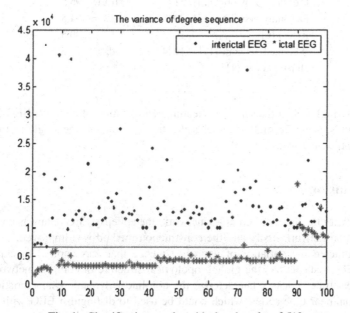

Fig. 1. Classification results with data lengths of 512

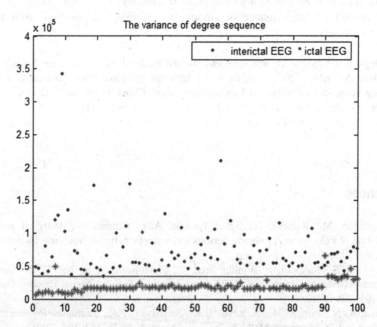

Fig. 2. Classification results with data lengths of 1024

Table 2. Result of the feature automatic detection algorithm

Method	Data length	Feature	ACC
Transition network	512	σ	98.5
Proximity networks [14]	512	NEED	96.5
Proximity networks [15]	2048	$Pclu$	94.5
RQA [16]	1024	DET	90.5
Weighted network [17]	1024	wd_r	94.5
Hurst + SVM [1]	****	****	87.5

significantly higher than that of single-feature classification of epileptic EEG based on other methods, achieved highest the classification accuracy of the single feature based on the complex network.

4 Conclusion

Combined with the theory of complex network, the epilepsy EEG data was constructed as a transition network firstly and the construction method was improved.

We extracted the variance of degree sequence that was applied to classify the epileptic EEG data set. As the global topological structure of complex network based on nonlinear time series σ, characterize the non-linear dynamic characteristics of the original nonlinear time series, which could be used to distinguish EEG with different nonlinear dynamic modes. Compared with the single feature classification accuracy of proximity networks, we got higher classification accuracy. The feature extracted in this paper improved the performance of automatic detection classification algorithm of epilepsy effectively.

Acknowlegement. This work was supported by the National Natural Science Foundation of China (Grant No. 61671220, 61640218, 61201428), the Shandong Distinguished Middleaged and Young Scientist Encourage and Reward Foundation, China (Grant No. ZR2016FB14), the Project of Shandong Province Higher Educational Science and Technology Program, China (Grant No. J16LN07), the Shandong Province Key Research and Development Program, China (Grant No. 2016GGX101022).

References

1. Nurujjaman, M., Ramesh, N., Sekar Iyengar, A.N.: Comparative study of nonlinear properties of EEG signals of normal persons and epileptic patients. Nonlinear Biomed. Phys. 3(1), 6–15 (2009)
2. Acharya, U.R., Molinari, F., Vinitha Sree, S., Chattopadhyay, S., Ng, K.-H., Suri, J.S.: Automated diagnosis of epileptic EEG using entropies. Biomed. Signal Process. Control 7(4), 401–408 (2012)
3. Kannathal, N., Lim, C.M., Acharya, U.R., Sadasivan, P.K.: Entropies for detection of epilepsy in EEG. Comput. Methods Programs Biomed. 80(3), 187–194 (2005)

4. Acharya, U., Vinitha Sree, S., Chattopadhyay, S., Wenwei, Y.U., Alvin, A.P.C.: Application of recurrence quantification analysis for the automated identification of epileptic EEG signal. Int. J. Neural Syst. **21**(3), 199–211 (2011)

5. Übeyli, E.D.: Combined neural network model employing wavelet coefficients for EEG signals classification. Digit. Signal Process. **19**(2), 297–308 (2009)

6. Gandhi, T., Panigrahi, B.K., Bhatia, M., Anand, S.: Expert model for detection of epileptic activity in EEG signature. Expert Syst. Appl. **37**(4), 3513–3520 (2010)

7. Song, Y., Liò, P.: A new approach for epileptic seizure detection sample entropy based feature extraction and extreme learning machine. J. Biomed. Sci. Eng. **3**(6), 556–567 (2010)

8. Yuan, Q., Zhou, W., Liu, Y.X., Wang, J.W.: Epileptic seizure detection with linear and nonlinear features. Epilepsy Behav. **24**(4), 415–421 (2012)

9. Yuan, Q., Zhou, W., Li, S., Cai, D.M.: Epileptic EEG classification based on extreme learning machine and nonlinear features. Epilepsy Res. **96**, 29–38 (2011)

10. Zhang, J., Small, M.: Complex network from pseudoperiodic time series: topology versus dynamics. Phys. Rev. Lett. **96** (2006)

11. Lacasa, L., Luque, B., Ballesteros, F., Luque, J., Nuno, J.C.: From time series to complex networks: the visibility graph. Proc. Natl. Acad. Sci. U.S.A. **105**, 4972–4975 (2008)

12. Sun, X.R., Small, M., Zhao, Y., Xue, X.P.: Characterizing system dynamics with a weighted and directed network constructed from time series data. Phys. A **24**(2), 1054–1500 (2013)

13. Wang, M., Tian, L.X.: From time series to complex networks: the phase space coarse graining. Phys. A **461**, 456–468 (2016)

14. Wang, F., Meng, Q., Zhou, W., Chen, S.: The feature extraction method of EEG signals based on degree distribution of complex networks from nonlinear time series. In: Huang, D.-S., Bevilacqua, V., Figueroa, J.C., Premaratne, P. (eds.) ICIC 2013. LNCS, vol. 7995, pp. 354–361. Springer, Heidelberg (2013). doi:10.1007/978-3-642-39479-9_42

15. Wang, F.L., Meng, Q.F., Chen, Y.H., Zhao, Y.Z.: Feature extraction method for epileptic seizure detection based on cluster coefficient distribution of complex network. WSEAS Trans. Comput., 351–360 (2014)

16. Meng, Q.F., Chen, S., Chen, Y.H.: Automatic detection of epileptic EEG based on recursive quantification analysis and support vector machine. Acta Phys. **6**(5) (2014)

17. Wang, F.L., Meng, Q.F., Xie, H.B., Chen, Y.H.: Novel feature extraction method based on weight difference of weighted network for epileptic seizure detection. In: 36th Annual International IEEE EMBS Conference, Chicago, Illinois, USA (2014)

Deep Belief Networks for EEG-Based Concealed Information Test

Qi Liu[1(✉)], Xiao-Guang Zhao[1], Zeng-Guang Hou[1],
and Hong-Guang Liu[2]

[1] The State Key Laboratory of Management and Control for Complex Systems,
Institute of Automation, University of Chinese Academy of Sciences,
Beijing 100190, China
popofay17@gmail.com
[2] Institute of Crime, Chinese People's Public Security University,
Beijing 100038, China

Abstract. This paper introduces a deep learning approach to the feature extraction of P300 cognitive component existing in electroencephalogram signals collected in an autobiographical paradigm test. A thorough belief mechanism is used for the extraction of deep characteristics rather than raw feature vectors to train the classifier. It is shown that the classification accuracy is satisfactory by learning deep from the experimental data. Experiments have validated the usefulness of the algorithm. The hidden information has been obtained accurately with a single electroencephalogram channel. Moreover, performances of support vector machine with different feature extraction methods are compared.

Keywords: Electroencephalogram · Concealed information test · Deep feature extraction · Deep belief networks

1 Introduction

In recent years, EEG-based concealed information test has drawn considerable attention in the field of criminal investigation. Many effective methods have been used for EEG signal analysis in Concealed Information Test (CIT) [1]. Compared to traditional methods based on physiological responses which are easily affected by emotions and stress, cognitive behavior based polygraph is considered more reliable and scientific that can reduce the risk in false positive errors [2]. In addition, EEG is more convenient, more harmless and more economical than other brain activity monitoring methods such as PET, MEG and fMRI [3].

Due to the complexity and particularity of actual criminal investigation tasks and poor ratio of signal intensity to noise intensity (SNR), increasing performance of recognition of raw EEG signals remains a live problem. In which, methods based on machine learning algorithms have achieved the most effective results. Numerous feature extraction approaches have been adopted in machine learning algorithms such as time or periodicity methods [4], model parameter methods [5], as well as methods on the basis of wavelet decomposition [6], etc. [7]. However, the distinguishability of a

F. Cong et al. (Eds.): ISNN 2017, Part II, LNCS 10262, pp. 498–506, 2017.
DOI: 10.1007/978-3-319-59081-3_58

certain feature is uncertain in different tasks, which may lead to a failure of recognition. Therefore, feature extraction methods which are capable of feature self-learning are necessary to be studied in this field. Recently, deep learning strategy has made great progress and the related algorithms have also been adopted in various fields such as EEG signal processing [8]. It can be viewed as a computational intelligent method since its similar mechanism to human brain. To improve the generalization performance of EEG feature, deep belief networks (DBN) is adopted to learn features automatically.

In this paper, we use the CIT technique and focus primarily on the feature extraction process of different brain waves evoked by relevant stimulus and control stimulus. DBN was applied to self-learn features of EEG signals. Then support vector machine (SVM) was implemented as the classifier. The classification performance is satisfactory and the runtime is acceptable.

2 Methods

2.1 Data Description

Data in this paper were from an autobiographical paradigm test [9]. There were 11 volunteering subjects in total participated who were all males at the age of between 22 and 35. They are all used to using the right hand and their vision are all normal or corrected to normal range. They have no idea what the test is based on and just know how to carry out the test. All the subjects were required to offer five numbers which all contained 4 digits and one of the numbers was the year of birth. The experimenter was not informed by the subjects of the birth date number until when the experiment ended. In the experiment, subject 11 took part in 3 runs while other subjects were involved in 2 runs. Due to wrong target stimulus counting (as was shown below), subject 1, 3, and 7 saw one of their runs vetoed. Finally, the study applied a total of 20 runs in the experiment. To achieve whole stimuli, in each run, the subject was exposed to each number with random for thirty times. Each number was revealed for one second and there was a two-second blank in the screen between the numbers. The experiment required the subjects to count how many times the number of the birth year was revealed instead of responding to the items. The subjects did not know that the entire target stimulus were displayed with 30 repetitions. EEG signals sampled at 256 Hz digitally were recorded at the Fz, Cz and Pz electrode positions of the 10–20 international electrode placement system (Fig. 1). The electrodes referred to linked mastoids. For the purpose of blink artifact detection, the experiment also recorded vertical EOG signals.

2.2 Methods

For the complexity and weak anti-interference capability of EEG, it is not easy to recognize effective data from raw signals. Figure 2 shows the raw waveforms. It is observed that the potential offset value of each sample belonging to the same category is quite different and there is no obvious distinction between samples belonging to separate categories.

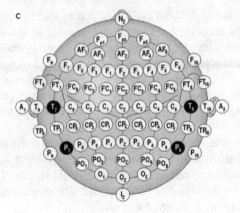

Fig. 1. 10–20 system of electrode placement

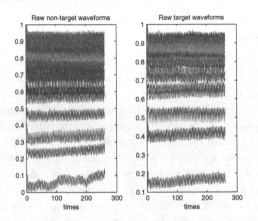

Fig. 2. Raw EEG waveforms

Figure 3 shows that the signal process mainly includes data collection, pre-processing, feature exaction and pattern classification.

(1) Pre-processing

This process consists of electrodes selection, segmentation of signals, superposition and filtering. For low SNR of EEG signals, the stimulations are repeated to remove unnecessary signals and enhance useful signals. Because the P300 frequency is primarily allocated in area with low frequency, the experiment designed a 6-order band pass Chebyshev Type I filter with cut-off frequencies 0.5 and 35 Hz to penetrate each epoch. Moreover, the data information matrix is designed into a range from 0 to 1 according to Eq. (1).

$$\mathbf{X}_{norm} = \mathbf{X} - \mathbf{X}_{min} / \mathbf{X}_{max} - \mathbf{X}_{min} \tag{1}$$

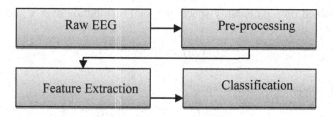

Fig. 3. The flowchart of signal processing

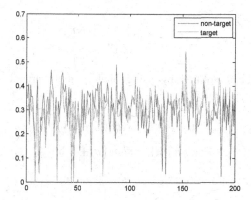

Fig. 4. Comparison of mean values of the two categories

(2) Deep Feature Extraction

To begin with, k-means method is adopted to represent features preliminary as described in [11]. Using subject 1 as an example, some differences between the two categories can be seen in Fig. 4 after the initial feature extraction. However, the difference is still too small to distinguish samples. Further feature extraction is implemented as following.

DBN could be considered to be a stack of RBMs (Restricted Boltz-man Machines), which are motivated from the idea of equilibrium from the statistical physics literature [12]:

$$E(\mathbf{v}, \mathbf{h}; \theta) = -\sum_j a_j v_j - \sum_i b_i h_i - \sum_{i,j} v_j h_i w_{ij} \qquad (2)$$

Where w_{ij} is the symmetric interaction term between distinct unit v_j and covered unit h_i, a_i as well as b_j are both the bias term. $\theta = \{\mathbf{w}, \mathbf{a}, \mathbf{b}\}$ is the model parameter need to be learned.

Equation (2) could be optimized in a tricky way by contrastive divergence that has been usually applied to border on the expectation by a sample deriving from a certain amount of Gibbs sampling iterations [13].

When defined on a probability space, the joint distribution over **v** and **h** is:

$$P(\mathbf{v}, \mathbf{h}) = \frac{1}{z} e^{-E(\mathbf{v}, \mathbf{h})} \tag{3}$$

where z is a standardized factor. Then

$$P(\mathbf{v}) = \sum_{\mathbf{h}} P(\mathbf{v}, \mathbf{h}) = \frac{e^{-F(\mathbf{v})}}{z} \tag{4}$$

in which

$$F(\mathbf{v}) = -\log \sum_{\mathbf{h}} e^{-E(\mathbf{v}, \mathbf{h})} \tag{5}$$

Model (2) can be simplified by using binary input variables. The conditional probabilities can be formulated as:

$$P(h_i = 1|\mathbf{v}) = sigm(b_i + w_i \mathbf{v})$$
$$P(v_j = 1|\mathbf{h}) = sigm\left(a_j + w_j' \mathbf{h}\right) \tag{6}$$

Then

$$F(\mathbf{v}) = -\mathbf{a}'\mathbf{v} - \sum_i \log\left(1 + e^{(c_i + w_i \mathbf{v})}\right) \tag{7}$$

$$-\frac{\partial \log P(\mathbf{v})}{\partial \theta} = \frac{\partial F(\mathbf{v})}{\partial \theta} - \sum_{\tilde{\mathbf{v}}} P(\tilde{\mathbf{v}}) \frac{\partial F(\tilde{\mathbf{v}})}{\partial \theta} \tag{8}$$

To make RBM stability, the energy of system should be the minimum. By the above formulas, $P(\mathbf{v})$ should be maximized. The partial derivative of loss function $-P(\mathbf{v})$ is calculated as:

$$-\frac{\partial \log P(\mathbf{v})}{\partial w_{ij}} = E_{\mathbf{v}}\left[P(h_i|\mathbf{v}) \cdot v_j\right]$$
$$- v_j^{(i)} \cdot sigm\left(w_i \cdot v^{(i)} + c_i\right)$$
$$-\frac{\partial \log P(\mathbf{v})}{\partial c_i} = E_{\mathbf{v}}[P(h_i|\mathbf{v})] - sigm\left(w_i \cdot v^{(i)}\right) \tag{9}$$
$$-\frac{\partial \log P(\mathbf{v})}{\partial b_j} = E_{\mathbf{v}}\left[P(v_j|\mathbf{h})\right] - v_j^{(i)}$$

Thus, the parameter θ corresponding to maximum $P(\mathbf{v})$ is obtained. DBN could then be trained with the greedy layer-wise method [12]. Each RBM is trained greedily and unsupervised [14]. The posterior distribution of the first RBM is used as the input distribution of the second RBM. Then the weights are fine-tuned by back propagation (BP) neural network. Figure 5 shows the architecture of DBN model and Fig. 6 displays the comparison of mean values between two categories. The difference is significant after feature learning by DBN.

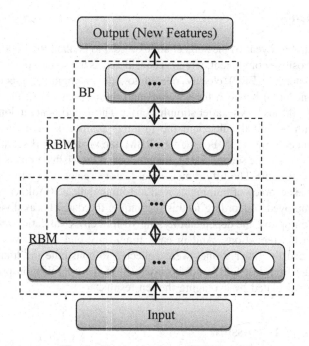

Fig. 5. The architecture of DBN model

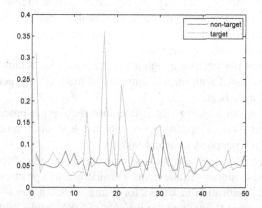

Fig. 6. Comparison of mean values of the two categories

(3) **Classification**

DBN model is viewed as a feature extraction system in this paper. Outputs of the last model were used as the new input feature vectors with labels of samples to train the SVM classifier.

3 Experiments

Responses to the birth year of the subject are expected to contain the P300 component, which is a late positive component, which is considered as the most typical and common event-related potential (ERP) closely related to human cognitive process. For the time-locked conception between the stimulus and the response [15], the value of the signal during 0–700 ms was set after stimulus onset. The experiment randomly assigned the weights with an initial value and the turning parameters were set as: learning rate = 0.07, momentum = 0.95. For the first RBM, the visible unit is set at 200 and the hidden unit is 100. For the second RBM, the number of the visible units is 100 and that of the hidden units is 50. The fifty-dimensional feature vector is input to libsvm.

To ensure the accuracy of training as well as testing data, a 10-fold cross-validation method was employed. According to this technique, the dataset was divided into ten subsets [16]. To improve the dependability, the 10-fold cross-validation procedure was performed with ten repetitions. And in each time, only one subset was used as the testing dataset and the other 9 ones were collected to constitute the training dataset. Particularly, data from test fold is not be involved in the optimization procedure. All final data were calculated by averaging the ten results.

4 Results and Discussion

This section made a test of performance of the DBN-SVM classification algorithm on the basis of the dataset presented in Sect. 2.1. Table 1 and Fig. 7 reveal the results. Specifically, Table 1 displays the recognition accuracy and runtime over all eleven subjects. Figure 7 compares performances of classifiers adopted different effective feature extraction methods for SVM classifier. All the experiments are repeated ten times, and the average results are reported.

From the effects of perspective, a high average accuracy is obtained. In addition, as shown in Fig. 7, compared with other features used methods, the performance of our approach is significantly better.

Moreover, it is worth noticing that it does not require pre-processing operations including artifact removal or bootstrapping which takes much time and allows the approach possible to be applied to actual tasks.

However, the complex application environment and unpredictable interference will definitely put forward higher requirements considering the practical applications in crime information identification tasks. As for future works, it would be interesting to investigate a way to overall fine-tune the weights of DBN model with regard to SVM learning rule [13].

Table 1. Performances of the algorithm over all subjects

Subject	Amount of samples	Accuracy (%)
S1	150	95.5
S2	300	98.9
S3	300	97.6
S4	150	96.7
S5	150	97.5
S6	300	98.0
S7	300	97.0
S8	150	96.3
S9	300	97.6
S10	300	96.2
S11	450	98.9
Average		97.3

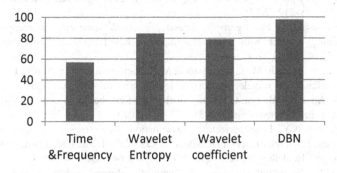

Fig. 7. Comparison of classification performances over different feature extraction methods

5 Conclusion

In this paper, deep learning strategy is applied for signal processing in concealed information test based on EEG. The introduction of DBN aims to better express characteristics of different signals. We choose SVM as the classifier which can avoid over-fitting effectively. According to the results, the method has been highly recognized. The study in this paper suggests that it is valuable to do further development on deep learning or other computational intelligence strategies applied in CIT based on EEG as well as provide reliable supports to actual future explorations.

Acknowledgement. Special thanks would be expressed to Dr. V. Abootalebi and the Research Center of Intelligent Signal Processing (RCISP), Iran, for the provision of the data. Besides, Dr. Deng Wang with department of Computer Science and Technology, Tongji University also deserves the appreciation, for the data support.

References

1. Abootalebi, V., Moradi, M.H., Khalilzadeh, M.A.: A comparison of methods for ERP assessment in a P300-based GKT. Int. J. Psychophysiol. **62**(2), 309–320 (2006)
2. Zhao, M., Zhang, C., Zhao, C.: New approach for concealed information identification based on ERP assessment. J. Med. Syst. **36**(4), 2401–2409 (2011)
3. Wolpaw, J.R., Birbaumer, N., McFarland, D.J., Pfortscheller, G., Vaughan, T.M.: Brain-computer interfaces for communication and control. Clin. Neurophysiol. **113**(6), 767–791 (2002)
4. Guo, X.J., Wu, X.P., Zhang, D.J.: Motor imagery EEG detection by empirical mode decomposition. In: International Joint Conference on Neural Networks, pp. 2619–2622 (2008)
5. Zhao, M.Y., Zhou, M.T., Zhu, Q.X.: Feature extraction and parameters selection of classification model on Brain-computer interface. In: IEEE 7th International Symposium on Bioinformatics and Bioengineering, pp. 1249–1253 (2007)
6. Sherwood, J., Derakhshani, R.: On classifiability of wavelet features for EEG-based Brain-computer interfaces. In: International Joint Conference on Neural Networks, pp. 2895–2902 (2009)
7. Subha, D.P., Joseph, P.K., Acharya, U.R., Lim, C.M.: EEG signal analysis: a survey. J. Med. Syst. **34**(2), 195–212 (2010)
8. Jirayucharoensak, S., Pan-Ngum, S., Israsena, P.: EEG-based emotion recognition using deep learning network with principal component based covariate shift adaptation. Sci. World J. **2014**, 1–10 (2014)
9. Abootalebi, V., Moradi, M.H., Khalilzadeh, M.A.: Detection of the cognitive components of brain potentials using wavelet coefficients. Iran. J. Biomed. Eng. **1**(1), 25–46 (2004)
10. Wang, D., Miao, D., Blohm, G.: A new method for EEG-based concealed information test. IEEE Trans. Inf. Forensics Secur. **8**(3), 520–527 (2013)
11. Coates, A., Ng, A.Y.: Learning feature representations with k-means. In: Montavon, G., Orr, G.B., Müller, K.-R. (eds.) Neural Networks: Tricks of the Trade. LNCS, vol. 7700, pp. 561–580. Springer, Heidelberg (2012). doi:10.1007/978-3-642-35289-8_30
12. Hinton, G.E., Salakhutdinov, R.R.: Reducing the dimensionality of data with neural network. Science **313**(5786), 504–507 (2006)
13. Hinton, G.E.: Training products of experts by minimizing contrastive divergence. Neural Comput. **14**(8), 1771–1800 (2002)
14. Bengio, Y., Lamblin, P., Popovici, D., Larochelle, H.: Greedy layer-wise training of deep networks. Adv. Neural. Inf. Process. Syst. **19**, 146–153 (2007)
15. Quiroga, R.Q.: Quantitative analysis of EEG signals: time-frequency methods and chaos theory. Institute of Physiology-Medical University Lubeck and Institute of Signal Processing-Medical University Lubeck (1998)
16. Ripley, B.D.: Pattern Recognition and Neural Networks. Cambridge University Press, Cambridge (1996)

Cluster Aggregation for Analyzing Event-Related Potentials

Reza Mahini[1], Tianyi Zhou[1,2], Peng Li[3], Asoke K. Nandi[4],
Huanjie Li[1], Hong Li[3], and Fengyu Cong[1,2(✉)]

[1] Department of Biomedical Engineering,
Faculty of Electronic Information and Electrical Engineering,
Dalian University of Technology, Dalian, China
r_mahini@mail.dlut.edu.cn, tianyi.zhou@foxmail.com,
{hj_li,cong}@dlut.edu.cn
[2] Department of Mathematical Information Technology,
University of Jyvaskyla, Jyväskylä, Finland
[3] College of Psychology and Sociology, Shenzhen University, Shenzhen, China
peng@szu.edu.cn, lihongwrm@vip.sina.com
[4] Department of Electronic and Computer Engineering,
Brunel University, London, UK
Asoke.Nandi@brunel.ac.uk

Abstract. Topographic analysis are references independent for Event-Related Potentials (ERPs), and thus render statistically unambiguous results. This drives us to develop an effective clustering approach to finding temporal samples possessing similar topographies for analysing the temporal-spatial ERPs data. The previous study called CARTOOL used single clustering method to cluster ERP data. Indeed, given a clustering method, the quality of clustering varies with data and the number of clusters, motivating us to implement and compare multiple clustering algorithms via using multiple similarity measurements. By finding the minimum distance among the various clustering methods and selecting the most selected clustering algorithms with other methods via voting the proposed method, a most suitable algorithm showing a considerable performance for a given dataset can be found. This cluster aggregation approach assists to use the most suitable founded cluster for each dataset. We demonstrated the effectiveness of the proposed method by using ERP data for cognitive neuroscience research.

Keywords: Cluster aggregation · Cognitive neuroscience · ERP data analysis · Spatial · Temporal

1 Introduction

Event-related potentials (ERPs) are important tools for cognitive neuroscience by analyzing peak measurements [1]. Usually, the mean amplitude of an ERP over a certain time range is measured as the peak amplitude for statistical analysis. The underlying assumption of this approach is that the topographies over that certain time range do not change. In order to validate the assumption, the clustering has been

© Springer International Publishing AG 2017
F. Cong et al. (Eds.): ISNN 2017, Part II, LNCS 10262, pp. 507–515, 2017.
DOI: 10.1007/978-3-319-59081-3_59

applied the temporal-spatial ERP data to find the temporal samples sharing the similar spatial topographies [2].

In the latest version of CARTOOL software 3.55 (2014), it is possible to use one of the two Clustering methods named, K-means and Hierarchical clustering with some good selective options [2]. It could be considerable that clustering algorithm selection and the quality of clustering would be affected by different conditions such as, dataset types, quality of data and etc. In this study, we demonstrate that for given dataset the proposed method can find a suitable clustering algorithm among various ones in a reliable way.

The following of the study is structured into 3 Sections; we start with clustering techniques for data analysis in Sect. 2 and the cluster aggregation method is described, Sect. 3 provides experimental results and discussion about the results and Sect. 4 includes conclusions and future works.

2 Method

Indeed, in clustering analysis, one solution to the question above is the use of numerical clustering validation algorithms and assessing the quality of clustering results in terms of many criteria. Since it is also true that no single clustering validation algorithm has been claimed to impartially evaluate the results of any clustering algorithm, the use of clustering validation is not an overwhelmingly reliable solution [3]. In this study, the two-way clustering is applied. Since the multi-way analysis is significant for ERP data analysis [4], it is worth extending the two-way clustering to the multi-way clustering. Consequently, we propose a new approach for that, how to use cluster aggregation aim to deal with uncertainty in datasets and clustering algorithms, using multiple clustering methods and multiple similarity measurements for cluster aggregation to achieve reliable analysis.

2.1 Clustering Methods

In this study, five popular standard clustering algorithms are used and they are briefly introduced as the following:

- **K-means Clustering.** Which for given a dataset with N data objects in an M-dimensional feature space, this algorithm determines a partition of K groups or clusters which detailed in [5].
- **Hierarchical Clustering.** Basically Hierarchical clustering algorithms are mainly classified into methods (bottom-up methods) and divisive agglomerative methods (top-down methods), based on how the hierarchical divide or merge is formed [6].
- **Fuzzy C-means (FCM).** Dunn developed fuzzy k-partition algorithms which minimize certain fuzzy extensions of the k-means least-squared-error criterion function [7]. Eventually, the generalised algorithm was named fuzzy c-means (FCM) [8].
- **Self-organizing Map (SOM).** Clustering in the neural network literature is generally based on competitive learning (CL) model, Kohonen made particularly strong implementation of CL in his work on learning vector quantisation (LVQ) and self-organising maps (SOM) also known as self-organising feature maps (SOFM) [9].

- **Diffusion Maps Spectral Clustering.** Spectral clustering is an algorithm which is very close to the graph cut clustering algorithm. It requires the computation of the first k eigenvectors of a Laplacian matrix; Diffusion map is a dimensionality reduction method that embeds the high-dimensional data to a low-dimensional space. Clustering is performed within the low-dimensional space [10].

2.2 Similarity Measurements

Partition–Partition (P–P) Comparison approaches, equivalent to median partition approaches, attempt to provide the solution of an optimization problem, which maximizes the total similarity to the given partitions [3]. In Eqs. 1 and 2, which R is the number of clustering algorithms and C^* is the clustering with maximum similarity with the other clusterings and minimum dissimilarity with them. In fact, here are several similarity measurements for measuring similarity or dissimilarity between partitions see [3].

$$C^* = argmax_{p \in \mathbb{P}_x} \sum_{j=1}^{R} \Gamma(C, C_j) \tag{1}$$

$$C^* = argmin_{p \in \mathbb{P}_x} \sum_{j=1}^{R} \mathcal{M}(C, C_j) \tag{2}$$

We used a number of similarity measurements aim to cluster aggregation. Briefly, we describe them in below,

Fowlkes and Mallows Distance Function
The Wallace distance of two clustering algorithms C, C' is,

$$W_I(C, C') = \frac{N_{11}}{\sum_k n_k(n_k - 1)/2} \tag{3}$$

$$W_{II}(C, C') = \frac{N_{11}}{\sum_k n'_k(n'_k - 1)/2} \tag{4}$$

Where N_{11} is the number of pairs of objects that were clustered in the same clusters in C and C'. They represent the probability that a pair of points which are in the same cluster C, C'.

$$\mathcal{F}(C, C') = \sqrt{W_I(C, C') W_{II}(C, C')} \tag{5}$$

The index is used by subtracting the base-line and normalizing by the range, so that the expected value of the normalized index is 0 while the maximum (attained for identical clustering algorithms) is [11].

Rand Distance Function

$$\mathcal{R}(C, C') = \frac{N_{11} + N_{00}}{n(n - 1)/2} \tag{6}$$

In the Eq. 6 N_{00} is the number of pairs of objects that were clustered in separate clusters in C and also C'. A similar transformation was introduced for Rand index and Adjusted Rand index.

Adjust Rand Distance Function

$$AR(C, C') = \frac{R(C, C') - E[R]}{1 - E[R]}$$

$$= \frac{\sum_{i,j}\binom{n_{i,j}}{2} - \left[\sum_i\binom{n_i}{2}\sum_j\binom{n_j}{2}\right] / \binom{n}{2}}{\frac{1}{2}\left[\sum_i\binom{n_i}{2} + \sum_j\binom{n_j}{2}\right] - \left[\sum_i\binom{n_i}{2}\sum_j\binom{n_j}{2}\right] / \binom{n}{2}} \tag{7}$$

The main motivation for adjusting indices like R and F is the observation that the unadjusted R, F do not range over the entire $[0, 1]$ interval (i.e. min $R > 0$, min $F > 0$). There are other criteria in the literature, to which the above discussion applies. For instance, the Jaccard index in [11].

Jaccard Distance Function

Jccard index for two clustering algorithms C, C' is,

$$\mathcal{J}(C, C') = \frac{N_{11}}{N_{11} + N_{01} + N_{10}} \tag{8}$$

2.3   Cluster Aggregation

Clustering comparison can be useful for examining whether the structures of the clusters match to some predefined classification of the instances. In fact, researchers use different distance algorithms, even clustering ensemble causes to obtain good results most of the time, in this study an aggregation method is used with acceptable workload. The consensus clustering problem is considered as an NP-hard problem [11] yet, we are still able to provide approximation guarantees for many of the algorithms, we propose via using a combination of different similarity measurements. Figure 1 illustrates the proposed aggregation algorithm model.

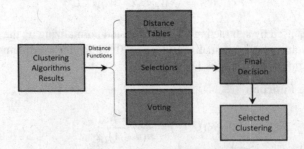

Fig. 1. The cluster aggregation model.

Clustering algorithms results can be compared together with the mentioned similarity measurements. In this study four distance functions are used to make a reliable comparison between clusterings. The idea is, the clustering algorithms are selected with a maximum value of similarity by the other algorithms in the distance tables. Next voting assisted to find which clustering has selected more than the others. The final decision is made based on the voting table to find a suitable clustering algorithm.

3 Experimental Results and Discussion

We implemented the proposed algorithm using by five clustering algorithms and four similarity measurements. First, because these algorithms are standard and also we need to control the number of clusters in proposed algorithm. In this study, ERP data which has been published in [12] form the gamboling task is used. We just selected one subject and one stimulus data randomly and the size of dataset is 500 temporal samples by 58 electrodes (features for each sample). Actually, we applied it to all the subjects' data to find group behavior which is very important in cognitive neuroscience. Figure 2 presents the 5 algorithms clustering performance for the sample dataset. Indeed, we considered the CARTOOL proposed algorithms (k-means and Hierarchical Clustering) as it is shown in this figure. We applied clustering algorithms for this dataset with 6 clusters according to eigenvalues distribution and the explained variance value diagram. Due to the limited space, they are not shown here.

Figure 3 demonstrates distance function tables based on 4 similarity measurements (Rand, Adjusted Rand, Fowlkes and Mallow and Jaccard indices) and for five clustering algorithms in the order with: K-means (1), Hierarchical (2), FCM (3), SOM (4) and Diff-Spec (5). Moreover, Tables 1 and 2 illustrate aggregation algorithm process results. Table 2 shows that the K-means have been selected 8 times by the other clustering methods, as a result K-means is used as the suitable method after comparison.

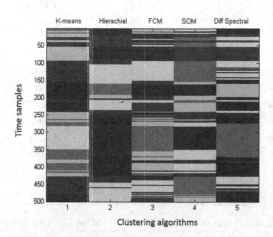

Fig. 2. The 5 Clustering algorithms performance for given dataset.

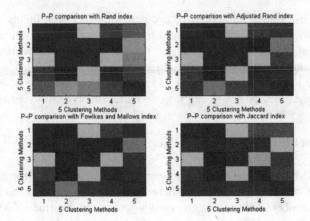

Fig. 3. Four distance function tables for 5 clustering algorithms.

Table 1. How the clustering algorithms select similar method based on similarity measurements, the numbers 1, 2, 3, 4, 5 indicate the selected clustering's code in selection.

No.	Table no.	Clustering meth.	Selected meth.	Similarity value
1	1	1	4	0.9912
2	1	2	5	0.8768
3	1	3	1	0.9162
4	1	4	1	0.9912
5	1	5	2	0.8768
6	2	1	4	0.9721
7	2	2	5	0.6136
8	2	3	1	0.7261
9	2	4	1	0.9721
10	2	5	2	0.6136
11	3	1	4	0.9776
12	3	2	5	0.6931
13	3	3	1	0.7781
14	3	4	1	0.9776
15	3	5	2	0.6931
16	4	1	4	0.9415
17	4	2	5	0.6221
18	4	3	1	0.6032
19	4	4	1	0.9712
20	4	5	3	0.4763

Table 2. Voting table for the selected methods.

K-means (1)	Hierarchical (2)	FCM (3)	SOM (4)	Diff_Spec (5)
8	3	1	4	4

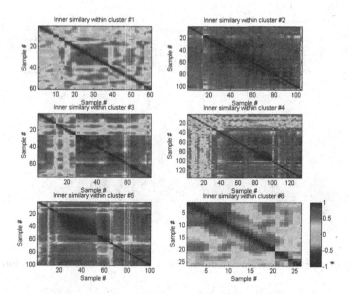

Fig. 4. Correlation coefficient among centroids in clusters.

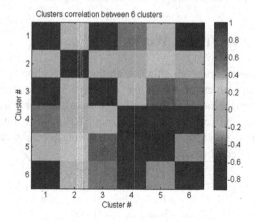

Fig. 5. Correlation between 6 clusters.

Inner similarity of objects inside clusters is presented in Fig. 4, reasonable correlation in order to objects in each cluster is appeared. Figure 5 shows the inter-cluster correlation and it means that the clusters are enough separated. Moreover, it would be very important to recognize important time windows corresponding to related ERP waveform with related topography maps based on 6 number of clusters, this concept is shown in Fig. 6. It is clearly seen that there are 6 different topography results for 6 clusters and it means that all the time points in each cluster have same topography and we could consider them as a point, this advantage assists us to find the reasonable time range to average the amplitudes for the ERP peak measurement, providing a reliable

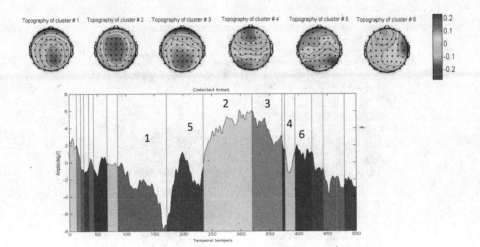

Fig. 6. Topographies of centroids of the six clusters and clustered ERP waveform into six clusters.

and objective way for cognitive neuroscience research. Due to the length limitation of the study, we will report the results of the full ERP dataset in the future study.

4   Conclusions and Future Works

In this study, we have proposed an effective approach to finding the temporal samples sharing similar topographies of ERPs for cognitive neuroscience research. Using several similarity measurements to find a better algorithm for clustering makes this method more reliable and suitable for processing ERP data. In future works, we are going to improve the method for brain signal processing by using clustering ensemble and consider other different datasets for processing.

Acknowledgements. This work was supported by the Fundamental Research Funds for the Central Universities [DUT16JJ(G)03] in Dalian University of Technology in China, and National Natural Science Foundation of China (Grant No. 81471742).

References

1. Luck, S.J.: An Introduction to the Event-Related Potential Technique. MIT Press, Cambridge (2014)
2. Brunet, D., Murray, M.M., Michel, C.M.: Spatiotemporal analysis of multichannel EEG: CARTOOL. Comput. Intell. Neurosci. **2**, 1–15 (2011)
3. Abu-Jamous, B., Fa, R., Nandi, A.K.: Integrative Cluster Analysis in Bioinformatics. Wiley, Hoboken (2015)

4. Cong, F., Lin, Q.-H., Kuang, L.-D., Gong, X.-F., Astikainen, P., Ristaniemi, T.: Tensor decomposition of EEG signals: a brief review. J. Neurosci. Methods **248**, 59–69 (2015)
5. Yu, S., Tranchevent, L., Liu, X., Glanzel, W., Suykens, J.A., De Moor, B., Moreau, Y.: Optimized data fusion for kernel k-means clustering. IEEE Trans. Pattern Anal. Mach. Intell. **34**, 1031–1039 (2012)
6. Tan, P.-N.: Introduction to Data Mining. Pearson Education, London (2006)
7. Dunn, J.C.: A fuzzy relative of the ISODATA process and its use in detecting compact well-separated clusters. J. Cybern. **3**, 32–57 (1973)
8. Bezdek, J.C.: Pattern Recognition with Fuzzy Objective Function Algorithms. Plenum Press, NewYork/London (2013)
9. Kohonen, T.: The self-organizing map. Proc. IEEE **78**, 1464–1480 (1990)
10. Sipola, T., Cong, F., Ristaniemi, T., Alluri, V., Toiviainen, P., Brattico, E., Nandi, A.K.: Diffusion map for clustering fMRI spatial maps extracted by independent component analysis. In: IEEE International Workshop on Machine Learning for Signal Processing (MLSP), pp. 1–6 (2013)
11. Meila, M.: Comparing clusterings - an information based distance. J. Multivar. Anal. **98**, 873–895 (2007)
12. Han, C., Li, P., Warren, C., Feng, T., Litman, J., Li, H.: Electrophysiological evidence for the importance of interpersonal curiosity. Brain Res. **1500**, 45–54 (2013)

Detection of Epileptic Seizure in EEG Using Sparse Representation and EMD

Qingfang Meng[1,2]([✉]), Shanshan Chen[1,2], Haihong Liu[1,2],
Yunxia Liu[1,2], and Dong Wang[1,2]

[1] School of Information Science and Engineering,
University of Jinan, Jinan 250022, China
ise_mengqf@ujn.edu.cn
[2] Shandong Provincial Key Laboratory of Network Based
Intelligent Computing, Jinan 250022, China

Abstract. Epileptic seizure detection is the most important part in the diagnosis of epilepsy. The automatic detection and classification of epileptic EEG signals has great clinical significance. This paper proposes a novel method for epileptic seizure detection using empirical mode decomposition (EMD) and sparse representation based classification (SRC). Firstly, EMD was used to decompose EEG into multiple Intrinsic Mode Function (IMF) components. Secondly, the features like variation coefficient, fluctuation index, relative energy and relative amplitude were extracted from the IMFs. Finally, in the framework of sparse representation based classification (SRC), the feature vectors of test sample were represented as a linear combination of the feature vector of training samples with sparse coefficients. Experimental results show that the time consumed by one epileptic EEG test sample is not more than 5.9 s, and the accuracy is up to 97.5%. In SRC, the raw EEG signals were replaced by extracted features, which could reduce data dimension and computational cost. The algorithm has a good performance in the recognition of ictal EEG. The higher recognition rate and fast speed make the method suit for the diagnosis of epilepsy in clinical application.

Keywords: Epileptic EEG signal · Sparse representation based classification (SRC) · Empirical mode decomposition (EMD)

1 Introduction

Epilepsy is a common neurological disorder characterized by the presence of recurring seizures. Brain activity during seizure differs greatly from that of normal state with respect to patterns of neuronal firing. The EEG (Electroencephalogram) has been a valuable clinical tool to monitor the epileptic seizures, which contains important information about the conditions and functions of the brain. Detection of epilepsy by visual inspection is very tedious and time-consuming, particularly for long-term EEG signals. In clinical practice, epileptic seizure detection is the most important part in the diagnosis of epilepsy.

Due to the scalp EEG signals are complex and nonstationary, many techniques have been developed for epileptic activity detection in several years. The methods include feature extraction and the design of classifiers mainly. Wavelet transform has

F. Cong et al. (Eds.): ISNN 2017, Part II, LNCS 10262, pp. 516–523, 2017.
DOI: 10.1007/978-3-319-59081-3_60

been used as a powerful signal processing technique since the EEG signals contain non-stationary or transitory characteristics [1, 2]. Reference [2] used the wavelet transform to analyze and characterize 3-Hz spike and wave complex in absence seizure.

Recently, the nonlinear dynamical methods have been widely applied since the EEG signals are considered as nonlinear. The nonlinear parameters, like the Fractal Intercept analysis [3], entropy [4, 5], recurrence quantification analysis [6], Hurst Exponent [7], Linear Discriminate Analysis [8] are usually used as the extracted feature value. Besides the feature extraction methods, the designs of classifiers also have an important effect on the epileptic EEG classification. Various effective classifiers such as artificial neural network [9], support vector machine [10, 11], relevance Vector Machine [12] and extreme learning machine [13], have been used to improve the performance of classification of epileptic EEG automatically. In Reference [10], EMD decomposes a raw signal into a set of complete and almost orthogonal components called intrinsic mode functions (IMFs). IMFs represent the natural oscillatory modes embedded in the raw signal. It has the benefit of self-adaptive capacity and has been widely used in physiological signals researches [14, 15].

Recently, sparse representation becomes a hot topic in pattern recognition. Sparse representation comes from compressed sensing, potentially using lower sampling rates than the Shannon-Nyquist bound. Sparse representation selects the most compact subset with a pursuit of the least number of base elements to express signals. In the scheme of the sparse representation based classification (SRC) developed by Wright et al. [16], a test EEG sample is sparsely represented on the training samples. It has been successfully used in lots of fields, such as blind speech signals separation [17], face recognition [16–18], EEG signals detection [19, 20]. The employed dictionary plays an important role in sparse representation. The history of dictionary design could range from the Fast Fourier Transform (FFT), Principal Component Analysis (PCA), wavelets to modern dictionary learning methods, such as KSVD, fisher discrimination dictionary learning (FDDL) model. Meng Yang et al. [21] proposed a fisher discrimination dictionary learning for sparse representation to improve the pattern classification performance. Reference [22] proposed a kernel sparse representation learning framework for time series classification with KSVD techniques.

In this paper, a novel method for epileptic seizure detection is presented based on SRC and EMD. Firstly, EMD was used to decompose EEG into multiple Intrinsic Mode Function (IMF) components. Secondly, the features like variation coefficient, fluctuation index, relative energy and relative amplitude were extracted from the IMFs. Finally, in the framework of sparse representation based classification (SRC), the feature vectors of test sample were represented as a linear combination of the feature vector of training samples by solving the l1-optimization problem.

2 Method

The technology of sparse representation originated from compressed sensing breaks through the sampling rate restriction of traditional Nyquist-Shannon theorem in the area of signal processing. Sparse representation has gained considerable attention in pattern classification recently.

2.1 Sparse Representation Based Classification (SRC)

Sparse representation based classification (SRC) presented by Wright et al. [16] have exhibited excellent performance for face recognition. The basic idea of SRC is to represent the input test sample as a sparse linear combination of the training samples with sparse coefficients. Given the training set of the i_{th} class as a matrix $A_i = [s_{i1}, s_{i2}, \ldots, s_{in}] \in R^{m \times n_i}$, where $v_{ij}, j = 1, 2, \ldots, n_i$ is an m-dimensional vector stretched by the j_{th} sample of the i_{th} class. For a test sample $y \in R^m$ from the same class, y could be well approximated by the linear combination of the training samples associated with the class i:

$$y = \sum_{j=1}^{n_i} \alpha_{ij} s_{ij} = A_i a_i \tag{1}$$

The $a_i = [a_{i1}, a_{i2}, \ldots, a_{in}]^T \in R^n$ are coefficients. If the test sample y belongs to the i_{th} class, the coefficient vector of all the training samples should be $a = [a_1; a_2; \ldots; a_k] = [0, \ldots, 0, \alpha_{i1}, \alpha_{i2}, \ldots, \alpha_{in_i}, 0, \ldots, 0]^T$.

When the solution a is sparse enough, the sparsest solution $y = Aa$ is NP-hard and difficult to approximate. Based on the development of sparse representation and compressed sensing theory, there is growing evidence that if the solution a is sparse enough, the sparsest solution can be formulated as the following l1-optimization problem:

$$\hat{a} = \arg \min \|a\|_1 \quad \text{subject to } Aa = y \tag{2}$$

Since real data are usually contaminated with some additive noise, sparse solution can be modified to account for an error tolerance ε by solving the following stable l1-minimization problem:

$$\hat{a} = \arg \min \|a\|_1 \quad \text{subject to } \|Aa - y\|_2 \leq \varepsilon \tag{3}$$

The solution can be obtained by using the MATLAB package provided by http://cvxr.com/cvx/. To identify a new test sample, the test sample y can be assigned to the object class with minimal reconstructed residual.

The SRC algorithm is as following:

Normalize the columns of A to have unit l2-norm.
Solve the sparse representation problems in (3) by convex optimization tool.
Reconstruct the test vector y by the coefficient vector associated with the i_{th} class as (1).
Compute residue $y - A\delta_i(\hat{a})$ of each class.
Identify test vector y to the class with the least residue.

2.2 Empirical Mode Decomposition (EMD)

EMD is an adaptive and efficient method applied to analysis non-stationary signals. The principle of empirical mode decomposition (EMD) technique is to decompose a signal automatically into a set of the band limited functions named Intrinsic Mode Functions (IMFs). Each IMF must satisfy two conditions: in the whole data set, the number of extreme and the number of zero crossings must either equal or differ at most by one; and at any point, the mean value of the upper envelope and lower envelope is zero. The EMD algorithm for the signal $x(t)$ can be summarized as follows:

(1) Identify the local maxima and minima of the original data $x(t)$, then connect respectively by a cubic spline line to produce the upper and lower envelops U_{max} and U_{min}.

(2) Obtain the mean value of corresponding data point

$$m_1 = \frac{U_{max} + U_{min}}{2} \tag{4}$$

(3) Define the difference between $x(t)$ and m_1 as the first component

$$h_1 = x(t) - m_1 \tag{5}$$

(4) Regard h_1 as new $x(t)$ and repeat the operation above until h_1 satisfies the IMF conditions, then obtain the first-order IMF, designate it as $c_1 = h_1$

(5) Defined the residue r_1 as $x(t)$ minus c_1

$$r_1 = x(t) - c_1 \tag{6}$$

(6) Taking the residue r_1 as a new data and repeating (1)–(5) and the second IMF component is obtained. If c_1 or r_1 is smaller than a predetermined value, or r_1 becomes a monotone function, the sifting process is stopped, or else repeated as the last step. Thus, a series of IMF can be obtained. The signal $x(t)$ can be expressed as

$$x(t) = \sum_{i=1}^{m} c_i + r_m \tag{7}$$

EMD decomposes each EEG signal into $(m = 1)$ frequency components. Here, m intrinsic mode functions (IMFs) represent the different higher frequency components of the original signals, while r_m corresponding to the lower frequency residue.

2.3   Feature Extraction of IMF Components

Variation coefficient, fluctuation index, relative energy and relative amplitude were extracted from the IMFs. The characteristic of ictal EEG signals are different from that of the interictal EEG signals. Variation coefficient measures the change of the signal's amplitude. Fluctuation index measures the intensity fluctuations of signals.

For epileptic EEG, the noise and trivial information in the raw EEGs can make the classification less effective and the complexity of sparse representation can be very high when the number of training samples is big. Feature extraction can reduce data dimension and computational cost. For long term EEG signals, the corresponding linear system is very large. For instance interictal EEG and ictal EEG signals contain 100 single-channel EEG epochs of 4097 points respectively. Each epoch is divided into four segments and the size is 400 × 1024. Due to the high-dimensional signal is sparse, selecting least than 4-D features can reduce the dimension of feature and computational complexity.

3 Results

The EEG signals used in this paper come from Bonn epileptic EEG. The Bonn dataset consists of five sets denoted as Z, O, N, F, and S, each containing 100 single-channel EEG epochs with duration of 23.6 s. In this study, Set F over interictal period and Set S over ictal period are examined. They all contain 100 epochs of 4097 points, and each epoch is divided into four segments with the same length of 1024 points. The classification of interictal (Set F) and ictal (Set S) EEGs is more difficult to solve but closer to clinical applications than the other classification problems based on this dataset.

For the recognition of ictal and interictal EEGs, half of the data per class (200 samples from each class) were chosen as the training samples, the rest samples from each class were used to test the performance of this method. For Bonn EEG data, EEG segments are decomposed by EMD firstly. By analyzing the spectrum of those IMFs with Fast Fourier Transform, the frequency ranges of IMF components are as follows: IMF1 (0–45 Hz), IMF2 (0–30 Hz), IMF3 (0–20 Hz), IMF4 (0–10 Hz), IMF5 (0–7 Hz), and IMF6 (0–3 Hz). Figure 1 shows the IMF components of an ictal EEG sample. From Fig. 1, the frequency of ictal EEG most concentrate on the IMF1–3 component (Table 2).

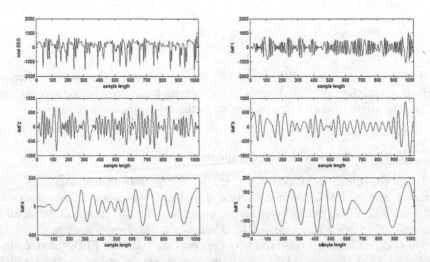

Fig. 1. The IMF components of an ictal EEG sample. The first one is the raw ictal EEG signal sample with 1024 points. The second one is the IMF1 frequency component. One by one, the last one is the fifth IMF component.

Form Table 1, it is found that based on SRC, the classification accuracy of IMF1–IMF3 component is better than the EEG sample without EMD. Especially the classification accuracy based on IMF1 linear feature can be up to 96.5%. In addition, the test time decreased obviously, which suit for the real time seizure detection in clinical application.

Table 1. Classification performance of IMF based on SRC

Component	Sensitivity	Specificity	Accuracy
Raw EEG	87.00	89.00	88.50
IMF1	97.00	95.00	96.50
IMF2	93.00	95.00	94.00
IMF3	88.00	91.50	89.50
IMF4	–	–	74.00
IMF5	–	–	66.50
IMF12	95.00	96.00	95.50
IMF123	99.00	96.00	97.50

Table 2. Comparision of the classification method

Method	Accuracy	Test time(s)
EMD+SVM	88.50	1.4
SRC	97.88 ± 2.87	40.0
Kernel SRC	98.63 ± 2.80	3.9
EMD + SRC	97.50 ± 2.50	0.5

4 Conclusions

The sparse representation methods have been widely applied to pattern classification in recent years. In the scheme of sparse representation based classification (SRC), a new test EEG sample is sparsely represented on the training dataset.

In this work, we propose a seizure detection method using SRC and EMD to classify the epileptic ictal EEG and interictal EEG. EMD is an adaptive and efficient method applied to analysis non-stationary signals. To make full use of subbands frequency characteristics, extracted the features such as variation coefficient, fluctuation index, relative energy and relative amplitude from the IMFs. The feature extraction can reduce the data dimension and computational cost. This method is faster than SVM, which have better performance in the real-time diagnosis of epilepsy in the future.

Acknowledgments. This work was supported by the National Natural Science Foundation of China (Grant No. 61671220, 61640218, 61201428), the Shandong Distinguished Middle-aged and Young Scientist Encourage and Reward Foundation, China (Grant No. ZR2016FB14), the Project of Shandong Province Higher Educational Science and Technology Program, China (Grant No. J16LN07), the Shandong Province Key Research and Development Program, China (Grant No. 2016GGX101022).

References

1. Subasi, A.: EEG signal classification using wavelet feature extraction and a mixture of expert model. Expert Syst. Appl. **32**(4), 1084–1093 (2007)
2. Adeli, H., Zhou, Z., Dadmehr, N.: Analysis of EEG records in an epileptic patient using wavelet transform. J. Neurosci. Methods **123**, 69–87 (2010)
3. Wang, Y., Zhou, W.D., Li, S.F., Yuan, Q., Geng, S.J.: Fractal intercept analysis of EEG and application for seizure detection. Chin. J. Biomed. Eng. **30**, 562–566 (2011)
4. Ocak, H.: Automatic detection of epileptic seizures in EEG using discrete wavelet transform and approximate entropy. Expert Syst. Appl. **36**, 2027–2036 (2009)
5. Achary, U.R., Molinari, F., Sree, S.V., Chattopadhyay, S., Ng, K.H., Suri, J.S.: Automated diagnosis of epileptic EEG using entropies. Biomed. Sig. Process. Control **7**, 401–408 (2012)
6. Acharya, U.R., Sree, V., Chattopadhyay, S., Yu, W.W., Alvin, P.C.: Application of recurrence quantification analysis for the automatic EEG signals. Int. J. Neural Syst. **21**, 199–211 (2011)
7. Cai, D.M., Zhou, W.D., Liu, K., Li, S.J., Geng, S.J.: Approach of eplileptic EEG detectionbased on hurst exponent and SVM. Chin. J. Biomed. Eng. **29**, 836–840 (2010)
8. Zhou, W., Liu, Y., Yuan, Q., Li, X.: Epileptic seizure detection using lacunarity and bayesian linear discriminant analysis in intracranial EEG. IEEE Trans. Biomed. Eng. **60**, 3375–3381 (2013)
9. Guo, L., Rivero, D., Seoane, J., Pazos, A.: Classification of EEG signals using relative wavelet energy and artificial neural networks. In: The First ACM/SIGEVO Summit on Genetic and Evolutionary Computation, pp. 177–183 (2009)
10. Li, S., Zhou, W., Yuan, Q., Geng, S., Cai, D.: Feature extraction and recognition of ictal EEG using EMD and SVM. Comput. Biol. Med. **43**, 807–816 (2013)
11. Zhao, J.L., Zhou, W.D., Liu, K., Cai, D.M.: EEG signal classification based on SVM and wavelet analysis. Comput. Appl. Softw. **28**, 114–116 (2011)
12. Lima, C.A.M., Coelho, A.L.V., Chagas, S.: Automatic EEG signal classification for epilepsy diagnosis with relevance vector machines. Expert Syst. Appl. **36**, 10054–10059 (2009)
13. Yuan, Q., Zhou, W., Li, S., Cai, D.: Epileptic EEG classification based on extreme learning machine and nonlinear features. Epilepsy Res. **96**, 29–38 (2011)
14. Tafreshi, A.K., Nasrabadi, A.M., Omidvarnia, A.H.: Epileptic seizure detection using empirical mode decomposition. In: 8th IEEE International Symposium on Signal Processing and Information Technology, pp. 238–242 (2008)
15. Braun, S., Feldman, M.: Decomposition of nonstationary signals into varying time scales: some aspects of the EMD and HVD methods. Mech. Syst. Sig. Process. **25**, 2608–2630 (2011)
16. Wright, J., Yang, A.Y., Ganesh, A., Sastry, S.S., Ma, Y.: Robust face recognition via sparse representation. IEEE Trans. Pattern Anal. Mach. Intell. **31**(2), 210–227 (2009)
17. Georgiev, P., Theis, F., Cichocki, A.: Sparse component analysis and blind source separation of underdetermined mixtures. IEEE Trans. Neural Netw. **16**(4), 992–996 (2005)
18. Yang, M., Zhang, L., Feng, X.C., Zhang, D.: Fisher discrimination dictionary learning for sparse representation. In: Proceedings of the International Conference Computer Vision (2011)
19. Ren, Y.F., Wu, Y., Ge, Y.B.: A co-training algorithm for EEG classification with biomimetic pattern recognition and sparse representation. Neurocomputing **137**, 212–222 (2014)

20. Faust, O., Acharya, U.R., Adeli, H., Adeli, A.: Wavelet-based EEG processing for computer-aided seizure detection and epilepsy diagnosis. SEIZURE: Eur. J. Epilepsy **26**, 56–64 (2015)
21. Yang, M., Zhang, L., Feng, X., Zhang, D.: Sparse representation based fisher discrimination dictionary learning for image classification. Int. J. Comput. Vis. **109**(3), 209–232 (2014)
22. Chen, Z.H., Zuo, W.M., Hu, Q.H., Lin, L.: Kernel sparse representation for time series classification. Inf. Sci. **292**(20), 15–26 (2015)

Scaling of Texture in Training Autoencoders for Classification of Histological Images of Colorectal Cancer

Tuan D. Pham[✉]

Department of Biomedical Engineering,
Linkoping University, 58183 Linkoping, Sweden
`tuan.pham@liu.se`

Abstract. Autoencoding in deep learning has been known as a useful tool for extracting image features in multiple layers, which are subsequently configured for classification by deep neural networks. A practical burden for the implementation of autoencoders is the time required for training a large number of artificial neurons. This paper shows the effects of scaling of texture in the histology of colorectal cancer, which can result in significant training time reduction being approximately to an exponential function, with improved classification rates.

Keywords: Deep neural networks · Image classification · Digital pathology · Colorectal cancer · Tissue types

1 Introduction

Tumors are well-known to be different among patients, and within the tumor itself at its tissue and its cell levels. In fact, tumor morphological heterogeneity has been recognized by pathologists, and used as the basis for many tumor grading prognostic classification systems [1]. In colorectal cancer (CRC), tumor architecture changes during tumor progression [2], and is related to patient prognosis [3]. Therefore, quantifying images of tissue types in CRC is important in digital histopathology [4]. In cancer research, the use of machine-learning techniques for computerized histological image analysis is a key factor for advancing methods of major diagnostic importance, minimizing human subjective error, and providing vital clinical information [5–8].

Deep learning [9] is a machine-learning method that operates on nonlinear processing layers known as autoencoders to extract useful features directly from the raw data for object classification. A deep-learning model is trained with a large set of data and a neural-network architecture of multiple layers. The accuracy of a deep leaning model largely depends on the amount of data used to train the model. In other words, to achieve an accurate deep-learning model, thousands or even millions of training samples are required, which can take a very large amount of time for the model training. Once a deep-learning model is appropriately trained, it can be applied in real-time applications.

© Springer International Publishing AG 2017
F. Cong et al. (Eds.): ISNN 2017, Part II, LNCS 10262, pp. 524–532, 2017.
DOI: 10.1007/978-3-319-59081-3_61

This study aims to tackle a state-of-the-art problem in translational colorectal cancer research with the application of deep learning and neural-network
classification in histological image analysis by effective handling of computational time requirement for the model training. The proposed texture scaling in
autoencoder training is promising for constructing a practical deep-NN-based
classifier that can provide precise and predictive diagnosis of the cancer with the
goal of providing a clinical decision support tool to help oncologists select the
best treatment plans for individual patients.

2   Scaling of Texture in CRC Histology

As texture can be synthesized to generate larger images of similar statistical
properties to improve texture retrieval [10], it can also be resampled to images
of smaller sizes that still preserve the spatial statistics of the same texture. By
resampling images of texture, the deep-learning training time required by the
autoencoders can be reduced. Here, bicubic interpolation, which is an extended
cubic interpolation for estimating data points on a two-dimensional regular grid,
is utilized for image resampling [11]. Bicubic interpolation is carried out using
cubic convolution algorithm, where the output pixel intensity is a weighted average of pixels in the nearest 4×4 neighborhood. In image processing, bicubic interpolation is preferred to bilinear interpolation. It is because images resampled
with bicubic interpolation are smoother and have fewer interpolation artifacts
than bilinear interpolation [11].

Figure 1 provides examples of histological images of CRC tumor and stroma
tissues, obtained from [7], that show visual appearance of textures. Figure 2
shows the experimental semivariograms [12] of CRC histological images of a
tumor tissue, a stroma tissue, and their resampled images. The experimental
semivariogram, denoted as $\gamma(h)$, is a function that represents the spatial correlation of spatial data measured with distances between all data pairs at sampled
locations, and mathematically defined as

$$\gamma(h) = \frac{1}{2N(h)} \sum_{i=1}^{N(h)} [Z(x_i) - Z(x_i + h)]^2, \tag{1}$$

where $Z(x_i)$, $i = 1, 2, \ldots, n$, be a sampling of size n, $N(h)$ is the number of
pairs of variables separated by distance h. In this study, the semi-variogram of
an image at a lag h is calculated by taking the sum of squared differences of the
intensity values of pixel pairs separated by h in both rows and columns, then
divided by the total number of the pixel pairs.

It can be observed from Fig. 2 that the semivariograms of the resampled textures preserve similar spatial structures to those of the original textures, where
the values of $\gamma(h = 1)$ for 50×50 and 20×20 images increase proportionally
to the scaling factors with respect to the original 150×150 images. The semi-
variogram shapes of the tumor-tissue texture follows a spherical function of the

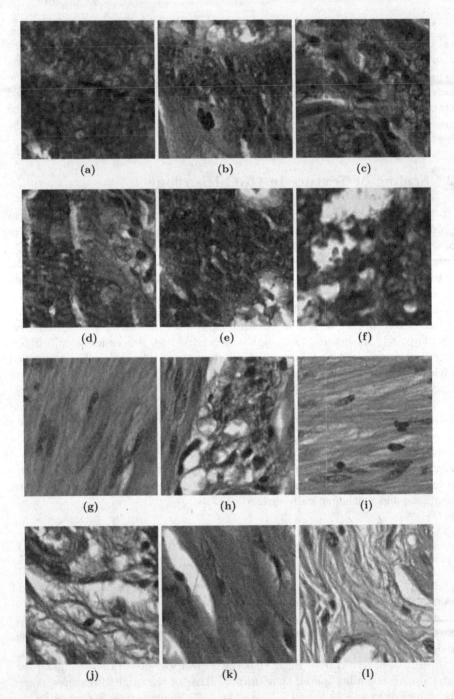

Fig. 1. CRC histological images [7]: (a)–(f) tumor tissues, and (g)–(l) stroma tissues.

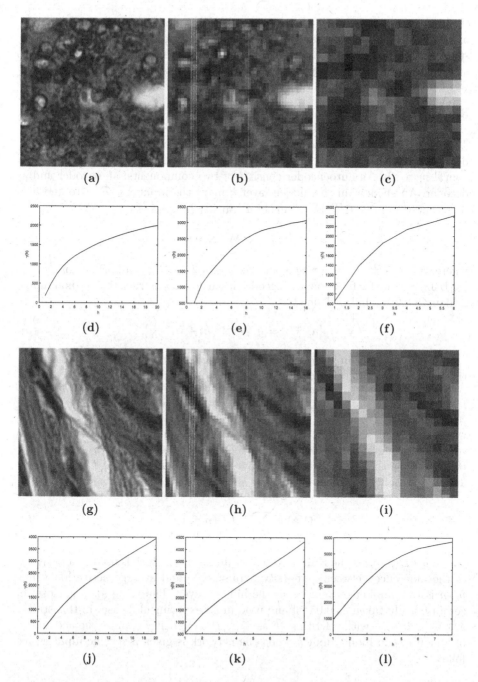

Fig. 2. (a)–(c) are 150 × 150, resized 50 × 50, and resized 20 × 20 images of the same CRC tumor tissue type, respectively, (d)–(f) are semivariograms of (a)–(c), respectively, (g)–(i) are 150 × 150, resized 50 × 50, and resized 20 × 20 images of the same stroma tissue type, respectively, and (j)–(l) are semivariograms of (g)–(i), respectively.

theoretical semivariogram, whereas those of the stroma-tissue texture approximately exhibit a linear curve [13]. Thus, the preservation of the spatial statistical structures of the scaled CRC histological images technically justify the notion of their texture scaling for feature learning using the autoencoders.

3   Configuration of Autoencoders

The discovery of hidden and effective features of an object can be obtained by deep learning with the implementation of an autoencoder that is an unsupervised neural network. An autoencoder consists of two components: an encoder and a decoder. An encoder in the hidden layer k maps the input $\mathbf{x} \in \mathcal{R}^{D_\mathbf{x}}$ to another representation $\mathbf{a}^k \in \mathcal{R}^{D^k}$ by means of a transfer function f:

$$\mathbf{a}^k = f(\mathbf{x}) = f(\mathbf{W}^k\,\mathbf{x} + \mathbf{b}^k), \tag{2}$$

where $\mathbf{W}^k \in \mathcal{R}^{D^k \times D_\mathbf{x}}$ is a weight matrix, and $\mathbf{b}^k \in \mathcal{R}^{D^k}$ is a bias vector.

Using a transfer function g, a decoder learns to reconstruct the original input $\mathbf{x}$, denoted as $\mathbf{y}$, and is defined as

$$\mathbf{y} = g[\mathbf{a}^{(k+1)}] = g[\mathbf{W}^{(k+1)}\,\mathbf{x} + \mathbf{b}^{(k+1)}]. \tag{3}$$

The optimal learning process of the autoencoder is performed by minimizing the following loss function $\mathcal{L}$ [9]:

$$\mathcal{L} = L(\mathbf{x},\mathbf{y}) + \lambda_s\,\Omega(\mathbf{a}) + \lambda_r\,\Omega(\mathbf{w}), \tag{4}$$

where $L(\mathbf{x},\mathbf{y})$ is a loss function such as the mean squared error that imposes a cost for the difference between $\mathbf{x}$ and its reconstructed signal $\mathbf{y}$, $\Omega(\mathbf{a})$ is a sparsity penalty, λ_s is the coefficient for the sparse penalty, $\Omega(\mathbf{w})$ is the L_2 regularizer and λ_r its coefficient. These terms are mathematically defined as follows.

$$\Omega(\mathbf{a}) = \sum_{i=1}^{n} KLD(p||\tilde{p}_i), \tag{5}$$

where $KLD(p||\tilde{p}_i)$ is the Kullback-Leibler divergence, which is used as a measure of the difference between a sparsity parameter p and average activation value $\tilde{p}_i$, n is the number of neurons in the hidden layer. The value for p is close to zero, typically taken as 0.05 [9] and used in all experimental cases in this study, and $\tilde{p}_i$ is defined with rewriting a_i^k as $a_i^k(x_j)$ to explicitly express the activation of a neuron i in hidden unit k when the network is given a specific input x_j as follows

$$\tilde{p}_i = \frac{1}{N} \sum_{j=1}^{N} a_i^k(x_j), \tag{6}$$

where N is the total number of training samples.

The L_2 regularizer attempts to add a regularization term on the weights to the loss function:

$$\Omega(\mathbf{w}) = \frac{1}{2} \sum_{k=1}^{L} \sum_{j=1}^{N} \sum_{i=1}^{V} (w_{ji}^k)^2 \tag{7}$$

where L, N, and V are the numbers of hidden layers, training samples, and training-data variables, respectively.

4   Results

The CRC histology image database [7], which is made publically available under a Creative Commons License, was used in this study. The database consists of RGB images of 8 texture classes in histological images of human colorectal cancer, each class has 625 images. The deep-NN based classification of tumor and stroma tissues with the configuration of the autoencoders described in the foregoing section was carried out in this experiment. A neural network with two hidden layers were trained individually in an unsupervised mode using the autoencoders. The feature vectors produced from the first autoencoder were passed to the second autoencoder to subsequently generate the second set of feature vectors that were used as the input to train the softmax layer in a supervised mode. The encoders from the autoencoders together with the softmax layer were stacked together to form a deep neural network. Finally, the fine tuning of the deep neural network was carried out to improve its classification power by performing the backpropagation to retrain the whole network in a supervised fashion. In the training of the two autoencoders, the sizes of the first and second hidden layers for the original images (150×150 pixels) and resampled images were chosen to be twice and half the number of image rows or columns, respectively.

Table 1 shows the average two-fold cross-validation results and associated computational times obtained from the deep neural networks for the two-class classification problem with various resamplings of textures, in which the scaling to 30×30 achieves the minimum classification error rates. Figure 3 shows the plots of texture scaling vs. computational time, and scaling vs. classification error, where the reduction in computational time for the autoencoder training in terms of image size approximately follows an exponential function, and the scaling of the CRC histology texture can be reduced 5 times with an improved classification rate. It was also of interest to see the effect of selecting various sizes for the hidden layers of the autoencoders using the resampled textures. Table 2 shows the average error rates obtained from the two-fold cross-validation using the resampled texture images of 30×30 and 20×20 pixels. All the results are also more favorable than the average error rate using the original images of a much larger size.

Table 1. Average classification error rates and computational times (seconds) of two-fold cross-validation using various texture scalings.

Image size	Hidden layer #1	Hidden layer #2	Error rate	Computational time
150 × 150	450	225	0.25	6403
130 × 130	390	195	0.21	2506
100 × 100	300	150	0.29	1848
80 × 80	240	120	0.28	1020
60 × 60	180	90	0.32	435
50 × 50	150	75	0.34	280
40 × 40	120	60	0.22	148
30 × 30	90	45	0.19	65
20 × 20	60	30	0.20	26

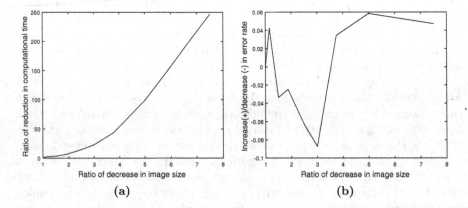

Fig. 3. Effects of CRC texture scaling in deep learning: (a) scaling vs. computational time, and (b) scaling vs. classification error rates.

Table 2. Average classification error rates of two-fold cross-validation using resampled images of smaller sizes.

Image size	Hidden layer #1	Hidden layer #2	Error rate
30 × 30	400	400	0.22
30 × 30	300	300	0.16
30 × 30	200	200	0.18
20 × 20	300	300	0.23
20 × 20	200	200	0.19
20 × 20	100	100	0.19

5   Conclusion

Large texture images in CRC histology are prone to excessive computational time for data training in deep neural-network classification, otherwise the performance of the classifier may suffer from achieving expected accuracy. This paper proposed texture scaling in CRC histology that can alleviate the computational burden in training the networks, while still produce better performance than the use of a much larger image size of texture. The experimental results suggest that the use of smaller texture image size with an appropriate selection of hidden layer sizes can be more favorable for the training of the autoencoders in terms of both time and accuracy. Furthermore, object recognition in 2D images is known to be well suited to the application of convolutional neural network (CNN) architectures, which use 2D convolutional layers to convolve deep-learning features with input image data for classification, and has become a popular deep-learning technique [14]. In this study, only two hidden layers were created for deep learning as an example, a larger number of hidden layers would be expected to show the higher performance of deep neural networks with respect to the increase in classification accuracy and reduction in machine-learning time. Furthermore, a future study will aim at the effective implementation of a CNN model for texture classification of CRC histological images with the feasibility of the computational speed.

References

1. Jesinghaus, M., et al.: Genetic heterogeneity in synchronous colorectal cancers impacts genotyping approaches and therapeutic strategies. Genes Chromosom. Cancer **55**, 268–277 (2016)
2. Burrell, R.A., McGranahan, N., Bartek, J., Swanton, C.: The causes and consequences of genetic heterogeneity in cancer evolution. Nature **501**, 338–345 (2013)
3. Lawrence, M.S., et al.: Mutational heterogeneity in cancer and the search for new cancer-associated genes. Nature **499**, 214–218 (2013)
4. The Cancer Genome Atlas Research Network: Comprehensive molecular characterization of human colon and rectal cancer. Nature **487**, 330–337 (2012)
5. Aerts, H.J., et al.: Decoding tumour phenotype by noninvasive imaging using a quantitative radiomics approach. Nat. Commun. **5**, Article no. 4006 (2014)
6. Janowczyk, A., Madabhushi, A.: Deep learning for digital pathology image analysis: a comprehensive tutorial with selected use cases. J. Pathol. Inform. **7**, 29 (2016)
7. Kather, J.N., et al.: Multi-class texture analysis in colorectal cancer histology. Sci. Rep. **6**, 27988 (2016)
8. Chaddad, A., et al.: Multi texture analysis of colorectal cancer continuum using multispectral imagery. PLoS ONE **11**, e0149893 (2016)
9. Goodfellow, I., Bengio, Y., Courville, A.: Deep Learning. MIT Press (2016, in preparation). http://www.deeplearningbook.org
10. Pham, T.D.: Enhancing texture characteristics with synthesis and noise for image retrieval. In: Proceedings of the 8th IEEE International Conference on Intelligent Systems, pp. 433–437. IEEE Press, New York (2016)
11. Keys, R.: Cubic convolution interpolation for digital image processing. IEEE Trans. Acoust. Speech Signal Process. **29**, 1153–1160 (1981)

12. Pham, T.D.: The semi-variogram and spectral distortion measures for image texture retrieval. IEEE Trans. Image Process. **5**, 1556–1565 (2016)
13. Olea, R.A.: Geostatistics for Engineers and Earth Scientists. Kluwer Academic Publishers, Boston (1999)
14. Nielsen, M.A.: Neural Networks and Deep Learning. Determination Press (2015). http://neuralnetworksanddeeplearning.com

Multi-channel EEG Classification Based on Fast Convolutional Feature Extraction

Qian Wang[1], Yongjun Hu[2(✉)], and He Chen[1]

[1] School of Electrical and Information Engineering,
University of Sydney, Sydney, NSW 2006, Australia
qwan9983@uni.sydney.edu.au, he.chen@sydney.edu.au
[2] School of Business, Guangzhou University, Guangzhou 510006, China
hyjsdu96@126.com

Abstract. In this paper, we develop a novel feature extraction approach for multi-channel electroencephalography (EEG) classification. Inspired by convolutional neural networks (CNNs), we devise a fast convolutional feature extraction approach for EEG classification. In our approach, convolutional filters are first applied to extract features of multi-channel EEG signals. Then weak classifier selection is adopted to adaptively choose important features, which will be used for final classification. After that, we evaluate the performance of selected features through classification accuracy. Experiments on BCI III IVa competition dataset demonstrate the superior performance of our method, compared with the same classifier without feature extraction and deep learning methods, such as CNNs and long short term memory (LSTM). This work can be used to form the framework of deep neural networks for EEG signal processing.

Keywords: EEG · Feature extraction · Convolutional filter · Classification

1 Introduction

Electroencephalography (EEG) is the recording of bio-potential, measured from the scalp of human brain. How to classify the EEG signals becomes a hot research issue [1, 2]. Classification of EEG signals reveals brain states and intention, which is one of the three major components of brain-computer interface (BCI) [3]. Researches in this field cannot fully satisfy the demand of application and there is only a few products of BCI in the market. This, to some extent, is due to the fact that most of EEG signal classification methods are traditional methods [1, 2], lagging behind the development of machine learning, such as deep learning. It is rare to apply deep learning methods in this area, which have brought fundamental breakthroughs in many other areas [4].

EEG signals are neither linear, nor stationary, and always interfered by the movement and potential of eye and muscle [5]. This makes it challenging to classify EEG signals using traditional methods. Directly inputting the raw data into classifier will not only take a long time for computation but also consume huge memory resources. In some cases, the signal-to-noise ratio can be increased by extracting features from auxiliary samples [6]. In this sense, it is essential to extract the features of EEG signals before their classification.

© Springer International Publishing AG 2017
F. Cong et al. (Eds.): ISNN 2017, Part II, LNCS 10262, pp. 533–540, 2017.
DOI: 10.1007/978-3-319-59081-3_62

Features of EEG signals can be extracted in time, space and frequency domains [3] or through other methods. In the open literature, there have been some existing methods for EEG feature extraction, such as Bayesian common spatial pattern (BCSP) [6], source analysis [7], time-frequency feature extraction [8], and discrete cosine transform (DCT) [9].

Existing methods for feature extraction, some focusing on single channel led to the loss of information, some using multi-channel but failed to insure the channel location. In this paper, we apply a new method of feature extraction to classify EEG signals based on convolutional filter design. The proposed approach is inspired by convolutional neural networks (CNNs) for sentence classification [10], which has already demonstrated excellent results on nature language processing. The work in [10] uses two-dimensional matrix to represent the one-dimension sentence based on word embedding. Inspired by this, we strategically regard multi-channel EEG signals as a two-dimensional matrix based on time dynamic and the location of electrodes. Then we follow the convolutional rule of two-dimensional processing to extract features from EEG signals. Several attempts are then made to find appropriate convolutional filters in order to capture important features of EEG samples. Later, we work on selecting representative features by adopting weak classifier selection algorithm implemented in [11]. Then artificial neural networks (ANNs) and linear discriminant analysis (LDA) are performed to classify EEG samples in order to evaluate the effectiveness of the selected features. Our work can be regard as an exploration of transferring deep learning methods into EEG signal processing to some extent.

Our experiments show that the proposed classifier with fast convolutional feature extraction achieves better performance, which is about 5% higher than the accuracy of the classifier without feature extraction. This shows that features extracted by convolutional filter are beneficial to EEG classification. Additionally, the result also illustrates that further work can be done on filter and architecture design to make convolutional neural networks suitable for EEG classification.

The rest of this paper is organized as follows. In Sect. 2 we present the proposed method used for feature extraction and classification. In Sect. 3, we describe how to carry out the experiment based on the BCI III IVa dataset to validate our method. We compare its performance with that of other classification methods as well. At last, we draw conclusions in Sect. 4.

2   Method

This section explains how our method works from feature extraction to classification and Fig. 1 provides a block diagram of our method. The raw EEG signals are filtered by band pass filter beforehand. Then, fast convolutional feature extraction is implemented to accomplish feature selection. Based on the most important and representative features, we choose ANNs as classifier to evaluate the performance of selected features.

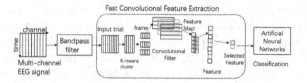

Fig. 1. The block diagram of classification method

2.1   Fast Convolutional Feature Extraction with Weak Classifier Selection

Two-Dimensional Processing

Based on the preprocessed data, we form a two-dimensional matrix using all channels. Then, we partition channels into several clusters to capture different channels' representative information. Considering time dependency of EEG signals, we divide EEG signals into M frames. Let $e_i^m \in \mathbb{R}^k$ be the k-dimensional vector corresponding to the clustered EEG signals of m-th frame at i-th time point from k clustered channels. Each frame consists of T time instants, and m-th frame is represented as

$$e_{1:T}^m = e_1^m \oplus e_2^m \oplus \ldots \oplus e_T^m, \tag{1}$$

where $\oplus$ represents the concatenation operator. And the length of each frame T can be adjusted to find the most suitable length to capture time dependency.

2D Convolutional Feature Extraction

Here is the way of using convolution for EEG feature extraction. This is analogous to the convolution used for sentence classification in [10]. Let $e_{i:i+j}^m$ refer to EEG signals of m-th frame from i-th time instant to $(i+j)$-th time instant. Convolution operation includes a filter $w \in \mathbb{R}^{hk}$, which filters a window of h time instants EEG signals to produce a new feature. For example, using

$$f_i^m = w e_{i:i+h-1}^m, \tag{2}$$

a feature f_i^m is generated. When apply this filter to each window of EEG signals from particular frame $\{e_{1:h}^m, e_{2:1+h}^m, \ldots, e_{T-h+1:T}^m\}$, *a feature map*

$$\mathbf{f}^m = [f_1^m, f_2^m, \ldots, f_{T-h+1}^m] \tag{3}$$

is produced with $\mathbf{f}^m \in \mathbb{R}^{T-h+1}$, followed by a max-over-time pooling operation [12] to take the maximum value of the feature map $\mathbf{f}^m = \max\{\mathbf{f}^m\}$ to represent the most important feature generated by corresponding convolution filter. Naturally, we regard it as the feature of current filter. This is to capture the most remarkable feature for each filter, i.e., the one with highest value from each *feature map*.

Filter Design

Apart from different channels' information, EEG signals are also a kind of time series. In this way, we design filters considering the time dynamic to capture both long-term dependency and short-term dependency of EEG signals.

Inspired by edge detection of image processing, we design 3 types of filters to capture different time characteristic shown in Fig. 2. Type 1 focuses on difference between current time instant and previous n instants while Type 2 captures changes in current instant and previous n instants as well as latter n instants. Type 3 detects the amplitude variance of EEG signal with n time instants. Here, we set $n = 1, 2, \ldots, 20$. And different lengths of filters aim to obtain both short-term and long-term dependency of EEG signals. Figure 2 shows columns from convolution filters (we display in row only to make it clear to understand). We expand each column into a two-dimensional convolutional filter to process all clusters.

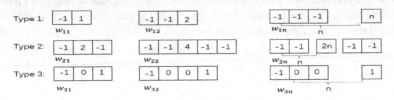

Fig. 2. Three types of convolution filters used for feature extraction. $n = 20$

Weak Classifier Selection

Before classification, we use weak classifier selection algorithm [11] to learn the importance of all features. Then, we choose the most representative features for later classifier.

2.2 Classifier

With the selected features, we use classification result to evaluate the effect of our algorithm. There are several popular ways for classification. Here, we focus on two widely used machine learning algorithms for EEG classification, linear discriminant analysis (LDA) [6] and artificial neural networks (ANNs) [9, 13].

3 Experiments and Results

3.1 Dataset and Evaluation

The dataset we used is BCI Competition III IVa[1] [14] dataset containing EEG signals from 5 subjects with 2 motor imagery movement classes, sampled at the rate of 100 Hz. There are totally 560 samples with labeled motor imagery movement class and each was cut to 3.5 s (350 time points) from the beginning of it. This makes it more challenging

---

[1] http://www.bbci.de/competition/iii/.

and drives us to classify. As for evaluation, we use the classification accuracy to evaluate the performance of feature extraction. Below is how accuracy is calculated

$$accuracy = \frac{n_{true}}{n_{test}} \times 100, \qquad (4)$$

where n_{test} denotes the number of total testing samples and n_{true} denotes the number of testing samples that are classified correctly.

3.2 Preparing Training Data

Before feature extraction, we applied bandpass filter to obtain the signal component in bandwidth from 7 Hz to 30 Hz. EEG signals from different channels was filtered. Then, we partitioned channels by k-means clustering and divided the data into several frames. By adjusting the number of clusters and the length of frames, we can find the most suitable combo of them.

3.3 Feature Extraction

Here, we applied 3 types of convolutional filters showed in Fig. 2 to capture the features of each frame. And each type of filters had 20 kinds of different lengths. In this way, each frame would have 60 features. After combining features from different frames of single sample, we ranked the importance of the features. Table 1 below shows the top 10 features in the important index when treating each sample as a frame.

Table 1. Top 10 important features treating each sample as a frame. $f_{i,j}$ denotes the feature generated by filter $w_{i,j}$ (i for type, j for length)

Ranking	Feature
1	$f_{2,1}$
2	$f_{3,19}$
3	$f_{3,8}$
4	$f_{3,6}$
5	$f_{3,18}$
6	$f_{2,2}$
7	$f_{3,20}$
8	$f_{3,17}$
9	$f_{1,3}$
10	$f_{3,16}$

3.4 Training Classifier

Initially, neural network is set with random weights and bias. With enough number of training epochs, parameters of neural network will converge to the best weights with highest accuracy. Considering the initialized random weight and bias, it is necessary to perform the experiment for several times and use the average accuracy to represent the overall performance. For a given number of hidden layers and same transfer function, both the size of the input vector and the representativeness of input features will influence the accuracy of classification.

With calculated importance index, we selected top m number of features and set m to 2, 3, 4, 5, 6, 7, 8, 9 and 10, respectively. We constructed the ANNs consisting of 10 hidden layers and 1 softmax layer, and adopted scaled conjugate gradient algorithm [15] for supervised learning. Before training, we randomly divided the dataset into two groups, 75% for training, the rest for testing, and used the testing group to calculate the accuracy of our algorithm. Then, we calculated the average classification accuracy. Base on that, we can find the best combo of features that is suitable for the classifier and can represent the EEG sample well.

3.5   Result

We evaluated the suitable number of features for classifier by the average accuracy. Figure 3 displays the performance of our algorithm, where the performance of LDA and ANNs were compared with different size of input feature vectors, length of frame and number of clusters.

Figure 3 below shows the accuracy of our methods with different number of input features. Plot (a) and (b) in Fig. 3 indicate that when channels of current EEG sample were partitioned into k clusters, classification accuracy of 32 clusters was the highest among the result of 32, 64, 96 clusters and that without clustering. Plot (c) and (d) in Fig. 3 demonstrate that when EEG sample was divided into several frames, dividing the sample into 7 frames at the length of 50 time instants achieved highest accuracy for classification.

When the channels of EEG signals were partitioned into 32 clusters with 7 frames, ANNs and LDA achieved their highest accuracy of 59.13% with top 4 features and 57.81% with top 3 features, respectively. A common trend was that both two method suffered performance degradation as the number of features used increasing. And it shows that our feature extraction has its advantage to some extent.

Besides, we directly applied the CNNs of one-layer architecture followed by max-time-over pool and softmax layer to realize the classification, and only got the accuracy of 44.76%. The accuracy of long short term memory (LSTM) was 52.32% which was only slightly higher than CNNs. ANNs and LDA were applied to EEG signals with single channel as well. The highest accuracy of both methods are displayed in Table 2. Among all methods, accuracy of classifier with our feature extraction method was up to 5% higher than those without feature extraction.

The dataset we used is relatively small and comes from 5 subjects with more individual difference. We randomly divided the data, which made the result more dependable as well as the feature extraction method. In addition, we focus on the performance of feature extraction and the result indicates the effectiveness of our method with accuracy improvement.

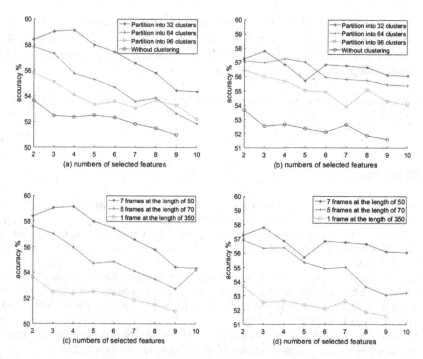

Fig. 3. Classification accuracy of ANNs and LDA with selected features. Plots (a) (c) are shown for accuracy of ANNs and plots (b) (d) are shown for accuracy of LDA

Table 2. Result of other methods

Method	CNN	LSTM	ANN with single channel	LDA with single channel	ANN with feature extraction	LDA with feature extraction
Accuracy %	44.76	52.32	54.36	52.93	59.13	57.81

4 Conclusion

We proposed a new feature extraction method for multi-channel electroencephalography (EEG) classification referring to convolutional neural networks (CNNs). Different from the use of normal CNNs in other areas, our work did not simply applying the CNNs to multi-channel EEG signals. We designed convolutional filters focusing on capturing characteristic of EEG signals.

In our method, based on convolutional filters, we extracted the remarkable features. Then, we applied weak classifier selection algorithm to obtain the importance index. Using classifier with same architecture but different from size of input vector, we found out best combo among all the features ranking in the index. And comparing with the other algorithms, such as traditional CNNs and classifier without feature extraction, our method showed its merits in terms of higher accuracy. Even though the improvement is not that significant, it paves the way for deep learning application in EEG signal processing. For future work, we will use the fast convolutional feature extraction to form the framework of deep neural networks to achieve higher performance in EEG signal processing.

Acknowledgements. The authors gratefully acknowledge financial support from China Scholarship Council.

References

1. Sonkin, K., Stankevich, L., Khomenko, Y., Nagornova, Z., Shemyakina, N., Koval, A., Perets, D.: Neurological classifier committee based on artificial neural networks and support vector machine for single-trial EEG signal decoding. In: Cheng, L., Liu, Q., Ronzhin, A. (eds.) ISNN 2016. LNCS, vol. 9719, pp. 100–107. Springer, Cham (2016). doi:10.1007/978-3-319-40663-3_12
2. Li, W., Zou, L., Zhou, T., Wang, C., Zhou, J.: A two-stage channel selection model for classifying EEG activities of young adults with internet addiction. In: Cheng, L., Liu, Q., Ronzhin, A. (eds.) ISNN 2016. LNCS, vol. 9719, pp. 66–73. Springer, Cham (2016). doi:10.1007/978-3-319-40663-3_8
3. Cichocki, A., Washizawa, Y., Rutkowski, T.M., et al.: Noninvasive BCIs: multiway signal-processing array decompositions. IEEE Comput. **41**(10), 34–42 (2008)
4. LeCun, Y., Bengio, Y., Hinton, G.: Deep learning. Nature **521**(7553), 436–444 (2015)
5. Klonowski, W.: Everything you wanted to ask about EEG but were afraid to get the right an-swer. Nonlinear Biomed. Phys. **3**(1), 2 (2009)
6. Kang, H., Choi, S.: Bayesian common spatial patterns for multi-subject EEG classification. Neural Netw. **57**, 39–50 (2014)
7. Qin, L., Ding, L., He, B.: Motor imagery classification by means of source analysis for brain–computer interface applications. J. Neural Eng. **1**(3), 135 (2004)
8. Boashash, B., Azemi, G., Khan, N.A.: Principles of time–frequency feature extraction for change detection in non-stationary signals: applications to newborn EEG abnormality detection. Pattern Recogn. **48**(3), 616–627 (2015)
9. Birvinskas, D., Jusas, V., Martišius, I., et al.: Data compression of EEG signals for artificial neural network classification. Inf. Technol. Control **42**(3), 238–241 (2013)
10. Kim, Y.: Convolutional neural networks for sentence classification. arXiv preprint arXiv: 1408.5882 (2014)
11. Viola, P., Jones, M.J.: Robust real-time face detection. Int. J. Comput. Vis. **57**(2), 137–154 (2004)
12. Collobert, R., Weston, J., Bottou, L., et al.: Natural language processing (almost) from scratch. J. Mach. Learn. Res. **12**(Aug), 2493–2537 (2011)
13. Lotte, F., Congedo, M., Lecuyer, A., Lamarche, F., Arnaldi, B.: A review of classification algorithms for EEG-based brain-computer interfaces. J. Neural Eng. **4**, 1–24 (2007)
14. Blankertz, B., Muller, K.R., Krusienski, D.J., et al.: The BCI competition III: validating alternative approaches to actual BCI problems. IEEE Trans. Neural Syst. Rehabil. Eng. **14** (2), 153–159 (2006)
15. Moller, M.F.: A scaled conjugate gradient algorithm for fast supervised learning. Neural Netw. **6**(4), 525–533 (1993)

Hearing Loss Detection in Medical Multimedia Data by Discrete Wavelet Packet Entropy and Single-Hidden Layer Neural Network Trained by Adaptive Learning-Rate Back Propagation

Shuihua Wang[1], Sidan Du[2], Yang Li[2], Huimin Lu[3], Ming Yang[4], Bin Liu[5], and Yudong Zhang[1,6(✉)]

[1] School of Computer Science and Technology,
Nanjing Normal University, Nanjing 210023, Jiangsu, China
zhangyudong@njnu.edu.cn
[2] School of Electronic Science and Engineering,
Nanjing University, Nanjing 210046, Jiangsu, China
[3] Department of Mechanical and Control Engineering,
Kyushu Institute of Technology, Fukuoka Prefecture 804-8550, Japan
[4] Department of Radiology, Children's Hospital of Nanjing Medical University,
Nanjing 210008, China
[5] Department of Radiology, Zhong-Da Hospital of Southeast University,
Nanjing 210009, China
[6] Translational Imaging Division, Columbia University,
New York, NY 10032, USA

Abstract. In order to develop an efficient computer-aided diagnosis system for detecting left-sided and right-sided sensorineural hearing loss, we used artificial intelligence in this study. First, 49 subjects were enrolled by magnetic resonance imaging scans. Second, the discrete wavelet packet entropy (DWPE) was utilized to extract global texture features from brain images. Third, single-hidden layer neural network (SLNN) was used as the classifier with training algorithm of adaptive learning-rate back propagation (ALBP). The 10 times of 5-fold cross validation demonstrated our proposed method yielded an overall accuracy of 95.31%, higher than standard back propagation method with accuracy of 87.14%. Besides, our method also outperforms the "FRFT + PCA (Yang, 2016)", "WE + DT (Kale, 2013)", and "WE + MRF (Vasta 2016)". In closing, our method is efficient.

Keywords: Hearing loss · Multimedia data · Discrete wavelet packet entropy · Single-hidden layer neural network

1 Introduction

Multimedia data is a combination content of different data forms: text, audio, image, animation, and video. In medical application, the multimedia data offer refers to 3D volumetric data obtained by different imaging techniques.

© Springer International Publishing AG 2017
F. Cong et al. (Eds.): ISNN 2017, Part II, LNCS 10262, pp. 541–549, 2017.
DOI: 10.1007/978-3-319-59081-3_63

The sensorineural hearing loss (SNHL) is a disease featuring in gradual deafness [1]. SNHL contains thee types: (i) sensory hearing loss (SHL), (ii) neural hearing loss (NHL), and (iii) both. SHL may be due to bad function of cochlear hair cell, and NHL may be because of impairment of cochlear nerve function.

In this study, we aimed to use multimedia data obtained by magnetic resonance imaging (MRI) scanning [2] to differentiate left-sided SNHL and right-sided SNHL. The detection basis is that SNHL patients will have slight to severe structural change in specific brain regions. Traditionally, the human eye-based detection is unreliable since the human eyes cannot perceive slight atrophy. Thus, artificial intelligence is employed in this study, which is aimed to develop a computer-aided diagnosis (CAD) system.

Traditional CAD systems mainly used discrete wavelet transform (DWT) [3–5] to learn global image features, and then employed latest pattern recognition tools. For example, Mao, Ma and Tian [6] used DWT to analyze the potential signals of local field. Ikawa [7] employed DWT to performance auditory brainstem response (ABR) operation. Nayak, Dash and Majhi [8] employed the DWT to identify brain images. They used AdaBoost with random forests as classifiers. Lahmiri [9] utilized three multi-resolution techniques: DWT, empirical mode decomposition (EMD), and variational mode decomposition (VMD). Chen and Chen [10] used principal component analysis (PCA) and generalized eigenvalue proximal support vector machine (GEPSVM). Gorriz and Ramírez [11] proposed a directed acyclic graph support vector machine method.

Nevertheless, DWT suffers from the disadvantage translational variance [12]. That means, even a slight translation may lead to different decomposition result [13]. Besides, the DWT decomposition will lead to larger dimension space ($\sim 10^6$) than original image ($\sim 10^5$) for a 256×256 size image, and it needs dimension reduction techniques, such as principal component analysis [14].

To solve this problem, we introduced a relatively new technique: discrete wavelet packet entropy (DWPE) [15–17] that can yield mere a few ($\sim 10^1$) translational invariant features. Besides, we used a single-hidden layer neural network as the classifier, which was trained by gradient descent with adaptive learning rate back propagation method.

2 Materials

Subjects were enrolled from outpatients of department of otorhinolaryngology and head-neck surgery and community. They were excluded if evidence existed of known psychiatric or neurological diseases, brain lesions, taking psychotropic medications, as well as contraindications to MR imaging.

Finally, the study collection includes 15 patients with left-sided SNHL (LSNHL), 14 patients with right-sided SNHL (RSNHL) and 20 age- and sex-matched healthy controls (HC), as shown in Table 1.

Preprocessing was implemented on the software platform of FMRIB Software Library (FSL) v5.0. The brain extraction tool (BET) was utilized to extract brain tissues. The results are shown in Fig. 1. Then, the extracted brains of all subjects were registered to MNI space. Three experienced radiologists were instructed to select the most distinctive (around 40-th) slice between SNHLs and HCs.

Table 1. Subject characteristics

	HC	LSNHL	RSNHL
Gender (f/m)	12/8	7/8	8/6
Education level (year)	11.5 ± 3.2	12.5 ± 1.7	12.1 ± 2.4
Age (year)	53.6 ± 5.4	51.7 ± 9.6	53.9 ± 7.6
Disease duration (year)	–	17.6 ± 17.3	14.2 ± 14.9
PTA of left ear (dB)	22.2 ± 2.1	78.1 ± 17.9	21.8 ± 3.2
PTA of right ear (dB)	21.3 ± 2.2	20.4 ± 4.2	80.9 ± 17.4

(PTA = pure tone average)

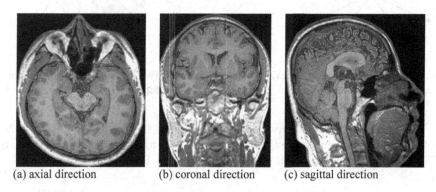

(a) axial direction   (b) coronal direction   (c) sagittal direction

Fig. 1. The green lines label the edge of BET result (Color figure online)

3 Methodology

3.1 Discrete Wavelet Packet Transform

In the field of signal processing, standard discrete wavelet transform (abbreviated as DWT) [18, 19] decomposes the given signal at each level, by submitting the previous approximation subband to the quadrature mirror filters (QMF) [20]. Its even-indexed downsampling causes the translational invariance problem [21].

On the other hand, discrete wavelet packet transform (DWPT) [22] is an improvement of standard DWT. DWPT passes both approximation and detail coefficients of previous decomposition level to QMF, so it can create a full binary tree [23]. In general, DWPT offers more features than DWT at the same decomposition levels [24].

Suppose x represents the original signal, c the channel index, d the decomposition level, p the position parameter, D the decomposition coefficients, and ψ the wavelet function, then DWPT is calculated as below:

$$D_p^{c,d} = \int_{-\infty}^{\infty} x(t)\psi_c(2^{-d}t - p)\mathrm{d}t \tag{1}$$

where. 2^d sequences will be yielded. Based on d-level decomposition, the decomposition results of $(d + 1)$ level is:

$$D_k^{2c,d+1} = \sum_{p\in Z} h(p - 2k) \times D_p^{c,d}, D_k^{2c+1,d+1} = \sum_{p\in Z} l(p - 2k) \times D_p^{c,d} \qquad (2)$$

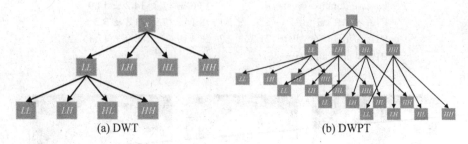

$$\text{(a) DWT} \qquad\qquad\qquad \text{(b) DWPT}$$

Fig. 2. Comparison between 2-level DWT and 2-level DWPT (x denotes for an image, H denotes the high-pass filter result, L denotes the low-pass filter result)

From Fig. 2, we can observe that for an image, DWT offer in total $(1 + 3d)$ coefficient subbands. In contrast, DWPT generates in total 4^d coefficients subbands. Thus, DWPT can provide much more information than DWT.

3.2 Shannon Entropy

Entropy was originally utilized to measure the system disorder degree [25]. It was generalized by Shannon to measure information contained in a given message [26]. Suppose m the index of grey level, h_m the probability of m-th grey level, and T the total number of grey levels, we have the Shannon entropy S as:

$$S = -\sum_{m=1}^{T} h_m \log_2(h_m) \qquad (3)$$

In the case of h_m equals to zero, the value of $0\log_2(0)$ is taken to 0 [27]. We calculated Shannon entropies of all subbands obtained from DWPT, and dubbed the results as discrete wavelet packet entropy (DWPE). For a brain image with size of 256×256, it has originally 65,536 features. A two-level DWPE can finally reduce the 65,536 features to only $2^4 = 16$ features.

3.3 Single-Hidden Layer Neural Network

The features were then presented into a classifier. There are many classifier in various fields, such as logistic regression [28], linear regression classifier [29, 30], extreme learning machine [31], decision tree [32, 33], etc.

In this study, we chose the classifier as a single-hidden layer neural-network (SLNN) [34] due to its superior performance. We did not employ multiple hidden layers [35], because one-hidden layer model is complicated enough to express our data. In a SLNN, the input nodes are connected to the hidden neuron layer, which is then connected to the output neuron layer.

The hidden neuron number is usually assigned with a large value. Afterwards, its value is decreased gradually till the classification performance reaches the peak result. The gradient descent with adaptive learning-rate back propagation (ALBP) algorithm [36] was employed to train the weights and biases of SLNN. Initial learning rate was set to 0.01. The increasing ratio and decreasing ratio of learning rate were set to 1.05 and 0.07, respectively. The maximum epoch is set to 5000.

4   Experiments and Results

4.1   DWPT Result

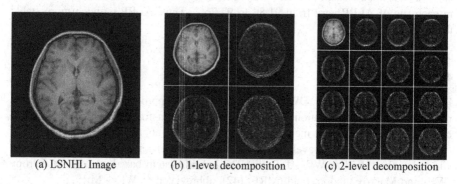

(a) LSNHL Image   (b) 1-level decomposition   (c) 2-level decomposition

Fig. 3. DWPT of a left-sided sensorineural hearing loss image

The 2-level DWPT result of a left-sided SNHL image is shown in Fig. 3. Here we can see in total 4 subbands are generated for 1-level decomposition, and 16 subbands are generated for 2-level decomposition.

4.2   Accuracy Performance

We repeated 5-fold cross validation [37] 10 times. The brief accuracy performance by BP algorithm is shown in left side in Table 2 with overall accuracy of 87.14%, and the accuracy performance by ALBP algorithm is shown in right side in Table 2 with overall accuracy of 95.31%. In these two tables, y/z represents y instances are successfully detected out of z instances.

The 10 repetition of 5-fold cross validation results indicate that this proposed ALBP performs better than classical BP algorithm. The reason lies in the adaptive learning-rate can accelerate the training procedure [38]. In standard BP, the learning

Table 2. Accuracy performance by BP and ALBP (R = Run; F = Fold; T = Total)

BP	F1	F2	F3	F4	F5	T	Acc.	ALBP	F1	F2	F3	F4	F5	T	Acc.
R1	10/10	9/10	7/9	10/10	8/10	44/49	89.80	R1	9/10	9/9	10/10	9/10	9/10	46/49	93.88
R2	10/10	10/10	9/10	7/9	8/10	44/49	89.80	R2	8/9	10/10	10/10	10/10	10/10	48/49	97.96
R3	9/9	9/10	9/10	8/10	10/10	45/49	91.84	R3	10/10	10/10	10/10	9/9	9/10	48/49	97.96
R4	6/10	10/10	7/9	9/10	7/10	39/49	79.59	R4	10/10	10/10	8/9	10/10	10/10	48/49	97.96
R5	9/10	9/10	8/9	8/10	8/10	42/49	85.71	R5	9/10	8/9	10/10	9/10	9/10	45/49	91.84
R6	9/10	9/9	8/10	7/10	9/10	42/49	85.71	R6	7/10	9/10	9/9	10/10	10/10	45/49	91.84
R7	8/10	8/10	9/10	9/9	10/10	44/49	89.80	R7	10/10	10/10	7/9	10/10	10/10	47/49	95.92
R8	10/10	10/10	9/10	7/9	9/10	45/49	91.84	R8	8/9	9/10	10/10	9/10	9/10	45/49	91.84
R9	7/10	9/10	9/10	9/9	6/10	40/49	81.63	R9	9/10	10/10	10/10	9/9	10/10	48/49	97.96
R10	9/9	8/10	9/10	9/10	7/10	42/49	85.71	R10	8/9	10/10	10/10	9/10	10/10	47/49	95.92
T							87.14	T							95.31

rate is unchanged, and thus the performance is sensitive to initial weight [39]. We see from left side of Table 2 that the accuracy in each run of BP vary from 79.59% to 91.84%. While the ALBP makes the learning rate responsive to the local error surface, and thus it is not as sensitive as BP. We see from right side of Table 2 that the accuracy in each run of ALBP vary from 91.84 to 97.76%. Thus, ALBP is much more stable than BP.

4.3 Comparison

Finally, we compared our DWPE + SLNN + ALBP approach with following three methods: (i) The combination of fractional Fourier transform (FRFT) and principal component analysis (PCA) method [40], which shall be abbreviated as FRFT + PCA. (ii) The combination of wavelet entropy (WE) and decision tree (DT) method [41], which is abbreviated as WE + DT. (iii) The hybrid system based on wavelet entropy (WE) and Markov random field (MRF) [42], abbreviated as WE + MRF.

Table 3. Comparison with state-of-the-art methods

Method	Overall accuracy
FRFT + PCA [40]	95.10%
WE + DT [41]	91.84%
WE + MRF [42]	91.02%
DWPE + SLNN + ALBP (Our)	95.31%

Table 3 shows that our method get superior overall accuracy of 95.31% to other three methods: FRFT + PCA [40], WE + DT [41], and WE + MRF [42]. The reason may be two folds: First, our method used DWPE, which combines two successful components, DWPT and Shannon entropy. Second, the wavelet packet transform is more efficient than fractional Fourier transform in image texture extraction. In the future, we shall try to use advanced classifiers, such as sparse autoencoder [43], convolutional neural network [44], and shared-weight neural network [45].

5 Conclusions

We developed a new computer-aided diagnosis system in this paper for detecting unilateral hearing loss, viz., left-sided or right-sided. The experiments gave promising results. In the future, we shall collect more data to further validate our method.

Acknowledgment. This study is supported by NSFC (61602250, 61271231), Program of Natural Science Research of Jiangsu Higher Education Institutions (16KJB520025, 15KJB470010), Natural Science Foundation of Jiangsu Province (BK20150983), Leading Initiative for Excellent Young Researcher (LEADER) of Ministry of Education, Culture, Sports, Science and Technology, Japan (16809746), Open Program of Jiangsu Key Laboratory of 3D Printing Equipment and Manufacturing (3DL201602), Open fund of Key Laboratory of Guangxi High Schools Complex System and Computational Intelligence (2016CSCI01), Open fund of Key Laboratory of Guangxi High Schools Complex System and Computational Intelligence (2016CSCI01).

References

1. Nakagawa, T., Yamamoto, M., Kumakawa, K., Usami, S., Hato, N., Tabuchi, K., Takahashi, M., Fujiwara, K., Sasaki, A., Komune, S., Yamamoto, N., Hiraumi, H., Sakamoto, T., Shimizu, A., Ito, J.: Prognostic impact of salvage treatment on hearing recovery in patients with sudden sensorineural hearing loss refractory to systemic corticosteroids: a retrospective observational study. Auris Nasus Larynx **43**, 489–494 (2016)
2. Prasad, A., Ghosh, P.K.: Information theoretic optimal vocal tract region selection from real time magnetic resonance images for broad phonetic class recognition. Comput. Speech Lang. **39**, 108–128 (2016)
3. Sun, P.: Preliminary research on abnormal brain detection by wavelet-energy and quantum-behaved PSO. Technol. Health Care **24**, S641–S649 (2016)
4. Pattanaworapan, K., Chamnongthai, K., Guo, J.M.: Signer-independence finger alphabet recognition using discrete wavelet transform and area level run lengths. J. Vis. Commun. Image Represent. **38**, 658–677 (2016)
5. Dong, Z., Phillips, P., Ji, G., Yang, J.: Exponential wavelet iterative shrinkage thresholding algorithm for compressed sensing magnetic resonance imaging. Inf. Sci. **322**, 115–132 (2015)
6. Mao, Y., Ma, M.F., Tian, X.: Phase Synchronization Analysis of theta-band of Local Field Potentials in the Anterior Cingulated Cortex of Rats under Fear Conditioning. In: International Symposium on Intelligent Information Technology Application, pp. 737–741. IEEE Computer Society (2008)
7. Ikawa, N.: Automated averaging of auditory evoked response waveforms using wavelet analysis, Int. J. Wavelets Multiresolut. Inf. Process. **11** (2013). Article ID: 1360009
8. Nayak, D.R., Dash, R., Majhi, B.: Brain MR image classification using two-dimensional discrete wavelet transform and AdaBoost with random forests. Neurocomputing **177**, 188–197 (2016)
9. Lahmiri, S.: Image characterization by fractal descriptors in variational mode decomposition domain: application to brain magnetic resonance. Phys. A **456**, 235–243 (2016)

10. Chen, Y., Chen, X.-Q.: Sensorineural hearing loss detection via discrete wavelet transform and principal component analysis combined with generalized eigenvalue proximal support vector machine and Tikhonov regularization. Multimedia Tools Appl. (2016). doi:10.1007/s11042-016-4087-6

11. Gorriz, J.M., Ramírez, J.: Wavelet entropy and directed acyclic graph support vector machine for detection of patients with unilateral hearing loss in MRI scanning, Frontiers Comput. Neurosci. **10** (2016) Article ID: 160

12. Liu, A.: Magnetic resonance brain image classification via stationary wavelet transform and generalized eigenvalue proximal support vector machine. J. Med. Imaging Health Inform. **5**, 1395–1403 (2015)

13. Zhou, X.-X., Yang, J.-F., Sheng, H., Wei, L., Yan, J., Sun, P.: Combination of stationary wavelet transform and kernel support vector machines for pathological brain detection. Simulation **92**, 827–837 (2016)

14. Ghods, A., Lee, H.H.: Probabilistic frequency-domain discrete wavelet transform for better detection of bearing faults in induction motors. Neurocomputing **188**, 206–216 (2016)

15. Sun, Y.X., Zhuang, C.G., Xiong, Z.H.: Real-time chatter detection using the weighted wavelet packet entropy. In: IEEE/ASME International Conference on Advanced Intelligent Mechatronics, pp. 1652–1657. IEEE, New York (2014)

16. Vyas, B., Maheshwari, R.P., Das, B.: Investigation for improved artificial intelligence techniques for thyristor-controlled series-compensated transmission line fault classification with discrete wavelet packet entropy measures. Electr. Power Compon. Syst. **42**, 554–566 (2014)

17. Yang, J.: Identification of green, oolong and black teas in China via wavelet packet entropy and fuzzy support vector machine. Entropy **17**, 6663–6682 (2015)

18. Arrais, E., Valentim, R.A.M., Brandao, G.B.: Real time QRS detection based on redundant discrete wavelet transform. IEEE Lat. Am. Trans. **14**, 1662–1668 (2016)

19. Hamzah, F.A.B., Yoshida, T., Iwahashi, M., Kiya, H.: Adaptive directional lifting structure of three dimensional non-separable discrete wavelet transform for high resolution volumetric data compression. IEICE Trans. Fundam. Electron. Commun. Comput. Sci. **E99A**, 892–899 (2016)

20. Kumar, A., Pooja, R., Singh, G.K.: Performance of different window functions for designing quadrature mirror filter bank using closed form method. Int. J. Signal Imaging Syst. Eng. **8**, 367–379 (2015)

21. Zhang, Y.D., Dong, Z.C., Ji, G.L., Wang, S.H.: An improved reconstruction method for CS-MRI based on exponential wavelet transform and iterative shrinkage/thresholding algorithm. J. Electromagn. Waves Appl. **28**, 2327–2338 (2014)

22. Baranwal, N., Singh, N., Nandi, G.C.: Indian sign language gesture recognition using discrete wavelet packet transform. In: International Conference on Signal Propagation and Computer Technology, pp. 573–577. IEEE (2014)

23. Gokmen, G.: The defect detection in glass materials by using discrete wavelet packet transform and artificial neural network. J. Vibroengineering **16**, 1434–1443 (2014)

24. Qin, Z.J., Wang, N., Gao, Y., Cuthbert, L.: Adaptive threshold for energy detector based on discrete wavelet packet transform. In: Wireless Telecommunications Symposium, pp. 171–177. IEEE (2012)

25. Ghafourian, M., Hassanabadi, H.: Shannon information entropies for the three-dimensional Klein-Gordon problem with the Poschl-Teller potential. J. Korean Phys. Soc. **68**, 1267–1271 (2016)

26. Phillips, P., Dong, Z., Yang, J.: Pathological brain detection in magnetic resonance imaging scanning by wavelet entropy and hybridization of biogeography-based optimization and particle swarm optimization. Prog. Electromagnet. Res. **152**, 41–58 (2015)

27. Alcoba, D.R., Torre, A., Lain, L., Massaccesi, G.E., Ona, O.B., Ayers, P.W., Van Raemdonck, M., Bultinck, P., Van Neck, D.: Performance of Shannon-entropy compacted N-electron wave functions for configuration interaction methods. Theor. Chem. Acc. **135** (11), 153 (2016)

28. Zhan, T.M., Chen, Y.: Multiple sclerosis detection based on biorthogonal wavelet transform, RBF kernel principal component analysis, and logistic regression. IEEE Access **4**, 7567–7576 (2016)

29. Du, S.: Alzheimer's disease detection by Pseudo Zernike moment and linear regression classification. CNS Neurol. Disord. - Drug Targets **16**, 11–15 (2017)

30. Chen, Y.: A feature-free 30-disease pathological brain detection system by linear regression classifier. CNS Neurol. Disord. - Drug Targets **16**, 5–10 (2017)

31. Lu, S., Qiu, X.: A pathological brain detection system based on extreme learning machine optimized by bat algorithm. CNS Neurol. Disord. - Drug Targets **16**, 23–29 (2017)

32. Zhou, X.-X.: Comparison of machine learning methods for stationary wavelet entropy-based multiple sclerosis detection: decision tree, k-nearest neighbors, and support vector machine. Simulation **92**, 861–871 (2016)

33. Zhang, Y.: Binary PSO with mutation operator for feature selection using decision tree applied to spam detection. Knowl.-Based Syst. **64**, 22–31 (2014)

34. Abbas, H.A., Belkheiri, M., Zegnini, B.: Feedback linearisation control of an induction machine augmented by single-hidden layer neural networks. Int. J. Control **89**, 140–155 (2016)

35. Sun, Y.: A multilayer perceptron based smart pathological brain detection system by fractional fourier entropy. J. Med. Syst. **40**, 173 (2016)

36. Hicham, A., Mohamed, B., Abdellah, E.F.: A model for sales forecasting based on fuzzy clustering and back-propagation neural networks with adaptive learning rate. In: International Conference on Complex Systems, pp. 111–115. IEEE (2012)

37. Lu, H.M.: Facial emotion recognition based on biorthogonal wavelet entropy, fuzzy support vector machine, and stratified cross validation. IEEE Access **4**, 8375–8385 (2016)

38. Iranmanesh, S.: A diffferential adaptive learning rate method for back-propagation neural networks. In: Proceedings of the 10th Wseas International Conference on Neural Networks, pp. 30–34. World Scientific And Engineering Acad And Soc (2009)

39. Murru, N., Rossini, R.: A Bayesian approach for initialization of weights in backpropagation neural net with application to character recognition. Neurocomputing **193**, 92–105 (2016)

40. Li, J.: Detection of left-sided and right-sided hearing loss via fractional fourier transform. Entropy **18**, 194 (2016)

41. Kale, M.C., Fleig, J.D., Imal, N.: Assessment of feasibility to use computer aided texture analysis based tool for parametric images of suspicious lesions in DCE-MR mammography, Comput. Math. Method Med. (2013). Article ID: 872676

42. Vasta, R., Augimeri, A., Cerasa, A., Nigro, S., Gramigna, V., Nonnis, M., Rocca, F., Zito, G., Quattrone, A.: ADNI: hippocampal subfield atrophies in converted and not-converted mild cognitive impairments patients by a markov random fields algorithm. Curr. Alzheimer Res. **13**, 566–574 (2016)

43. Hou, X.: Seven-layer deep neural network based on sparse autoencoder for voxelwise detection of cerebral microbleed. Multimedia Tools Appl. (2017). doi:10.1007/s11042-017-4554-8

44. Nogueira, R.F., Lotufo, R.D., Machado, R.C.: Fingerprint liveness detection using convolutional neural networks. IEEE Trans. Inf. Forensic Secur. **11**, 1206–1213 (2016)

45. Chen, M., Li, Y., Han, L.: Detection of dendritic spines using wavelet-based conditional symmetric analysis and regularized morphological shared-weight neural networks, Comput. Math. Method Med. (2015). Article ID: 454076

Study on Differences of Early-Mid ERPs Induced by Emotional Face and Scene Images

Xin Wang[1], Jingna Jin[1,2], Zhipeng Liu[1,2], and Tao Yin[1,2(✉)]

[1] Institute of Biomedical Engineering,
Chinese Academy of Medical Sciences & Peking Union Medical College,
Tianjin 300192, People's Republic of China
wangxin121926@163.com, jjn_902@163.com,
lzpeng67@163.com, bme500@163.com
[2] Neuroscience Center, Chinese Academy of Medical Sciences,
Beijing 100005, People's Republic of China

Abstract. Researches about the cortical processing mechanisms of emotions have important scientific significance and application value. To probe cortical processing differences of emotional face and scene images, electroencephalogram (EEG) of sixteen volunteers was recorded while they watching emotional images. Early-mid occipital ERPs (Event Related Potentials) under different images were compared, and RMS (Root Mean Square) was calculated to analyze activities of the whole brain. Results showed that the N1 (170 ms) amplitudes and P2 (250 ms) amplitudes induced by face images were respectively larger and smaller than that induced by scene images, which embodied specific processing of faces and reprocessing of complex scenes. Negative scene images were processed preferentially and induced more obvious N1 than positive or neutral scene images. Comparisons of ERPs among the whole brain displayed that occipital lobe was the main active region and frontal lobe responsible for emotional regulation was activated mainly at moments of N1 and P2. The early-mid ERPs comparisons explicitly showed cortical processing differences of emotional face and scene images, which deserved further studies.

Keywords: Event Related Potentials (ERPs) · Emotional face · Emotional scene

1 Introduction

Emotion is a comprehensive state when people perceiving the outside world, which includes psychological reaction, physiological reaction, related thinking and behaviors [1–3]. In scientific researches, emotions were usually induced by images, videos or recalling emotional events. Among them, images with specific meanings would do better to induce unitary and specific emotion, such as face and scene images.

Face images not only provide specific identity information, but also contain rich social information such as age, gender and emotion [4]. Cortical processing of face

Research supported by the Special Funds of the National Natural Science Foundation of China (No. 81127003), CAMS Initiative for Innovative Medicine(CAMS-I2M-1004).

F. Cong et al. (Eds.): ISNN 2017, Part II, LNCS 10262, pp. 550–558, 2017.
DOI: 10.1007/978-3-319-59081-3_64

images was different from non-face images, which was embodied by prosopagnosia in clinical researches, effects of inverted faces in behavioral researches and the fusiform gyrus in functional magnetic resonance imaging studies [5, 6]. Face images with specific expressions would produce special visual stimulation and emotion induction. According to the mirror neurons empathy principles [7], participants' emotions would be induced partly when they viewing others' emotional faces, along with similar brain activities and biochemical reactions. Therefore, cortical processing of emotional face images contains both emotional perception and specific processing of faces [8].

Scene images generally contain abundant contents of landscape, people and situations, which provides highly immersion to the viewer. Intense emotions usually could be induced by emotional scene images with specific meanings [8], such as positive family reunion images and negative disaster casualty images. Negative scenes are often related to risk factors in the environment. In order to eliminate or avoid this situation, the cerebral cortex tends to process negative emotions preferentially [9–11]. As emotional scene images often contain faces, both emotion induction and specific processing of faces were involved in the cortical processing.

Early-mid cortical processing of emotional images mainly includes the demand and allocation of attention and emotion perception [1]. Previous ERP researches showed that: faces evoked greater N170 [4] (a negative wave with the incubation period of 170 ms, mainly distributed in the right side of the occipito-temporal lobe) than non-face objects, which was known as the specific ERP component of faces. As stronger immersive and more occupied attention resources, emotional scene images usually evoked more obvious ERP components than face images, such as EPN (Early Posterior Negativity) and LPP (Late Positive Potential) [8].

However, when the complexity of scenes and the specificity of faces doing efforts together on emotion induction, the cortical processing is unclear [11]. Even though scene images provided much stronger immersion, the emotion area presented by face images is much bigger. Therefore, the emotional processing activated by scene images and face images need intensive study. The strength and progress of brain activities induced by scene images and face images can't be easily speculated. This study intends to study temporal-spatial differences of ERP signals induced by emotional face and scene images, taking advantages of non-invasive and real-time of ERP. Further, this study aims to obtain more detailed and in-depth understanding about attention resource allocation and emotional perception among scenes, faces, and emotions.

2 Materials and Methods

2.1 Experiments Protocols

Sixteen young students (7 males, average age 27 ± 3) were organized in the experiments, with good mood and no history of mental illness.

Totally 45 scene images and 60 face images were selected from the web photo gallery, and both were divided in three types of positive, negative and neutral. Among scene images, specific emotional meanings were presented in positive (happy family gatherings and travels) and negative (disaster, hunger and death) scenes, but not in neutral scenes (common daily items). Emotional face images contain complete faces with positive (happy and joyful), negative (sad and disgusted) or neutral expressions. The size and resolution of all images were adjusted to the most appropriate parameters.

The face and scene images were divided into two groups in the experiments. Images appeared randomly and each image was presented for 4 s. Subjects kept feeling emotions as much as possible when images were presenting, so as to form certain emotional experiences. When images disappeared, subjects pressed corresponding keys to judge emotions into positive, negative or neutral. In order to calm the emotional state caused by the previous image, a resting period of 4 s was set before the image presenting and was prompted as the "+".

2.2   Signal Acquisition and Pre-processing

The experiments were carried out in a private room with soft light and good sound insulation. Subjects sat before the screen comfortably, and the signal acquisition devices were placed on the right side of the subjects, as shown in Fig. 1. The EEG of 32 leads were recorded by ActiveTwo (Biosemi, Netherlands) with the sampling rate of 1024 Hz. In addition, perpendicular and horizontal electrooculogram (EOG) was recorded for eliminating interference of eye movements in EEG.

Band-pass filtering of 0.1–60 Hz was applied to remove interference with high frequency and drift with low frequency in EEG. Then, EEG of 32 leads was averaged as the reference for data conversion. At last, independent component analysis (ICA) was used to eliminate interference of EOG.

Fig. 1. Scene of the experiments

2.3 Data Analysis

Firstly, accuracies and reaction time of all subjects when they performing emotion judgment tasks were calculated, and the reaction time was defined as the interval between the image disappearing and pressing the key.

Then, EEG induced by images of the same emotion was averaged and the mean amplitude of EEG within 200 ms before image appearing was calculated as the baseline for calibration. ERP signal was obtained as the Eq. (1).

$$S_t = x_t - \frac{1}{200} \sum\nolimits_{j=-200}^{0} x_j \qquad (1)$$

Where, S is the computed ERP signal, $t = -200–1000$ ms, which represents the time interval from 200 ms before image appearing to 1000 ms after image appearing; x_t is the EEG signal.

ERP analysis examined changes of the ERP waveforms and components induced by emotional images during a period of time with the time resolution of a few milliseconds. As the images belong to visual stimulation essentially, this study mainly compared ERP waveforms, amplitudes and incubation period of the occipital area (primary visual cortex), focusing on ERP components of P1 (incubation period within 90–110 ms), N1 (incubation period within 160–180 ms) and P2 (incubation period within 235–255 ms).

Finally, although ERP waveforms of all leads were roughly similar, peaks of ERP components among all leads didn't appear at the same time. Therefore, root mean square (RMS) of four ERP components was calculated, including P1, N1, P2 and P3 of all leads, to examine cortical neural activities of the corresponding brain regions in the process of emotion induction.

3   Results

3.1   Behavior Analysis

Behavioral results showed that accuracies of all participants were above 95%, reaction time was around 800 ms and no significant statistical difference was found among different judgment tasks of emotions.

3.2   ERP Amplitude Comparisons of Oz

ERP waveforms induced by emotional face and scene images were similar on the whole, as shown in Fig. 2. To quantify differences among the ERP waveforms, amplitudes and incubation period of three ERP components were extracted, including P1, N1 and P2. Paired sample T test was applied to do statistics significant analysis.

ERP amplitude comparisons of Oz between face and scene images were shown in Fig. 3(a). Amplitudes of N1 under scene images were smaller than that under face images ($p = 0.0311$), which was usually regarded as the specific ERP component of faces [5]. Even though faces were often existed in scene images, they were not as

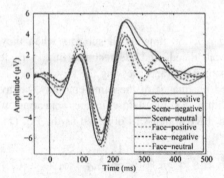

Fig. 2. ERP waveforms of Oz under face and scene images

obvious as that in face images. Amplitudes of P2 under scene images were bigger than that under face images (p = 0.0104), which reflected obvious cortical reprocessing on complex scene images. As lots of information was involved in the scene images, more resources were needed for cortical reprocessing.

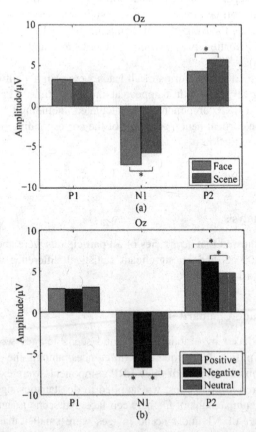

Fig. 3. ERP amplitude comparisons of Oz under different images. (a) Comparisons of face and scene images (b) Comparisons of scene images with different emotions

ERP amplitude comparisons of Oz among positive, negative and neutral scene images were shown in Fig. 3(b). Amplitudes of N1 under negative scene images were bigger than that under positive or neutral scene images (p = 0.0193). Incubation period of N1 under negative scene images (154 ms) were earlier than that under positive scene images (162 ms, p = 0.0519) or neutral scene images (164 ms, p = 0.0049). As negative emotions were usually related with dangerous hazards, preferential cortical processing and more attention was applied, which was accorded with the evolutionary psychology [10, 11]. Amplitudes of P2 under neutral scene images were smaller than that under positive (p = 0.0080) or negative (p = 0.0078). Therefore, emotional scene images would evoke more cortical reprocessing than neutral scene images, though the same complexity.

No significant statistics differences were found among ERP amplitudes of positive, negative and neutral face images. As subjects mainly focused on judging emotions of face images, they paid less attention on feeling emotions.

3.3   ERP Amplitudes Comparisons of the Whole Brain

To study activities of the whole brain under emotional images, RMS of P1, N1, P2 and P3 was calculated as the Eq. (2).

$$\text{RMS} = \sqrt{\frac{S_1^2 + S_2^2 + \cdots + S_n^2}{n}} = \sqrt{\frac{\sum_{i=1}^{n} S_i^2}{n}} \tag{2}$$

Where, n is the length of the period and S is the computed ERP signal based on Eq. (1), $i = 1, 2, ..., n$.

Periods of P1, N1, P2 and P3 were set as 72–112 ms, 148–180 ms, 234–262 ms and 335–365 ms successively. Topographic maps based on RMS were shown in Fig. 4.

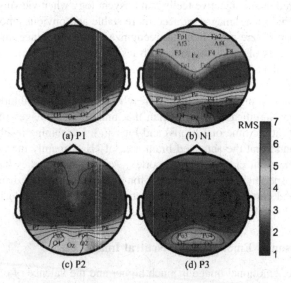

Fig. 4. Topographic maps based on RMS of four ERP components

The occipital cortex as the primary visual cortex was activated in the whole process of viewing images, especially at moments of N1 and P2. The frontal cortex was activated mainly at the moment of N1, which was related with emotion perception and regulation. In addition, as emotional judgment tasks in the experiments were easier, cortical activity at the moment of P3 was not obvious.

4   Discussion

4.1   Emotional Differences of Scene and Face Images

Emotional images are common stimulation materials to induce certain emotions, such as scene images and face images. Scene images generally contain abundant contents of landscape, people and situations, which provides highly immersion to the viewer. For face images with specific expressions, special visual stimulation and emotion induction would be induced according to the mirror neurons empathy principles [7]. That means, participants' emotions would be induced partly when they viewing others' emotional faces, along with similar brain activities and biochemical reactions. Even though scene images provide much stronger immersion, the emotional area presented by face images is much bigger. This study has compared the strength and progress of brain activities induced by scene images and face images to probe corresponding cortical processing mechanisms.

In results of ERP comparisons, the amplitudes of P2 under scene images were much bigger than that under face images ($p = 0.0104$), especially for positive and negative scene images, as shown in Fig. 2 (at the time of 240 ms) and Fig. 3(a). That is to say, scene images could induce much stronger emotions than face images, at least in the present experiments. Thus we speculate that immersion of emotional images is much more important than the exposed emotional area. In the callback after experiments, participants shared their subjective feeling and psychology when viewing images. They all agreed with that scene images induced more stable and obvious emotions. For face images, participants were more likely to recognize emotions of face images according to the slight changes of facial structure with less feeling and experience of emotions. Therefore, hand-picked images, optimized tasks and procedures are all needed to induce stable and obvious emotions by facial expression images.

When considering mainly face images, results showed that amplitudes of N1 under face images were significantly bigger than that under scene images ($p = 0.0311$), as shown in Fig. 2 (at the time of 170 ms) and Fig. 3(a). Combining results in Figs. 3(a) and 4(b), we found that the activated brain area of N1 is mainly the occipital cortex, especially for the right cipito-temporal cortex. As previous studies have shown that N170 was the specific ERP component for faces and mainly disturbed in right oc-cipito-temporal cortex [4], results in this study verified that conclusion once again, which suggested specific processing of face images.

4.2   Comparison of Emotional and Neutral Images

As the arousal of emotional image is much higher and the valence of emotional image is much lower than that of neutral image, the corresponding brain activaty is much stronger. In the results of ERP comparison, amplitudes of P2 under neutral scene images were much smaller than that under positive (p = 0.0080) or negative (p = 0.0078) scene images, as shown in Fig. 2 (at the time of 240 ms) and Fig. 3(b). Thus, emotional scene images do evoke more cortical reprocessing than neutral scene images, which reflected the greater experience of emotional stimuli.

Previous studies about emotions induced by images showed that the LPP was more obvious for emotional images than that for neutral images [8, 12–14]. In this study, the trend of difference between emotional and neutral scene images is the same with previous studies [8, 12]. But the waveform of ERP is different from the LPP in previous studies [8, 12], which may be related with the stimuli-presenting mode of images. Images were presented randomly in this study and intervals of two successive images were relatively longer. However, in previous studies [8, 12], lots of images were presented rapidly and continuously, which evoked accumulated emotional effects and were shown as LPP. In addition, the P1 was usually related with physical properties of images, including brightness and contrast. In this study, no significant statistics difference was found among face and scene images, which suggested good consistency on the physical properties of selected images.

4.3 Relevance of Attention and Negative Emotion

Emotional images usually imply certain environmental factors, which would induce specific action or action inclination [15]. From the evolutionary psychological perspective, the positive scene is beneficial to people's survival, which guides people to approach. On the contrary, negative scenes may endanger personal survival. It is very important to quickly capture the source and orientation of the negative factors in the environment [16]. Thus, negative scenes often attract more attention to guide the corresponding physiological and behavioral changes, compared with the pleasure and neutral scenes.

According the results in Fig. 4, brain activities during the time of T1 mainly occurred in the occipital lobe, which showed the allocation of attention. As shown in Fig. 3(b), amplitudes of N1 under negative scene images were bigger than that under positive or neutral scene images (p = 0.0193). Thus, the occipital lobe of negative scene picture was more activated than that of positive and neutral scene images, which implied that more attention resources were occupied by negative scene images, which is in line with the precious study.

5 Conclusion

This study confirmed differences on cortical emotional processing of face and scene emotional images, based on the early-mid ERP comparisons (P1, N1 and P2) and whole brain topography analysis. Especially, the results revealed that the immersion of emotional images is much more important than the exposed emotional area. Furthermore, this study figured out the significant relevance between attention and negative emotion, and explained the physiological basis. The results of this study would provide important reference for further exploration of the cortical emotional processing mechanism.

However, the activated intensity, time course and spatial distribution is still not enough. In next work, we will involve more nonlinear analysis methods to reveal the internal relation and essence of emotion perception, such as the brain network analysis and the EEG source analysis.

References

1. Viviani, R.: Emotion regulation, attention to emotion, and the ventral attentional network. Hum. Neurosci. **13**, 359–365 (2013)
2. Esslen, M., Pascual-Marqui, R.D., Hell, D., Kochi, K., Lehmann, D.: Brain areas and time course of emotional processing. NeuroImage **21**, 1189–1203 (2004)
3. Jaeger, A., Rugg, M.D.: Implicit effects of emotional contexts: an ERP study. Cogn. Affect. Behav. Neurosci. **12**, 748–760 (2012)
4. Jiang, Y., Shannon, R.W., Vizueta, N., Bernat, E.M., Patrick, C.J., He, S.: Dynamics of processing invisible faces in the brain: automatic neural encoding of facial expression information. Neuroimage **44**, 1171–1177 (2009)
5. Blau, V.C., Maurer, U., Tottenham, N., McCandliss, B.D.: The face-specific N170 component is modulated by emotional facial expression. Behav. Brain Funct. **3**, 7 (2007)
6. Morelli, S.A., Lieberman, M.D.: The role of automaticity and attention in neural processes underlying empathy for happiness, sadness and anxiety. Front. Hum. Neurosci. **7**, 1–15 (2013)
7. Kokal, L., Gazzola, V., Keysers, C.: Acting together in and beyond the mirror neuron system. NeuroImage **47**, 2046–2056 (2009)
8. Adolphs, R.: Recognizing emotion from facial expressions: psychological and neurological mechanisms. Behav. Cogn. Neurosci. Rev. **1**, 21–62 (2002)
9. Thom, N., Knight, J., Dishman, R., Sabatinelli, D., Johnson, D.C., Clementz, B.: Emotional scenes elicit more pronounced self-reported emotional experience and greater EPN and LPP modulation when compared to emotional faces. Cogn. Affect. Behav. Neurosci. **14**, 849–860 (2014)
10. Moser, J.S., Hajcak, G., Bukay, E., Simons, R.F.: Intentional modulation of emotional responding to unpleasant pictures: an ERP study. Psychophysiology **43**, 292–296 (2006)
11. Schacht, A., Sommer, W.: Emotions in word and face processing: early and late cortical responses. Brain Cogn. **69**, 538–550 (2009)
12. Olofsson, J.K., Nordin, S., Sequeira, H., Polich, J.: Affective picture processing: an integrative review of ERP findings. Biol. Psychol. **77**, 247–265 (2008)
13. Luo, W., Feng, W., He, W., Wang, N.Y., Luo, Y.J.: Three stages of facial expression processing: ERP study with rapid serial visual presentation. Neuroimage **49**, 1857–1867 (2010)
14. Liu, Y., Huang, H., McGinnis-Deweese, M., Keil, A., Ding, M.: Neural substrate of the late positive potential in emotional processing. J. Neurosci. **32**, 14563–14572 (2012)
15. Cosmides, L., Tooby, J.: Evolutionary psychology and the emotions. Handb. Emot. **2**, 91–115 (2000)
16. Oaten, M., Stevenson, R.J., Case, T.I.: Disgust as a disease-avoidance mechanism. Psychol. Bull. **135**, 303–321 (2009)

Comparison of Functional Network Connectivity and Granger Causality for Resting State fMRI Data

Ce Zhang[1], Qiu-Hua Lin[1(✉)], Chao-Ying Zhang[1,2,3,4,5],
Ying-Guang Hao[1], Xiao-Feng Gong[1], Fengyu Cong[2,3],
and Vince D. Calhoun[4,5]

[1] School of Information and Communication Engineering,
Dalian University of Technology, Dalian 116024, China
qhlin@dlut.edu.cn
[2] Department of Biomedical Engineering,
Dalian University of Technology, Dalian, China
[3] Department of Mathematical Information Technology,
University of Jyvaskyla, Jyväskylä, Finland
[4] The Mind Research Network, Albuquerque, NM 87106, USA
[5] Department of Electrical and Computer Engineering,
University of New Mexico, Albuquerque, NM 87131, USA

Abstract. Functional network connectivity (FNC) and Granger causality have been widely used to identify functional and effective connectivity for resting functional magnetic resonance imaging (fMRI) data. However, the relationship between these two approaches is still unclear, making it difficult to compare results. In this study, we investigate the relationship by constraining the FNC lags and the causality coherences for analyzing resting state fMRI data. The two techniques were applied respectively to examine the connectivity within default mode network related components extracted by group independent component analysis. The results show that FNC and Granger causality provide complementary results. In addition, when the temporal delays between two nodes were larger and the causality coherences were distinct, the two approaches exhibit consistent functional and effective connectivity. The consensus between the two approaches provides additional confidence in the results and provides a link between functional and effective connectivity.

Keywords: Functional network connectivity · Granger causality · Resting state fMRI · Group ICA · Default mode network

1 Introduction

Over the past decades, an increasing number of analytical methods have been introduced to explore the functional and effective connectivity among brain functional networks [1, 2]. Functional network connectivity (FNC) is a powerful functional connectivity approach for assessing temporal coherence among brain networks by utilizing lag shift correlations between nodes [3]. On the other side, as a typical method for effective

© Springer International Publishing AG 2017
F. Cong et al. (Eds.): ISNN 2017, Part II, LNCS 10262, pp. 559–566, 2017.
DOI: 10.1007/978-3-319-59081-3_65

connectivity, Granger causality is a statistical method for exploring the predictability and dependencies to establish causal relationships between brain networks [4].

FNC and Granger causality have been separately applied to fMRI data for identifying typical resting connectivity networks. In particular, FNC had been used to distinguish the abnormal relationships among several specific networks in psychiatric patients from the normal controls [5, 6]. Comparisons of functional network connectivity during resting and task conditions showed that functional network connectivity was stronger during rest compared to task [7]. As for Granger causality, its investigation for functional brain organization also found that schizophrenia patients exhibited significantly enhanced causal influence between specific regions [4, 8, 9].

Quite recently, FNC and Granger causality have both been utilized for analyzing connectivity changes among different age stages based on resting and task fMRI data [10]. FNC was employed to detect internetwork connectivity between the salience network, executive control networks and default mode networks (DMNs), while Granger causality was used to analyze the effective connectivity. In [10], FNC and Granger causality were used as two entirely different approaches with no analysis about connections between their results. As such, this study aims to directly compare the two approaches and examine how to leverage any complementary information they provide about the data. Eighty-two subjects of resting state fMRI data were used in the comparative analyses.

The rest of this paper is organized as follows. Section 2 introduces the resting state fMRI data we used, the components we extracted, and the two key algorithms: FNC and Granger causality. In Sect. 3, we presented the results of the two approaches, and compared the results of FNC under different time-lags and those of Granger causality with causality coherence constraint. Section 4 has the conclusions.

2   Methods

2.1   Materials

The fMRI Data from 82 subjects were collected using a 3T Siemens Trio scanner with the parameters: repeat time (TR) = 2 s, echo time = 29 ms, field of view = 240 mm, flip angle = 75°, slice thickness = 3.5 mm, gap = 1.05 mm, matrix size = 64 × 64 33, voxel size = 3.75 × 3.75 × 4.55 mm^3, number of timepoints = 150. All subjects were instructed to do nothing but keep their eyes open during the scan. Data were preprocessed using the statistical parametric mapping (SPM) software package. After motion correction, spatial normalization with isotropic resampling to voxels of 4 × 4×4 mm^3 in standard Montreal Neurological Institute brain space, and spatial smoothing with an 8 mm full width at half maximum Gaussian kernel, we obtained fMRI datasets with dimension 53 × 53 × 46 × 150 for each subject.

2.2   Extraction of Components

Spatial group independent component analysis (ICA), which has been widely used for extracting components from fMRI data, was performed for all 82 subjects using the toolbox GIFT (http://mialab.mrn.org/software/gift) [11]. Since high model order enables us to evaluate multiple sub-networks within each network domain [12], we separated 120 independent components (IC) using the Infomax algorithm. After ICA separation, we further extracted seven DMN-related components based on their spatial map references [13]. These components, as shown in Fig. 1, were medial prefrontal cortex (MPFC) corresponding to IC10, left and right inferior parietal lobule (IPL) corresponding to IC80, IC16, and posterior cingulate cortex (PCC) corresponding to IC22, IC52, IC66, and IC71. Prior to being applied to FNC and Granger causality, the time courses of these seven DMN-related components were low-pass filtered (Butterworth, cutoff frequency 0.15 Hz).

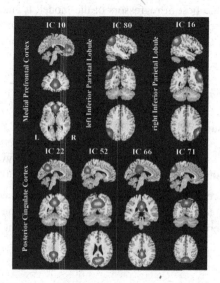

Fig. 1. Spatial maps of the seven DMN-related components. The final spatial maps were z-scored and thresholded at $|Z| \geq 2$ and displayed at the three most informative slices.

2.3 Functional Network Connectivity (FNC)

FNC computes the lag-shift Pearson's correlation coefficient between pairs of time courses using the FNC toolbox (http://mialab.mrn.org/software):

$$\rho_{\Delta t} = \frac{X_{t_0}^T Y_{t_0 + \Delta t}}{\sqrt{X_{t_0}^T X_{t_0}} \times \sqrt{Y_{t_0 + \Delta t}^T Y_{t_0 + \Delta t}}} \tag{1}$$

where $\rho_{\Delta t}$ represents correlation between two time courses X and Y, while Δt stands for the time shifting from the initial reference point t_0. FNC recorded the maximal lagged correlation $\rho_{max}^{(k)} = \max\{\rho_{\Delta t}\}$ and its corresponding lag $\Delta t^{(k)}$ for a single subject k, $k = 1, \ldots, K$ (K is the number of subjects), and then averaged across all subjects. The statistical significance of these correlations and lags was finally calculated by using one sample t-test at $p < 0.05$ corrected by false discovery rate (FDR), respectively [3].

2.4   Granger Causality

Granger causality relies on linear regression models of a stochastic process. Specifically, if the information in the past of a time series can be used to improve the prediction accuracy of the future of another time series, then the former is the Granger cause of the latter. Let X_t, Y_t be two stationary variables, i.e., the two time courses here, the autoregressive model can be described as:

$$X_t = \sum_{j=1}^{m} a_j X_{t-j} + \sum_{j=1}^{m} b_j Y_{t-j} + \varepsilon_t, \quad Y_t = \sum_{j=1}^{m} c_j X_{t-j} + \sum_{j=1}^{m} d_j Y_{t-j} + \eta_t \tag{2}$$

where a_j, b_j, c_j, and d_j are best fit regressors of the model, ε_t and η_t are two zero-mean uncorrelated white-noise series. The model order m can be determined by MDL criterion. The measure of the strength of the causality $X \to Y$ can be defined as,

$$C_{\overrightarrow{xy}} = \frac{\sigma_\varepsilon^4 |(1-d)c|^2}{(\sigma_\varepsilon^2 |(1-d)|^2 + \sigma_\eta^2 |(b)|^2)(\sigma_\varepsilon^2 |(c)|^2 + \sigma_\eta^2 |(1-a)|^2)} \tag{3}$$

Similarly, the measure of the strength of the causality $Y \to X$ ($C_{\overrightarrow{yx}}$) can be defined with another numerator $\sigma_\eta^4 |(1-a)b|^2$. We call ($C_{\overrightarrow{xy}}$) and ($C_{\overrightarrow{yx}}$) the causality coherences. It should be noted that $0 < (C_{\overrightarrow{xy}}) < 1$ and similarly for ($C_{\overrightarrow{yx}}$) [4, 14]. The causality coherences were computed for all subjects and then averaged. After performing one sample t-test ($p < 0.05$, corrected by FDR), the directional influence between two components adopted the statistically significant causality with larger values.

3   Results

3.1   FNC and Granger Without Lag and Causality Coherence Constraints

Figure 2 demonstrates the results for FNC and Granger causality. The direction of an arrow in FNC results indicates the time delay between two components. For example, in Fig. 2(a), an arrow from IC10 to IC71 represents that IC10 precedes IC71 by certain time units. Meanwhile, in Granger causality, Fig. 2(b), an arrow IC10 → IC71 represents IC10 is the Granger cause of IC71.

For the FNC results shown in Fig. 2(a), the temporal correlations existed in every pair of components and the time-lags varied in a large range from 0.0015 s to 0.650 s. Causality was found in a subset of the components, as shown in Fig. 2(b). IC80 and IC16 (bilateral IPL) were the Granger cause of IC66 and IC71 (PCC), and IC10 (MPFC) caused IC71 (PCC), which suggests that in the DMN, the PCC may work as a special node that seldom generate but mostly receives Granger connections. The internal connection of PCC shows that IC22 caused IC71, and IC52 was the Granger cause of IC66 and IC71.

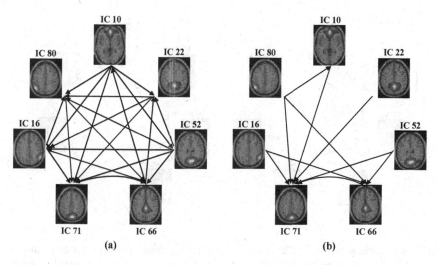

Fig. 2. Results for FNC and Granger. (a) Functional connectivity detected by FNC. (b) Effective connectivity detected by Granger causality.

When we focused on the common connectivity in both, it was not difficult to find most connections showed the same directionality. For example, IC71 and IC66 which lagged to IC80 and IC16 (bilateral IPL) in functional network connectivity were also identified as caused by them in the Granger causality approach. This was also true of connections within the PCC.

Nevertheless, it is obvious that the number of significant connections identified by FNC was much larger than that in Granger causality, and there were also a few discrepancies. The arrow direction was reverse for connections IC80 - IC10 as well as IC66 - IC71. As such, we next added constraints to the lags and causality coherences to investigate their influence on the connectivity.

3.2 FNC with Lag Constraints

In the analysis of FNC, correlation and lag values were simultaneously examined for possible combinations, and the lag markers reflected the chronological order of the related components. Considering that a small latency in FNC may be influenced by noise in the time courses and thus hard to show precedence relationship in time, we ignored lags less than 0.05 s in functional network connectivity. Figure 3 shows the results. In Fig. 3(a), the values on the line represent the 'lag (second)/correlation coefficient' between two connected components. Moreover, in Fig. 3(b) the arrow A → B is also expressed by the values of '$C_{\overrightarrow{AB}}/C_{\overrightarrow{BA}}$'.

After omitting the connections with small lags, we found that there were only ten connections left in functional network connectivity and each of them corresponded to a specific connectivity in Granger causality. The directional connectivity between components obtained from FNC was quite similar to the directionality obtained from

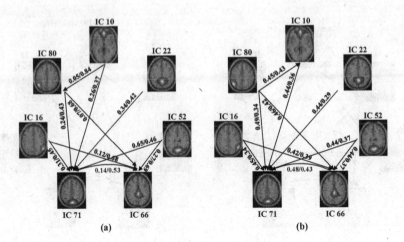

Fig. 3. Comparison of FNC and Granger with lag constraints. (a) Functional connectivity detected by FNC ignoring lags less than 0.05 s. (b) Effective connectivity detected by Granger causality. Values on the line represent 'lag (second)/correlation coefficient' in (a), while on the arrow A → B also express '$C_{\overrightarrow{AB}}/C_{\overrightarrow{BA}}$' in (b).

Granger causality: there were 8/10 connections exhibiting a close correspondence between time delay and causality. Two of the ten connections, those between IC80 - IC10 and IC66 - IC71, showed reverse directions.

It was easy to see that the time delays between IC80 - IC10 and IC66 - IC71 were also small (0.05 s for IC80 - IC10 and 0.14 s for IC66 - IC71). In addition, the difference between the causality coherences '$C_{\overrightarrow{AB}}/C_{\overrightarrow{BA}}$' was also small (less than or equal to 0.05). We further constrained lags or causality coherences in order to focus on consistent connectivity.

3.3   FNC and Granger with Both Lag and Causality Coherence Constraints

Figure 4 illustrates the results of FNC and Granger causality with both lag and causality coherence constraints. We ignored FNC connections with lags less than 0.15 s and kept Granger causality connections with '$C_{\overrightarrow{AB}}/C_{\overrightarrow{BA}}$' difference greater than 0.05. Taking the causality from A to B as an example, we find the causality is distinct when $C_{\overrightarrow{AB}} - C_{\overrightarrow{BA}} > 0.05$.

After selecting the connections with large lags and distinct causality, we found that the directional functional connectivity obtained from FNC had the same directionality obtained from Granger causality. As shown in Fig. 4, the bilateral IPL (IC16, IC80) and the MPFC (IC10) actually preceded and caused IC71 of PCC. The direction within the PCC showed consistent results in both FNC and Granger causality, which implied in the DMN some specific components of PCC were in charge of receiving connections from others. Ultimately, six connections with effective lags or causality were left, in

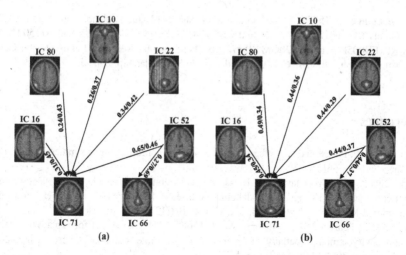

Fig. 4. Comparison of FNC and Granger with lag constraints. (a) Functional connectivity ignoring lags less than 0.15 s. (b) Effective connectivity with '$C_{\overrightarrow{AB}}/C_{\overrightarrow{BA}}$' difference greater than 0.05. Values on the line represent 'lag (second)/correlation coefficient' in (a), while on the arrow A → B also express '$C_{\overrightarrow{AB}}/C_{\overrightarrow{BA}}$' in (b).

which the two approaches can give consistent results. These results show that for these cases the lags and causality coherences are more reliable for assessing temporal and causal relationship between components than small ones which may be influenced by noise. The consistent connections identified by both methods provide reliable and stable results for estimating functional and effective connectivity.

4   Conclusions

The FNC approach detects the maximal shift lagged correlations between all pair-wise components, while the Granger causality analyzes the causal relationship between components. Previously, the two approaches have been used to detect distinct functional connectivity and effective connectivity. In this study we compared the two approaches by constraining the FNC lags and the causality coherences. When we removed small FNC lags and causality coherences, we obtained consistent functional and effective connectivity based on resting state fMRI data. The results support the conclusion that time delay has a specific meaning to the Granger causality and is a main factor driving the results. Our results also suggest that additional advantages can be gained by using FNC and Granger causality in combination. We obtain unique information from each approach, i.e., the correlation structure detected by FNC when the lags are small, and the causal relationship found by Granger causality when the difference between two reversing causality coherences is small, but also the convergent information identified by both methods provides reliable and stable information for enhancing the analysis of functional and effective connectivity. In the future, we will test how the two approaches are connected in task fMRI data.

Acknowledgments. This work was supported by National Natural Science Foundation of China under Grants 61379012, 61671106, and 81471742, NSF grants 0840895 and 0715022, NIH grants R01EB005846 and 5P20GM103472, the Fundamental Research Funds for the Central Universities (China, DUT14RC(3)037), and the China Scholarship Council.

References

1. Friston, K.J.: Functional and effective connectivity in neuroimaging: a synthesis. Hum. Brain Mapp. **2**(1–2), 56–78 (1994)
2. Greicius, M.D., Srivastava, G., Reiss, A.L., Menon, V., Raichle, M.E.: Default-mode network activity distinguishes Alzheimer's disease from healthy aging: evidence from functional MRI. Proc. Natl. Acad. Sci. U.S.A. **101**(13), 4637–4642 (2004)
3. Jafri, M.J., Pearlson, G.D., Stevens, M., Calhoun, V.D.: A method for functional network connectivity among spatially independent resting-state components in schizophrenia. NeuroImage **39**(4), 1666–1681 (2008)
4. Demirci, O., Stevens, M.C., Andreasen, N.C., Michael, A., Liu, J., While, T., Pearlson, G. D., Clark, V.P., Calhoun, V.D.: Investigation of relationships between fMRI brain networks in the spectral domain using ICA and Granger causality reveals distinct differences between schizophrenia patients and healthy controls. NeuroImage **46**(2), 419–431 (2009)
5. Allen, E.A., Erhardt, E.B., Damaraju, E., Gruner, W., Segall, J.M., Silva, R.F., et al.: A baseline for the multivariate comparison of resting-state networks. Front. Syst. Neurosci. **5**, 2 (2011)
6. Das, P., Calhoun, V.D., Malhi, G.S.: Bipolar and borderline patients display differential patterns of functional connectivity among resting state networks. NeuroImage **98**, 73–81 (2014)
7. Arbabshirani, M.R., Havlicek, M., Kiehl, K.A., Pearlson, G.D., Calhoun, V.D.: Functional network connectivity during rest and task conditions: a comparative study. Hum. Brain Mapp. **34**(11), 2959–2971 (2013)
8. Palaniyappan, L., Simmonite, M., White, T.P., Liddle, E.B., Liddle, P.F.: Neural primacy of the salience processing system in schizophrenia. Neuron **79**(4), 814–828 (2013)
9. Potvin, S., Lungu, O., Tikàsz, A., Mendrek, A.: Abnormal effective fronto-limbic connectivity during emotion processing in schizophrenia. Prog. Neuropsychopharmacol. Biol. Psychiatry **72**, 1–8 (2017)
10. Archer, J.A., Lee, A., Qiu, A., Chen, S.H.A.: A comprehensive analysis of connectivity and aging over the adult life span. Brain Connect. **6**(2), 169–185 (2016)
11. Calhoun, V.D., Adali, T., Pearlson, G.D., Pekar, J.J.: A method for making group inferences from functional MRI data using independent component analysis. Hum. Brain Mapp. **14**(3), 140–151 (2001)
12. Abou-Elseoud, A., Starck, T., Remes, J., Nikkinen, J., Tervonen, O., Kiviniemi, V.: The effect of model order selection in group PICA. Hum. Brain Mapp. **31**(8), 1207–1216 (2010)
13. Smith, S.M., Fox, P.T., Miller, K.L., Glahn, D.C., Fox, P.M., Mackay, C.E., et al.: Correspondence of the brain's functional architecture during activation and rest. Proc. Natl. Acad. Sci. U.S.A. **106**(31), 13040–13045 (2009)
14. Granger, C.W.J.: Investigating causal relations by econometric models and cross-spectral methods. Econometrica **37**(3), 424–438 (1969)

Neural Oscillations as a Bridge Between GABAergic System and Emotional Behaviors

Tao Zhang[1(✉)], Qun Li[1], and Zhuo Yang[2]

[1] College of Life Sciences and State Key Laboratory of Medicinal Chemical
Biology, Nankai University, Tianjin 300071, People's Republic of China
zhangtao@nankai.edu.cn
[2] School of Medicine, Nankai University, Tianjin 300071,
People's Republic of China

Abstract. This study aims to investigate if neural oscillations can play a role as
a bridge between GABAergic systems and emotional behaviors. Mice were
divided into streptozotocin-induced diabetes mellitus (DM) group and control
(CON) group. After 8 weeks of successfully modeling, the diabetic mice
exhibited depression and anxiety, while the $GABA_A R$ α_1 subunit expression and
the GABA level were significantly changed as well. There were increase power
percent at gamma (50–80 Hz) band in the power spectrum between these two
groups. However, DM attenuated the identical-frequency strength of phase
synchronization and information flow at both theta (8–13 Hz) and gamma
rhythms, and reduced the theta-gamma phase-amplitude coupling strength either
in the PP & DG regions or on PP-DG pathway. It suggests that DM changes the
pattern of neural oscillations by modulating the $GABA_A R$ α_1 subunit expression
and the GABA level in depression and anxiety in diabetic mice.

Keywords: Neural oscillations · GABA · Depression · Anxiety

1 Introduction

Neural oscillations are generated by interaction and influence of neurons with each
other through excitatory and inhibitory synaptic connections. As known, the classification for oscillations is usually identified with different frequencies, such as theta (3–
13 Hz) and gamma (30–100 Hz) [1]. Each frequency has its own distinctive effect on
corresponding to different physiological functions. GABA and its receptors, play an
key role in neural oscillations and fundamentally implicate in modulating all of frequency bands [2]. However, the role of neural oscillations, in which depressive-like
behaviors are induced by GABAergic deficits in diabetic rodents, is yet to be eluci-
dated. In the study, a hypothesis was raised that neural synchronization as the critical
"middle ground" connecting GABAergic system with emotional behaviors could play a
role of bridge between behaviors and molecular biology. Accordingly, the diabetic
mouse-model was established and the alterations of emotional behaviors were examined. The protein expression of GABAAR α_1 subunit and the concentration of GABA
were determined. In order to represent a potential neural mechanism, the functional role
of neural oscillations was extensively investigated.

© Springer International Publishing AG 2017
F. Cong et al. (Eds.): ISNN 2017, Part II, LNCS 10262, pp. 567–574, 2017.
DOI: 10.1007/978-3-319-59081-3_66

2   Materials and Methods

2.1   Animals and Treatment

In the study, eighteen male C57BL/6 J mice (22–24 g, around 7–8 week-old) were used, which were purchased from the Laboratory Animal Center, Academy of Military Medical Science of People's Liberation Army. Mice were reared in a specific pathogen free house with a 12 h light-dark cycle (8 am–8 pm), and they were free to food and water under a constant temperature (22 ± 1 °C) and about 50% humidity. Protocols were approved by the Animal Research Ethics Committee, School of Medicine, Nankai University. And all efforts were made to minimize the number of animals used and their suffering. Eighteen mice were randomly divided into control group (CON group, n = 10) and diabetes mellitus group (DM group, n = 8). Mice in the DM group were intraperitoneally injected 55 mg/kg/day streptozotocin (STZ) for consecutive five days. At the same time, mice in the CON group were intraperitoneally injected the same volume of sterile citrate buffer. On the 8th day, the glucose concentration of blood was detected. If the glucose level was not lower than 16.7 mM, mice were identified as hyperglycemia.

2.2   Biological Experiments

Behavioral tests including elevated plus maze, tail suspension and forced swimming were performed. Then, electrophysiological experiments were carried out at the end of the behavioral experiments. The signals of local field potential (LFPs) were recorded from both the hippocampal perforant path (PP) and dentate gyrus (DG) regions. At last, molecular biological tests including Western and HPLC were performed. The more details cold be seen.

2.3   Neurodynamic Analysis

The following approaches provide the neural oscillations assessment at theta frequency band (8–13 Hz) and gamma frequency band (50–80 Hz).

Power Spectrum Density (PSD) Analysis. PSD analysis was used to test the power changes of different frequency in both the hippocampal PP and DG areas. Multitaper Spectral Estimation is one of the most classical methods to perform PSD analysis. Its advantage is that sequence data analysis can be performed by using an optimal set of orthogonal localization windows (Slepian windows) to obtain direct spectrum and average spectral estimation, thereby achieving a smaller variance and bias properties for the estimation result.

Phase Locking Value (PLV). Phase synchronization of oscillations at an identical frequency could promote neural communication [3]. PLV is a great identical frequency algorithm to measure the strength of phase synchronization by the mean value of a composite of phase difference between two brain areas. Before extracting the phase of two signals, the original LFPs were filtered into theta and gamma frequency bands by the mean of eegfilt.m from EEGLAB toolbox [4]. The bandwidth was set to 1 Hz incremented in 1 Hz steps. Afterward, the instantaneous phases of filtered LFPs (theta, gamma) were obtained by Hilbert transform from the hippocampal PP and DG areas in all these frequency bands. N was for the length of the signal. Then PLV was given as

$$PLV = \left| \frac{1}{N} \sum_{t=1}^{N} \exp(i[\phi_{PP}(t) - \phi_{DG}(t)]) \right| \tag{1}$$

Generalized Partial Directed Coherence (gPDC). The gPDC algorithm has been developed to assess the directional coupling in an identical frequency based on multivariate vector autoregressive model [5]. Actually, its meaning is the Granger causality in the frequency domain. The method is characterized by analysis based on the signal itself and direction recognition of information flow.

Mean Vector Length (MVL). Phase-amplitude coupling (PAC) has been by far the most commonly reported in recent years as one of cross-frequency coupling types. In order to assess the strength of PAC, the MVL method was employed in both the hippocampal PP and DG areas as well as PP-DG pathway [6].

3   Results

3.1   DM Induced Depression and Changed GABAAR α_1 and GABA

There was significant difference of FST immobility between these two groups ($t_{8.43} = -4.804$, $P = 0.001$). Moreover, the time of immobility in the DM group was higher than that in the CON group ($t_{8.84} = -3.810, P = 0.004$), which was consistent with the results obtained from the FST. DM markedly decreased the expression of GABAAR α_1 subunit to 0.390 ± 0.028. The level of GABA in the hippocampus showed that there was significant difference between the CON group and the DM group ($Z = -39.273$, $P < 0.001$). The level of GABA in the CON group was 1.28 ± 0.23 µg/mg protein. However, the level of GABA was increased to 5.63 ± 0.13 µg/mg protein in the DM group.

3.2   Power Percent Increased at Gamma Band but Not at Theta Band

PSD analysis showed that DM decreased the power percent of gamma ($t_{16} = -2.787$, $P = 0.013$) bands while there were no statistical changes of power percent at theta ($t_{16} = 1.857$, $P = 0.082$) bands in the hippocampal PP area (Fig. 1C Left). The similar results were also obtained from PSD estimating in the DG area at theta rhythm ($t_{8.85} = 0.174$, $P = 0.866$) and gamma rhythm ($Z = -2.488$, $P = 0.012$) frequency bands (Fig. 1C Right).

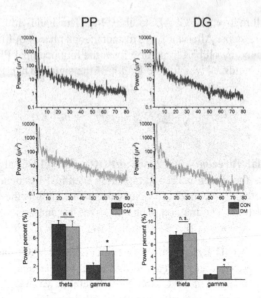

Fig. 1. There were significant changes of power at gamma frequency bands but not at theta bands between the CON group and the DM group. (A) and (B) were representative power spectrum in the CON and DM groups resulting from 40 s window at 1 Hz to 80 Hz, respectively. (C) The statistical results of power spectrum distribution in PP (left) and DG (right) areas. The n.s. represents no significant differences of power between the CON group and the DM group.

3.3   Strength of Phase Synchronization Was Impaired by DM on PP-DG Pathway

PLV measurement showed that the strength of phase synchronization was slightly decreased at theta frequency band in the DM group compared to that in the CON group ($t_{9.02} = 2.105$, $P = 0.065$). However, there was significant difference of the strength of phase synchronization between these two groups in gamma band ($t_{16} = 2.468$, $P = 0.025$, Fig. 2).

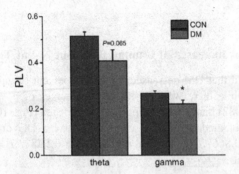

Fig. 2. The impairment of phase synchronization measured by PLV in an identical frequency on PP-DG pathway in DM mice. *$P < 0.05$ and **$P < 0.01$ represent significant differences of PLV between the CON group and the DM group.

3.4 Direction of Coupling Was Significantly Altered by DM on PP-DG Pathway

Figure 3 showed the gPDC analysis of PP-DG pathway in the CON and DM groups over theta, gamma bands. The group values for the unidirectional influence from the PP to DG were presented in Fig. 3A. It was found that the unidirectional influence indexes in all three frequency bands were considerable decreased in the diabetic state, and the statistical differences between the CON group and the DM group were found in the theta band ($Z = -3.554$, $P < 0.001$) and gamma band ($t_{16} = 7.825$, $P < 0.001$), respectively (Fig. 3A). Interestingly, bidirectional coupling indexes between PP and the DG were greatly altered from positive values to negative values at the same bands (theta: $t_{7.79} = 9.168$, $P < 0.001$; gamma: $t_{16} = 7.935$, $P < 0.001$, Fig. 3B). Together with above indexes, it could be inferred that DM significantly attenuated unidirectional information flow from PP to DG, and further altered the direction of bidirectional information flow.

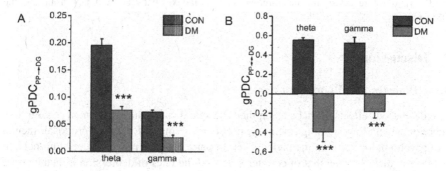

Fig. 3. The gPDC index represented the changes of coupling direction in an identical frequency band on PP-DG pathway in DM mice. (A) Decreased unidirectional coupling gPDC index from PP to DG in DM mice. (B) Altered bidirectional gPDC index in DM mice at these three frequency bands. **$P < 0.01$ and ***$P < 0.001$ represent significant differences between the CON group and the DM group.

3.5 Strength of Phase-Amplitude Coupling Was Significantly Impaired by DM

In order to measure cross-frequency coupling between PP and DG areas, MVL algorithm was employed. There was a clear and strong theta-gamma PAC in the CON group. (Figure 4A). Moreover, the strength of PAC was obviously decreased in the DM group (Fig. 4A-right column). Again the similar changes of PAC were in either PP or DG area. Statistical analysis showed that the coupling strength between theta and gamma was significantly reduced in the PP area ($Z = -2.132$, $P = 0.033$, Fig. 4B), and the DG area ($Z = -2.932$, $P = 0.003$, Fig. 4D). Most importantly, the strength of PAC between PP and DG areas was also considerably decreased ($Z = -3.465$, $P = 0.001$, Fig. 4C).

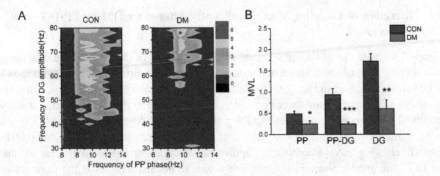

Fig. 4. Reduced MVL index at cross-frequency (theta-gamma) between the hippocampal PP and DG regions as well as PP-DG pathway in DM mice. (A) Representative PAC of PP-DG pathway between PP low frequency phase (6–14 Hz) and DG gamma band amplitude (30–80 Hz) in a 40 s window of one mouse in the CON group (left) and another in the DM group (right). (B–D) The statistical results of theta and gamma cross-frequency measured by MVL in PP (B), PP-DG pathway (C), and DG (D), respectively. *$P < 0.05$, **$P < 0.01$ and ***$P < 0.001$ represent significant differences of the indexes between the CON group and the DM group.

4   Discussion

4.1   Diabetes and Depression

The deleterious effects of diabetes on the CNS could cause the impairment of cognitive functions, including learning, memory, problem solving, informative proposal, mental and psychomotor speed in the patients [7]. In particular, the risk of depression incident is twice in diabetes than that of counterparts, and the rates of depression in diabetes are in the range of 11% to 13.5% a share. A battery of neuropsychological and neurobehavioral changes in diabetic subjects have been discovered on the experimental animals. In the present study, it was found that mice spend less time in the open arms, presented less enteries into open arms, and had a lower percent of enteries into open arms of EPM in the DM group compared to that in the CON group. It indicated that mice became higher anxiety in DM group, which was consistent with Tang's results. Moreover, our data showed that mice in the DM group spent much longer time of immobility in FST and TST, which was in accordance with the results of other research groups [8].

4.2   GABAergic System and Diabetic Encephalopathy-Related Depression

GABA, as one of the most important inhibitory amino acid neurotransmitters of CNS, almost affects all neuronal activities, and about 1/3 of the neurons employ GABA as their primary neurotransmitter. The relationship between GABA and depression is widely investigated in both human and animals. In comparable animal models, rats exposed to CMS had lower hippocampal GABA levels accompanied by depression-like symptoms [9]. A previous study reported that there was lower basal striatal GABA levels in the experiment of STZ-induced diabetic rats [10]. GABAergic drugs were able to reverse these depressive-like behaviors, suggesting that GABA treatment protected against the development of diabetic complications in STZ-induced diabetic rats [11]. On the contrary, GABA level is obviously increased in the cerebrospinal fluid of diabetic rats. Consequently, it is speculated that DM and depression induced the species-specific, cerebral area-specific, and tissues-specific abnormal concentration of GABA. As we known, most ionotropic $GABA_ARs$ are comprised of two α, one β, and two γ subunits. GABAergic effects are mediated via ionotropic $GABA_ARs$, which are functionally defined by their α subunits $(\alpha_1 - \alpha_6)$. Repeated swim-stress reduces $GABA_AR$ α subunit mRNAs in the mouse hippocampus [12]. Furthermore, decreased $GABA_AR$ clustering results in enhanced anxiety in mice heterozygous for the γ_2 subunit. But how about the protein expression of $GABA_AR$ α_1 subunit in diabetic mice still poorly understood. In this study, the protein expression of $GABA_AR$ α_1 subunit was dramatically reduced in hippocampus of diabetic mice compared to that of the non-diabetic mice.

4.3 A Role of Neural Oscillations as a Bridge Between GABAergic System and Emotional Behaviors

If the well-studied of GABAergic system was viewed as a "small ground" with molecular level, and emotional behavior was viewed as a "large ground" with behavioral level, then neural oscillations may be as a bridge between them. Based on this hypothesis, both theta and gamma frequencies in the hippocampus were chosen to be the bonds investigating mechanism in diabetic encephalopathy. Neural oscillations can be measured at different levels or scales, such as large-scale electroencephalographic (EEG), electrocorticogram (ECoG), and magnetoencephalography (MEG) recordings, medium-scale LFPs, and small-scale action potential recordings. All of them are viewed as the critical "middle ground" linking molecule to behavior. In the study, the data showed that both the strength of phase coupling (Fig. 2) and the strength of unidirectional coupling (Fig. 3A) on gamma between the hippocampal PP and DG was considerably decreased by DM, suggesting that both low and high gamma rhythms were implicated in anxiety-like and depression-like emotional behavior in DM mice. Moreover, gamma-band rhythmogenesis is a synchronous activity of fast-spiking inhibitory interneurons, with the resulting rhythmic inhibition producing neural ensemble synchrony by generating a narrow window for effective excitation. When applying the GABAAR antagonist bicuculline to the hippocampal slice in vitro, gamma rhythmic would be blocked. Our study also provided an evidence that the decreased expression of GABAAR α_1 subunit and the compensative increased GABA level were able to generate above kind of change at gamma frequency band in the hippocampus of DM mice.

Acknowledgement. This work was supported by grants from the National Natural Science Foundation of China (11232005, 31171053).

References

1. Lakatos, P., et al.: An oscillatory hierarchy controlling neuronal excitability and stimulus processing in the auditory cortex. J. Neurophysiol. **94**, 1904–1911 (2005)
2. Vijayan, S., et al.: Thalamocortical mechanisms for the anteriorization of alpha rhythms during propofol-induced unconsciousness. J. Neurosci. Off. J. Soc. Neurosci. **33**, 11070–11075 (2013)
3. Fell, J., Axmacher, N.: The role of phase synchronization in memory processes. Nat. Rev. Neurosci. **12**, 105–118 (2011)
4. Delorme, A., Makeig, S.: EEGLAB: an open-source toolbox for analysis of EEG dynamics. J. Neurosci. Methods **134**, 9–21 (2004)
5. Mi, X.C., et al.: Performance comparison between gPDC and PCMI for measuring directionality of neural information flow. J. Neurosci. Methods **227**, 57–64 (2014)
6. Canolty, R.T., et al.: High gamma power is phase-locked to theta oscillations in human neocortex. Science **313**, 1626–1628 (2006)
7. Wayhs, C.A.Y., et al.: GABAergic modulation in diabetic encephalopathy-related depression. Curr. Pharm. Des. **21**, 4980–4988 (2015)
8. Tang, Z.J., et al.: Antidepressant-like and anxiolytic-like effects of hydrogen sulfide in streptozotocin-induced diabetic rats through inhibition of hippocampal oxidative stress. Behav. Pharmacol. **26**, 427–435 (2015)
9. Grønli, J., et al.: Extracellular levels of serotonin and GABA in the hippocampus after chronic mild stress in rats. A microdialysis study in an animal model of depression. Behav. Brain Res. **181**, 42–51 (2007)
10. Gomez, R., et al.: Lower in vivo brain extracellular GABA concentration in diabetic rats during forced swimming. Brain Res. **968**, 281–284 (2003)
11. Nakagawa, T., et al.: Protective effects of γ-aminobutyric acid in rats with streptozotocin-induced diabetes. J. Nutr. Sci. Vitaminol. **51**, 278–282 (2005)
12. Montpied, P., et al.: Repeated swim-stress reduces GABAA receptor α subunit mRNAs in the mouse hippocampus. Mol. Brain Res. **18**, 267–272 (1993)

Impacts of Working Memory Training
on Brain Network Topology

Dongping Zhao[1], Qiushi Zhang[2,3], Li Yao[2], and Xiaojie Zhao[2(✉)]

[1] School of Electronics Engineering and Computer Science, Peking University,
Beijing, China
[2] College of Information Science and Technology, Beijing Normal University,
Beijing, China
`zhaox86@163.com`
[3] Department of Computer Science, University of Texas at Dallas, Richardson,
TX, USA

Abstract. A variety of network analysis methods that can reveal the neural mechanism underlying the course of dealing information in the brain by characterizing the topology and properties of brain networks have been applied to investigate the complexity of brain activities.. Working memory refers to the maintaining and handling of information in high-level cognition. It has been demonstrated that working memory performances can be enhanced by training. However, how working memory training affects the brain network topology and behavioral performance remains unclear. In this study, independent component analysis and graph theory were applied to the study of brain networks during real time fMRI based working memory training. The results showed that the training not only recruited the central execution network, the default-mode network, and the salience network, but also exerted lasting effects on the brain minimum spanning tree structure. These results demonstrated that the organization and working pattern of brain networks were altered by the training and provide new insights into the neural mechanisms underlying working memory training.

Keywords: Working memory · Independent component analysis · Minimum spanning tree · Network topology · Training

1 Introduction

Working memory (WM) refers to a cognitive system which provides temporary holding and manipulating of the important information for a range of complex cognitive activities, and WM training is deemed to an effective approach to improve an individual's cognition [1]. Previous studies have found WM behavior training can recruit and modulate several networks, which included the central executive network (CEN) [2], the default-mode network (DMN) [3], and the salience network (SN) [4]. Previous study has demonstrated that WM training significantly changed the connection strength intra- and inter-network in these three networks, and there was a significant correlation between the strength alteration and behavioral improvement [5]. Further studies reported that WM training also affected the topological relationships of these brain networks [6].

© Springer International Publishing AG 2017
F. Cong et al. (Eds.): ISNN 2017, Part II, LNCS 10262, pp. 575–582, 2017.
DOI: 10.1007/978-3-319-59081-3_67

The recently emerged neuro-feedback training based on real time functional magnetic resonance imaging (fMRI) technology introduces a new method to feed back and regulate the localized dynamic brain activities in individuals to achieve the behavioral performance improvement [7]. Recent studies have demonstrated that the brain activities in some localized regions can be regulated, such as insula [8], amygdala [9], primary motor cortex (PMA) [10], and the left dorsal lateral prefrontal cortex (DLPFC) [11]. Further studies found that the self-regulation of a localized region may impact on its connection with other regions [12, 13]. However, these studies focused on the impact of real time fMRI on the connection intra- and inter- networks, and few studies involved the effect of real time fMRI on the topological property of networks, which also play an important role in WM training and behavioral improvements.

Minimum spanning tree (MST) is an approach in graph theory to study the network topology which can be used to reveal the evolution of brain disease or cognitive function. Lee et al. found that the developmental stages of epilepsy can be classified using the MST structure [14]. Ciftci reported the topological changes of the DMN network in Alzheimer's patients by constructing the MST of DMN network [15]. Boersma et al. indicated the measurement of MST has a high sensitivity in detection the topological alterations of functional network during childhood brain development [16]. It is worth noting that the construction of the MST requires all network nodes, which guarantees the integrity of information when researching the interaction among multiple brain networks [17]. However, the application of MST in fMRI study is still relatively limited, especially in the application of cognitive function training.

In this study, independent component analysis (ICA) and MST were applied to explore the brain network topology during WM training by real time fMRI. Based on our previous work [11], ICA was firstly performed to determine the CEN, DMN and SN that were recruited in the WM training. Then, the MST was used to investigate the evolution of the brain network topology during the training. On the basis of previous findings, we expect to explore the WM training effect on the organization and working pattern of brain networks.

2 Methods and Materials

2.1 Subjects and Experimental Procedure

Thirty healthy right hand subjects took part in the real time fMRI based WM training included the experimental group (8 males and 7 females) and the control group (8 males and 7 females). All subjects signed the informed consent before the experiment. The experiment was approved by the Institutional Review Board of the State Key Laboratory of Cognitive Neuroscience and Learning in Beijing Normal University.

The experimental procedure included two real time fMRI feedback sessions, at an interval of seven days. Each session included a T1-weighted scan, a digital 3-back localizer run for selecting the target ROI to feedback, and four training runs. In each training run, subjects were instructed to use a cognitive strategy to increase the number of bars in the thermometer that represented the activities of target ROI, as high as possible. The subjects in the control group performed the same experimental procedure with the same instructions, except that a sham feedback signal from the other subject was presented to them. On the days before and after each session, all the subjects performed behavioral tests including the forward digit span, backward digit span and letter memory tasks. More details about the experimental procedure have been reported in our previous study [11].

2.2 fMRI Data Acquisition

All images data were acquired in a SIEMENS 3.0 T scanner at the MRI center of Beijing Normal University. A single-shot T2*-weighed echo-planar imaging (EPI) sequence (TE = 30 ms, TR = 2000 ms, flip angle = 90°, In-plane resolution = 3.125 × 3.125 mm^2, matrix size = 64 × 64, slice = 33; slice thickness = 3.60 mm) was used for fMRI image acquisition. A T1-weighted magnetization-prepared rapid gradient echo (MPRAGE) sequence (matrix size = 256 × 256, 176 partitions, 1 mm^3 isotropic voxels, TR = 2530 ms, TE = 3.45 ms, flip angle = 7°) was used for structural image acquisition.

2.3 Independent Component Analysis

Firstly, the fMRI images were preprocessed using SPM8 software which included slice timing, head motion correction, normalization to the Montreal Neurological Institute (MNI) space, reslicing into a resolution of 3 × 3 × 4 mm^3, and spatial smoothing using a Gaussian kernel with full-width at half maximum (FWHM) of 8 mm (http://www.fil.ion.ucl.ac.uk/spm). Then, ICA was performed using GIFT 2.0 (http://mialab.mrn.org/software/gift/) on the fMRI images from all training runs of all subjects in the experimental and control groups to determine the brain networks involved in the WM training. Using the minimum description length (MDL) criteria, the optimal number of independent components was selected to 22. The Infomax ICA algorithm was used to calculate the individual spatial maps and time courses for each subject. According to the spatial patterns in previous studies [3, 18], CEN, DMN and SN were determined. The ROIs in each network were defined as a spherical region centered on the local activation maximum with a radius of 6 mm.

2.4 Minimum Spanning Tree Analysis

Minimum spanning tree (MST) is a kind of tree structure connecting all the nodes by acyclic way with minimum cost. For constructing the MST, the first step was to extract the time series of all nodes in the network, calculated their correlation coefficient for each run of every subject and used the reciprocal of correlation coefficient as the weights to build a connection graph. Then, Kruskal's algorithm in Python 2.7.11 (https://www.python.org) was applied to carry out MST analysis. In each run, the average connection graph across subjects was built for the experimental group and control group respectively and adopted for generating the connection graph on the group level.

The difference degree between two MST X and Y is defined as $D_{(X,Y)} = D_{(Y|X)}$ + $D_{(X|Y)}$. The $D_{(Y|X)}$ represents the sum of distance between node i and all its neighbors in Y as follows.

$$D_{(Y|X)} = \frac{1}{N} \sum_{i=1}^{N} \log \left| \frac{D_{Y(i)}}{D_{X(i)}} \right| \qquad (1)$$

$D_x(i)$ is the sum of distance between node i and others in X. For each subject, the difference degree between the first run and the others were calculated to evaluate the influence of the training on the MST structure.

3 Results

The spatial distribution of twenty two independent components (brain network) on the group level was acquired by the ICA analysis. Comparing with the spatial patterns in previous studies [3, 18], the three network components of CEN, DMN, and SN were determined by visual inspection. According to the local activation maximum, the coordinates of all the nodes we choose are shown in Table 1.

Table 1. MNI coordinates of the ROIs of the three networks.

Network	Region	No.	[x, y, z]	T_{max}
CEN	Left DLPFC	B	−42, 26, 38	30.5
	Right DLPFC	G	45, 23, 38	28.79
	Left PPC	E	−45, −52, 46	33.95
	Right PPC	J	42, −52, 50	35.19
DMN	vMPFC	K	−3, 47, −10	25.8
	PCC	F	−3, −55, 22	23.45
	Left MTG/AG	D	−45, −61, 26	18.77
	Right MTG/AG	I	51, −64, 26	12.69
SN	ACC	A	6, 23, 26	17.61
	Left insula	C	−33, 20, 6	15.35
	Right insula	H	36, 20, 6	13.57

The MST structure of experimental group and control group are shown in Fig. 1. It revealed that the MST structures changed following the training stages which suggested the brain network topological structures could be influenced by the training. Based on the statistical analysis of difference degree, all MST structures of the brain network from run 2 to run 8 were different with that in run 1, and the difference degree to run 1 significantly increased in the experimental group ($p = 0.027$). However, there was no same obvious changes in the control group ($p = 0.33$) (Fig. 2).

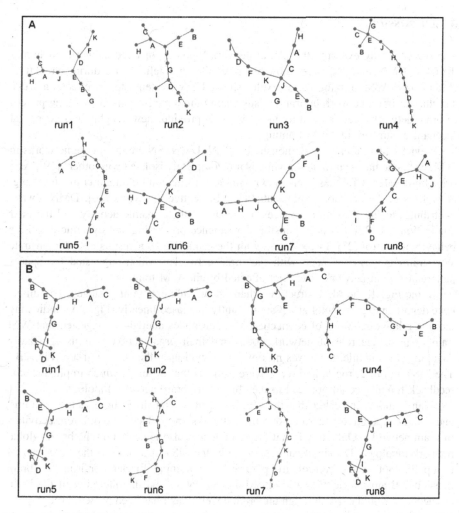

Fig. 1. The MST of each training run in the experimental group (A) and the control group (B). The letters from A to H represent the brain network nodes (See Table 1).

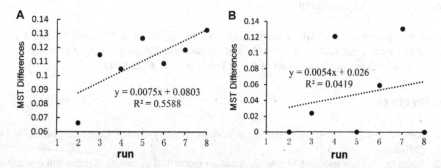

Fig. 2. The trend of difference degree in the experimental group (A) and the control group (B) as the training progressed. The scattered points stand for the difference degree of MST between the run 1 and the other runs.

4 Discussion

The present study explored the global network topology of three networks including the CEN, DMN, and SN, affected by the up-regulation of left DLPFC during a real time fMRI based WM training. The results showed a persistent steady effect on MST structure of brain network by training suggesting the organization and working pattern of brain networks were altered by the training. It provides new insights into the neural mechanisms underlying WM training.

Previous literatures have demonstrated CEN, DMN, SN are three core networks in WM, each one has own function while is not completely isolated with others [19]. As a prior knowledge, CEN has been shown vital to information holding and manipulating, as well as to the decision making about goal-directive behaviors [20]. DMN always contributes to the WM performances by means of the dynamic activity and the inter connection [21, 22]. SN is responsible for salience processing and dynamic cognitive behavioral control [23]. In consistent with these findings, our results also demonstrate WM training was associated with the three networks, which suggested the spatial patterns of brain networks were not affected by the WM training.

In the Fig. 1, the MST structures changed on different training stages. The difference degree in experimental group significantly increased linearly (Fig. 2A), indicating that the relative relations of connection between nodes altered. It suggested that WM training made the role of network nodes and their organization patterns constantly changing. Meanwhile, there was no obvious alteration in the control group (Fig. 2B). The MST in run 2, run 5, and run 8 were same as that in run 1, which implied shame feedback training could not lead to the changes in brain network topology.

Many studies indicated that the regulation of the localized brain activities influenced the cooperative pattern among the networks and induced the different activities of brain networks. Our study found these influences also appeared on the brains global network topology. The significant changing of the MST structures in the experimental group but not in the control group during the training further confirms the brain plasticity. Brain plasticity is a kind of inherent nature and the alteration of the brain connections would exist although the brain development has been passed through [24]. In our study, the self-regulating to the target region using various memory strategies by subjects can be regarded as a process to re-configure the brain network structure, which suggested the feedback training could be applied in the cognitive improvement and cognitive disorder rehabilitation.

Acknowledgments. This study is supported by the Funds of National Natural Science Foundation of China (grant number 61473044), and International Cooperation and Exchange of the National Natural Science Foundation of China (grant number 61210001).

References

1. Klingberg, T.: Training and plasticity of working memory. Trends Cogn. Sci. **14**, 317–324 (2010)
2. Koechlin, E., Summerfield, C.: An information theoretical approach to prefrontal executive function. Trends Cogn. Sci. **11**, 229–235 (2007)

3. Raichle, M.E., Macleod, A.M., Snyder, A.Z., Powers, W.J., Gusnard, D.A., Shulman, G.L.: A default mode of brain function. Proc. Natl. Acad. Sci. **98**, 676–682 (2001)

4. Critchley, H.D., Wiens, S., Rotshtein, P., Ohman, A., Dolan, R.J.: Neural systems supporting interoceptive awareness. Nat. Neurosci. **7**, 189–195 (2004)

5. Zhang, Q.S., Zhang, G.Y., Yao, L., Zhao, X.J.: Impact of real-time fMRI working memory feedback training on the interactions between three core brain networks. Front. Behav. Neurosci. **9**(244), 1–9 (2015)

6. Langer, N., von Bastian, C.C., Wirz, H., Oberauer, K., Jäncke, L.: The effects of working memory training on functional brain network efficiency. Cortex **49**, 2424–2438 (2013)

7. Caria, A., Sitaram, R., Birbaumer, N.: Real-time fMRI: a tool for local brain regulation. Neuroscientist **18**, 487–501 (2012)

8. Caria, A., Sitaram, R., Veit, R., Begliomini, C., Birbaumer, N.: Volitional control of anterior insula activity modulates the response to aversive stimuli. A real-time functional magnetic resonance imaging study. Biol. Psychiatry **68**, 425–432 (2010)

9. Zotev, V., Krueger, F., Phillips, R., Alvarez, R.P., Simmons, W.K., Bellgowan, P., Drevets, W.C., Bodurka, J.: Self-regulation of amygdala activation using real-time fMRI neurofeedback. PLoS One **6**, e24522 (2011)

10. Zhao, X.J., Zhang, H., Song, S.T., Ye, Q., Guo, J., Yao, L.: Causal interaction following the alteration of target region activation during motor imagery training using real-time fMRI. Front. Hum. Neurosci. **7**, 866 (2013)

11. Zhang, G.Y., Yao, L., Zhang, H., Long, Z.Y., Zhao, X.J.: Improved working memory performance through self-regulation of dorsal lateral prefrontal cortex activation using real-time fMRI. PLoS One **8**, e73735 (2013)

12. Hampson, M., Scheinost, D., Qiu, M., Bhawnani, J., Lacadie, C.M., Leckman, J.F., Constable, R.T., Papademetris, X.: Biofeedback of real-time functional magnetic resonance imaging data from the supplementary motor area reduces functional connectivity to subcortical regions. Brain Connect. **1**, 91–98 (2011)

13. Lee, J.H., Kim, J., Yoo, S.S.: Real-time fMRI-based neurofeedback reinforces causality of attention networks. Neurosci. Res. **72**, 347–354 (2012)

14. Lee, U., Kim, S., Jung, K.Y.: Classification of epilepsy types through global network analysis of scalp electroencephalograms. Phys. Rev. E Stat. Nonlinear Soft. Matter. Phys. **73**, 041–920 (2006)

15. Ciftci, K.: Minimum spanning tree reflects the alterations of the default mode network during Alzheimer's disease. Ann. Biomed. Eng. **39**(5), 1493–1504 (2011)

16. Boersma, M., Smit, D.J., Boomsma, D.I., de Geus, E.J., Delemarre-van de Waal, H.A., Stam, C.J.: Growing trees in child brains: graph theoretical analysis of electroencephalography-derived minimum spanning tree in 5- and 7-year-old children reflects brain maturation. Brain Connect. **3**(1), 50–60 (2013)

17. Tewarie, P., van Dellen, E., Hillebrand, A., Stam, C.J.: The minimum spanning tree: an unbiased method for brain network analysis. Neuroimage **104**, 177–188 (2015)

18. Uddin, L.Q., Supekar, K.S., Ryali, S., Menon, V.: Dynamic reconfiguration of structural and functional connectivity across core neurocognitive brain networks with development. J. Neurosci. **31**, 18578–18589 (2011)

19. Menon, V.: Large-scale brain networks and psychopathology: a unifying triple network model. Trends Cogn. Sci. **15**, 483–506 (2011)

20. Sridharan, D., Levitin, D.J., Menon, V.: A critical role for the right fronto-insular cortex in switching between central-executive and default-mode networks. Proc. Natl. Acad. Sci. **105**, 12569–12574 (2008)

21. Esposito, F., Aragri, A., Latorre, V., Popolizio, T., Scarabino, T., Cirillo, S., Marciano, E., Tedeschi, G., Di Salle, F.: Does the default-mode functional connectivity of the brain correlate with working-memory performances? Arch. Ital. Biol. **147**, 11–20 (2009)
22. Pyka, M., Beckmann, C.F., Schoning, S., Hauke, S., Heider, D., Kugel, H., Arolt, V., Konrad, C.: Impact of working memory load on FMRI resting state pattern in subsequent resting phases. PLoS One **4**(9), e7198 (2009)
23. Menon, V., Uddin, L.Q.: Saliency, switching, attention and control: a network model of insula function. Brain Struct. Funct. **214**, 655–667 (2010)
24. Pascual-Leone, A., Amedi, A., Fregni, F., Merabet, L.B.: The plastic human brain cortex. Annu. Rev. Neurosci. **28**, 377–401 (2005)

A Novel Biologically Inspired Hierarchical Model for Image Recommendation

Yan-Feng Lu[1(✉)], Hong Qiao[2,3], Yi Li[4], Li-Hao Jia[1], and Ai-Xuan Zhang[5]

[1] Research Center for Brain-Inspired Intelligence, Institute of Automation, Chinese Academy of Sciences, Beijing, China
{yanfeng.lv,lihao.jia}@ia.ac.cn
[2] The State Key Laboratory of Management and Control for Complex Systems, Institute of Automation, Chinese Academy of Sciences, Beijing, China
hong.qiao@ia.ac.cn
[3] CAS Center for Excellence in Brain Science and Intelligence Technology, Shanghai, China
[4] School of Information Engineering, Nanchang University, Nanchang, China
littlepear@ncu.edu.cn
[5] School of Mechanical and Material Engineering, North China University of Technology, Beijing, China
xyeahr@gmail.com

Abstract. The biologically inspired model (BIM) for invariant feature representation has attracted widespread attention recently, which approximately follows the organization of cortex visuel. BIM is a computational architecture with four layers. With the image data size increases, the four-layer framework is prone to be overfitting, which limits its application. To address this issue, motivated by biology, we propose a biologically inspired hierarchical model (BIHM) for image feature representation, which adds two more discriminative layers upon the conventional four-layer framework. In contrast to the conventional BIM that mimics the inferior temporal cortex, which corresponds to the low level feature invariance and selectivity, the proposed BIHM adds two more layers upon the conventional BIM framework to simulate inferotemporal cortex, exploring higher level feature invariance and selectivity. Furthermore, we firstly utilize the BIHM in the image recommendation. To demonstrate the effectiveness of proposed model, we use it in image recommendation task and perform experiment on CalTech5 datasets. The experiment results show that BIHM exhibits higher performance than conventional BIM and is very comparable to existing architectures.

Keywords: Visual cortex · Biologically inspired model · Classification · Image recommendation

1 Introduction

Recently, there are a large number of digital images made every day in the internet. Effective image recommendation, and retrieval tools are strongly demanded by consumers, including fashion, shopping online, sensing, medicine, and so on [1–4].

© Springer International Publishing AG 2017
F. Cong et al. (Eds.): ISNN 2017, Part II, LNCS 10262, pp. 583–590, 2017.
DOI: 10.1007/978-3-319-59081-3_68

Efficient image retrieval techniques play important roles in the image recommendation system. Among them, content based image retrieval (CBIR) has drawn widespread research attentiveness in the last decade. The retrieval performance of a CBIR system essentially depends on the feature representation, which has been carried out extensive research by researchers for decades [5–7]. In the visual system, humans trend to adopt high-level features (concepts) to represent images and measure their similarity. By contrast, while the most computer vision techniques extracted features with low-level features (color, shape, structure, texture and so on.) [8, 9].

In recent years, important developments have been made in the research of brain science [10–13]. The findings in the primary visual cortex V1 area are of significance. While researching the V1 area, Hubel and Wiesel discovered that the visual cortex analyzes features into various ways with different spatial orientations and frequencies [12]. The discovery gives an important support to early neuroscience theories. Based on these theories, Riesenhuber and Poggio described an original calculation framework for object recognition, called biologically inspired model (BIM) that tends to model the cognitive mechanism of the visual cortex [13]. Serre et al. upgraded the original BIM model and presented the standard BIM [16], which shows that the visual framework significantly improve the performance of object recognition.

The standard BIM made a big step forward for extending neurobiological models to deal with real-world vision tasks, and many enhanced models have been proposed about on the basic framework of it. Mutch and Lowe modified the model by constrain-ing the number of feature input, inhibiting S1/C1 outputs, and increasing feature selection [17]. Qiao et al. introduced some preliminary cognition and active attention mechanism based on the BIM framework [14, 15]. Lu et al. proposed a novel receptive field in the S1 layers and upgraded the framework by novel patch selection and matching processes [18–21]. These approaches obtain better performances by incorporating some biologically motivated properties in addition. In conclusion, all these systems have shown that the relevant biological findings are helpful for constructing more robust computer vision algorithms.

BIM is a calculation model with S1, C1, S2, and C2 four layers, which concentrates on the invariance and selectivity of features [16]. With the image data size increases, the four-layer framework is prone to be overfitting in the big data cases, which limits its application. To address this weakness, we describe a biologically inspired hierarchical model (BIHM), which adds two more discriminative layers upon the conventional four-layer framework. In contrast to the conventional BIM that successfully mimics the inferior temporal cortex (from V1 to V4) of the human visual system, which corresponds to the low level feature invariance and selectivity, the proposed BIHM adds two more layers based on the conventional BIM framework to simulate up to inferotemporal cortex (i.e., PIT and AIT), exploring higher level feature invariance and selectivity.

The remaining part of the article is organized as follows: in Sect. 2, we give an introduction about the conventional BIM; in Sect. 3, we propose the BIHM method and utilize the BIHM in image recommendation; in Sect. 4, we show experimental results based on Caltech05 database; finally, in Sect. 5, we concludes this paper.

2   Biologically Inspired Model Review

The biologically inspired model (BIM) presented by Serre et al. has attracted wide-spread attention recently. It has been successfully applied in various recognition tasks: From single object recognition in the clutter condition to multi-categorization as well as the understanding of complex scene.

The units in S1 layer are corresponding to the simple cells of the visual cortex, which compute the response for input image by Gabor filter bank. The C1 layer is corresponding to cortical complex cell layer and shows robustness to scale and shift transformations by pooling the afferent S1 units with the MAX operation in the same scale and orientation band. It tends to have larger receptive fields and increases the robustness to deformations from layer S1 to C1.

In the S2 stage, units pool C1 units from a local neighborhood across all orientations. A mass of patches are chose from the C1 layers of training images at random before that. Then the S2 units behave as radial basis function and pool over the C1 units in a Gaussian-like tuning way on the Euclidean distance between an input patch and a stored prototype. The C2 layer pools the S2 units over all scales and positions with a global maximum operation, which devotes to obtain shift- and scale-invariant responses. Therefore, for the sampled k prototype patches, a k-dimensional feature vector is finally obtained after the four-stage features extraction, which has larger receptive fields and shows shift-invariant and scale-invariant properties.

3   Biologically Inspired Hierarchical Model

Build upon Hubel and Wiesel's theories on visual cortex [10–12], the simple cells are not sensitive to illumination and need edge-like response at a particular phase, and position. The complex cells respond well to bars with a particular phase, while they are not sensitive to both phase and position of the bar in the receptive fields (RFs). At the upper layers the hypercomplex cells not only respond to bars in the phase and position invariant way, but also are selective to the bars with a specific length. Hubel and Wiesel suggested that this increasingly invariant and complex feature representations should be built by integrating the inputs from lower levels.

Two sort of functional layers exist in the visual cortex framework: the S layers consisted of simple cells are interleaved with the C layers consisted of complex cells. In brief, along the hierarchy, each S layer increase the feature selectivity by tuning to features of increasing complexity, and each C layers increase the invariance to 2D transformations such as slight changes in position and scale by a max pooling operation [16, 22].

S3 stage: In the S3 stage, the similar step is iterated one more time to increase the complexity of the prior response at the C2 level. The S3 unit presents the similarity between a sampled patch and the previous C2 layer in a Gaussian-like tuning way by Euclidean distance. These S3 units are interleaved with the C2 layers consisted of complex cells. The mathematical equation of the corresponding S3 layers is given by:

$$S3 = \exp(-\beta * \sum_{i=1}^{N} (C2(j,k) - F_i)^2), \tag{1}$$

where β defines the sharpness of the exponential function, $C2(j,k)$ is the afferent C2 layer with a specific scale j and orientation k, and F_i denotes a sampled patch from the prior C2 layers, N is the number of the sampled patches.

C3 stage: In the C3 stage, the C3 units acquired by pooling the S3 units with the same selectivity at adjacent scales and positions. The C3 units show the similar selectivity to complex features with the S3 units, but with a broader size of invariance. The S3 and C3 layers give a description of largely tuned shape. The set of invariant C3 responses can be calculated by doing a global maximum value of inputting S3 units across all positions and scales. The responses of the C3 layers are given by:

$$C3 = \max_{(} i,j,\sigma)(S3(i,j,\sigma)). \tag{2}$$

where (i,j) denotes the position of S3 units and σ is the corresponding scale. The export is a feature vector with C3 values. The vector is used as the C3 features in the experiment tasks.

The S3 and C3 stages follow the operations in S2 and C2, their further interleaving and max-like pooling with the inferior layers introduce better selectivity and invariance. The deeper features with more robust and discriminative information benefit the BIHM model in the cluttered recommendation tasks.

4 Experiments

In this part, we design an experiment to do the evaluation of BIHM in the image recommendation tasks. We compare the BIHM with BIM and SIFT on Caltech05 database [16]. Figure 1 shows the sample images of the CalTech5 datasets. Given the

Fig. 1. Sample images of the CalTech5 datasets.

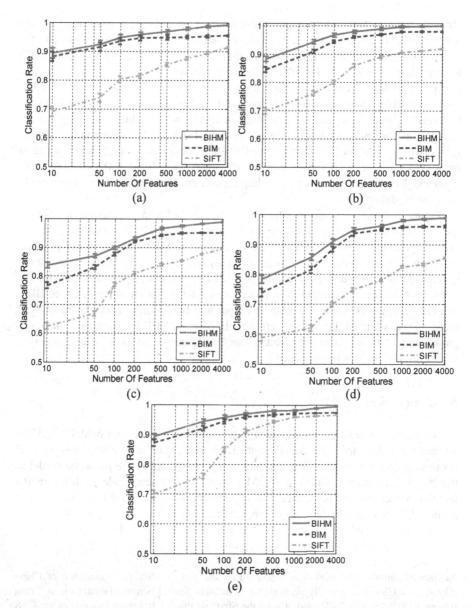

Fig. 2. Comparison of BIHM with standard BIM and SIFT on the CalTech5 dataset: (a) aeroplanes, (b) cars, (c) leaves, (d) faces, and (e) motorcycles.

various appearance transformation of the images, we applied the position-scale-invariant C3 features of BIHM, and passed the features to a classifier to execute classification. (In the experiments of this article, we select the linear Lib-SVM [23] as the classifier).

To make the experiment at a feature level and ensure a fair comparison between the methods, we neglected the position information from SIFT, because it was shown in

[16, 24] that structural information does not appear to improve classification perfor-mance. We use the same experimental setting in [16]. We compared the position-scale-invariant C2 features from the conversional BIM and the updated C3 features from BIHM with the SIFT features. We pass these features to an SVM and perform a present/absent classification task.

In this experiment, we chose 25 images at random from each category from the CaltTech5 database as positive training images and 25 different images from back-grounds as the negative training set. In the test stage, 100 different images (every category of the CaltTech5 database) and 100 other background images were chosen at random as a testing set. A various number of features (i.e., 10, 50, 100, 200, 500, 1000, 2000 and 4000) were obtained by selecting them from the 5,000 available at random to train the models.

Figure 2 shows the experimental results of the CalTech5 database with different numbers of features. Generally, it was demonstrated that BIHM exceeded BIM and SIFT in precision in most classes in this database. The SIFT features are adept at the detection of a transformed seen image, but they may lack discriminability in a more general classification task [16]. BIHM and BIM significantly exceeded SIFT for the faces, cars, leaves, and airplanes; in most cases, BIHM was clearly superior to BIM, especially, when the feature number is bigger and bigger, the improvement of BIM performance is limited due to the overfitting. In contrast, BIHM performance maintains growth trend with the increasing feature number. Therefore, in most image categories, BIHM shows superior performance.

5   Conclusion

In the article, we describe a novel biologically inspired hierarchical model by adding S3 and C3 layers upon the conventional BIM framework to represent a high level invariance and selectivity of features, and successfully applied the proposed model in the image recommendation. The BIHM provides a balanced trade-off between the features' selectivity and invariance. The experiment on Caltech05 datasets showed our proposed model qualifies in image recommendation tasks. Our research by far has mainly concentrated on the performance upgrading of BIM. The improvement of calculating speed upon BIM will be our future work.

Acknowledgment. This work was supported by the National Science Foundation of China (Grant 61603389) and partially supported by National Natural Science Foundation of China (Grants 61210009, 61502494) and also by the Strategic Priority Research Program of the CAS (Grant XDB02080003).

References

1. Wu, L., Jin, R., Jain, A.K.: Tag completion for image retrieval. IEEE Trans. Pattern Anal. Mach. Intell. **35**, 716–727 (2013)

2. Kim, M., Park, S.O.: Group affinity based social trust model for an intelligent movie recommender system. Multimed. Tools Appl. **64**, 505–516 (2013)
3. Wang, J., Shim, B.: On the recovery limit of sparse signals using orthogonal matching pursuit. IEEE Trans. Signal Process. **60**, 4973–4976 (2012)
4. Viana, W., Braga, R., Lemos, F.D.A., Souza, J.M.O., Carmo, R.A.F., Andrade, R.M.C., et al.: Mobile photo recommendation and logbook generation using context-tagged images. IEEE Multimed. **21**, 24–34 (2014)
5. Wan, J., Wang, D., Hoi, S.C.H., et al.: Deep learning for content-based image retrieval: a comprehensive study. In: 22nd ACM International Conference on Multimedia, pp. 157–166. ACM, Orlando (2014)
6. Babenko, A., Slesarev, A., Chigorin, A., Lempitsky, V.: Neural codes for image retrieval. In: Fleet, D., Pajdla, T., Schiele, B., Tuytelaars, T. (eds.) ECCV 2014. LNCS, vol. 8689, pp. 584–599. Springer, Cham (2014). doi:10.1007/978-3-319-10590-1_38
7. Donahue, J., Anne, H.L., Guadarrama, S., Rohrbach, M., Venugopalan, S., Saenko, K., et al.: Long-term recurrent convolutional networks for visual recognition and description, pp. 2625–2634 (2015)
8. Habibian, A., van de Sande, K.E.A., Snoek, C.G.M.: Recommendations for video event recognition using concept vocabularies. In: 3rd ACM Conference on International Conference on Multimedia Retrieval, pp. 89–96. ACM, New York (2013)
9. Tam, K.P.: Concepts and measures related to connection to nature: similarities and differences. J. Environ. Psychol. **34**, 64–78 (2013)
10. Hubel, D.H., Wiesel, T.N.: Receptive fields of single neurones in the cat's striate cortex. J. Physiol. **148**, 574–591 (1959)
11. Hubel, D.H., Wiesel, T.N.: Receptive fields, binocular interaction and functional architecture in the cat's visual cortex. J. Physiol. **160**(1), 106–154 (1962)
12. Poggio, T., Girosi, F.: Networks for approximation and learning. Proc. IEEE **78**, 1481–1497 (1990)
13. Riesenhuber, M., Poggio, T.: Hierarchical models of object recognition in cortex. Nat. Neurosci. **2**, 1019–1025 (1999)
14. Qiao, H., Xi, X., Li, Y., Wu, W., Li, F.: Biologically inspired visual model with preliminary cognition and active attention adjustment. IEEE Trans. Cybern. **45**, 2612–2624 (2015)
15. Qiao, H., Li, C., Yin, P., Wu, W., Liu, Z.Y.: Human-inspired motion model of upper-limb with fast response and learning ability – a promising direction for robot system and control. Assem. Autom. **36**, 97–107 (2016)
16. Serre, T., Wolf, L., Bileschi, S., Riesenhuber, M., Poggio, T.: Robust object recognition with cortex-like mechanisms. IEEE Trans. Pattern Anal. Mach. Intell. **29**, 411–426 (2007)
17. Mutch, J., Lowe, D.G.: Multiclass object recognition with sparse, localized features. In: IEEE Computer Society Conference on Computer Vision and Pattern Recognition, New York, vol. 1, pp. 11–18 (2006)
18. Lu, Y.F., Zhang, H.Z., Kang, T.K., Choi, I.H., Lim, M.T.: Extended biologically inspired model for object recognition based on oriented Gaussian-Hermite moment. Neurocomputing **139**, 189–201 (2014)
19. Lu, Y.F., Kang, T.K., Zhang, H.Z., Lim, M.T.: Enhanced hierarchical model of object recognition based on a novel patch selection method in salient regions. IET Comput. Vis. **9**, 663–672 (2015)
20. Lu, Y.F., Zhang, H.Z., Kang, T.K., Lim, M.T.: Dominant orientation patch matching for HMAX. Neurocomputing **193**, 155–166 (2016)
21. Zhang, H.Z., Lu, Y.F., Kang, T.K., Lim, M.T.: B-HMAX: a fast binary biologically inspired model for object recognition. Neurocomputing **218**, 242–250 (2016)

22. Serre, T., Riesenhuber, M.: Realistic modeling of simple and complex cell tuning in the HMAX model, and implications for invariant object recognition in cortex (2004)
23. Chang, C.C., Lin, C.J.: LIBSVM: a library for support vector machines. ACM Trans. Intell. Syst. Technol. **2**, 1–27 (2011)
24. Lowe, D.G.: Distinctive image features from scale-invariant keypoints. Int. J. Comput. Vis. **60**, 91–110 (2004)

Author Index

the United States
okmasters